T0176425

Sampling and Estimation from Finite Populations

Sampling and Estimation from Finite Populations

Yves Tillé
Université de Neuchâtel
Switzerland

Most of this book has been translated from French by **Ilya Hekimi**

Original French title: *Théorie des sondages : Échantillonnage et estimation en populations finies*

This edition first published 2020
© 2020 John Wiley & Sons Ltd

The right of Yves Tillé to be identified as the author of this work has been asserted in accordance with law.

Registered Offices
John Wiley & Sons, Inc., 111 River Street, Hoboken, NJ 07030, USA
John Wiley & Sons Ltd, The Atrium, Southern Gate, Chichester, West Sussex, PO19 8SQ, UK

Editorial Office
9600 Garsington Road, Oxford, OX4 2DQ, UK

For details of our global editorial offices, customer services, and more information about Wiley products visit us at www.wiley.com.

Wiley also publishes its books in a variety of electronic formats and by print-on-demand. Some content that appears in standard print versions of this book may not be available in other formats.

Library of Congress Cataloging-in-Publication Data

Names: Tillé, Yves, author. | Hekimi, Ilya, translator.
Title: Sampling and estimation from finite populations / Yves Tillé ; most
    of this book has been translated from French by Ilya Hekimi.
Other titles: Théorie des sondages. English
Description: Hoboken, NJ : Wiley, [2020] | Series: Wiley series in
    probability and statistics applied. Probability and statistics section |
    Translation of: Théorie des sondages : échantillonnage et estimation
    en populations finies. | Includes bibliographical references and index.
Identifiers: LCCN 2019048451 | ISBN 9780470682050 (hardback) | ISBN
    9781119071266 (adobe pdf) | ISBN 9781119071273 (epub)
Subjects: LCSH: Sampling (Statistics) | Public opinion polls – Statistical
    methods. | Estimation theory.
Classification: LCC QA276.6 .T62813 2020 | DDC 519.5/2 – dc23
LC record available at https://lccn.loc.gov/2019048451

Cover Design: Wiley
Cover Image: © gremlin/Getty Images

Set in 10/12pt WarnockPro by SPi Global, Chennai, India

Printed and bound by CPI Group (UK) Ltd, Croydon, CR0 4YY

10 9 8 7 6 5 4 3 2 1

Contents

15.10   Linearization by the Sample Indicator Variables   *324*
15.10.1 The Method   *324*
15.10.2 Linearization of a Quantile   *326*
15.10.3 Linearization of a Calibrated Estimator   *327*
15.10.4 Linearization of a Multiple Regression Estimator   *328*
15.10.5 Linearization of an Estimator of a Complex Function with Calibrated
         Weights   *329*
15.10.6 Linearization of the Gini Index   *330*
15.11   Discussion on Variance Estimation   *331*
         Exercises   *331*

16   **Treatment of Nonresponse**   *333*
16.1   Sources of Error   *333*
16.2   Coverage Errors   *334*
16.3   Different Types of Nonresponse   *334*
16.4   Nonresponse Modeling   *335*
16.5   Treating Nonresponse by Reweighting   *336*
16.5.1   Nonresponse Coming from a Sample   *336*
16.5.2   Modeling the Nonresponse Mechanism   *337*
16.5.3   Direct Calibration of Nonresponse   *339*
16.5.4   Reweighting by Generalized Calibration   *341*
16.6   Imputation   *342*
16.6.1   General Principles   *342*
16.6.2   Imputing From an Existing Value   *342*
16.6.3   Imputation by Prediction   *342*
16.6.4   Link Between Regression Imputation and Reweighting   *343*
16.6.5   Random Imputation   *345*
16.7   Variance Estimation with Nonresponse   *347*
16.7.1   General Principles   *347*
16.7.2   Estimation by Direct Calibration   *348*
16.7.3   General Case   *349*
16.7.4   Variance for Maximum Likelihood Estimation   *350*
16.7.5   Variance for Estimation by Calibration   *353*
16.7.6   Variance of an Estimator Imputed by Regression   *356*
16.7.7   Other Variance Estimation Techniques   *357*

17   **Summary Solutions to the Exercises**   *359*

         Bibliography   *379*

         Author Index   *405*

         Subject Index   *411*

List of Figures

List of Tables

List of Algorithms

Preface

The first version of this book was published in 2001, the year I left the Ecole Nationale de la Statistique et de l'Analyse de l'Information (ENSAI) in Rennes (France) to teach at the University of Neuchâtel in Switzerland. This version came from several course materials of sampling theory that I had taught in Rennes. At the ENSAI, the collaboration with Jean-Claude Deville was particularly stimulating.

The editing of this new edition was laborious and was done in fits and starts. I thank all those who reviewed the drafts and provided me with their comments. Special thanks to Monique Graf for her meticulous re-reading of some chapters.

The almost 20 years I spent in Neuchâtel were dotted with multiple adventures. I am particularly grateful to Philippe Eichenberger and Jean-Pierre Renfer, who successively headed the Statistical Methods Section of the Federal Statistical Office. Their trust and professionalism helped to establish a fruitful exchange between the Institute of Statistics of the University of Neuchâtel and the Swiss Federal Statistical Office.

I am also very grateful to the PhD students that I have had the pleasure of mentoring so far. Each thesis is an adventure that teaches both supervisor and doctoral student. Thank you to Alina Matei, Lionel Quality, Desislava Nedyalkova, Erika Antal, Matti Langel, Toky Randrianasolo, Eric Graf, Caren Hasler, Matthieu Wilhelm, Mihaela Guinand-Anastasiade, and Audrey-Anne Vallée who trusted me and whom I had the pleasure to supervise for a few years.

Neuchâtel, 2018   *Yves Tillé*

Preface to the First French Edition

This book contains teaching material that I started to develop in 1994. All chapters have indeed served as a support for teaching, a course, training, a workshop or a seminar. By grouping this material, I hope to present a coherent and modern set of results on the sampling, estimation, and treatment of nonresponses, in other words, on all the statistical operations of a standard sample survey.

In producing this book, my goal is not to provide a comprehensive overview of survey sampling theory, but rather to show that sampling theory is a living discipline, with a very broad scope. If, in several chapters demonstrations have been discarded, I have always been careful to refer the reader to bibliographical references. The abundance of very recent publications attests to the fertility of the 1990s in this area. All the developments presented in this book are based on the so-called "design-based" approach. In theory, there is another point of view based on population modeling. I intentionally left this approach aside, not out of disinterest, but to propose an approach that I deem consistent and ethically acceptable to the public statistician.

I would like to thank all the people who, in one way or another, helped me to make this book: Laurence Broze, who entrusted me with my first sampling course at the University Lille 3, Carl Särndal, who encouraged me on several occasions, and Yves Berger, with whom I shared an office at the Université Libre de Bruxelles for several years and who gave me a multitude of relevent remarks. My thanks also go to Antonio Canedo who taught me to use LaTeX, to Lydia Zaïd who has corrected the manuscript several times, and to Jean Dumais for his many constructive comments.

I wrote most of this book at the *École Nationale de la Statistique et de l'Analyse de l'Information*. The warm atmosphere that prevailed in the statistics department gave me a lot of support. I especially thank my colleagues Fabienne Gaude, Camelia Goga, and Sylvie Rousseau, who meticulously reread the manuscript, and Germaine Razé, who did the work of reproduction of the proofs. Several exercises are due to Pascal Ardilly, Jean-Claude Deville, and Laurent Wilms. I want to thank them for allowing me to reproduce them. My gratitude goes particularly to Jean-Claude Deville for our fruitful collaboration within the Laboratory of Survey Statistics of the Center for Research in Economics and Statistics. The chapters on the splitting method and balanced sampling also reflect the research that we have done together.

Bruz, 2001   *Yves Tillé*

Table of Notations

#	cardinal (number of elements in a set)
$\ll$	much less than
$\backslash$	$A \backslash B$ complement of B in A
$'$	function $f'(x)$ is the derivative of $f(x)$
!	factorial: $n! = n \times (n-1) \times \cdots \times 2 \times 1$
$\binom{N}{n}$	$\frac{N!}{n!(N-n)!}$ number of ways to choose k units from N units
$[a \pm b]$	interval $[a - b, a + b]$
$\approx$	is approximately equal to
$\propto$	is proportional to
$\sim$	follows a specific probability distribution (for a random value)
$\mathbb{1}\{A\}$	equals 1 if A is true and 0 otherwise
a_k	number of times unit k is in the sample
$\mathbf{a}$	vector of a_k
$B_0, B_1, B_2, \ldots$	population regression coefficients
$\mathbf{B}$	vector of population regression coefficients
$\beta_0, \beta_1, \beta_2, \ldots$	regression coefficients for model M
β	vector of regression coefficients of model M
$\widehat{\mathbf{B}}$	vector of estimated regression coefficients
$\widehat{\beta}$	vector of estimated regression coefficients of the model
C	cube whose vertices are samples
$\mathrm{cov}_p(X, Y)$	covariance between random variables X and Y
$cov(X, Y)$	estimated covariance between random variables X and Y
CV	population coefficient of variation
$\widehat{\mathrm{CV}}$	estimated coefficient of variation
d_k	$d_k = 1/\pi_k$ expansion estimator survey weights
$\mathrm{E}_p(\widehat{Y})$	mathematical expectation under the sampling design $p(.)$ of estimator $\widehat{Y}$
$\mathrm{E}_M(\widehat{Y})$	mathematical expectation under the model M of estimator $\widehat{Y}$
$\mathrm{E}_q(\widehat{Y})$	mathematical expectation under the nonresponse mechanism q of estimator $\widehat{Y}$
$\mathrm{E}_I(\widehat{Y})$	mathematical expectation under the imputation mechanism I of estimator $\widehat{Y}$
MSE	mean square error
f	sampling fraction $f = n/N$

$g_k(.,.)$	pseudo-distance derivative for calibration
g_k	adjustment factor after calibration called g-weight $g_k = \pi_k w_k = \pi_k/d_k$
$G_k(.,.)$	pseudo-distance for calibration
h	strata or post-strata index
$\mathrm{IC}(1-\alpha)$	confidence interval with confidence level $1-\alpha$
k ou ℓ	indicates a statistical unit, $k \in U$ or $\ell \in U$
K	$C \cap Q$ intersection of the cube and constraint space for the cube method
m	number of clusters or primary units in the sample of clusters or primary units
M	number of clusters or primary units in the population
n	Sample size (without replacement)
n_i	number of secondary units sampled in primary unit i
n_S	size of the sample in S if the size is random
N	population size
n_h	Sample size in stratum or post-stratum U_h
N_h	number of units in stratum or post-stratum U_h
N_i	number of secondary units in primary unit i
N_{ij}	population totals when (i,j) is a contingency table
$\mathbb{N}$	set of natural numbers
$\mathbb{N}_+$	set of positive natural numbers with zero
$p(s)$	probability of selecting sample s
p_i	probability of sampling unit i for sampling with replacement
P or P_D	proportion of units belonging to domain D
$\Pr(A)$	probability that event A occurs
$\Pr(A|B)$	probability that event A occurs, given B occurred
Q	subspace of constraints for the cube method
r_k	response indicator
$\mathbb{R}$	set of real numbers
$\mathbb{R}_+$	set of positive real numbers with zero
$\mathbb{R}_+^*$	set of strictly positive real numbers
s	Sample or subset of the population, $s \subset U$
s_y^2	Sample variance of variable y
s_{yh}^2	Sample variance of y in stratum or post-stratum h
s_{xy}	covariance between variables x and y in the sample
S	random sample such that $Pr(S = s) = p(s)$
s_y^2	variance of variance y in the population
S_{xy}	covariance between variables x and y in the population
S_h	random sample selected in stratum or post-stratum h
s_{yh}^2	population variance of y in the stratum or post-stratum h
$\top$	vector $\mathbf{u}^\top$ is the transpose of vector $\mathbf{u}$
U	finite population of size N
U_h	stratum or post-stratum h, where $h = 1, \dots, H$
v_k	linearized variable
$v_{HT}(\widehat{Y})$	Horvitz–Thompson estimator of the variance of estimator $\widehat{Y}$
$v_{SYG}(\widehat{Y})$	Sen–Yates–Grundy estimator of the variance of estimator $\widehat{Y}$

$\mathrm{var}_p(\hat{Y})$	variance of estimator $\hat{Y}$ under the survey design
$\mathrm{var}_M(\hat{Y})$	variance of estimator $\hat{Y}$ under the model
$\mathrm{var}_q(\hat{Y})$	variance of estimator $\hat{Y}$ under the nonresponse mechanism
$\mathrm{var}_I(\hat{Y})$	variance of estimator $\hat{Y}$ under the imputation mechanism
$v(\hat{Y})$	variance estimator of estimator $\hat{Y}$
w_k or $w_k(S)$	weight associated with individual k in the sample after calibration
x	auxiliary variable
x_k	auxiliary variable value of unit k
$\mathbf{x}_k$	vector in $\mathbb{R}^p$ of the p values taken by the auxiliary variables on k
X	total value of the auxiliary variable over all the units of U
$\hat{X}$	expansion estimator of X
$\overline{X}$	mean value of the auxiliary variables over all the units of U
$\hat{\overline{X}}$	expansion estimator of $\overline{X}$
y	variable of interest
y_k	value of the variable of interest for unit k
y_k^*	imputed value of y for k (treating nonresponse)
Y	total value of the variable of interest over all the units of U
Y_h	total value of the variable of interest over all the units in stratum or post-stratum U_h
Y_i	total of y_k in primary unit or cluster i
$\hat{Y}$	expansion estimator of Y
$\overline{Y}$	mean value of the variable of interest over all the units of U
$\overline{Y}_h$	mean value of the variable of interest over all units of stratum or post-stratum U_h
$\hat{\overline{Y}}_h$	estimator of the mean value of the variable of interest over all units of stratum or post-stratum U_h
$\hat{\overline{Y}}$	expansion estimator of $\overline{Y}$
$\hat{Y}_{\mathrm{BLU}}$	best unbiased linear estimator under the model of total Y
$\hat{Y}_{\mathrm{CAL}}$	calibrated estimator of total Y
$\hat{Y}_D$	difference estimator of total Y
$\hat{Y}_h$	estimator of total Y_h in stratum or post-stratum U_h
$\hat{Y}_{\mathrm{HAJ}}$	Hájek estimator of Y
$\hat{Y}_{\mathrm{HH}}$	Hansen–Hurwitz estimator of Y
$\hat{Y}_{\mathrm{IMP}}$	estimator used when missing values are imputed
$\hat{Y}_{\mathrm{OPT}}$	expansion estimator of the total in an optimal stratified design
$\hat{Y}_{\mathrm{POST}}$	post-stratified estimator of the total Y
$\hat{Y}_{\mathrm{PROP}}$	expansion estimator of the total in a stratified design with proportional allocation
$\hat{Y}_{\mathrm{REG}}$	regression estimator of total Y
$\hat{Y}_{\mathrm{REGM}}$	multiple regression estimator of total Y
$\hat{Y}_{\mathrm{REG\text{-}OPT}}$	optimal regression estimator of total Y
$\hat{Y}_{\mathrm{RB}}$	Rao–Blackwellized estimator of Y
$\hat{Y}_Q$	ratio estimator of the total
$\hat{Y}_{\mathrm{STRAT}}$	expansion estimator of the total in a stratified design

z_p	quantile of order p of a standardized normal random variable
α	probability that the parameter is outside the interval
$\Delta_{k\ell}$	$\pi_{k\ell} - \pi_k \pi_\ell$
π_k	inclusion probability of unit k
$\pi_{k\ell}$	second-order inclusion probabilities for units k and ℓ $\pi_{k\ell} = \Pr(k \text{ and } \ell \in S)$
ϕ_k	response probability of unit k
σ^2	variance of an infinite population or variable of y or variance under the model
ρ	correlation between x and y in the population

1

A History of Ideas in Survey Sampling Theory

1.1 Introduction

Looking back, the debates that animated a scientific discipline often appear futile. However, the history of sampling theory is particularly instructive. It is one of the specializations of statistics which itself has a somewhat special position, since it is used in almost all scientific disciplines. Statistics is inseparable from its fields of application since it determines how data should be processed. Statistics is the cornerstone of quantitative scientific methods. It is not possible to determine the relevance of the applications of a statistical technique without referring to the scientific methods of the disciplines in which it is applied.

Scientific truth is often presented as the consensus of a scientific community at a specific point in time. The history of a scientific discipline is the story of these consensuses and especially of their changes. Since the work of Thomas Samuel Kuhn (1970), we have considered that science develops around paradigms that are, according to Kuhn (1970, p. 10), "models from which spring particular coherent traditions of scientific research." These models have two characteristics: "Their achievement was sufficiently unprecedented to attract an enduring group of adherents away from competing modes of scientific activity. Simultaneously, it was sufficiently open-ended to leave all sorts of problems for the redefined group of practitioners to resolve." (Kuhn, 1970, p. 10).

Many authors have proposed a chronology of discoveries in survey theory that reflect the major controversies that have marked its development (see among others Hansen & Madow, 1974; Hansen et al., 1983; Owen & Cochran, 1976; Sheynin, 1986; Stigler, 1986). Bellhouse (1988a) interprets this timeline as a story of the great ideas that contributed to the development of survey sampling theory. Statistics is a peculiar science. With mathematics for tools, it allows the methodology of the other disciplines to be finalized. Because of the close correlation between a method and the multiplicity of its fields of action, statistics is based on a multitude of different ideas from the various disciplines in which it is applied.

The theory of survey sampling plays a preponderant role in the development of statistics. However, the use of sampling techniques has been accepted only very recently. Among the controversies that have animated this theory, we find some of the classical debates of mathematical statistics, such as the role of modeling and a discussion of estimation techniques. Sampling theory was torn between the major currents of statistics and gave rise to multiple approaches: design-based, model-based, model-assisted, predictive, and Bayesian.

Sampling and Estimation from Finite Populations, First Edition. Yves Tillé.
© 2020 John Wiley & Sons Ltd. Published 2020 by John Wiley & Sons Ltd.

1.2 Enumerative Statistics During the 19th Century

In the Middle Ages, several attempts to extrapolate partial data to an entire population can be found in Droesbeke et al. (1987). In 1783, in France, Pierre Simon de Laplace (see 1847) presented to the Academy of Sciences a method to determine the number of inhabitants from birth registers using a sample of regions. He proposed to calculate, from this sample of regions, the ratio of the number of inhabitants to the number of births and then to multiply it by the total number of births, which could be obtained with precision for the whole population. Laplace even suggested estimating "the error to be feared" by referring to the central limit theorem. In addition, he recommended the use of a ratio estimator using the total number of births as auxiliary information. Survey methodology as well as probabilistic tools were known before the 19th century. However, never during this period was there a consensus about their validity.

The development of statistics (etymologically, from German: analysis of data about the state) is inseparable from the emergence of modern states in the 19th century. One of the most outstanding personalities in the official statistics of the 19th century is the Belgian Adolphe Quételet (1796–1874). He knew of Laplace's method and maintained a correspondence with him. According to Stigler (1986, pp. 164–165), Quételet was initially attracted to the idea of using partial data. He even tried to apply Laplace's method to estimate the population of the Netherlands in 1824 (which Belgium was a part of until 1830). However, it seems that he then rallied to a note from Keverberg (1827) which severely criticized the use of partial data in the name of precision and accuracy:

> In my opinion, there is only one way to arrive at an exact knowledge of the population and the elements of which it is composed: it is that of an actual and detailed enumeration; that is to say, the formation of nominative states of all the inhabitants, with indication of their age and occupation. Only by this mode of operation can reliable documents be obtained on the actual number of inhabitants of a country, and at the same time on the statistics of the ages of which the population is composed, and the branches of industry in which it finds the means of comfort and prosperity.[1]

In one of his letters to the Duke of Saxe-Coburg Gotha, Quételet (1846, p. 293) also advocates for an exhaustive statement:

> La Place had proposed to substitute for the census of a large country, such as France, some special censuses in selected departments where this kind of operation might have more chances of success, and then to carefully determine the ratio of the population either at birth or at death. By means of these ratios of the births and deaths of all the other departments, figures which can be ascertained with sufficient accuracy, it is then easy to determine the population of

---

1 Translated from French: "À mon avis, il n'existe qu'un seul moyen de parvenir à une connaissance exacte de la population et des élémens dont elle se compose : c'est celle d'un dénombrement effectif et détaillé ; c'est-à-dire, de la formation d'états nominatifs de tous les habitans, avec indication de leur âge et de leur profession. Ce n'est que par ce mode d'opérer, qu'on peut obtenir des documens dignes de confiance sur le nombre réel d'habitans d'un pays, et en même temps sur la statistique des âges dont la population se compose, et des branches d'industrie dans lesquelles elle trouve des moyens d'aisance et de prospérité."

the whole kingdom. This way of operating is very expeditious, but it supposes an invariable ratio passing from one department to another. [···] This indirect method must be avoided as much as possible, although it may be useful in some cases, where the administration would have to proceed quickly; it can also be used with advantage as a means of control.[2]

It is interesting to examine the argument used by Quételet (1846, p. 293) to justify his position.

> To not obtain the faculty of verifying the documents that are collected is to fail in one of the principal rules of science. Statistics is valuable only by its accuracy; without this essential quality, it becomes null, dangerous even, since it leads to error.[3]

Again, accuracy is considered a basic principle of statistical science. Despite the existence of probabilistic tools and despite various applications of sampling techniques, the use of partial data was perceived as a dubious and unscientific method. Quételet had a great influence on the development of official statistics. He participated in the creation of a section for statistics within the British Association of the Advancement of Sciences in 1833 with Thomas Malthus and Charles Babbage (see Horvàth, 1974). One of its objectives was to harmonize the production of official statistics. He organized the International Congress of Statistics in Brussels in 1853. Quételet was well acquainted with the administrative systems of France, the United Kingdom, the Netherlands, and Belgium. He has probably contributed to the idea that the use of partial data is unscientific.

Some personalities, such as Malthus and Babbage in Great Britain, and Quételet in Belgium, contributed greatly to the development of statistical methodology. On the other hand, the establishment of a statistical apparatus was a necessity in the construction of modern states, and it is probably not a coincidence that these personalities come from the two countries most rapidly affected by the industrial revolution. At that time, the statistician's objective was mainly to make enumerations. The main concern was to inventory the resources of nations. In this context, the use of sampling was unanimously rejected as an inexact and fundamentally unscientific procedure. Throughout the 19th century, the discussions of statisticians focused on how to obtain reliable data and on the presentation, interpretation, and possibly modeling (adjustment) of these data.

---

2 Translated from French: "La Place avait proposé de substituer au recensement d'un grand pays, tel que la France, quelques recensements particuliers dans des départements choisis, où ce genre d'opération pouvait avoir plus de chances de succès, puis d'y déterminer avec soin le rapport de la population soit aux naissances soit aux décès. Au moyen de ces rapports des naissances et des décès de tous les autres départements, chiffres qu'on peut constater avec assez d'exactitude, il devient facile ensuite de déterminer la population de tout le royaume. Cette manière d'opérer est très expéditive, mais elle suppose un rapport invariable en passant d'un département à un autre. [···] Cette méthode indirecte doit être évitée autant que possible, bien qu'elle puisse être utile dans certains cas, où l'administration aurait à procéder avec rapidité ; on peut aussi l'employer avec avantage comme moyen de contrôle."
3 Translated from French: "Ne pas se procurer la faculté de vérifier les documents que l'on réunit, c'est manquer à l'une des principales règles de la science. La statistique n'a de valeur que par son exactitude ; sans cette qualité essentielle, elle devient nulle, dangereuse même puisqu'elle conduit à l'erreur."

1.3   Controversy on the use of Partial Data

In 1895, the Norwegian Anders Nicolai Kiær, Director of the Central Statistical Office of Norway, presented to the Congress of the International Statistical Institute of Statistics (ISI) in Bern a work entitled *Observations et expériences concernant des dénombrements représentatifs* (*Observations and experiments on representative enumeration*) for a survey conducted in Norway. Kiær (1896) first selected a sample of cities and municipalities. Then, in each of these municipalities, he selected only some individuals using the first letter of their surnames. He applied a two-stage design, but the choice of the units was not random. Kiær argues for the use of partial data if it is produced using a "representative method". According to this method, the sample must be a representation with a reduced size of the population. Kiær's concept of representativeness is linked to the quota method. His speech was followed by a heated debate, and the proceedings of the Congress of the ISI reflect a long dispute. Let us take a closer look at the arguments from two opponents of Kiær's method (see ISI General Assembly Minutes, 1896).

> Georg von Mayr (Prussia)[· · ·] It is especially dangerous to call for this system of representative investigations within an assembly of statisticians. It is understandable that for legislative or administrative purposes such limited enumeration may be useful – but then it must be remembered that it can never replace complete statistical observation. It is all the more necessary to support this point, that there is among us in these days a current among mathematicians who, in many directions, would rather calculate than observe. But we must remain firm and say: no calculation where observation can be done.[4]

> Guillaume Milliet (Switzerland). I believe that it is not right to give a congressional voice to the representative method(which can only be an expedient) an importance that serious statistics will never recognize. No doubt, statistics made with this method, or, as I might call it, statistics, *pars pro toto*, has given us here and there interesting information; but its principle is so much in contradiction with the demands of the statistical method that as statisticians, we should not grant to imperfect things the same right of bourgeoisie, so to speak, that we accord to the ideal that scientifically we propose to reach.[5]

---

4 Translated from French: "C'est surtout dangereux de se déclarer pour ce système des investigations représentatives au sein d'une assemblée de statisticiens. On comprend que pour des buts législatifs ou administratifs un tel dénombrement restreint peut être utile – mais alors il ne faut pas oublier qu'il ne peut jamais remplacer l'observation statistique complète. Il est d'autant plus nécessaire d'appuyer là-dessus, qu'il y a parmi nous dans ces jours un courant au sein des mathématiciens qui, dans de nombreuses directions, voudraient plutôt calculer qu'observer. Mais il faut rester ferme et dire : pas de calcul là où l'observation peut être faite."

5 Translated from French: "Je crois qu'il n'est pas juste de donner par un vœu du congrès à la méthode représentative (qui enfin ne peut être qu'un expédient) une importance que la statistique sérieuse ne reconnaîtra jamais. Sans doute, la statistique faite avec cette méthode ou, comme je pourrais l'appeler, la statistique, *pars pro toto*, nous a donné ça et là des renseignements intéressants ; mais son principe est tellement en contradiction avec les exigences que doit avoir la méthode statistique, que, comme statisticiens, nous ne devons pas accorder aux choses imparfaites le même droit de bourgeoisie, pour ainsi dire, que nous accordons à l'idéal que scientifiquement nous nous proposons d'atteindre."

The content of these reactions can again be summarized as follows: since statistics is by definition exhaustive, renouncing complete enumeration denies the very mission of statistical science. The discussion does not concern the method proposed by Kiaer, but is on the definition of statistical science. However, Kiaer did not let go, and continued to defend the representative method in 1897 at the congress of the ISI at St. Petersburg (see Kiær, 1899), in 1901 in Budapest, and in 1903 in Berlin (see Kiær, 1903, 1905). After this date, the issue is no longer mentioned at the ISI Congress. However, Kiær obtained the support of Arthur Bowley (1869–1957), who then played a decisive role in the development of sampling theory. Bowley (1906) presented an empirical verification of the application of the central limit theorem to sampling. He was the true promoter of random sampling techniques, developed stratified designs with proportional allocations, and used the law of total variance. It will be necessary to wait for the end of the First World War and the emergence of a new generation of statisticians for the problem to be rediscussed within the ISI. On this subject, we cannot help but quote Max Plank's reflection on the appearance of new scientific truths: "a new scientific truth does not triumph by convincing its opponents and making them see the light, but rather because its opponents eventually die, and a new generation grows up that is familiar with it" (quoted by Kuhn, 1970, p. 151).

In 1924, a commission (composed of Arthur Bowley, Corrado Gini, Adolphe Jensen, Lucien March, Verrijn Stuart, and Frantz Zizek) was created to evaluate the relevance of using the representative method. The results of this commission, entitled "Report on the representative method of statistics", were presented at the 1925 ISI Congress in Rome. The commission accepted the principle of survey sampling as long as the methodology is respected. Thirty years after Kiær's communication, the idea of sampling was officially accepted. The commission laid the foundation for future research. Two methods are clearly distinguished: "random selection" and "purposive selection". These two methods correspond to two fundamentally different scientific approaches. On the one hand, the validation of random methods is based on the calculation of probabilities that allows confidence intervals to be build for certain parameters. On the other hand, the validation of the purposive selection method can only be obtained through experimentation by comparing the obtained estimations to census results. Therefore, random methods are validated by a strictly mathematical argument while purposive methods are validated by an experimental approach.

1.4 Development of a Survey Sampling Theory

The report of the commission presented to the ISI Congress in 1925 marked the official recognition of the use of survey sampling. Most of the basic problems had already been posed, such as the use of random samples and the calculation of the variance of the estimators for simple and stratified designs. The acceptance of the use of partial data, and especially the recommendation to use random designs, led to a rapid mathematization of this theory. At that time, the calculation of probabilities was already known. In addition, statisticians had already developed a theory for experimental statistics. Everything was in place for the rapid progress of a fertile field of research: the construction of a statistical theory of survey sampling.

Jerzy Neyman (1894–1981) developed a large part of the foundations of the probabilistic theory of sampling for simple, stratified, and cluster designs. He also determined
the optimal allocation of a stratified design. The optimal allocation method challenges
the basic idea of the quota method, which is the "representativeness". Indeed, depending on the optimal stratification, the sample should not be a miniature of the population
as some strata must be overrepresented. The article published by Neyman (1934) in the
Journal of the Royal Statistical Society is currently considered one of the founding texts
of sampling theory. Neyman identified the main fields of research and his work was
to have a very important impact in later years. We now know that Tschuprow (1923)
had already obtained some of the results that were attributed to Neyman, but the latter
seems to have found them independently of Tschuprow. It is not surprising that such a
discovery was made simultaneously in several places. From the moment that the use of
random samples was considered a valid method, the theory would arise directly from
the application of the theory of probability.

1.5 The US Elections of 1936

During the same period, the implementation of the quota method contributed much
more to the development of the use of survey sampling methods than theoretical
studies. The 1936 US election marked an important turning point in the handling of
questionnaire surveys. The facts can be summarized as follows. The major American
newspapers used to publish, before the elections, the results of empirical surveys
produced from large samples (two million people polled for the *Literary Digest*) but
without any method to select individuals. While most polls predicted Landon's victory,
Roosevelt was elected. Surveys conducted by Crossley, Roper, and Gallup on smaller
samples but using the quota method gave a correct prediction. This event helped to
confirm the validity of the data provided by opinion polls.

This event, which favored the increase in the practice of sample surveys, was made
without reference to the probabilistic theory that had already been developed. The
method of Crossley, Roper, and Gallup is indeed not probabilistic but empirical,
therefore validation of the adequacy of the method is experimental and absolutely not
mathematical.

1.6 The Statistical Theory of Survey Sampling

The establishment of a new scientific consensus in 1925 and the identification of major
lines of research in the following years led to a very rapid development of survey theory.
During the Second World War, research continued in the United States. Important contributions are due to Deming & Stephan (1940), Stephan (1942, 1945, 1948) and Deming
(1948, 1950, 1960), especially on the question of adjusting statistical tables to census
data. Cornfield (1944) proposed using indicator variables for the presence of units in the
sample. Cochran (1939, 1942, 1946, 1961) and Hansen & Hurwitz (1943, 1949) showed
the interest of unequal probability sampling with replacement. Madow (1949) proposed

unequal probability systematic sampling (see also Hansen et al., 1953a,b). This is quickly established that an unequal probability sampling with fixed size without replacement is a complex problem. Narain (1951), Horvitz & Thompson (1952), Sen (1953), and Yates & Grundy (1953) presented several methods with unequal probabilities in two articles that are certainly among the most cited in this field. Devoted to the examination of several designs with unequal probabilities, these texts are mentioned for the general estimator (expansion estimator) of the total, which is also proposed and discussed. The expansion estimator is, in fact, an unbiased general estimator applicable to any sampling design without replacement. However, the proposed estimator of variance has a default. Yates & Grundy (1979) showed that the variance estimator proposed by Horvitz and Thompson can be negative. They proposed a valid variant when the sample is of fixed sample size and gives sufficient conditions for it to be positive. As early as the 1950s, the problem of sampling with unequal probabilities attracted considerable interest, which was reflected in the publication of more than 200 articles. Before turning to rank statistics, Hájek (1981) discussed the problem in detail. A book of synthesis by Brewer & Hanif (1983) was devoted entirely to this subject, which seems far from exhausted, as evidenced by regular publications.

The theory of survey sampling, which makes abundant use of the calculation of probabilities, attracted the attention of university statisticians and very quickly they reviewed all aspects of this theory that have a mathematical interest. A coherent mathematical theory of survey sampling was constructed. The statisticians very quickly came up against a difficult problem: surveys with finite populations. The proposed model postulated the identifiability of the units. This component of the model makes irrelevant the application of the reduction by sufficiency and the maximum likelihood method. Godambe (1955) states that there is no optimal linear estimator. This result is one of the many pieces of evidence showing the impossibility of defining optimal estimation procedures for general sampling designs in finite populations. Next, Basu (1969) and Basu & Ghosh (1967) demonstrated that the reduction by sufficiency is limited to the suppression of the information concerning the multiplicity of the units and therefore of the nonoperationality of this method. Several approaches were examined, including one from the theory of the decision. New properties, such as hyperadmissibility (see Hanurav, 1968), are defined for estimators applicable in finite populations.

A purely theoretical school of survey sampling developed rapidly. This theory attracted the attention of researchers specializing in mathematical statistics, such as Debabrata Basu, who was interested in the specifics of the theory of survey sampling. However, many of the proposed results were theorems of the nonexistence of optimal solutions. Research on the question of the foundations of inference in survey theory was becoming so important that it was the subject of a symposium in Waterloo, Canada, in 1971. At this symposium, the intervention of Calyampudi Radhakrishna Rao (1971, p. 178), began with a very pessimistic statement:

> I may mention that in statistical methodology, the existence of uniformly optimum procedures (such as UMV unbiased estimator, uniformly most powerful critical region for testing a hypothesis) is a rare exception rather than a rule. That is the reason why ad hoc criteria are introduced to restrict the class of procedures

in which an optimum may be sought. It is not surprising that the same situation is obtained in sampling for a finite situation. However, it presents some further complications which do not seem to exist for sampling from infinite populations.

This introduction announced the direction of current research.

In survey sampling theory, there is no theorem showing the optimality of an estimation procedure for general sampling designs. Optimal estimation methods can only be found by restricting them to particular classes of procedures. Even if one limits oneself to a particular class of estimators (such as the class of linear or unbiased estimators), it is not possible to obtain interesting results. One possible way out of this impasse is to change the formalization of the problem, for example by assuming that the population itself is random.

1.7 Modeling the Population

The absence of tangible general results concerning certain classes of estimators led to the development of population modeling by means of a model called "superpopulation". In this model-based approach, it is assumed that the values taken by the variable of interest on the observation units of the population are the realizations of random variables. The superpopulation model defines a class of distributions to which these random variables are supposed to belong. The sample is then derived from a double random experiment: a realization of the model that generates the population and then the choice of the sample. The idea of modeling the population was present in Brewer (1963a), but it was developed by Royall (1970b, 1971, 1976b) (see also Valliant et al., 2000; Chambers & Clark, 2012).

Drawing on the fact that the random sample is an "ancillary" statistic, Royall proposed to work conditionally on it. In other words, he considered that once the sample is selected, the choice of units is no longer random. This new modeling allowed the development of a particular research school. The model must express a known and previously accepted relationship. According to Royall, if the superpopulation model "adequately" describes the population, the inference can be conducted only with respect to the model, conditional to the sample selection. The use of the model then allows us to determine an optimal estimator.

One can object that a model is always an approximate representation of the population. However, the model is not built to be tested for data but to "assist" the estimation. If the model is correct, then Royall's method will provide a powerful estimator. If the model is false, the bias may be so important that the confidence intervals built for the parameter are not valid. This is essentially the critique stated by Hansen et al. (1983).

The debate is interesting because the arguments are not in the domain of mathematical statistics. Mathematically, these two theories are obviously correct. The argument relates to the adequacy of formalization to reality and is therefore necessarily external to the mathematical aspect of statistical development. In addition, the modeling proposed by Royall is particular. Above all, it makes it possible to break a theoretical impasse and therefore provide optimal estimators. However, the relevance of modeling is questionable and will be considered in a completely different way depending on whether one takes the arguments of sociology, demography or econometrics, three disciplines that are intimately related to the methodology of statistics. A comment from Dalenius (see Hansen et al., 1983, p. 800) highlights this problem:

> That is not to say that the arguments for or against parametric inference in the usual statistical theory are not of interest in the context of the theory of survey sampling. In our assessment of these arguments, however, we must pay attention to the relevant specifics of the applications.

According to Dalenius, it is therefore in the discipline in which the theory of survey sampling is applied that useful conclusions should be drawn concerning the adequacy of a superpopulation model.

The statistical theory of surveys mainly applies in official statistics institutes. These institutes do not develop a science but have a mission from their states. There is a fairly standard argument by the heads of national statistical institutes: the use of a superpopulation model in an estimation procedure is a breach of a principle of impartiality which is part of the ethics of statisticians. This argument comes directly from the current definition of official statistics. The principle of impartiality is part of this definition as the principle of accuracy was part of it in the 19th century. If modeling a population is easily conceived as a research tool or as a predictive tool, it remains fundamentally questionable in the field of official statistics.

1.8 Attempt to a Synthesis

The "superpopulation" approach has led to extremely fruitful research. The development of a hybrid approach called the model-assisted approach allows valid inferences to be provided under the model but is also robust when the model is wrong. This view was mainly developed by a Swedish school (see Särndal et al., 1992). The model allows us to take into account auxiliary information at the time of estimation while preserving properties of robustness for the estimators in the event of nonadequacy of the model. It is actually very difficult to construct an estimator that takes into account a set of auxiliary information after the selection of the sample without making a hypothesis, even a simple one, on the relation existing between the auxiliary information and the variable of interest. The modeling allows a conceptualization of this type of presumption. The model-assisted approach allows us to construct interesting and practical estimators. It is now clear that the introduction of a model is a necessity for dealing with some nonresponse and estimation problems in small areas. In this type of problem, whatever the technique used, one always postulates the existence of a model even if sometimes this is implicit. The model also deserves to be clearly determined in order to explain the underlying ideas that justify the application of the method.

1.9 Auxiliary Information

The 1990s were marked by the emergence of the concept of auxiliary information. This relatively general notion includes all information external to the survey itself used to increase the accuracy of the results of a survey. This information can be the knowledge of

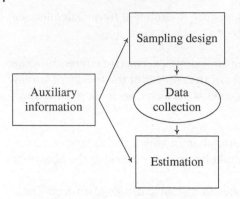

Figure 1.1 Auxiliary information can be used before or after data collection to improve estimations.

the values of one or more variables on all the units of the population or simply a function of these values. For most surveys, auxiliary information is available. It can be given by a census or simply by the sampling frame. Examples of auxiliary information include the total of a variable on the population, subtotals according to subpopulations, averages, proportions, variances, and values of a variable on all the units of the sampling frame. Therefore, the notion of auxiliary information encompasses all data from censuses or administrative sources.

The main objective is to use all this information to obtain accurate results. As shown in Figure 1.1, the auxiliary information can be used during the conception of the sampling design and at the time of the estimation of the parameters. When auxiliary information is used to construct the sampling design, a design is sought that provides accurate estimators for a given price or is inexpensive for given accuracy criteria. For these reasons, one can use unequal, stratified, balanced, clustered or multi-stage probability designs. When information is used in the estimation stage, it is used to "calibrate" the results of the census auxiliary information to the survey. The general method of calibration (in French, *calage*) of Deville & Särndal (1992) allows auxiliary information to be used without explicit references to a model.

Any sampling problem deals with how to use the available information. With the idea of auxiliary information, one frees oneself from the modeling of the population. This concept, which will be the main theme of this book, allows us to think of the problem of planning and estimation in an integrated way.

1.10   Recent References and Development

The books of precursors are Yates (1946, 1949, 1960, 1979), Deming (1948, 1950, 1960), Thionet (1953), Sukhatme (1954), Hansen et al. (1953a,b), Cochran (1953, 1963, 1977), Dalenius (1957), Kish (1965, 1989, 1995), Murthy (1967), Raj (1968), Johnson & Smith (1969), Sukhatme & Sukhatme (1970), Konijn (1973), Lanke (1975), Cassel et al. (1977, 1993), Jessen (1978), Hájek (1981), and Kalton (1983). These books are worth consulting because many modern ideas, especially on calibration and balancing, are discussed in them.

Important reference works include Skinner et al. (1989), Särndal et al. (1992), Lohr (1999, 2009b), Thompson (1997), Brewer (2002), Ardilly & Tillé (2006), and Fuller (2011). The series *Handbook of Statistics*, which is devoted to sampling, was published at 15-year intervals. First, a volume headed by Krishnaiah & Rao (1994), then two volumes headed by Pfeffermann & Rao (2009a,b). There is also a recent collective work led by Wolf et al. (2016).

The works of Thompson (1992, 1997, 2012) and Thompson & Seber (1996) are devoted to sampling in space. Methods for environmental sampling are developed in Gregoire & Valentine (2007) and for forestry in Mandallaz (2008). Several books are dedicated to unequal probability sampling and sampling algorithms. One can cite Brewer & Hanif (1983), Gabler (1990), and Tillé (2006). The model-based approach is clearly described in Valliant et al. (2000), Chambers & Clark (2012), and Valliant et al. (2013).

Many relevant books have been published and are still available in French. One can cite Thionet (1953), Desabie (1966), Deroo & Dussaix (1980), Gouriéroux (1981), Grosbras (1987), Dussaix & Grosbras (1992), Dussaix & Grosbras (1996), Ardilly (1994, 2006), Ardilly & Tillé (2003), and Ardilly & Lavallée (2017). In Italian, one can consult the works of Cicchitelli et al. (1992, 1997), Frosini et al. (2011), and Conti & Marella (2012). In Spanish, there exist also the books of Pérez López (2000), Tillé (2010), and Gutiérrez (2009) as well as a translation of the book of Sharon Lohr (2000). In German, one finds the books of Stenger (1985) and of Kauermann & Küchenhoff (2010). Finally, in Chinese there is a book by Ren & Ma (1996) and in Korean by Kim (2017).

Recently, new research fields have been opened. Small area estimation from survey data has became a major research topic (Rao, 2003; Rao & Molina, 2015). Recent developments in survey methodology are described in Groves (2004b) and Groves et al. (2009). Indirect sampling involves the selection of samples from a population that is not the population of interest but has links to it (Lavallée, 2002, 2007), for example new sampling algorithms have been developed to select balanced samples (Tillé, 2006). Adaptive sampling consists of completing the initial sample based on preliminary results (Thompson, 1992; Thompson & Seber, 1996). Capture–recapture methods are used to estimate the size of animal populations. Variations of these methods sometimes allow rare population sizes to be estimated or coverage surveys to be carried out (Pollock, 2000; Seber, 2002).

Resampling methods have been developed for finite populations (Shao & Tu, 1995; Groves, 2004b). Of course measurement errors will always remain a major research topic (Fuller, 1987; Groves, 2004a). Finally, substantial progress has been made in nonresponse methods: reweighting methods or imputation techniques (Särndal & Lundström, 2005; Bethlehem et al., 2011; De Waal et al., 2011; Kim & Shao, 2013).

One of the challenges that is currently emerging is the integration of data from multiple sources: administrative files, registers, and samples. In a thoughtful article entitled *Big data: are we making a big mistake?*, Tim Harford (2014) reminds us that the abundance of data is never a guarantee of quality. Access to new sources of data should not make us fall back into the mistakes of the past, as was the case during the 1936 US presidential election (see Section 1.5, page 6).

There have been methods for decades to integrate data from different sources. However, the multiplication of available sources makes these integration issues more and more complex. There is still a lot of research and development work needed to define the methods for integrating data from multiple sources by appropriately addressing the different measurement errors.

2

Population, Sample, and Estimation

2.1 Population

The theory of survey sampling is a set of tools for studying a population by examining only a part of it. The population denoted by U is assumed to be discrete and composed of a finite number of elements, e.g. companies, individuals, and households. Each unit can be identified by a label or an identification number. For simplicity, we will consider that the population is a set of identification numbers, in other words, $U = \{1, \dots, k, \dots, N\}$. There is also a formalization for continuous populations (Deville, 1989; Cordy, 1993) that will not be developed below.

In population U, we are interested in variables (or characters) x or y. The values taken by these variables on the observation unit k are denoted by x_k and y_k. In the design-based approach, x_k and y_k are not random. Indeed, under this approach, the only source of randomness is the way of selecting the sample.

The objective is to estimate parameters in this population. These parameters are also called functions of interest because they do not correspond to the usual definition of parameter used in inferential statistics for a parametric model. Parameters are simply functions of x_k or y_k. For example, the goal may be to estimate totals,

$$X = \sum_{k \in U} x_k, \quad Y = \sum_{k \in U} y_k,$$

means,

$$\overline{X} = \frac{1}{N} \sum_{k \in U} x_k, \quad \overline{Y} = \frac{1}{N} \sum_{k \in U} y_k,$$

or population variances,

$$S_x^2 = \frac{1}{N-1} \sum_{k \in U} (x_k - \overline{X})^2, \quad S_y^2 = \frac{1}{N-1} \sum_{k \in U} (y_k - \overline{Y})^2.$$

The population size N is not necessarily known and therefore can also be an estimation objective. However, as one can write

$$N = \sum_{k \in U} 1,$$

the estimation of the population size is a problem of the same nature as the estimation of X or Y.

Sampling and Estimation from Finite Populations, First Edition. Yves Tillé.
© 2020 John Wiley & Sons Ltd. Published 2020 by John Wiley & Sons Ltd.

Some parameters may depend jointly on two variables, such as the ratio of two totals,

$$R = \frac{Y}{X},$$

the population covariance,

$$S_{xy} = \frac{1}{N-1} \sum_{k \in U} (x_k - \overline{X})(y_k - \overline{Y})$$

or the correlation coefficient,

$$\rho_{xy} = \frac{S_{xy}}{S_x S_y}.$$

These parameters are unknown and will therefore be estimated using a sample.

2.2 Sample

A sample without replacement s is simply a subset of the population $s \subset U$. We also consider the set $\mathcal{S} = \{s | s \subset U\}$ of all the possible samples. For instance, if $U = \{1, 2, 3\}$, then

$$\mathcal{S} = \{\emptyset, \{1\}, \{2\}, \{3\}, \{1,2\}, \{1,3\}, \{2,3\}, U\},$$

where $\emptyset = \{\}$ denotes the empty set.

Definition 2.1 *A sampling design without replacement $p(\cdot)$ is a probability distribution on $\mathcal{S}$ such that*

$$p(s) \geq 0 \ \text{for all } s \subset U \ \text{and} \ \sum_{s \subset U} p(s) = 1.$$

Definition 2.2 *A random sample S is a random variable whose values are the samples:*

$$\Pr(S = s) = p(s) \ \text{for all } s \in U.$$

A random sample can also be defined as a discrete random vector composed of non-negative integer variables $\boldsymbol{a} = (a_1, \dots, a_k, \dots, a_N)^\top$. The variable a_k represents the number of times unit k is selected in the sample. If the sample is without replacement then variable a_k can only take the values 0 or 1 and therefore has a Bernoulli distribution. In general, random variables $a_1, \dots, a_N$ are not independent except in very special cases. The use of indicator variables a_k was introduced by Cornfield (1944) and greatly simplified the notation in survey sampling theory because it allows us to clearly separate the values of the variables x_k or y_k from the source of randomness $\boldsymbol{a}$.

Often, we try to select the sample as randomly as possible. The usual measure of randomness of a probability distribution is the entropy.

Definition 2.3 *The entropy of a sampling design is the quantity*

$$I(p) = - \sum_{s \subset U} p(s) \log p(s).$$

We suppose that $0 \log 0 = 0$.

We can search for sampling designs that maximize the entropy, with constraints such as a fixed sample size or given inclusion probabilities (see Section 2.3). A very random sampling design has better asymptotic properties and allows a more reliable inference (Berger, 1996, 1998a; Brewer & Donadio, 2003).

The sample size n_S is the number of units selected in the sample. We can write

$$n_S = \#S = \sum_{k \in U} a_k.$$

When the sample size is not random, we say that the sample is of fixed sample size and we simply denote it by n.

The variables are observed only on the units selected in the sample. A statistic $T(S)$ is a function of the values y_k that are observed on the random sample: $T(S) = f(y_k, k \in S)$. This statistic takes the value $T(s)$ on the sample s. The expectation under the design is defined from the sampling design:

$$E_p[T(S)] = \sum_{s \subset U} p(s) T(s).$$

The variance operator is defined using the expectation operator:

$$\mathrm{var}_p[T(S)] = E_p\{T(S) - E_p[T(S)]\}^2 = E_p[T(S)^2] - E_p^2[T(S)].$$

2.3 Inclusion Probabilities

The inclusion probability π_k is the probability that unit k is selected in the sample. This probability is, in theory, derived from the sampling design:

$$\pi_k = \Pr(k \in S) = E_p(a_k) = \sum_{\substack{s \subset U \\ s \ni k}} p(s),$$

for all $k \in U$. In sampling designs without replacement, the random variables a_k have Bernoulli distributions with parameter π_k. There is no particular reason to select units with equal probabilities. However, it will be seen below that it is important that all inclusion probabilities be nonzero.

The second-order inclusion probability (or joint inclusion probability) $\pi_{k\ell}$ is the probability that units k and ℓ are selected together in the sample:

$$\pi_{k\ell} = \Pr(k \in S \text{ and } \ell \in S) = E_p(a_k a_\ell) = \sum_{\substack{s \in U \\ s \supset \{k, \ell\}}} p(s),$$

for all $k, \ell \in U$. In sampling designs without replacement, when $k = \ell$, the second-order inclusion probability is reduced to the first-order inclusion probability, in other words $\pi_{kk} = \pi_k$, for all $k \in U$.

The variance of the indicator variable a_k is denoted by

$$\Delta_{kk} = \mathrm{var}_p(a_k) = E_p(a_k^2) - E_p^2(a_k) = E_p(a_k) - E_p^2(a_k) = \pi_k - \pi_k^2 = \pi_k(1 - \pi_k),$$

which is the variance of a Bernoulli variable. The covariances between indicators are

$$\Delta_{k\ell} = \mathrm{cov}_p(a_k, a_\ell) = E_p(a_k, a_\ell) - E_p(a_k)E_p(a_\ell) = \pi_{k\ell} - \pi_k \pi_\ell.$$

One can also use a matrix notation. Let

$$a = (a_1, \dots, a_N)^{\mathsf{T}}$$

be a column vector. The vector of inclusion probabilities is

$$\pi = E_p(a) = (\pi_1, \dots, \pi_N)^{\mathsf{T}}.$$

Define also the symmetric matrix:

$$\Pi = E_p(a\,a^{\mathsf{T}}) = \begin{pmatrix} \pi_1 & \cdots & \pi_{1k} & \cdots & \pi_{1N} \\ \vdots & \ddots & \vdots & & \vdots \\ \pi_{k1} & \cdots & \pi_k & \cdots & \pi_{kN} \\ \vdots & & \vdots & \ddots & \vdots \\ \pi_{N1} & \cdots & \pi_{Nk} & \cdots & \pi_N \end{pmatrix}$$

and the variance–covariance matrix

$$\Delta = \mathrm{var}_p(a) = E_p(a\,a^{\mathsf{T}}) - E_p(a)\,E_p(a^{\mathsf{T}})$$

$$= \Pi - \pi\pi^{\mathsf{T}} = \begin{pmatrix} \Delta_{11} & \cdots & \Delta_{1k} & \cdots & \Delta_{1N} \\ \vdots & \ddots & \vdots & & \vdots \\ \Delta_{k1} & \cdots & \Delta_{kk} & \cdots & \Delta_{kN} \\ \vdots & & \vdots & \ddots & \vdots \\ \Delta_{N1} & \cdots & \Delta_{Nk} & \cdots & \Delta_{NN} \end{pmatrix}.$$

Matrix Δ is a variance–covariance matrix which is therefore semi-definite positive.

Result 2.1 *The sum of the inclusion probabilities is equal to the expected sample size.*

Proof:

$$E_p(n_S) = E_p\left(\sum_{k \in U} a_k\right) = \sum_{k \in U} E_p(a_k) = \sum_{k \in U} \pi_k.$$

∎

If the sample size is fixed, then n_S is not random. In this case, the sum of the inclusion probabilities is exactly equal to the sample size.

Result 2.2 *If the random sample is of fixed sample size, then*

$$\sum_{\ell \in U} \pi_{k\ell} = n\pi_k \quad \text{and} \quad \sum_{k \in U} \Delta_{k\ell} = 0.$$

Proof: Let $\mathbf{1}$ be a column vector consisting of N ones and $\mathbf{0}$ a column vector consisting of N zeros, the sample size can be written as $n = \mathbf{1}^{\mathsf{T}} a$. We directly have

$$\Pi\mathbf{1} = E_p(a\,a^{\mathsf{T}})\mathbf{1} = E_p(a\,a^{\mathsf{T}}\mathbf{1}) = E_p(an) = n\pi$$

and

$$\Delta\mathbf{1} = (\Pi - \pi\,\pi^{\mathsf{T}})\mathbf{1} = n\pi - n\pi = \mathbf{0}.$$

∎

If the sample is of fixed sample size, the sum of all rows and all columns of Δ is zero. Therefore, matrix Δ is singular. Its rank is then less than or equal to $N - 1$.

Example 2.1 Let $U = \{1, 2, 3\}$ be a population of size $N = 3$. The sampling design is defined as follows:

$$p(\{1,2\}) = \frac{1}{2}, p(\{1,3\}) = \frac{1}{4}, p(\{2,3\}) = \frac{1}{4},$$

and $p(s) = 0$ for all other samples. The random sample is of fixed sample size $n = 2$.

$$\pi_1 = \frac{1}{2} + \frac{1}{4} = \frac{3}{4}, \ \pi_2 = \frac{1}{2} + \frac{1}{4} = \frac{3}{4}, \ \pi_3 = \frac{1}{4} + \frac{1}{4} = \frac{1}{2}.$$

The sum of the inclusion probabilities is equal to the sample size. Indeed,

$$\sum_{k \in U} \pi_k = \frac{3}{4} + \frac{3}{4} + \frac{1}{2} = 2.$$

The joint inclusion probabilities are

$$\pi_{12} = \frac{1}{2}, \pi_{13} = \frac{1}{4}, \pi_{23} = \frac{1}{4}.$$

Therefore, the matrices are

$$\mathbf{\Pi} = \begin{pmatrix} 3/4 & 1/2 & 1/4 \\ 1/2 & 3/4 & 1/4 \\ 1/4 & 1/4 & 1/2 \end{pmatrix}$$

and

$$\mathbf{\Delta} = \mathbf{\Pi} - \boldsymbol{\pi}\boldsymbol{\pi}^{\mathsf{T}} = \begin{pmatrix} 3/4 & 1/2 & 1/4 \\ 1/2 & 3/4 & 1/4 \\ 1/4 & 1/4 & 1/2 \end{pmatrix} - \begin{pmatrix} 9/16 & 9/16 & 3/8 \\ 9/16 & 9/16 & 3/8 \\ 3/8 & 3/8 & 1/4 \end{pmatrix}$$

$$= \begin{pmatrix} 3/16 & -1/16 & -1/8 \\ -1/16 & 3/16 & -1/8 \\ -1/8 & -1/8 & 1/4 \end{pmatrix}.$$

We find that the sums of all the rows and all the columns of $\mathbf{\Delta}$ are null because the sampling design is of fixed sample size $n = 2$.

2.4 Parameter Estimation

A parameter or a function of interest $\theta = \theta(y_k, k \in U)$ is a function of the values taken by one or more variables in the population. A statistic is a function of the data observed in the random sample $T(S) = f(y_k, k \in S) = f(a_k y_k, k \in U)$.

Definition 2.4 *An estimator $\hat{\theta}$ is a statistic used to estimate a parameter.*

If $\hat{\theta}(s)$ denotes the value taken by the estimator on sample s, the expectation of the estimator is

$$E_p(\hat{\theta}) = \sum_{s \subset U} p(s)\hat{\theta}(s).$$

Definition 2.5 *An estimator $\hat{\theta}$ is said to be unbiased if* $E_p(\hat{\theta}) = \theta$, *for all* $y \in \mathbb{R}^N$, *where* $y = (y_1, \dots, y_N)^{\mathsf{T}}$.

Definition 2.6 *The bias of an estimator $\hat{\theta}$ is the difference between its expectation and the parameter it estimates:*

$$B_p(\hat{\theta}) = E_p(\hat{\theta}) - \theta.$$

From the expectation, we can define the variance of the estimator:

$$\text{var}_p(\hat{\theta}) = E_p[\hat{\theta} - E_p(\hat{\theta})]^2 = E_p(\hat{\theta}^2) - E_p^2(\hat{\theta})$$

and the mean squared error (MSE):

$$\text{MSE}_p(\hat{\theta}) = E_p(\hat{\theta} - \theta)^2.$$

Result 2.3 *The mean squared error is the sum of the variance and the square of the bias:*

$$\text{MSE}_p(\hat{\theta}) = \text{var}_p(\hat{\theta}) + B_p^2(\hat{\theta}).$$

Proof:

$$\text{MSE}_p(\hat{\theta}) = E_p(\hat{\theta} - \theta)^2 = E_p[\hat{\theta} - E_p(\hat{\theta}) + E_p(\hat{\theta}) - \theta]^2$$

$$= E_p[\hat{\theta} - E_p(\hat{\theta})]^2 + [E_p(\hat{\theta}) - \theta]^2 + 2[E_p(\hat{\theta}) - \theta] \underbrace{E_p[\hat{\theta} - E_p(\hat{\theta})]}_{=0}$$

$$= \text{var}_p(\hat{\theta}) + B_p^2(\hat{\theta}). \qquad \blacksquare$$

2.5 Estimation of a Total

For estimating the total

$$Y = \sum_{k \in U} y_k,$$

the basic estimator is the expansion estimator:

$$\hat{Y} = \sum_{k \in S} \frac{y_k}{\pi_k} = \sum_{\substack{k \in U \\ \pi_k > 0}} \frac{y_k a_k}{\pi_k}.$$

This estimator was proposed by Narain (1951) and Horvitz & Thompson (1952). It is often called the Horvitz–Thompson estimator, but Narain's article precedes that of Horvitz–Thompson. It may also be referred to as the Narain or Narain–Horvitz–Thompson estimator or π-estimator or estimator by dilated values.

Often, one writes

$$\hat{Y} = \sum_{k \in U} \frac{y_k a_k}{\pi_k},$$

but this is correct only if $\pi_k > 0$, for all $k \in U$. If any inclusion probabilities are zero, then y_k is divided by 0. Of course, if an inclusion probability is zero, the corresponding unit is never selected in the sample.

In this estimator, the values taken by the variable are weighted by the inverse of their inclusion probabilities. The inverse of π_k is often denoted by $d_k = 1/\pi_k$ and can be interpreted as the number of units that unit k represents in the population. The value d_k is the weight assigned to unit k. The expansion estimator can also be written as

$$\hat{Y} = \sum_{\substack{k \in U \\ \pi_k > 0}} d_k y_k a_k.$$

Result 2.4 *A necessary and sufficient condition for the expansion estimator $\hat{Y}$ to be unbiased is that $\pi_k > 0$, for all $k \in U$.*

Proof: Since

$$E_p(\hat{Y}) = E_p\left(\sum_{\substack{k \in U \\ \pi_k > 0}} \frac{y_k a_k}{\pi_k} \right) = \sum_{\substack{k \in U \\ \pi_k > 0}} \frac{y_k E_p(a_k)}{\pi_k}$$

$$= \sum_{\substack{k \in U \\ \pi_k > 0}} \frac{y_k \pi_k}{\pi_k} = \sum_{\substack{k \in U \\ \pi_k > 0}} y_k = \sum_{k \in U} y_k - \sum_{\substack{k \in U \\ \pi_k = 0}} y_k,$$

the bias of the estimator is

$$B_p(\hat{Y}) = E_p(\hat{Y}) - Y = - \sum_{\substack{k \in U \\ \pi_k = 0}} y_k.$$

The bias is zero if and only if $\pi_k > 0$, for all $k \in U$. ∎

If some inclusion probabilities are zero, it is impossible to estimate Y without bias. It is said that the sampling design does not cover the population or is vitiated by a coverage problem. We sometimes hear that a sample is biased, but this terminology should be avoided because bias is a property of an estimator and not of a sample. In what follows, we will consider that all the sampling designs have nonzero first-order inclusion probabilities.

2.6 Estimation of a Mean

For estimating the mean

$$\overline{Y} = \frac{1}{N} \sum_{k \in U} y_k,$$

we can simply divide the expansion estimator of the total by the size of the population. We obtain

$$\hat{\overline{Y}} = \frac{\hat{Y}}{N} = \frac{1}{N} \sum_{k \in S} \frac{y_k}{\pi_k}.$$

However, the population size is not necessarily known.

Estimator $\widehat{\overline{Y}}$ is unbiased, but it suffers from a serious problem. If the variable of interest y_k is constant, in other words, $y_k = C$, then

$$\widehat{\overline{Y}} = \frac{1}{N} \sum_{k \in S} \frac{C}{\pi_k} = C \frac{1}{N} \sum_{k \in S} \frac{1}{\pi_k} \neq C. \tag{2.1}$$

Indeed, there is no reason that

$$\frac{1}{N} \sum_{k \in S} \frac{1}{\pi_k}$$

is equal to 1, even if the expectation of this quantity is equal to 1. The expansion estimator of the mean is thus vitiated by an error that does not depend on the variability of variable y. Even if $y_k = C$, $\widehat{\overline{Y}}$ remains random.

We refer to the following definition:

Definition 2.7 *An estimator of the mean $\widehat{\overline{Y}}$ is said to be linearly invariant if, for all $a, b \in \mathbb{R}$, when $z_k = a + b y_k$ then $\widehat{\overline{Z}} = a + b \widehat{\overline{Y}}$.*

For this reason, even when the size of the population N is known, it is recommended to use the estimator of Hájek (1971) (HAJ) which consists of dividing the total by the sum of the inverses of the inclusion probabilities:

$$\widehat{\overline{Y}}_{\text{HAJ}} = \frac{\sum_{k \in S} y_k / \pi_k}{\sum_{k \in S} 1 / \pi_k}. \tag{2.2}$$

When $y_k = C$, then $\text{var}_p(\widehat{\overline{Y}}_{\text{HAJ}}) = 0$. Therefore, the bad property of the expansion estimator is solved because the Hájek estimator is linearly invariant. However, $\widehat{\overline{Y}}_{\text{HAJ}}$ is usually biased because it is a ratio of two random variables. In some cases, such as simple random sampling without replacement with fixed sample size, the Hájek ratio is equal to the expansion estimator.

2.7 Variance of the Total Estimator

Result 2.5 *The variance of the expansion estimator of the total is*

$$\text{var}_p(\widehat{Y}) = \sum_{k \in U} \sum_{\ell \in U} \frac{y_k y_\ell}{\pi_k \pi_\ell} \Delta_{k\ell}. \tag{2.3}$$

Proof:

$$\text{var}_p(\widehat{Y}) = \text{var}_p \left(\sum_{k \in U} \frac{y_k a_k}{\pi_k} \right) = \sum_{k \in U} \sum_{\ell \in U} \frac{y_k y_\ell \, \text{cov}_p(a_k, a_\ell)}{\pi_k \pi_\ell} = \sum_{k \in U} \sum_{\ell \in U} \frac{y_k y_\ell}{\pi_k \pi_\ell} \Delta_{k\ell}. \qquad \blacksquare$$

It is also possible to write the expansion estimator with a vector notation. Let $\check{y}$ be the vector of dilated values:

$$\check{y} = \left(\frac{y_1}{\pi_1}, \cdots, \frac{y_k}{\pi_k}, \cdots, \frac{y_N}{\pi_N} \right)^{\mathsf{T}}.$$

This vector only exists if all the inclusion probabilities are nonzero. We can then write

$$\widehat{Y} = \check{y}^\top a.$$

The calculation of the variance is then immediate:

$$\mathrm{var}_p(\widehat{Y}) = \mathrm{var}_p(\check{y}^\top a) = \check{y}^\top \mathrm{var}_p(a)\check{y} = \check{y}^\top \Delta \check{y}.$$

If the sample is of fixed sample size, Sen (1953) and Yates & Grundy (1953) have shown the following result:

Result 2.6 *If the sampling design is of fixed sample size, then the variance of the expansion estimator can also be written as*

$$\mathrm{var}_p(\widehat{Y}) = \frac{-1}{2} \sum_{k \in U} \sum_{\ell \in U} \left(\frac{y_k}{\pi_k} - \frac{y_\ell}{\pi_\ell} \right)^2 \Delta_{k\ell}. \tag{2.4}$$

Proof: By expanding the square of Expression (2.4), we obtain

$$\frac{-1}{2} \sum_{k \in U} \sum_{\ell \in U} \left(\frac{y_k}{\pi_k} - \frac{y_\ell}{\pi_\ell} \right)^2 \Delta_{k\ell} = -\frac{1}{2} \sum_{k \in U} \sum_{\ell \in U} \left(\frac{y_k^2}{\pi_k^2} + \frac{y_\ell^2}{\pi_\ell^2} - 2\frac{y_k}{\pi_k}\frac{y_\ell}{\pi_\ell} \right) \Delta_{k\ell}$$

$$= -\frac{1}{2} \sum_{k \in U} \frac{y_k^2}{\pi_k^2} \sum_{\ell \in U} \Delta_{k\ell} - \frac{1}{2} \sum_{\ell \in U} \frac{y_\ell^2}{\pi_\ell^2} \sum_{k \in U} \Delta_{k\ell} + \sum_{k \in U} \sum_{\ell \in U} \frac{y_k}{\pi_k}\frac{y_\ell}{\pi_\ell} \Delta_{k\ell}. \tag{2.5}$$

Under a design with fixed sample size, it has been proved in Result 2.2, page 16, that

$$\sum_{k \in U} \Delta_{k\ell} = 0, \text{ for all } \ell \in U \quad \text{and} \quad \sum_{\ell \in U} \Delta_{k\ell} = 0, \text{ for all } k \in U.$$

The first two terms of Expression (2.5) are therefore null and we find the expression of Result 2.5. ∎

In order to estimate the variance, we use the following general result:

Result 2.7 *Let $g(.,.)$ be a function from $\mathbb{R}^2$ to $\mathbb{R}$. A necessary and sufficient condition for*

$$\sum_{k \in S} \sum_{\ell \in S} \frac{g(y_k, y_\ell)}{\pi_{k\ell}}$$

to unbiased ly estimate

$$\sum_{k \in U} \sum_{\ell \in U} g(y_k, y_\ell)$$

is that $\pi_{k\ell} > 0$ for all $k, \ell \in U$.

Proof: Since

$$\mathrm{E}_p \left[\sum_{k \in S} \sum_{\ell \in S} \frac{g(y_k, y_\ell)}{\pi_{k\ell}} \right] = \mathrm{E}_p \left[\sum_{\substack{k \in U \\ \pi_{k\ell} > 0}} \sum_{\ell \in U} \frac{g(y_k, y_\ell) a_k a_\ell}{\pi_{k\ell}} \right]$$

$$= \sum_{\substack{k \in U}} \sum_{\substack{\ell \in U \\ \pi_{k\ell} > 0}} \frac{g(y_k, y_\ell)}{\pi_{k\ell}} E_p(a_k a_\ell) = \sum_{\substack{k \in U}} \sum_{\substack{\ell \in U \\ \pi_{k\ell} > 0}} g(y_k, y_\ell)$$

$$= \sum_{k \in U} \sum_{\ell \in U} g(y_k, y_\ell) - \sum_{\substack{k \in U}} \sum_{\substack{\ell \in U \\ \pi_{k\ell} = 0}} g(y_k, y_\ell),$$

the estimator is unbiased if and only if $\pi_{k\ell} > 0$ for all $k, \ell \in U$. ∎

This result enables us to construct two variance estimators. The first one is called the Horvitz-Thompson estimator. It is obtained by applying Result 2.7 to Expression (2.3):

$$v_{HT}(\widehat{Y}) = \sum_{k \in S} \sum_{\ell \in S} \frac{y_k y_\ell}{\pi_k \pi_\ell} \frac{\Delta_{k\ell}}{\pi_{k\ell}}. \tag{2.6}$$

This estimator can take negative values, but is unbiased.

The second one is called Sen–Yates–Grundy estimator (see Sen, 1953; Yates & Grundy, 1953). It is unbiased only for designs with fixed sample size s. It is obtained by applying Result 2.7 to Expression (2.4):

$$v_{SYG}(\widehat{Y}) = \frac{-1}{2} \sum_{k \in S} \sum_{\ell \in S} \left(\frac{y_k}{\pi_k} - \frac{y_\ell}{\pi_\ell} \right)^2 \frac{\Delta_{k\ell}}{\pi_{k\ell}}. \tag{2.7}$$

This estimator can also take negative values, but when $\Delta_{k\ell} \leq 0$, for all $k \neq \ell \in U$, then the estimator is always positive. This condition is called the Yates–Grundy condition.

The Yates–Grundy condition is a special case of the negative correlation property defined as follows:

Definition 2.8 *A sampling design is said to be negatively correlated if, for all $G \subset U$,*

$$E_p \left(\prod_{k \in G} a_k \right) \leq \prod_{k \in G} E_p(a_k).$$

2.8 Sampling with Replacement

Sampling designs with replacement should not be used except in very special cases, such as in indirect sampling, described in Section 8.4, page 187 (Deville & Lavallée, 2006; Lavallée, 2007), adaptive sampling, described in Section 8.4.2, page 188 (Thompson, 1988), or capture–recapture techniques, also called capture–mark sampling, described in Section 8.5, page 191 (Pollock, 1981; Amstrup et al., 2005).

In a sampling design with replacement, the same unit can be selected several times in the sample. The random sample can be written using a vector $a = (a_1, \ldots, a_N)^\top$, where a_k represents the number of times that unit k is selected in the sample. The a_k can therefore take any non-negative integer value.

The vector a is therefore a positive discrete random vector. The value $\mu_k = E_p(a_k)$ is the expectation under the design of the number of times that unit k is selected. We also write $\mu_{k\ell} = E_p(a_k a_\ell)$, $\Sigma_{kk} = \text{var}_p(a_k)$, and $\Sigma_{k\ell} = \text{cov}_p(a_k, a_\ell)$. The expectation

$\mu_k = E_p(a_k) \neq \mu_{kk} = E_p(a_k^2)$, since a_k can take values larger than 1. We assume that $\mu_k > 0$ for all $k \in U$. Under this assumption, the Hansen–Hurwitz (HH) estimator is

$$\hat{Y}_{HH} = \sum_{k \in U} \frac{y_k a_k}{\mu_k}$$

and is unbiased for Y (Hansen & Hurwitz, 1949). The demonstration is the same as for Result 2.4.

The variance is

$$\text{var}_p(\hat{Y}_{HH}) = \sum_{k \in U} \sum_{\ell \in U} \frac{y_k y_\ell}{\mu_k \mu_\ell} \Sigma_{k\ell}.$$

If $\mu_{k\ell} > 0$ for all $k, \ell \in U$, this variance can be unbiasedly estimated by

$$v(\hat{Y}_{HH}) = \sum_{k \in U} \sum_{\ell \in U} \frac{a_k y_k a_\ell y_\ell}{\mu_k \mu_\ell} \frac{\Sigma_{k\ell}}{\mu_{k\ell}}.$$

Indeed,

$$E_p \, v(\hat{Y}_{HH}) = \sum_{k \in U} \sum_{\ell \in U} \frac{y_k y_\ell}{\mu_k \mu_\ell} \frac{\Sigma_{k\ell} E_p(a_k a_\ell)}{\mu_{k\ell}} = \text{var}_p(\hat{Y}_{HH}).$$

There are two other possibilities for estimating the total without bias. To do this, we use the reduction function $r(\cdot)$ from $\mathbb{N}$ to $\{0, 1\}$:

$$r(a_k) = \begin{cases} 0 \text{ if } a_k = 0 \\ 1 \text{ if } a_k \geq 1. \end{cases} \tag{2.8}$$

This function removes the multiplicity of units in the sense that units selected more than once in the sample are kept only once.

We then write π_k, the first-order inclusion probability

$$\pi_k = E_p[r(a_k)] = \Pr(a_k \geq 1),$$

and $\pi_{k\ell}$, the second-order inclusion probability

$$\pi_{k\ell} = E_p[r(a_k)r(a_\ell)] = \Pr(a_k \geq 1 \text{ and } a_\ell \geq 1).$$

By keeping only the distinct units, we can then simply use the expansion estimator:

$$\hat{Y} = \sum_{k \in U} \frac{y_k r(a_k)}{\pi_k}.$$

Obviously, if the design with replacement is of fixed sample size, in other words, if

$$\text{var}_p \left(\sum_{k \in U} a_k \right) = 0,$$

the sample of distinct units does not necessarily have a fixed sample size. The expansion estimator is not necessarily more accurate than the Hansen–Hurwitz estimator.

A third solution consists of calculating the so-called Rao–Blackwellized estimator. Without going into the technical details, it is possible to show that in the design-based theory, the minimal sufficient statistic can be constructed by removing the information concerning the multiplicity of units. In other words, if a unit is selected several times in the sample with replacement, it is conserved only once (Basu & Ghosh, 1967; Basu, 1969; Cassel et al., 1977, 1993; Thompson & Seber, 1996, p. 35). Knowing a minimal sufficient statistic, one can then calculate the augmented estimator (also called the Rao–Blackwellized estimator) by conditioning an estimator with respect to the minimal sufficient statistic.

Concretely, we calculate the conditional expectation $v_k = E_p[a_k|r(a)]$, for all $k \in U$, where $r(a) = (r(a_1), \dots, r(a_N))^\top$. Since $a_k = 0$ implies that $v_k = 0$, we can define the Rao–Blackwellized estimator (RB):

$$\hat{Y}_{\mathrm{RB}} = \sum_{k \in U} \frac{y_k v_k}{\mu_k}. \tag{2.9}$$

This estimator is unbiased because $E_p(v_k) = E_p E_p[a_k|r(a)] = \mu_k$. Moreover, since

$$\hat{Y}_{\mathrm{RB}} = E_p[\hat{Y}_{\mathrm{HH}}|r(a)],$$

and

$$\mathrm{var}_p(\hat{Y}_{\mathrm{HH}}) = \mathrm{var}_p \, E_p[\hat{Y}_{\mathrm{HH}}|r(a)] + E_p \, \mathrm{var}_p[\hat{Y}_{\mathrm{HH}}|r(a)],$$

we have

$$\mathrm{var}_p(\hat{Y}_{\mathrm{RB}}) \leq \mathrm{var}_p(\hat{Y}_{\mathrm{HH}}).$$

The Hansen–Hurwitz estimator should therefore in principle never be used. It is said that the Hansen–Hurwitz estimator is not admissible in the sense that it can always be improved by calculating its conditional expectation. However, this conditional expectation can sometimes be very complex to calculate. Rao–Blackwellization is at the heart of the theory of adaptive sampling, which can lead to multiple selections of the same unit in the sample (Thompson, 1990; Félix-Medina, 2000; Thompson, 1991a; Thompson & Seber, 1996).

Exercises

2.1 Show that

$$S_y^2 = \frac{1}{N-1} \sum_{k \in U} (y_k - \overline{Y})^2 = \frac{1}{2N(N-1)} \sum_{k \in U} \sum_{\substack{\ell \in U \\ \ell \neq k}} (y_k - y_\ell)^2.$$

2.2 Let $U = \{1, 2, 3\}$ be a population with the following sampling design:

$$p(\{1,2\}) = \frac{1}{2}, p(\{1,3\}) = \frac{1}{4}, p(\{2,3\}) = \frac{1}{4}.$$

Give the first-order inclusion probabilities. Give the variance–covariance matrix of the indicator variables.

2.3 Let $U = \{1, 2, 3\}$ and have the following sampling design:

$$p(\{1,2\}) = \frac{1}{2}, p(\{1,3\}) = \frac{1}{4}, p(\{2,3\}) = \frac{1}{4}.$$

1. Give the probability distributions of the expansion estimator and the Hájek estimator of the mean. Give the probability distributions of the two variance estimators of the expansion estimator and calculate their bias.
2. Give the probability distributions of the two variance estimators of the expansion estimator of the mean in the case where $y_k = \pi_k, k \in U$.

2.4 Let $p(s)$ be a sampling design without replacement applied to a population U of size N. Let $\pi_k, k \in U$, and $\pi_{k\ell}, k, \ell \in U, k \neq \ell$ denote the first- and second-order inclusion probabilities, respectively, and S is the random sample. Consider the following estimator:

$$\hat{\theta} = \frac{1}{N^2} \sum_{k \in S} \frac{y_k}{\pi_k} + \frac{1}{N^2} \sum_{k \in S} \sum_{\substack{\ell \in S \\ \ell \neq k}} \frac{y_\ell}{\pi_{k\ell}}.$$

For which function of interest is this estimator unbiased?

2.5 For a design without replacement with strictly positive inclusion probabilities, construct an unbiased estimator for S_y^2.

2.6 Let U be a finite population and let S be the random sample of U obtained by means of a design with inclusion probabilities π_k and $\pi_{k\ell}$. We suppose that this design is balanced on a variable z. In other words,

$$\sum_{k \in S} \frac{z_k}{\pi_k} = \sum_{k \in U} z_k. \tag{2.10}$$

The total of the variable of interest y is

$$Y = \sum_{k \in U} y_k$$

and can be unbiasedly estimated by

$$\hat{Y} = \sum_{k \in S} \frac{y_k}{\pi_k}.$$

1. Show that

$$\sum_{\substack{k \in U \\ k \neq \ell}} \frac{z_k \pi_{k\ell}}{\pi_k} = \pi_\ell \sum_{k \in U} z_k - z_\ell. \tag{2.11}$$

2. What particular result do we obtain when $z_k = \pi_k, k \in U$?
3. Show that

$$\operatorname{var}_p(\hat{Y}) = \frac{1}{2} \sum_{k \in U} \sum_{\substack{\ell \in U \\ \ell \neq k}} \left(\frac{y_k}{z_k} - \frac{y_\ell}{z_\ell} \right)^2 z_k z_\ell \frac{\pi_k \pi_\ell - \pi_{k\ell}}{\pi_k \pi_\ell}. \tag{2.12}$$

4. What result is generalized by Expression (2.12)?
5. Construct an unbiased variance estimator from Expression (2.12).

2.7 Let $\Delta = (\Delta_{k\ell})$ be the variance–covariance matrix of the indicators of the presence of the units in the sample for a design $p(s)$:

$$\Delta = \begin{pmatrix} 1 & 1 & 1 & -1 & -1 \\ 1 & 1 & 1 & -1 & -1 \\ 1 & 1 & 1 & -1 & -1 \\ -1 & -1 & -1 & 1 & 1 \\ -1 & -1 & -1 & 1 & 1 \end{pmatrix} \times \frac{6}{25}.$$

1. Is this design of fixed sample size?
2. Does this design satisfy the Yates–Grundy conditions?
3. Calculate the inclusion probabilities of this design knowing that

$$\pi_1 = \pi_2 = \pi_3 > \pi_4 = \pi_5.$$

4. Give the matrix of the second-order inclusion probabilities.
5. Give the probabilities associated with all possible samples.

3

Simple and Systematic Designs

Simple designs form the basis of sampling theory. They serve as a reference for defining more complex designs, such as stratified, cluster, and multi-stage designs. The study of simple designs is not limited to the classic description of simple random sampling without replacement (Section 3.1), Bernoulli sampling (Section 3.2), or simple random sampling with replacement (Sections 3.3 and 3.4). Under simple random sampling with replacement, it is only possible to keep distinct units (Section 3.5). Simple random sampling with replacement enables the use of the results of Section 2.8, page 22, for calculating the Rao–Blackwellized estimator. Inverse designs may be of practical interest, such as solving the "coupon collector's problem", also called "inverse sampling" (Section 3.6). The estimation of other functions of interest (Section 3.7) are examined such as counts, proportions or ratios. We then solve the problem of determining the size of the sample (Section 3.8) for simple designs with and without replacement. We give several algorithms (Section 3.9) to implement the designs described in this chapter. Finally, we describe systematic sampling with equal probabilities (Section 3.10), which can be an interesting alternative to simple random sampling without replacement. Section 3.11 is dedicated to calculating the entropy for simple, Bernoulli, and systematic designs.

3.1 Simple Random Sampling without Replacement with Fixed Sample Size

3.1.1 Sampling Design and Inclusion Probabilities

A sampling design without replacement is said to be simple if all samples with the same size have the same probability of being selected. The number of subsets of size n in a population of N is the number of combinations of n units among N. There is only one sampling design without replacement with fixed sample size n:

Definition 3.1 *A sampling design with fixed sample size n is said to be simple without replacement if and only if*

$$p(s) = \begin{cases} \binom{N}{n}^{-1} & \text{if } \#s = n \\ 0 & \text{otherwise,} \end{cases}$$

Sampling and Estimation from Finite Populations, First Edition. Yves Tillé.
© 2020 John Wiley & Sons Ltd. Published 2020 by John Wiley & Sons Ltd.

where $n \in \{1, \dots, N\}$.

The inclusion probabilities of all orders can be derived from the design. For the first-order inclusion probabilities, we have

$$\pi_k = \sum_{\substack{s \subset U \\ s \ni k}} p(s) = \sum_{\substack{s \in \mathcal{S}_n \\ s \ni k}} \binom{N}{n}^{-1},$$

where $\mathcal{S}_n = \{s \subset U | \#s = n\}$. There are $\binom{N-1}{n-1}$ samples of size n containing unit k:

$$\#\{s \subset U | \#s = n, k \in s\} = \binom{N-1}{n-1}.$$

Therefore, we obtain

$$\pi_k = \binom{N-1}{n-1}\binom{N}{n}^{-1} = \frac{n}{N}, \text{ for all } k \in U.$$

For the joint inclusion probabilities, we have

$$\pi_{k\ell} = \sum_{\substack{s \subset U \\ s \ni k,\ell}} p(s) = \sum_{\substack{s \in \mathcal{S}_n \\ s \ni k,\ell}} \binom{N}{n}^{-1},$$

$k \neq \ell \in U$. Since there exists $\binom{N-2}{n-2}$ samples of size n that jointly contain units k and ℓ, we obtain

$$\pi_{k\ell} = \binom{N-2}{n-2}\binom{N}{n}^{-1} = \frac{n(n-1)}{N(N-1)},$$

for all $k \neq \ell \in U$. Finally, we obtain

$$\Delta_{k\ell} = \begin{cases} \pi_{k\ell} - \pi_k \pi_\ell = \dfrac{n(n-1)}{N(N-1)} - \dfrac{n^2}{N^2} = -\dfrac{n}{N(N-1)}\left(1 - \dfrac{n}{N}\right) & \text{if } k \neq \ell \\[2ex] \pi_k(1 - \pi_k) = \dfrac{n}{N}\left(1 - \dfrac{n}{N}\right) & \text{if } k = \ell. \end{cases} \tag{3.1}$$

3.1.2 The Expansion Estimator and its Variance

Under simple random sampling without replacement, the expansion estimator of the total is simplified considerably because the inclusion probabilities are all equal. Since $\pi_k = n/N$, we have $d_k = 1/\pi_k = N/n$ and therefore

$$\hat{Y} = \sum_{k \in S} \frac{y_k}{\pi_k} = \sum_{k \in S} \frac{N}{n} y_k = \frac{N}{n} \sum_{k \in S} y_k = N\hat{\bar{Y}},$$

where $\hat{\bar{Y}}$ is the simple mean calculated on the sample:

$$\hat{\bar{Y}} = \frac{1}{n} \sum_{k \in S} y_k.$$

Therefore, to estimate the total, it is necessary to know the size of the population N. However, it is not possible to select a sample using simple random sampling without replacement without knowing the population size (see Section 3.9).

Since the design has a fixed sample size and the first- and second-order inclusion probabilities are strictly positive, one can use either the expression of the variance given in Expression (2.3) or the expression of the variance given in Expression (2.4).

Result 3.1 *Under simple random sampling without replacement with fixed sample size,*

$$\text{var}_p(\widehat{Y}) = N^2 \frac{N-n}{N} \frac{S_y^2}{n} = N^2 \left(1 - \frac{n}{N}\right) \frac{S_y^2}{n} = N^2 \left(\frac{1}{n} - \frac{1}{N}\right) S_y^2, \qquad (3.2)$$

where

$$S_y^2 = \frac{1}{N-1} \sum_{k \in U} (y_k - \overline{Y})^2.$$

Proof: From Expression (2.4), we obtain

$$\text{var}_p(\widehat{Y}) = -\frac{1}{2} \sum_{k \in U} \sum_{\substack{\ell \in U \\ \ell \neq k}} \left(\frac{y_k}{\pi_k} - \frac{y_\ell}{\pi_\ell}\right)^2 \Delta_{k\ell}$$

$$= \frac{1}{2} \sum_{k \in U} \sum_{\substack{\ell \in U \\ \ell \neq k}} \left(\frac{y_k N}{n} - \frac{y_\ell N}{n}\right)^2 \left(1 - \frac{n}{N}\right) \frac{n}{N(N-1)}$$

$$= \frac{N^2}{n} \left(1 - \frac{n}{N}\right) \frac{1}{2N(N-1)} \sum_{k \in U} \sum_{\substack{\ell \in U \\ \ell \neq k}} (y_k - y_\ell)^2.$$

However,

$$S_y^2 = \frac{1}{2N(N-1)} \sum_{k \in U} \sum_{\substack{\ell \in U \\ \ell \neq k}} (y_k - y_\ell)^2 = \frac{1}{N-1} \sum_{k \in U} (y_k - \overline{Y})^2.$$

We therefore obtain Expression (3.2). ∎

If the sampling fraction is denoted by $f = n/N$, we can also write

$$\text{var}_p(\widehat{Y}) = N^2 (1 - f) \frac{S_y^2}{n}.$$

If all the units in the population are selected, the variance $\widehat{Y}$ is zero.

The variance can be estimated either by using the Horvitz–Thompson estimator given in Expression (2.6) or by using the Sen–Yates–Grundy estimator given in Expression (2.7).

Result 3.2 *Under simple random sampling without replacement, the Sen–Yates–Grundy variance estimator can be written as*

$$v_{\text{SYG}}(\widehat{Y}) = N^2 \left(1 - \frac{n}{N}\right) \frac{s_y^2}{n}, \qquad (3.3)$$

where s_y^2 is the variance calculated on the sample:

$$s_y^2 = \frac{1}{n-1} \sum_{k \in S} (y_k - \widehat{\overline{Y}})^2.$$

Proof: The Sen–Yates–Grundy estimator can be written as

$$v_{SYG}(\widehat{Y}) = -\frac{1}{2} \sum_{k \in S} \sum_{\substack{\ell \in S \\ \ell \neq k}} \frac{\Delta_{k\ell}}{\pi_{k\ell}} \left(\frac{y_k}{\pi_k} - \frac{y_\ell}{\pi_\ell} \right)^2$$

$$= -\frac{1}{2} \sum_{k \in S} \sum_{\substack{\ell \in S \\ \ell \neq k}} -\frac{n}{N(N-1)} \left(1 - \frac{n}{N} \right) \frac{N(N-1)}{n(n-1)} \left(\frac{N}{n} y_k - \frac{N}{n} y_\ell \right)^2$$

$$= \frac{N^2}{n} \left(1 - \frac{n}{N} \right) \frac{1}{2n(n-1)} \sum_{k \in S} \sum_{\substack{\ell \in S \\ \ell \neq k}} (y_k - y_\ell)^2.$$

Now, the sample variance s_y^2 can be written as

$$s_y^2 = \frac{1}{2n(n-1)} \sum_{k \in S} \sum_{\substack{\ell \in S \\ \ell \neq k}} (y_k - y_\ell)^2 = \frac{1}{n-1} \sum_{k \in S} (y_k - \widehat{\overline{Y}})^2,$$

which gives Expression (3.3). ∎

Result 3.3 *Under simple random sampling without replacement, the Horvitz–Thompson variance estimator can be written as*

$$v_{SYG}(\widehat{Y}) = N^2 \left(1 - \frac{n}{N} \right) \frac{s_y^2}{n}$$

and is therefore equal to the Sen–Yates–Grundy estimator.

Proof: The Horvitz–Thompson estimator can be written as

$$v_{HT}(\widehat{Y}) = \sum_{k \in S} \sum_{\ell \in S} \frac{y_k y_\ell}{\pi_k \pi_\ell} \frac{\Delta_{k\ell}}{\pi_{k\ell}}$$

$$= \sum_{k \in S} \frac{y_k^2}{\pi_k^2} \frac{\Delta_{kk}}{\pi_{kk}} + \sum_{k \in S} \sum_{\substack{\ell \in S \\ \ell \neq k}} \frac{y_k y_\ell}{\pi_k \pi_\ell} \frac{\Delta_{k\ell}}{\pi_{k\ell}}$$

$$= \sum_{k \in S} \left(1 - \frac{n}{N} \right) \frac{n}{N} \frac{N}{n} \frac{N^2}{n^2} y_k^2$$

$$\quad - \sum_{k \in S} \sum_{\substack{\ell \in S \\ \ell \neq k}} \frac{n}{N(N-1)} \left(1 - \frac{n}{N} \right) \frac{N(N-1)}{n(n-1)} \frac{N^2}{n^2} y_k y_\ell$$

$$= \frac{N^2}{n} \left(1 - \frac{n}{N} \right) \left[\frac{1}{n} \sum_{k \in S} y_k^2 - \frac{1}{n(n-1)} \sum_{k \in S} \sum_{\substack{\ell \in S \\ \ell \neq k}} y_k y_\ell \right].$$

Since

$$\frac{1}{n} \sum_{k \in S} y_k^2 - \frac{1}{n(n-1)} \sum_{k \in S} \sum_{\substack{\ell \in S \\ \ell \neq k}} y_k y_\ell = \frac{1}{n-1} \sum_{k \in S} y_k^2 - \frac{1}{n(n-1)} \sum_{k \in S} \sum_{\ell \in S} y_k y_\ell$$

$$= \frac{1}{n-1} \sum_{k \in S} y_k^2 - \frac{n}{n-1} \widehat{\overline{Y}}^2 = \frac{1}{n-1} \sum_{k \in S} (y_k - \widehat{\overline{Y}})^2 = s_y^2,$$

we finally obtain

$$v_{HT}(\widehat{Y}) = N^2 \left(1 - \frac{n}{N}\right) \frac{s_y^2}{n}.$$

■

Under simple random sampling without replacement, the Sen–Yates–Grundy and Horvitz–Thompson estimators are equal, but in most other designs they take different values. The Sen–Yates–Grundy and Horvitz–Thompson estimators unbiasedly estimate the variance of the expansion estimator:

$$E_p \left[N^2 \left(1 - \frac{n}{N}\right) \frac{s_y^2}{n} \right] = N^2 \left(1 - \frac{n}{N}\right) \frac{S_y^2}{n}.$$

It follows that s_y^2 is an unbiased estimator of S_y^2.

The estimator of the mean $\overline{Y} = Y/N$ is obtained directly by dividing the estimator of the total $\widehat{Y}$ by N:

$$\widehat{\overline{Y}} = \frac{\widehat{Y}}{N} = \frac{1}{N} \sum_{k \in S} \frac{y_k}{\pi_k} = \frac{1}{N} \sum_{k \in S} \frac{N y_k}{n} = \frac{1}{n} \sum_{k \in S} y_k,$$

which is the simple mean on the sample. The variance of the mean estimator is

$$\mathrm{var}_p(\widehat{\overline{Y}}) = \mathrm{var}_p\left(\frac{\widehat{Y}}{N}\right) = \frac{1}{N^2}\mathrm{var}_p(\widehat{Y}) = \left(1 - \frac{n}{N}\right) \frac{S_y^2}{n},$$

and can be simply estimated by

$$v(\widehat{\overline{Y}}) = \left(1 - \frac{n}{N}\right) \frac{s_y^2}{n}.$$

The quantity

$$1 - \frac{n}{N} = \frac{N-n}{N}$$

is always between 0 and 1 and is called "finite population correction". This amount is the gain obtained by realizing a simple random sampling design without replacement with respect to a simple random sampling design with replacement.

3.1.3 Comment on the Variance–Covariance Matrix

Matrix $\boldsymbol{\Delta}$ of the $\Delta_{k\ell}$ is the operator that allows the calculation of the variance. Indeed, if

$$\check{\mathbf{y}} = (\check{y}_1, \dots, \check{y}_k, \dots, \check{y}_N)^{\mathsf{T}} = (y_1/\pi_1, \dots, y_k/\pi_k, \dots, y_N/\pi_N)^{\mathsf{T}},$$

the variance of the expansion estimator can be written as a quadratic form. If we refer to Expression (3.1), one can write

$$\mathrm{var}_p(\widehat{Y}) = \check{\mathbf{y}}^{\mathsf{T}} \boldsymbol{\Delta} \check{\mathbf{y}}.$$

The analysis of the properties of matrix $\boldsymbol{\Delta}$ allows us to better understand the value of simple random sampling.

Under simple random sampling without replacement, this matrix can be written as

$$\boldsymbol{\Delta} = \left(1 - \frac{n}{N}\right)\frac{nN}{N-1}\mathbf{P},$$

where

$$\mathbf{P} = \begin{pmatrix} \frac{N-1}{N} & \cdots & -\frac{1}{N} & \cdots & -\frac{1}{N} \\ \vdots & \ddots & \vdots & & \vdots \\ -\frac{1}{N} & \cdots & \frac{N-1}{N} & \cdots & -\frac{1}{N} \\ \vdots & & \vdots & \ddots & \vdots \\ -\frac{1}{N} & \cdots & -\frac{1}{N} & \cdots & \frac{N-1}{N} \end{pmatrix}.$$

Since $\mathbf{PP} = \mathbf{P}$, matrix $\mathbf{P}$ is idempotent. We can check that $\mathbf{P1} = \mathbf{0}$, where $\mathbf{1} = (1, \ldots, 1, \ldots, 1)^\top \in \mathbb{R}^N$. Matrix $\mathbf{P}$ therefore has an eigenvalue equal to 0 which is associated with the eigenvector $\mathbf{1}$.

Let us consider all normalized and centered vectors $\boldsymbol{\breve{u}}$, that is to say vectors $\boldsymbol{\breve{u}}$ such that

$$\sum_{k \in U} \breve{u}_k = 0 \quad \text{and} \quad \sum_{k \in U} \breve{u}_k^2 = 1,$$

then, we check that

$$\boldsymbol{\breve{u}}^\top \mathbf{P} \boldsymbol{\breve{u}} = 1 \quad \text{and} \quad \boldsymbol{\breve{u}}^\top \boldsymbol{\Delta} \boldsymbol{\breve{u}} = \left(1 - \frac{n}{N}\right)\frac{nN}{N-1}.$$

The $N - 1$ eigenvalues of $\mathbf{P}$ associated with normal orthogonal vectors to $\mathbf{1}$ are therefore equal to 1.

Simple random sampling therefore gives the same importance to all the centered vectors. It does not favor a particular direction. In this sense, the use of simple random sampling is a way of selecting a sample without preconceptions on the studied population. However, if auxiliary information is known and the variable of interest y is suspected of being linked to this auxiliary information, then it will be more interesting to use a design that integrates this auxiliary information, such as stratified (Chapter 4) or balanced (Chapter 6) designs, with possibly unequal inclusion probabilities (Chapter 5).

3.2 Bernoulli Sampling

3.2.1 Sampling Design and Inclusion Probabilities

In Bernoulli sampling, the indicator variables a_k are N independent Bernoulli random variables with the same parameter $\pi \in]0, 1[$:

$$\Pr(a_k = 1) = \pi \quad \text{and} \quad \Pr(a_k = 0) = 1 - \pi, \text{for all } k \in U.$$

This design is without replacement, but the sample size is random. Bernoulli sampling has limited appeal for selecting a sample, as this design does not allow the cost of the survey to be controlled. Bernoulli sampling can nevertheless be interesting when the units are presented successively, for example at a counter, without knowing the size of the population *a priori*. The main interest in the Bernoulli design lies in the treatment of nonresponse (see Chapter 16, page 333). This design is then used to model the response propensity of the individuals in a survey. Under this model, each individual decides whether or not to respond, independently of the others, with the same probability π.

More formally, we refer to the following definition:

Definition 3.2 *A sampling design is said to be a Bernoulli design if $p(s) = \pi^{n_s}(1 - \pi)^{N-n_s}$, for all $s \subset U$, where n_s is the size of sample s.*

In Bernoulli designs, all the possible samples have a nonzero probability of being selected, including the empty sample $\emptyset$ and the entire population U. Bernoulli sampling is a simple design because all samples with the same size have the same probability of being selected.

The sum of $p(s)$ is equal to 1. Indeed,

$$\sum_{s \subset U} p(s) = \sum_{r=0}^{N} \sum_{s \in \mathcal{S}_r} p(s) = \sum_{r=0}^{N} \sum_{s \in \mathcal{S}_r} \pi^r(1 - \pi)^{N-r}$$

$$= \sum_{r=0}^{N} \binom{N}{r} \pi^r(1 - \pi)^{N-r} = [\pi + (1 - \pi)]^N = 1.$$

We can derive the inclusion probabilities of the sampling design. Indeed,

$$\Pr(k \in S) = \sum_{\substack{s \subset U \\ s \ni k}} p(s) = \sum_{r=1}^{N} \sum_{\substack{s \in \mathcal{S}_r \\ s \ni k}} p(s) = \sum_{r=1}^{N} \sum_{\substack{s \in \mathcal{S}_r \\ s \ni k}} \pi^r(1 - \pi)^{N-r}$$

$$= \sum_{r=1}^{N} \binom{N-1}{r-1} \pi^r(1 - \pi)^{N-r} = \pi \sum_{z=0}^{N-1} \binom{N-1}{z} \pi^z(1 - \pi)^{N-1-z} = \pi.$$

For the second-order inclusion probabilities, we have for $k \neq \ell$:

$$\Pr(k \text{ and } \ell \in S) = \sum_{\substack{s \subset U \\ s \ni k,\ell}} p(s) = \sum_{r=2}^{N} \sum_{\substack{s \in \mathcal{S}_r \\ s \ni k,\ell}} p(s) = \sum_{r=2}^{N} \sum_{\substack{s \in \mathcal{S}_r \\ s \ni k,\ell}} \pi^r(1 - \pi)^{N-r}$$

$$= \sum_{r=2}^{N} \binom{N-2}{r-2} \pi^r(1 - \pi)^{N-r} = \pi^2 \sum_{z=0}^{N-2} \binom{N-2}{z} \pi^z(1 - \pi)^{N-2-z} = \pi^2.$$

Therefore, we have

$$\Delta_{k\ell} = \begin{cases} \pi_{k\ell} - \pi_k\pi_\ell = 0 & \text{if } k \neq \ell \\ \pi(1 - \pi) & \text{if } k = \ell. \end{cases}$$

In a Bernoulli design, the sample size n_S is a random variable whose distribution is binomial: $n_S \sim \text{Bin}(N, \pi)$. We can therefore write

$$\Pr(n_S = r) = \binom{N}{r} \pi^r(1 - \pi)^{N-r}, r = 0, \dots, N.$$

There is a nonzero probability of selecting an empty sample, $\Pr(n_S = 0) = (1 - \pi)^N$, or selecting the entire population, $\Pr(n_S = N) = \pi^N$. However, these probabilities become small when N is large.

If the Bernoulli design is conditioned on the sample size, then the Bernoulli design becomes a simple random sampling design without replacement and with fixed sample size. Indeed,

$$p(s|n_S = r) = \frac{p(s \text{ and } n_S = r)}{\Pr(n_S = r)} = \frac{\pi^r(1 - \pi)^{N-r}}{\binom{N}{r} \pi^r(1 - \pi)^{N-r}} = \binom{N}{r}^{-1}.$$

We therefore have the conditional inclusion probabilities:

$$E_p(a_k|n_S) = \frac{n_S}{N} \quad \text{and} \quad E_p(a_k a_\ell|n_S) = \frac{n_S(n_S - 1)}{N(N - 1)}, k \neq \ell \in U.$$

In practice, inference can be done conditioning on n_S, which consists of considering that n_S is not random but fixed. The sample is then treated as if it came from a simple random sample without replacement with fixed sample size. This result also suggests a way to draw a simple random sample of fixed sample size. A sequence of samples is selected using Bernoulli sampling until a sample of size r is obtained. However, this way may require significant calculation time. More efficient sampling algorithms are presented in Section 3.9.

3.2.2 Estimation

In Bernoulli sampling, one can use the expansion estimator:

$$\hat{Y} = \sum_{k \in S} \frac{y_k}{\pi} = \frac{1}{\pi} \sum_{k \in S} y_k.$$

In the case where $n_S = 0$, the expansion estimator is zero.

Due to the independence of the a_k, its variance is

$$\text{var}_p(\hat{Y}) = \sum_{k \in U} \frac{y_k^2}{\pi^2} \text{var}_p(a_k) = \sum_{k \in U} \frac{y_k^2}{\pi^2} \pi(1 - \pi) = \frac{1 - \pi}{\pi} \sum_{k \in U} y_k^2,$$

and it can be unbiasedly estimated by

$$v(\hat{Y}) = \frac{1 - \pi}{\pi^2} \sum_{k \in S} y_k^2.$$

However, in a Bernoulli design, the expansion estimator suffers from a significant problem. If the variable of interest is constant, in other words, $y_k = c$ for all $k \in U$, then

$$\hat{Y} = \frac{cn_S}{\pi}$$

and

$$\text{var}_p(\hat{Y}) = \frac{(1 - \pi)Nc^2}{\pi}.$$

Therefore, the variance of the estimator is not zero when the variable is constant, which means that the expansion estimator is tainted with an error that comes from the variance of n_S and not from the dispersion of y_k.

To remedy this problem, we can use the following alternative estimator (ALT) of the total:

$$\hat{Y}_{\text{ALT}} = \begin{cases} \dfrac{N}{n_S} \sum_{k \in S} y_k & \text{if } n_S \geq 1 \\ 0 & \text{if } n_S = 0. \end{cases}$$

When $n_S \geq 1$, its conditional expectation is

$$E_p(\hat{Y}_{\text{ALT}} | n_S = r) = Y, \text{ with } r = 1, \dots, N.$$

The unconditional expectation is then:

$$E_p E_p(\hat{Y}_{\text{ALT}} | n_S) = Y \times \Pr(n_S \geq 1) + 0 \times \Pr(n_S = 0)$$
$$= Y[1 - \Pr(n_S = 0)] = Y[1 - (1 - \pi)^N].$$

Since we condition on n_S, the design is simple without replacement with fixed sample size and the conditional variance is

$$\text{var}_p(\hat{Y}_{\text{ALT}} | n_S = r) = N^2 \frac{N - r}{Nr} S_y^2, \text{ with } r = 1, \dots, N.$$

Its unconditional variance is then:

$$\text{var}_p(\hat{Y}_{\text{ALT}}) = E_p \text{var}_p(\hat{Y}_{\text{ALT}} | n_S) + \text{var}_p E_p(\hat{Y}_{\text{ALT}} | n_S).$$

Since

$$E_p \text{var}_p(\hat{Y}_{\text{ALT}} | n_S) = 0 \times \Pr(n_S = 0) + E_p \left(N^2 \frac{N - n_S}{N n_S} S_y^2 \middle| n_S \geq 1 \right)$$
$$= N^2 \left[E_p \left(\frac{1}{n_S} \middle| n_S \geq 1 \right) - \frac{1}{N} \right] S_y^2,$$

and

$$\text{var}_p E_p(\hat{Y}_{\text{ALT}} | n_S) = (0 - Y)^2 \Pr(n_S = 0) + (Y - Y)^2 \Pr(n_S \geq 1) = Y^2 (1 - \pi)^N,$$

we obtain

$$\text{var}_p(\hat{Y}_{\text{ALT}}) = N^2 \left[E_p \left(\frac{1}{n_S} \middle| n_S \geq 1 \right) - \frac{1}{N} \right] S_y^2 + Y^2 (1 - \pi)^N.$$

The calculation of the inverse moment of the truncated binomial distribution

$$E_p(1/n_S | n_S \geq 1)$$

is somewhat delicate and has been studied among others by Kabe (1976). It is possible to show that $\text{var}_p(\hat{Y}_{\text{ALT}})$ is not necessarily smaller than $\text{var}_p(\hat{Y})$. However, we prefer to use $\hat{Y}_{\text{ALT}}$ rather than the expansion estimator because this estimator is zero when variable y_k is constant, which is not the case for the expansion estimator. Under Bernoulli designs, it is generally recommended to conduct the inference conditioning on n_S, which amounts to considering that the design is simple without replacement with fixed sample size. The inference is then carried out as described in Section 3.1.

3.3 Simple Random Sampling with Replacement

Simple random sampling with replacement with fixed sample size m is obtained by implementing the following sampling procedure: a unit is selected m times with equal probabilities $1/N$. Each drawing is done independently on the same population U. We can then define the random variable a_k which counts the number of times that unit k is selected in the sample. The vector $\mathbf{a} = (a_1, \ldots, a_k, \ldots, a_N)^\mathsf{T}$ has a multinomial distribution and each of the a_k follow a binomial distribution with parameter $1/N$ and exponent m, i.e. $a_k \sim \mathrm{Bin}(m, 1/N)$. Therefore, we have

$$E_p(a_k) = \frac{m}{N} \quad \text{and} \quad \mathrm{var}_p(a_k) = \frac{m}{N}\left(1 - \frac{1}{N}\right).$$

Moreover, since $\mathbf{a}$ follows a multinomial distribution, we have that

$$\mathrm{cov}_p(a_k, a_\ell) = -\frac{m}{N^2}.$$

The simple sample mean takes into account information about the multiplicity of the units:

$$\widehat{\overline{Y}}_{\mathrm{HH}} = \frac{1}{m} \sum_{k \in U} y_k a_k.$$

The simple sample mean is a special case of the Hansen & Hurwitz (1943) estimator (see Section 2.8, page 22) under simple random sampling with replacement.

The same unit of the population can be taken into account several times in the calculation of the mean. In the worst case, only one unit is selected m times. One can also select a sample whose size is larger than that of the population.

Result 3.4 *The Hansen–Hurwitz estimator unbiasedly estimates $\overline{Y}$ if and only if $\mu_k > 0$ for all $k \in U$ and its variance is*

$$\mathrm{var}_p(\widehat{\overline{Y}}_{\mathrm{HH}}) = \frac{N-1}{N}\frac{S_y^2}{m}.$$

Proof:

$$E_p(\widehat{\overline{Y}}_{\mathrm{HH}}) = \frac{1}{m} \sum_{k \in U} y_k E_p(a_k) = \frac{1}{m} \sum_{k \in U} y_k \frac{m}{N} = \overline{Y}.$$

Moreover,

$$\mathrm{var}_p(\widehat{\overline{Y}}_{\mathrm{HH}}) = \frac{1}{m^2} \sum_{k \in U} y_k^2 \mathrm{var}_p(a_k) + \frac{1}{m^2} \sum_{k \in U} \sum_{\substack{\ell \in U \\ \ell \neq k}} y_k y_\ell \mathrm{cov}_p(a_k, a_\ell)$$

$$= \frac{1}{m^2} \sum_{k \in U} y_k^2 \frac{m}{N}\left(1 - \frac{1}{N}\right) - \frac{1}{m^2} \sum_{k \in U} \sum_{\substack{\ell \in U \\ \ell \neq k}} y_k y_\ell \frac{m}{N^2}$$

$$= \frac{1}{Nm} \sum_{k \in U} y_k^2 - \frac{1}{mN^2} \sum_{k \in U} \sum_{\ell \in U} y_k y_\ell = \frac{1}{mN} \sum_{k \in U} (y_k - \overline{Y})^2 = \frac{N-1}{N}\frac{S_y^2}{m}. \qquad \blacksquare$$

By using a matrix notation, we have

$$E_p(\mathbf{a}) = \boldsymbol{\mu} = (\mu_1, \dots, \mu_k, \dots, \mu_N)^\top,$$

and

$$\text{var}_p(\mathbf{a}) = \frac{m}{N}\left(\mathbf{I} - \frac{\mathbf{1}\mathbf{1}^\top}{N}\right),$$

where $\mathbf{I}$ is an identity matrix of dimension $N \times N$ and $\mathbf{1}$ is a vector of N ones. Therefore, we can write

$$\widehat{\overline{Y}}_{\text{HH}} = \frac{1}{m}\mathbf{a}^\top\mathbf{y}.$$

and

$$\text{var}_p(\widehat{\overline{Y}}_{\text{HH}}) = \frac{1}{m^2}\mathbf{y}^\top\text{var}_p(\mathbf{a})\mathbf{y} = \frac{1}{mN}\mathbf{y}^\top\left(\mathbf{I} - \frac{\mathbf{1}\mathbf{1}^\top}{N}\right)\mathbf{y} = \frac{1}{m}\left(\frac{\mathbf{y}^\top\mathbf{y}}{N} - \overline{Y}^2\right).$$

In order to estimate $\text{var}_p(\widehat{\overline{Y}}_{\text{HH}})$, we use the following result:

Result 3.5 *In a simple random sampling with replacement, the uncorrected variance of the population:*

$$\frac{N-1}{N}S_y^2 = \frac{1}{N}\sum_{k\in U}(y_k - \overline{Y})^2$$

can be unbiasedly estimated by the sample variance and calculated by

$$s_y^2 = \frac{1}{m-1}\sum_{k\in U}(y_k - \widehat{\overline{Y}}_{\text{HH}})^2 a_k.$$

Proof: Since

$$s_y^2 = \frac{1}{m-1}\sum_{k\in U}(y_k - \widehat{\overline{Y}}_{\text{HH}})^2 a_k = \frac{1}{2m(m-1)}\sum_{k\in U}\sum_{\substack{\ell\in U\\\ell\neq k}}(y_k - y_\ell)^2 a_k a_\ell,$$

we have

$$E_p(s_y^2) = \frac{1}{2m(m-1)}\sum_{k\in U}\sum_{\substack{\ell\in U\\\ell\neq k}}(y_k - y_\ell)^2 E_p(a_k a_\ell).$$

In addition, if $k \neq \ell$, then

$$E_p(a_k a_\ell) = \text{cov}_p(a_k, a_\ell) + E_p(a_k)E_p(a_\ell) = -\frac{m}{N^2} + \frac{m^2}{N^2} = \frac{m(m-1)}{N^2}.$$

Therefore,

$$E_p(s_y^2) = \frac{1}{2m(m-1)}\sum_{k\in U}\sum_{\substack{\ell\in U\\\ell\neq k}}(y_k - y_\ell)^2\frac{m(m-1)}{N^2}$$

$$= \frac{1}{2N^2}\sum_{k\in U}\sum_{\substack{\ell\in U\\\ell\neq k}}(y_k - y_\ell)^2 = \frac{1}{N}\sum_{k\in U}(y_k - \overline{Y})^2 = \frac{N-1}{N}S_y^2.$$

■

The same statistic s_y^2 therefore does not estimate the same parameter depending on whether the design is with or without replacement.

3.4 Comparison of the Designs with and Without Replacement

Simple random sampling without replacement is always more interesting than sampling with replacement. The ratio of the variances of the mean estimators for the designs without and with replacement for the same sample size is always strictly less than 1 as soon as $n \geq 2$:

$$\frac{\text{var}_p(\widehat{\overline{Y}})}{\text{var}_p(\widehat{\overline{Y}}_{\text{HH}})} = \frac{\left(1 - \frac{n}{N}\right)\frac{S_y^2}{n}}{\frac{N-1}{N}S_y^2} = \frac{N-n}{N-1} < 1.$$

In Table 3.1, the different estimators are presented. The variance calculated in the sample does not estimate the same function of interest depending on whether the simple design is with or without replacement.

When N tends to infinity, the two variances of the mean estimator are equal. In practice, there is no interest in selecting the same unit in the sample several times. In the case of a simple random sampling with replacement, it is always more interesting to conserve only distinct units.

3.5 Sampling with Replacement and Retaining Distinct Units

3.5.1 Sample Size and Sampling Design

In this section, we show that three unbiased estimators can be defined in a design with replacement: the expansion estimator, the Hansen–Hurwitz estimator, which retains information about the multiplicity of units, and the Rao–Blackwellized estimator. We can show the interest of working only on distinct individuals, but it is impossible to define an optimal estimator in the sense of having minimum variance (see Basu, 1958; Raj & Khamis, 1958; Pathak, 1961, 1962, 1988; Chikkagoudar, 1966; Konijn, 1973). Simple designs show the difficulties inherent in finite population estimation. However, we will see that these results are sometimes of practical interest.

Table 3.1 Simple designs: summary table.

Sampling design	Without replacement	With replacement
Sample size	n	m
Mean estimator	$\widehat{\overline{Y}} = \frac{1}{n}\sum_{k \in S} y_k$	$\widehat{\overline{Y}}_{\text{HH}} = \frac{1}{m}\sum_{k \in U} a_k y_k$
Variance of the mean estimator	$\text{var}_p(\widehat{\overline{Y}}) = \left(1 - \frac{n}{N}\right)\frac{1}{n}S_y^2$	$\text{var}_p(\widehat{\overline{Y}}_{\text{HH}}) = \frac{N-1}{Nm}S_y^2$
Expectation of the sample variance	$E_p(s_y^2) = S_y^2$	$E_p(s_y^2) = \frac{N-1}{N}S_y^2$
Variance estimator of the mean estimator	$v(\widehat{\overline{Y}}) = \left(1 - \frac{n}{N}\right)\frac{s_y^2}{n}$	$v(\widehat{\overline{Y}}_{\text{HH}}) = \frac{s_y^2}{m}$

Consider a simple random sampling with replacement of size m in a population of N. Also, let a_k denote the number of times unit k is selected and let

$$n_S = \sum_{k \in U} r(a_k)$$

be the number of distinct units, where the function $r(\cdot)$ is defined in Expression (2.8), page 23. The random variable n_S can take values between 1 and m. Its probability distribution is given by the following result:

Result 3.6 *Consider a simple random sampling with replacement of size m in a population of N, the probability distribution of the number of distinct units n_S is*

$$Pr(n_S = u) = \frac{N!}{(N-u)!N^m} \left\{ \begin{matrix} u \\ m \end{matrix} \right\}, u = 1, \dots, \min(m, N), \tag{3.4}$$

where $\left\{ \begin{matrix} u \\ m \end{matrix} \right\}$ is a Stirling number of the second kind:

$$\left\{ \begin{matrix} u \\ m \end{matrix} \right\} = \frac{1}{u!} \sum_{i=1}^{u} \binom{u}{i} i^m (-1)^{u-i}.$$

Proof: A Stirling number of the second kind $\left\{ \begin{matrix} u \\ m \end{matrix} \right\}$ represents the number of ways to partition a set of m elements into u nonempty sets. These numbers can be constructed through a recurrence relation:

$$\left\{ \begin{matrix} u \\ m \end{matrix} \right\} = \left\{ \begin{matrix} u-1 \\ m-1 \end{matrix} \right\} + u \left\{ \begin{matrix} u \\ m-1 \end{matrix} \right\}, \tag{3.5}$$

with initial values $\left\{ \begin{matrix} 1 \\ 1 \end{matrix} \right\} = 1$ and $\left\{ \begin{matrix} u \\ 1 \end{matrix} \right\} = 0, u \neq 1$. An inventory of the properties of second-order Stirling numbers is given by Abramowitz & Stegun (1964, pp. 824–825.). For the proof, define $n_S(m)$ as the number of distinct units obtained by selecting at random with replacement m units in a population of size N. We can write the recurrence equation:

$$Pr[n_S(m) = u]$$
$$= Pr[n_S(m-1) = u-1]$$
$$\quad \times Pr[\text{select at the } m\text{th drawing a nonselected unit}$$
$$\quad\quad \text{knowing that } n_S(m-1) = u-1]$$
$$\quad + Pr[n_S(m-1) = u]$$
$$\quad \times Pr[\text{select at the } m\text{th drawing a unit already selected}$$
$$\quad\quad \text{knowing that } n_S(m-1) = u]$$
$$= Pr[n_S(m-1) = u-1]\frac{N-u+1}{N} + Pr[n_S(m-1) = u]\frac{u}{N}.$$

In addition, we have the initial conditions:

$$Pr[n_S(1) = u] = \begin{cases} 1 & \text{if } u = 1 \\ 0 & \text{otherwise.} \end{cases}$$

If we assume that Result 3.6 is true at step $m - 1$ and if we consider property (3.5), we can show that the recurrence equation is satisfied by Expression (3.4). Indeed,

$$\Pr[n_S(m-1) = u - 1]\frac{N-u+1}{N} + \Pr[n_S(m-1) = u]\frac{u}{N}$$

$$= \frac{N!}{(N-u+1)!N^{m-1}}\left\{{u-1 \atop m-1}\right\}\frac{N-u+1}{N} + \frac{N!}{(N-u)!N^{m-1}}\left\{{u \atop m-1}\right\}\frac{u}{N}$$

$$= \frac{N!}{(N-u)!N^m}\left(\left\{{u-1 \atop m-1}\right\} + u\left\{{u \atop m-1}\right\}\right)$$

$$= \frac{N!}{(N-u)!N^m}\left\{{u \atop m}\right\} = \Pr[n_S(m) = u].$$

We can check that the initial conditions are also satisfied. ∎

The design without replacement defined on U by selecting a simple random sample with replacement and retaining only the distinct units can therefore be fully specified. Conditionally to n_S, the sampling design is simple without replacement, which is written as

$$p(s|n_S = u) = \begin{cases} \left(\dbinom{N}{u}\right)^{-1} & \text{if } u = 1, \dots, \min(m, N) \\ 0 & \text{otherwise.} \end{cases}$$

Therefore, the unconditional design can be written as

$$p(s) = \begin{cases} \left(\dbinom{N}{u}\right)^{-1}\Pr(n_S = u) & \text{if } u = 1, \dots, \min(m, N) \\ 0 & \text{otherwise.} \end{cases}$$

In order to determine the inclusion probabilities of this design, it is necessary to know the moments of n_S. They can be deduced thanks to the following result:

Result 3.7 *If n_S is the number of distinct units obtained by selecting m units with replacement in a population of N, then*

$$E_p\left[\prod_{i=0}^{j-1}(N - n_S - i)\right] = \frac{(N-j)^m N!}{N^m (N-j)!}, j = 1, \dots, N. \tag{3.6}$$

Proof:

$$E_p\left[\prod_{i=0}^{j-1}(N - n_S - i)\right] = \sum_{u=1}^{\min(m,N)}\left[\prod_{i=0}^{j-1}(N - u - i)\right]\frac{N!}{(N-u)!N^m}\left\{{r \atop m}\right\}$$

$$= \sum_{u=1}^{\min(m,N-j)}\frac{(N-u)!}{(N-u-j)!}\frac{N!}{(N-u)!N^m}\left\{{u \atop m}\right\}$$

$$= \frac{(N-j)^m N!}{N^m (N-j)!}\sum_{u=1}^{\min(m,N-j)}\frac{(N-j)!}{(N-u-j)!(N-j)^m}\left\{{u \atop m}\right\}. \tag{3.7}$$

Since

$$\frac{(N-j)!}{(N-u-j)!(N-j)^m} \left\{ \begin{matrix} u \\ m \end{matrix} \right\}$$

is the probability of obtaining exactly u distinct units by selecting exactly m units with replacement in a population of size $N - j$. Therefore,

$$\sum_{u=1}^{\min(m,N-j)} \frac{(N-j)!}{(N-u-j)!(N-j)^m} \left\{ \begin{matrix} u \\ m \end{matrix} \right\} = 1,$$

and by (3.7), we obtain (3.6). ∎

By Expression (3.6) and taking $j = 1$, we can derive the expectation of the sample size:

$$E_p(n_S) = N \left[1 - \left(\frac{N-1}{N} \right)^m \right]. \tag{3.8}$$

Using Expression (3.6) again and taking $j = 2$, we can derive the variance of n_S:

$$\text{var}_p(n_S) = \frac{(N-1)^m}{N^{m-1}} + (N-1)\frac{(N-2)^m}{N^{m-1}} - \frac{(N-1)^{2m}}{N^{2m-2}}.$$

3.5.2 Inclusion Probabilities and Estimation

If we consider that, conditioning on n_S, the sampling design is simple without replacement, the inclusion probabilities can be deduced from (3.8):

$$\pi_k = E_p E_p[r(a_k)|n_S)] = E_p \left(\frac{n_S}{N} \right) = 1 - \left(\frac{N-1}{N} \right)^m, k \in U.$$

The inclusion probability can also be obtained by direct reasoning. Indeed,

$$\pi_k = E_p[r(a_k)] = 1 - \Pr(a_k = 0).$$

However, the probability that k is never selected in m independent draws is

$$\Pr(a_k = 0) = \left(\frac{N-1}{N} \right)^m,$$

and therefore

$$\pi_k = 1 - \left(\frac{N-1}{N} \right)^m.$$

Knowing the inclusion probabilities, we can calculate the expansion estimator of the mean, which is

$$\widehat{\overline{Y}} = \frac{1}{N \left[1 - \left(\frac{N-1}{N} \right)^m \right]} \sum_{k \in S} y_k. \tag{3.9}$$

The expansion estimator here has the bad property developed in Expression (2.1), page 20. If $y_k = C$, for all $k \in U$, we have

$$\widehat{\overline{Y}} = \frac{C n_S}{N \left[1 - \left(\frac{N-1}{N} \right)^m \right]}.$$

In this case, the expansion estimator is then a random variable of mean C and nonzero variance. For this reason, we prefer the Rao–Blackwellized estimator of $\widehat{Y}_{\mathrm{HH}}$ given in Expression (2.9):

Result 3.8 *Consider a simple random sampling with replacement and the Hansen–Hurwitz estimator*

$$\widehat{\overline{Y}}_{\mathrm{HH}} = \frac{1}{m} \sum_{k \in U} y_k a_k$$

of $\overline{Y}$, the Rao–Blackwellized estimator of $\widehat{\overline{Y}}_{\mathrm{HH}}$ is

$$\widehat{\overline{Y}}_{\mathrm{RB}} = \frac{1}{n_S} \sum_{k \in S} y_k. \tag{3.10}$$

Proof: The Rao–Blackwellized estimator is

$$\widehat{\overline{Y}}_{\mathrm{RB}} = \frac{1}{m} \sum_{k \in S} y_k E_p(a_k | r(\mathbf{a})),$$

where $r(\mathbf{a}) = (r(a_1), \dots, r(a_k), \dots, r(a_N))^{\mathsf{T}}$. However, $E_p[a_k | r(\mathbf{a})] = 0$ if $k \notin S$ and

$$\sum_{k \in S} a_k = m.$$

Since, for reasons of symmetry, $E_p[a_k | r(\mathbf{a})] = E_p[a_\ell | r(\mathbf{a})], k, \ell \in S$, we have

$$E_p[a_k | r(\mathbf{a})] = \frac{m}{n_S}, k \in S,$$

from which one directly derives (3.10). ∎

Estimator $\widehat{\overline{Y}}_{\mathrm{RB}}$ is, in this case, equal to the Hájek estimator (see Expression (2.2)). Indeed, since the π_k are constant for all k, we have

$$\widehat{\overline{Y}}_{\mathrm{HAJ}} = \frac{\sum_{k \in S} y_k / \pi_k}{\sum_{k \in S} 1 / \pi_k} = \frac{1}{n_S} \sum_{k \in S} y_k = \widehat{\overline{Y}}_{\mathrm{RB}}.$$

Since n_S is random, the determination of the expectation and variance of $\widehat{\overline{Y}}_{\mathrm{RB}}$ is somewhat delicate. However, conditioning on n_S, the design is simple without replacement, which directly gives the expectation and the conditional variance of this estimator:

$$E_p(\widehat{\overline{Y}}_{\mathrm{RB}} | n_S) = \overline{Y}$$

and

$$\mathrm{var}_p(\widehat{\overline{Y}}_{\mathrm{RB}} | n_S) = \left(1 - \frac{n_S}{N}\right) \frac{S_y^2}{n_S}.$$

The Rao–Blackwellized estimator of the mean is conditionally unbiased and, therefore, it is also unconditionally unbiased:

$$E_p(\widehat{\overline{Y}}_{\mathrm{RB}}) = E_p[E_p(\widehat{\overline{Y}}_{\mathrm{RB}} | n_S)] = E_p(\overline{Y}) = \overline{Y}.$$

The variance of $\widehat{\overline{Y}}_{RB}$ can be determined by using the law of total variance:

$$\text{var}_p(\widehat{\overline{Y}}_{RB}) = \text{var}_p[E_p(\widehat{\overline{Y}}_{RB}|n_S)] + E_p[\text{var}_p(\widehat{\overline{Y}}_{RB}|n_S)]$$

$$= \text{var}_p(\overline{Y}) + E_p\left(\frac{N - n_S}{N}\frac{S_y^2}{n_S}\right) = \left[E_p\left(\frac{1}{n_S}\right) - \frac{1}{N}\right]S_y^2. \qquad (3.11)$$

To determine the variance of $\widehat{\overline{Y}}_{RB}$, we must know the expectation of $1/n_S$ given by the following result:

Result 3.9 *If n_S is the number of distinct units obtained by selecting m random units with replacement in a population of size N, then*

$$E_p\left(\frac{1}{n_S}\right) = \frac{1}{N^m}\sum_{j=1}^{N} j^{m-1}. \qquad (3.12)$$

Proof: For the proof, let $n_S(N)$ denote the number of distinct units obtained by selecting m units according to simple random sampling with replacement in a population of size N. We will first show that

$$E_p\left[\frac{N^m}{n_S(N)}\right] - E_p\left[\frac{(N-1)^m}{n_S(N-1)}\right] = N^{m-1}. \qquad (3.13)$$

Indeed,

$$E_p\left[\frac{N^m}{n_S(N)} - \frac{(N-1)^m}{n_S(N-1)}\right]$$

$$= \sum_{r=1}^{\min(m,N)} \frac{1}{r}\frac{N!}{(N-r)!}\left\{\begin{matrix}r\\m\end{matrix}\right\} - \sum_{r=1}^{\min(m,N-1)} \frac{1}{r}\frac{(N-1)!}{(N-1-r)!}\left\{\begin{matrix}r\\m\end{matrix}\right\}$$

$$= N^{m-1}\sum_{r=1}^{\min(m,N)} \frac{N!}{(N-r)!N^m}\left\{\begin{matrix}r\\m\end{matrix}\right\} = N^{m-1}.$$

By (3.13), we obtain a recurrence equation:

$$E_p\left[\frac{1}{n_S(N)}\right] = \frac{1}{N} + \frac{(N-1)^m}{N^m}E_p\left[\frac{1}{n_S(N-1)}\right].$$

The initial condition is immediate:

$$E_p\left[\frac{1}{n_S(1)}\right] = 1.$$

We then can check that (3.12) satisfies the recurrence equation and the initial condition. ■

Finally, this result gives the variance of the Rao–Blackwellized estimator, considering (3.12) and (3.11), we obtain

$$\text{var}_p(\widehat{\overline{Y}}_{RB}) = \frac{S_y^2}{N^m}\sum_{j=1}^{N-1} j^{m-1}.$$

3.5.3 Comparison of the Estimators

The Rao–Blackwell theorem shows that

$$\operatorname{var}_p(\widehat{\overline{Y}}_{\mathrm{RB}}) \le \operatorname{var}_p(\widehat{\overline{Y}}_{\mathrm{HH}}).$$

Jensen's inequality shows that a harmonic mean is always less than an arithmetic mean. So we have

$$\frac{1}{E_p(1/n_S)} \le E_p(n_S) = N\left[1 - \left(\frac{N-1}{N}\right)^m\right] \le m,$$

and therefore

$$E_p\left(\frac{1}{n_S}\right) \ge \frac{1}{E_p(n_S)} = \frac{1}{N\left[1 - \left(\frac{N-1}{N}\right)^m\right]} \ge \frac{1}{m}.$$

We obtain

$$\begin{aligned}
\operatorname{var}_p(\widehat{\overline{Y}}_{\mathrm{RB}}) &= \left[E_p\left(\frac{1}{n_S}\right) - \frac{1}{N}\right] S_y^2 \\
&\ge \left[1 - \left(\frac{N-1}{N}\right)^m\right]^{-1}\left(\frac{N-1}{N}\right)^m S_y^2 \\
&\ge \left(\frac{1}{m} - \frac{1}{N}\right) S_y^2 = \frac{N-m}{Nm} S_y^2 = \operatorname{var}_p(\widehat{\overline{Y}}_{\mathrm{SRS}}),
\end{aligned}$$

where $\widehat{\overline{Y}}_{\mathrm{SRS}}$ is the estimator of the mean in a simple random sampling without replacement with fixed sample size. For the same sample size, we have

$$\operatorname{var}_p(\widehat{\overline{Y}}_{\mathrm{SRS}}) \le \operatorname{var}_p(\widehat{\overline{Y}}_{\mathrm{RB}}) \le \operatorname{var}_p(\widehat{\overline{Y}}_{\mathrm{HH}}).$$

Simple random sampling without replacement is always more accurate. However, if the design is with replacement, it is always better to only retain the distinct units.

Finally, we can examine the variance of the expansion estimator and compare it with other estimators.

Result 3.10 *Under simple random sampling with replacement and with distinct units, the variance of the expansion estimator given in Expression (3.9) is*

$$\operatorname{var}_p(\widehat{\overline{Y}}) = \left[\frac{N}{E_p(n_S)} - \frac{\operatorname{var}_p(n_S)}{[E_p(n_S)]^2} - 1\right]\frac{S_y^2}{N} + \frac{\overline{Y}^2}{[E_p(n_S)]^2}\operatorname{var}_p(n_S).$$

Proof: Since

$$\widehat{\overline{Y}} = \frac{n_S}{E_p(n_S)}\widehat{\overline{Y}}_{\mathrm{RB}},$$

the variance of $\widehat{\overline{Y}}$ is

$$\begin{aligned}
\operatorname{var}_p(\widehat{\overline{Y}}) &= E_p\,\operatorname{var}_p(\widehat{\overline{Y}}|n_S) + \operatorname{var}_p\,E_p(\widehat{\overline{Y}}|n_S) \\
&= E_p\left[\frac{n_S^2}{[E_p(n_S)]^2}\operatorname{var}_p(\widehat{\overline{Y}}_{\mathrm{RB}}|n_S)\right] + \operatorname{var}_p\left[\frac{n_S}{E_p(n_S)}E_p(\widehat{\overline{Y}}_{\mathrm{RB}}|n_S)\right]
\end{aligned}$$

$$= E_p \left[\frac{n_S^2}{[E_p(n_S)]^2} \frac{N - n_S}{N n_S} S_y^2 \right] + \frac{\overline{Y}^2 \text{var}_p(n_S)}{[E_p(n_S)]^2}$$

$$= \left[\frac{N}{E_p(n_S)} - \frac{\text{var}_p(n_S)}{[E_p(n_S)]^2} - 1 \right] \frac{S_y^2}{N} + \frac{\overline{Y}^2 \text{var}_p(n_S)}{[E_p(n_S)]^2}.$$

∎

Comparing $\widehat{\overline{Y}}$ with $\widehat{\overline{Y}}_{RB}$, we obtain

$$\text{var}_p(\widehat{\overline{Y}}) - \text{var}_p(\widehat{\overline{Y}}_{RB})$$

$$= \left\{ \frac{N}{E_p(n_S)} - E_p \left(\frac{N}{n_S} \right) - \frac{\text{var}_p(n_S)}{[E_p(n_S)]^2} \right\} \frac{S_y^2}{N} + \frac{\overline{Y}^2}{[E_p(n_S)]^2} \text{var}_p(n_S).$$

A necessary and sufficient condition for

$$\text{var}_p(\widehat{\overline{Y}}) \le \text{var}_p(\widehat{\overline{Y}}_{RB})$$

is therefore that

$$\overline{Y}^2 \le \left\{ \frac{E_p(n_S)}{\text{var}_p(n_S)} \left[E_p \left(\frac{1}{n_S} \right) E_p(n_S) - 1 \right] + \frac{1}{N} \right\} S_y^2.$$

For example, taking $\overline{Y} = 0$ is sufficient for the expansion estimator to be better than the Rao–Blackwellized estimator. On the other hand, if $\overline{Y}^2$ is large, the Rao–Blackwellized estimator has a smaller variance. This example illustrates a difficulty inherent in survey sampling. Even for an application as basic as simple random sampling with replacement, there is no optimal estimator. The choice of the estimator cannot be made solely on the basis of its variance. In the current case, we will of course choose the Rao–Blackwellized estimator, since it is linearly invariant and conditionally unbiased, two properties that the expansion estimator does not have.

3.6 Inverse Sampling with Replacement

Inverse sampling, studied by among others Basu (1958) and Chikkagoudar (1966), consists of selecting units with equal probability and with replacement until a fixed number r of distinct units is obtained. This problem is also known as the coupon collector's problem. Indeed, the collector tries to complete a series of objects by taking them at random with replacement in a finite population. If X is the number of units selected with replacement, we have

$$\Pr[X = m]$$

$= \Pr[\text{select } r - 1 \text{distinct units in } m - 1 \text{drawing with replacement}]$

$\times \Pr[\text{select at the } mith \text{ drawing a unit not already selected}$

knowing that $r - 1$ distinct units were already selected]

$$= \frac{N!}{(N - r + 1)! N^{m-1}} \left\{ \begin{matrix} r - 1 \\ m - 1 \end{matrix} \right\} \times \frac{N - r + 1}{N} = \frac{N!}{(N - r)! N^m} \left\{ \begin{matrix} r - 1 \\ m - 1 \end{matrix} \right\},$$

for $m = r, r + 1, \ldots$ We can show that (see, for instance, Tillé et al., 1996)

$$E_p(X) = \sum_{j=0}^{r-1} \frac{N}{N-j}, \tag{3.14}$$

and that

$$\mathrm{var}_p(X) = N \sum_{j=0}^{r-1} \frac{j}{(N-j)^2}.$$

This distribution has many practical applications ranging from the price evaluation of a child's picture collection to the estimation of an animal population. Several applications in ethology are also given in Tillé et al. (1996). If $N - r$ is large, we can construct an approximation of (3.14). Indeed, if a is large, we have

$$\sum_{j=1}^{a} \frac{1}{j} \approx \log a + \gamma,$$

where $\gamma \approx 0.57721$ is the Euler-Mascheroni constant (see Abramowitz & Stegun, 1964, p. 255).

If we want to obtain the whole collection (case $r = N$), the expectation can be approximated by

$$E_p(X) = \sum_{j=0}^{N-1} \frac{N}{N-j} = N \sum_{j=1}^{N} \frac{1}{j} \approx N \log N + N\gamma.$$

If we want to obtain a part of the collection (case $r < N$), the expectation can be approximated by

$$E_p(X) = \sum_{j=0}^{r-1} \frac{N}{N-j} = N \sum_{j=1}^{N} \frac{1}{j} - N \sum_{j=1}^{N-r} \frac{1}{j}$$

$$\approx N \log N - N \log(N - r) = N \log \frac{N}{N-r}.$$

Example 3.1 To complete a collection of 200 images, assuming that they are selected by means of simple random sampling with replacement, the expectation of the average number of images that needs to be bought is

$$E_p(X) = \sum_{j=0}^{N-1} \frac{N}{N-j} = N \sum_{j=1}^{N} \frac{1}{j} \approx 1175.606.$$

Using the approximation, we obtain

$$E_p(X) \approx N \log N + N\gamma \approx 1175.105.$$

If, for a collection of 200 images, we want only 100 distinct images:

$$E_p(X) = \sum_{j=0}^{r-1} \frac{N}{N-j} = \sum_{j=0}^{99} \frac{200}{200-j} \approx 138.1307.$$

Using the approximation, we find that

$$E_p(X) \approx N \log \frac{N}{N-r} \approx 138.6294.$$

3.7 Estimation of Other Functions of Interest

3.7.1 Estimation of a Count or a Proportion

Often, the variables of interest are qualitative. The variable of interest is then a characteristic that may or may not have a unit. We will therefore define the variable z as follows:

$$z_k = \begin{cases} 1 & \text{if unit } k \text{ has the characteristic} \\ 0 & \text{otherwise.} \end{cases}$$

Therefore, characteristic z is dichotomous. However, it can be treated in exactly the same way as a quantitative variable.

Indeed, the number of units that have the characteristic, or the count, is none other than the total Z of the z_k values. Likewise, the proportion P can be defined simply as the average of the z_k. The usual functions of interest become

$$\bar{Z} = \frac{\#\{z_k \in U | z_k = 1\}}{N} = P, \quad Z = \#\{z_k \in U | z_k = 1\} = NP$$

and

$$S_z^2 = \frac{N}{N-1} P(1-P).$$

The problem of estimating a proportion is only a special case of the estimation of the mean and that of the estimation of the population size is a special case of estimating a total. The expressions of variance simplify considerably, however. In simple random sampling design without replacement, we have

$$\hat{Z} = \frac{N}{n} \sum_{k \in S} z_k = N\hat{P}, \quad \hat{P} = \frac{1}{n} \sum_{k \in S} z_k,$$

$$s_z^2 = \frac{1}{n-1} \sum_{k \in S} (z_k - \hat{P})^2 = \frac{n}{n-1} \left(\frac{1}{n} \sum_{k \in S} z_k^2 - \hat{P}^2 \right) = \frac{n}{n-1} \hat{P}(1-\hat{P}),$$

$$\text{var}_p(\hat{P}) = \left(1 - \frac{n}{N}\right) \frac{N}{(N-1)n} P(1-P),$$

and

$$v(\hat{P}) = \left(1 - \frac{n}{N}\right) \frac{1}{n-1} \hat{P}(1-\hat{P}).$$

Under simple random sampling with replacement, we obtain

$$\hat{Z}_{HH} = \frac{N}{m} \sum_{k \in U} a_k z_k, \quad \hat{P}_{HH} = \frac{1}{m} \sum_{k \in U} a_k z_k,$$

$$s_z^2 = \frac{1}{m-1}\sum_{k\in U} a_k(z_k - \widehat{P}_{HH})^2 = \frac{m}{m-1}\left(\frac{1}{m}\sum_{k\in U} a_k z_k^2 - \widehat{P}_{HH}^2\right)$$

$$= \frac{m}{m-1}\widehat{P}_{HH}(1 - \widehat{P}_{HH}),$$

$$\mathrm{var}_p(\widehat{P}_{HH}) = \frac{1}{m}P(1-P),$$

and

$$v(\widehat{P}_{HH}) = \frac{1}{m}\frac{m}{m-1}\widehat{P}_{HH}(1-\widehat{P}_{HH}) = \frac{1}{m-1}\widehat{P}_{HH}(1-\widehat{P}_{HH}).$$

The peculiarity of the estimator of a proportion is that only P must be estimated and then it is possible to directly determine the variance estimator of $\widehat{P}$.

3.7.2 Estimation of a Ratio

Another function of interest that we often want to estimate is the ratio. We call a ratio any quantity that can be written in the form

$$R = \frac{Y}{X},$$

where Y and X are the totals of the values taken by the variables y and x on the units of the population. Under simple random sampling without replacement, we can estimate R by

$$\widehat{R} = \frac{\widehat{Y}}{\widehat{X}},$$

where

$$\widehat{Y} = \frac{N}{n}\sum_{k\in S} y_k \quad \text{and} \quad \widehat{X} = \frac{N}{n}\sum_{k\in S} x_k.$$

This estimator is a ratio of two statistics. The determination of its variance is relatively delicate, so we will be satisfied with an approximate value of the bias. To do this, we use the following result:

$$\widehat{R} - R = \frac{\widehat{Y} - R\widehat{X}}{\widehat{X}} = \frac{\widehat{Y} - R\widehat{X}}{X(1-\varepsilon)}$$

$$= \frac{\widehat{Y} - R\widehat{X}}{X}(1 + \varepsilon + \varepsilon^2 + \varepsilon^3 + \cdots), \tag{3.15}$$

where

$$\varepsilon = -\frac{\widehat{X} - X}{X}.$$

Note that $E_p(\varepsilon) = 0$ and that

$$\mathrm{var}_p(\varepsilon) = \left(1 - \frac{n}{N}\right)\frac{S_x^2}{\overline{X}^2 n}.$$

If n is large, we can consider that ε is a small variation. We can therefore construct an approximation by using a limiting development with respect to ε:

$$B_p(\widehat{R}) = E_p(\widehat{R} - R) = E_p\left[\frac{\widehat{Y} - R\widehat{X}}{X}(1 + \varepsilon + \varepsilon^2 + \varepsilon^3 + \cdots)\right]$$

$$\approx E_p\left[\frac{\widehat{Y} - R\widehat{X}}{X}(1 + \varepsilon)\right] = E_p\left[\frac{\widehat{Y} - R\widehat{X}}{X}\left(1 - \frac{\widehat{X} - X}{X}\right)\right].$$

Since

$$E_p\left(\frac{\widehat{Y} - R\widehat{X}}{X}\right) = 0,$$

we obtain

$$B_p(\widehat{R}) \approx E_p\left[\frac{(R\widehat{X} - \widehat{Y})(\widehat{X} - X)}{X^2}\right] = E_p\left[\frac{(R\widehat{X} - XR - \widehat{Y} + Y)(\widehat{X} - X)}{X^2}\right]$$

$$= \frac{R\,E_p(\widehat{X} - X)^2 - E_p(\widehat{Y} - Y)(\widehat{X} - X)}{X^2} = \frac{R\,\mathrm{var}_p(\widehat{X}) - \mathrm{cov}_p(\widehat{Y}, \widehat{X})}{X^2},$$

which gives

$$B_p(\widehat{R}) \approx \frac{N^2}{X^2}\left(1 - \frac{n}{N}\right)\frac{1}{n}(R\,S_x^2 - S_{xy}),$$

where

$$S_{xy} = \frac{1}{N-1}\sum_{k\in U}(y_k - \overline{Y})(x_k - \overline{X}).$$

The bias is negligible when the sample size is large. The bias is also approximately zero when $A = \overline{Y} - \overline{X}S_{xy}/S_x^2$ is zero. Parameter A can be interpreted as the intercept of the regression line from y to x. The bias is therefore approximately zero if the regression line passes through the origin.

It is possible to approximate the mean squared error of the ratio estimator using the linearization technique (see Section 15.4.1). We can also construct an approximation of the mean squared error by using only the first term of the development of Expression (3.15). We obtain the approximation

$$\mathrm{MSE}_p(\widehat{R}) \approx E_p(\widehat{R} - R)^2 \approx E_p\left(\frac{\widehat{Y} - R\widehat{X}}{X}\right)^2$$

$$= \frac{1}{X^2}[\mathrm{var}_p(\widehat{Y}) + R^2\,\mathrm{var}_p(\widehat{X}) - 2R\mathrm{cov}_p(\widehat{X}, \widehat{Y})]$$

$$= \left(1 - \frac{n}{N}\right)\frac{1}{n}\frac{1}{X^2}(S_y^2 + R^2\,S_x^2 - 2RS_{xy}).$$

This mean squared error can be estimated by

$$v(\widehat{R}) = \left(1 - \frac{n}{N}\right)\frac{1}{n}\frac{1}{\widehat{X}^2}(s_y^2 + \widehat{R}^2\,s_x^2 - 2\widehat{R}\,s_{xy}),$$

where

$$s_{xy} = \frac{1}{n-1}\sum_{k\in S}(y_k - \widehat{\overline{Y}})(x_k - \widehat{\overline{X}}).$$

3.8   Determination of the Sample Size

One of the tricky questions to address before starting a survey is the choice of sample size. However, it is very rare that the size of a survey is fixed for purely statistical reasons. Budgets, often allocated before the start of the survey, directly determine the sample size. Nevertheless, it is important to consider whether it is possible to achieve the targets set within the budget.

The objective is to reach a fixed precision for the studied parameter θ which is determined from the context of the study. Therefore, we want θ to be in the confidence interval defined around $\hat{\theta}$ with a probability of $1 - \alpha$.

$$\Pr\{\theta \in [\hat{\theta} - b, \hat{\theta} + b]\} = 1 - \alpha. \tag{3.16}$$

Assuming the normality of $\hat{\theta}$, this confidence interval is

$$\Pr\{\theta \in [\hat{\theta} - z_{1-\alpha/2}\sqrt{v(\hat{\theta})}, \hat{\theta} + z_{1-\alpha/2}\sqrt{v(\hat{\theta})}]\} = 1 - \alpha,$$

where $z_{1-\alpha/2}$ is the $1 - \alpha/2$ quantile of a standardized normal random variable. For the estimation of a mean in simple random sampling without replacement, we obtain

$$\Pr\left\{\overline{Y} \in \left[\hat{\overline{Y}} - z_{1-\alpha/2}\, s_y\, \sqrt{\left(1 - \frac{n}{N}\right)\frac{1}{n}},\right.\right.$$

$$\left.\left. \hat{\overline{Y}} + z_{1-\alpha/2}\, s_y\, \sqrt{\left(1 - \frac{n}{N}\right)\frac{1}{n}}\right]\right\} = 1 - \alpha,$$

However, for this interval to correspond to (3.16), it is necessary that

$$z_{1-\alpha/2}^2 s_y^2 \left(1 - \frac{n}{N}\right)\frac{1}{n} \leq b^2,$$

which means that

$$n \geq \frac{N s_y^2 z_{1-\alpha/2}^2}{b^2 N + s_y^2 z_{1-\alpha/2}^2}. \tag{3.17}$$

Of course, we need to know the variance s_y^2 or at least an estimate of it to be able to determine that size. The problem is easier to solve when we are interested in a proportion. In that case,

$$s_y^2 = \frac{n}{n-1}\hat{P}(1 - \hat{P}).$$

By (3.17), we obtain

$$n \geq \frac{N\hat{P}(1 - \hat{P})\frac{n}{n-1}z_{1-\alpha/2}^2}{b^2 N + \hat{P}(1 - \hat{P})\frac{n}{n-1}z_{1-\alpha/2}^2},$$

which gives

$$n \geq \frac{N[b^2 + \hat{P}(1 - \hat{P})z_{1-\alpha/2}^2]}{b^2 N + \hat{P}(1 - \hat{P})z_{1-\alpha/2}^2}.$$

Table 3.2 Example of sample sizes required for different population sizes and different values of b for $\alpha = 0.05$ and $\hat{P} = 1/2$.

	Population size N			
b	10 000	100 000	1 000 000	10 000 0000
0.1	97	97	98	98
0.05	371	384	386	386
0.025	1 333	1 515	1 536	1 538
0.01	4 900	8 764	9 514	9 596
0.005	7 935	27 755	36 996	38 270
0.0025	9 390	60 579	133 198	151 340
0.0001	9 999	99 896	989 695	9 056 960

If $\hat{P}$ is not known, we can always use the worst case, when $\hat{P} = 1/2$, and then use this overestimate for the size of the sample. Table 3.2 gives the sample sizes required for different population sizes with $\hat{P} = 1/2$, $\alpha = 0.05$, and $z_{1-\alpha/2} = 1.96$. When the size of the population is large, it is found that a halving of the length of the confidence interval requires a four-fold increase in the size of the sample.

3.9 Implementation of Simple Random Sampling Designs

3.9.1 Objectives and Principles

One of the crucial steps in the implementation of a sampling design is the selection of the observation units. These must be able to be formulated in the form of a simple algorithm, which is efficient, fast, and consumes little memory space. Ideally, an algorithm should be able to be applied in a single pass of a data file. Sometimes the size of the population is not known in advance, for example when selecting customers coming to a counter as they come through.

The sampling algorithm must not be confused with the sampling design. An algorithm is a procedure that can be programmed and allows a sample to be selected, while a sampling design is given by the definition of the probabilities $p(s)$ to select the samples $s \subset U$. Several fundamentally different algorithms can be used to implement the same sampling design. The most intuitive algorithm is usually not the most effective.

3.9.2 Bernoulli Sampling

The Bernoulli design described in Section 3.2 is certainly not a very adequate method to select a sample, but its implementation is extremely simple because of the independence of the a_k. The file is read sequentially and the units are selected independently with an inclusion probability π. The implementation of this design is described in Algorithm 1.

Bernoulli sampling is usually a bad solution, especially when we want to select few units. There exists an algorithm (see Algorithm 2) of equivalent complexity to Algorithm 1 which ensures a fixed sample size.

Algorithm 1 Bernoulli sampling

Definition u : real ; k : integer ;

$\qquad\qquad\qquad\qquad$ $u =$ uniform random variable in $[0, 1[$;

Repeat for $k = 1, \dots, N$ $\Big|$ $\Big|$ if $u < \pi$ select unit k;

$\qquad\qquad\qquad\qquad$ else do not select unit k.

3.9.3 Successive Drawing of the Units

The classical way of implementing a simple random sampling design is the following: one selects randomly a unit with equal probabilities among the N units of U. Then, the selected unit is removed from the population and a new individual is selected from the $N - 1$ units not yet selected. This operation is repeated n times. At each step, the selected unit is removed from the population. Therefore, the probability of selecting the n individuals $k_1, \dots, k_n$ in the order is $\{N(N - 1)(N - 2) \times \cdots \times (N - n + 1)\}^{-1} = (N - n)!/N!$. Since there are $n!$ ways to order $\{k_1, \dots, k_n\}$, the probability of selecting a sample s of n units is

$$p(s) = \frac{(N - n)!}{N!} n! = \binom{N}{n}^{-1}.$$

Therefore, these probabilities effectively define simple random sampling without replacement.

However, this method has a flaw. Its implementation in algorithmic form is complex; it requires n reads from the data file without counting the operations of sorting. This operation can take a considerable amount of time when N is large.

3.9.4 Random Sorting Method

A second, conceptually very simple, fixed method is to randomly sort the data file containing the population. In practice, we assign a uniform random number in $[0, 1[$ to each unit in the population. The file is then sorted in ascending (or descending) order of the random numbers. We finally choose the first n (or the last n) units from the ordered file. This sampling procedure is very easy to implement. However, the entire data file must be sorted. This operation can be very long when the file is large. The following result shows that this method gives simple random sampling:

Result 3.11 *(Sunter, 1977) [p. 263]* *If S is the random sample obtained by selecting the first n individuals in a file randomly sorted in descending order, then*

$$\Pr(S = s) = \begin{cases} \binom{N}{n}^{-1} & \text{if } n_s = n \\ 0 & \text{otherwise.} \end{cases}$$

Proof: Consider the distribution function of the largest random number generated for $N - n$ units of the file. If u_i is the uniform random variable for the ith unit, this function is

$$F(x) = \Pr\left[\bigcap_{i=1}^{N-n}(u_i \leq x)\right] = \prod_{i=1}^{N-n} \Pr(u_i \leq x) = x^{N-n}, 0 < x < 1.$$

The probability that the remaining n units are all larger than x is $(1-x)^n$, so the probability of selecting a particular sample s is

$$p(s) = \int_0^1 (1-x)^n dF(x) = (N-n) \int_0^1 (1-x)^n x^{N-n-1} dx. \tag{3.18}$$

Now Expression (3.18) is an Euler integral of the second kind, which is

$$B(x,y) = \int_0^1 t^x (1-t)^y dt = \frac{x!y!}{(x+y+1)!}, x, y \text{ integer.} \tag{3.19}$$

From (3.18) and (3.19), we directly obtain

$$p(s) = \binom{N}{n}^{-1}.$$

∎

A variant of this method allows us to avoid sorting the data file: the first n individuals of the file are selected from the outset in the starting sample by assigning each of them a realization of a uniform $[0,1[$ random variable. Next, the following $N - n$ individuals are examined. Whenever a new individual is taken into consideration, one generates for him a uniform $[0,1[$ random variable. If this is less than the largest variable generated for the individuals included in the sample, this individual is selected in the sample instead of the unit with the largest random number. It will be seen later that this method can still be improved (see the reservoir method, Section 3.9.6).

3.9.5 Selection–Rejection Method

The selection–rejection method allows the selection of a sample in a single reading of the data file. It was described by Fan et al. (1962), Bebbington (1975), and Devroye (1986, p. 620). This method is called Algorithm S by Knuth (1981, pp. 136–138). To implement it, we need to know the population size N and that of the sample n. Algorithm 2 is certainly the best solution when the size of the population is known.

Algorithm 2 Selection–rejection method

Definition k, j: integer; u: real;
$k = 0$;
$j = 0$;

Repeat while $j < n$ u = uniform random variable $[0, 1[$;
     If $u < \dfrac{n-j}{N-k}$ then   select unit $k + 1$;
                     $j = j + 1$;
     else do not select unit $k + 1$;
     $k = k + 1$.

Algorithm 2 uses variable k to count the number of units of the file already examined and variable j to count the number of units already selected. The sample must contain

n units. If n_k denotes the number of selected individuals after passing k units of the file, we deduce from the algorithm that

$$\Pr(k + 1 \in S | n_k = j) = \frac{n - j}{N - k}, k = 0, \dots, N - 1,$$

which can also be written as

$$\Pr(k \in S | n_{k-1} = j) = \frac{n - j}{N - (k - 1)}, k = 1, \dots, N,$$

and

$$\Pr(k \notin S | n_{k-1} = j) = 1 - \frac{n - j}{N - (k - 1)}, k = 1, \dots, N.$$

We will show that this algorithm results in a simple random sampling without replacement.

Result 3.12 *Let S be the random sample obtained by means of Algorithm 2. Then,*

$$\Pr(S = s) = \begin{cases} \binom{N}{n}^{-1} & \text{if } n_s = n \\ 0 & \text{otherwise.} \end{cases}$$

Proof: From Algorithm 2, we directly obtain

$$\Pr(S = s)$$

$$= \left[\prod_{k \in s} \Pr(k \in s | n_{k-1} = \#\{j \in s \,|\, j < k\}) \right]$$

$$\times \prod_{k \in U \setminus s} \Pr(k \notin s | n_{k-1} = \#\{j \in s \,|\, j < k\})$$

$$= \left[\prod_{k \in s} \frac{n - \#\{j \in s \,|\, j < k\}}{N - (k - 1)} \right] \times \prod_{k \in U \setminus s} \left[1 - \frac{n - \#\{j \in s \,|\, j < k\}}{N - (k - 1)} \right]$$

$$= \left[\prod_{k \in U} (N - k + 1) \right]^{-1} \times \left[\prod_{k \in s} (n - \#\{j \in s \,|\, j < k\}) \right]$$

$$\times \left[\prod_{k \in U \setminus s} (N - k + 1 - n + \#\{j \in s \,|\, j < k\}) \right]$$

$$= \frac{n!(N - n)!}{N!} = \binom{N}{n}^{-1}.$$

∎

3.9.6 The Reservoir Method

The reservoir method, described by Knuth (1981, p. 144), McLeod & Bellhouse (1983), Vitter (1985) and Devroye (1986, pp. 638–640), is also called the "sample update method" or the " R " algorithm. We immediately select the first n units from the file in the reservoir. Then, the reservoir is updated by successively examining the last $N - n$ individuals in the file.

Algorithm 3 is a little more time and memory size consuming than Algorithm 2. Indeed, it is necessary to at least define a vector of size n to run it. However, it has an advantage: the size of the population must not be known beforehand. We just have to stop at the last unit of the file.

Algorithm 3 Reservoir method

Definition : k integer; u real;
Select as initial reservoir the first n units of the file;
Repeat for $k = n + 1, \dots, N$
$\quad u$ = uniform random variable $[0, 1]$;
$\quad$ If $u < \dfrac{n}{k}$ $\begin{array}{l} \text{select unit } k; \\ \text{a unit is removed with equal probabilities} \\ \qquad\qquad\qquad \text{from the reservoir;} \\ \text{the selected unit } k \text{ takes the place of the removed unit;} \end{array}$
$\quad$ else do not select unit k.

Again, this algorithm provides a simple random sampling design.

Result 3.13 *(McLeod & Bellhouse, 1983)* *Let S be the random sample obtained by means of Algorithm 3, then*

$$
\Pr(S = s) = \begin{cases} \dbinom{N}{n}^{-1} & \text{if } n_s = n \\ 0 & \text{otherwise.} \end{cases}
\tag{3.20}
$$

Proof: Let U_k denote the subpopulation of the first k elements of U with $k = n, n + 1, \dots, N$, and S_k a random sample of size n selected in U_k. All $s_k \subset U_k$ of size n are possible values of the random sample S_k obtained in step k. We will show that

$$
\Pr(S_k = s_k) = \binom{k}{n}^{-1}, \quad \text{for all } s_k \subset U_k \text{ such that } \#s_k = n.
\tag{3.21}
$$

Expression (3.21) is true when $k = n$. Indeed, in this case, $\Pr(S_n = U_n) = 1$. When $k > n$, we will show (3.21) by induction, by assuming (3.21) true for $k - 1$. Consider a possible sample s_k of U_k. It is necessary to distinguish two cases:

- Case 1. $k \notin s_k$
  We then have

$$
\Pr(S_k = s_k)
$$
$$
= \Pr(S_{k-1} = s_k) \times \Pr(\text{unit } k \text{ is not selected in step } k)
$$
$$
= \binom{k-1}{n}^{-1} \left(1 - \frac{n}{k}\right) = \binom{k}{n}^{-1}.
\tag{3.22}
$$

- Case 2. $k \in s_k$

  In this case, the probability of selecting sample s_k at step k is the probability of
  1) selecting at step $k - 1$ a sample $(s_k \setminus \{k\}) \cup \{i\}$, where i is any element of U_{k-1} that is not in s_k;
  2) at step k, select unit k;
  3) precisely replace unit i by unit k.

This probability can therefore be written as

$$\Pr(S_k = s_k) = \sum_{i \in U_{k-1} \setminus s_k} \Pr\{S_{k-1} = (s_k \setminus \{k\}) \cup \{i\}\} \times \frac{n}{k} \times \frac{1}{n}$$

$$= (k - n) \binom{k-1}{n}^{-1} \frac{1}{k} = \binom{k}{n}^{-1}. \tag{3.23}$$

Results (3.22) and (3.23) show (3.21) by induction. Next we obtain (3.20) by taking $k = N$ in (3.21). ∎

For all the algorithms used to select a random sample by means of simple random sampling design without replacement, one must know the population size N before selecting the sample except for the reservoir method. For the latter, the size of the population is known only at the end of the application of the algorithm.

3.9.7 Implementation of Simple Random Sampling with Replacement

The most intuitive way of selecting a sample with a simple random sampling design with replacement is to repeat n times the selection of an individual in the population without changing the population. These n selections must be made independently of each other. In this case, the vector $\mathbf{a} = (a_1, \dots, a_k, \dots, a_N)$ follows a multinomial distribution.

However, it is possible to implement this design in a simpler way by using a sequential algorithm. The method is based on the following result: If $\mathbf{a}$ follows a multinomial distribution, then the conditional distribution of vector $(a_{k+1}, a_{k+2}, \dots, a_N)$ conditioning on $(a_1, \dots, a_k)$ is also multinomial (on this topic, see Kemp & Kemp, 1987; Bol'shev, 1965; Brown & Bromberg, 1984; Dagpunar, 1988; Devroye, 1986; Davis, 1993; Loukas & Kemp, 1983; Johnson et al., 1997; Tillé, 2006). This result enables us to construct Algorithm 4, which is the best solution to select a sample by means of a simple random sampling design with replacement.

Algorithm 4 Sequential algorithm for simple random sampling with replacement

---

Definition k: integer;

For $k = 1, \dots, N$

select a_k times unit k according to the binomial distribution:

$$a_k \sim \mathcal{B}\left(n - \sum_{i=1}^{k-1} a_i, \frac{1}{N - k + 1}\right).$$

3.10 Systematic Sampling with Equal Probabilities

In order to select a sample by means of systematic sampling, $G = 1/\pi$, the inverse of the inclusion probability, must be an integer. We proceed according to Algorithm 5.

Algorithm 5 Systematic sampling with equal probabilities

- Generate an integer random number h with equal probabilities between 1 and G.
- Select units with order numbers

$$h, h + G, h + 2G, \dots, h + (n-1)G, \dots$$

The support of the sampling design is very small. A systematic design has only G distinct samples. If N is a multiple of G, then the sample size is $n = \pi N = N/G$. The samples are

$$\{1, 1 + G, 1 + 2G, \dots, 1 + (n-1)G\},$$

$$\{2, 2 + G, 2 + 2G, \dots, 2 + (n-1)G\}, \dots,$$

$$\{h, h + G, h + 2G, \dots, h + (n-1)G\}, \dots,$$

$$\{G, 2G, 3G, \dots, nG\}.$$

If N is not a multiple of G, then the size of the sample n_S is random and is

$$n_S = \begin{cases} \lfloor N/G \rfloor & \text{with probability } 1 - N/G + \lfloor N/G \rfloor \\ \lfloor N/G \rfloor + 1 & \text{with probability } N/G - \lfloor N/G \rfloor, \end{cases}$$

where $\lfloor N/G \rfloor$ is the largest integer less than or equal to N/G. In all cases,

$$E_p(n_S) = \lfloor N/G \rfloor (1 - N/G + \lfloor N/G \rfloor) + (\lfloor N/G \rfloor + 1)(N/G - \lfloor N/G \rfloor) = \frac{N}{G}.$$

Algorithm 5 produces a sampling design with equal first-order inclusion probabilities. Indeed,

$$\pi_k = \frac{1}{G}, \text{ for all } k \in U.$$

The joint inclusion probabilities are

$$\pi_{k\ell} = \begin{cases} 1/G & \text{if } k \bmod G = \ell \bmod G \\ 0 & \text{if } k \bmod G \neq \ell \bmod G. \end{cases}$$

Therefore, most second-order inclusion probabilities are zero, which makes it impossible to unbiasedly estimate the variance.

Therefore, the expansion estimator of the total is

$$\hat{Y} = G \sum_{k \in S} y_k,$$

and the estimator of the mean is

$$\widehat{\overline{Y}} = \frac{G}{N} \sum_{k \in S} y_k.$$

The variance of this estimator is

$$\mathrm{var}_p(\widehat{Y}) = \sum_{h=1}^{G} [\widehat{Y}(h) - Y]^2,$$

where $\widehat{Y}(h)$ is the estimator of the total obtained for the hth sample.

However, it is not possible to unbiasedly estimate this variance, since many second-order inclusion probabilities are zero. This is a major flaw of systematic sampling. The application of systematic sampling after sorting the frame randomly is equivalent to simple random sampling without replacement. This method is then equivalent to the random sorting method (see Section 3.9.4). Systematic sampling is sometimes equated with simple random sampling without replacement for calculating variance. However, the systematic design is sometimes more or less precise than a simple design.

On one hand, if the data file is sorted according to an order that is related to the variables of interest, we obtain a gain of precision compared to simple random sampling because a stratification effect is generated automatically by systematic sampling. For example, suppose that a business register is sorted by establishment size, systematic sampling automatically respects the proportions of each size category. If the variables of interest are related to the size, the gain in precision can be important.

On the other hand, the sampling period may correspond to a periodicity naturally present in the sampling frame. For example, suppose that we select one in ten dwellings in a district where all buildings are five stories high. We can then select only ground floors. If this problem arises, the variance of the estimator can be considerably increased.

Although it is impossible to unbiasedly estimate the variance, we can construct variance estimators that can take into account the correlation between the file order and the variable of interest. Many publications deal with this issue (Cochran, 1946; Bellhouse & Rao, 1975; Iachan, 1982, 1983; Wolter, 1984; Murthy & Rao, 1988; Bellhouse, 1988b; Berger, 2003; Bartolucci & Montanari, 2006).

It is possible to select a fixed number n of units even when N/G is not an integer. This generalization is obtained by applying the inclusion probabilities $\pi_k = n/N, k \in U$, to systematic sampling with unequal probabilities (see Section 5.6, page 89). Systematic sampling may also be considered as a cluster sampling where only one cluster is selected (see Section 7.1, page 143).

3.11 Entropy for Simple and Systematic Designs

3.11.1 Bernoulli Designs and Entropy

Bernoulli designs maximize the entropy on the set of all of the possible samples:

Result 3.14 *The Bernoulli design with $\pi = 1/2$ is the maximum entropy design on $\mathcal{S} = \{s|s \subset U\}$ and is $p(s) = 2^{-N}$, for all $s \subset U$.*

Proof: By canceling the derivative of the Lagrange function

$$\mathcal{L}(p, \lambda) = -\sum_{s \in \mathcal{S}} p(s) \log p(s) + \lambda \left(\sum_{s \in \mathcal{S}} p(s) - 1 \right),$$

with respect to $p(s)$, we obtain

$$\frac{\partial \mathcal{L}(p, \lambda)}{\partial p(s)} = -1 - \log p(s) + \lambda = 0.$$

Therefore, $p(s) = \exp(\lambda - 1)$, for all $s \in \mathcal{S}$. Since

$$\sum_{s \in \mathcal{S}} p(s) = \sum_{s \in \mathcal{S}} \exp(\lambda - 1) = 2^N \exp(\lambda - 1) = 1.$$

We obtain $p(s) = 1/2^N$, for all $s \in \mathcal{S}$. ∎

In this case, all the samples have exactly the same probability of being selected. The entropy of a Bernoulli design with $\pi = 1/2$ is

$$I(p_{bern}) = -\sum_{s \in \mathcal{S}} p(s) \log p(s) = -\sum_{s \in \mathcal{S}} 2^{-N} \log 2^{-N} = N \log 2. \qquad (3.24)$$

This design is the most random among all sampling designs on population U.

If we now search for a design that maximizes the entropy by fixing the expectation of n_S, we also obtain a Bernoulli design.

Result 3.15 *The maximum entropy design on $\mathcal{S} = \{s|s \subset U\}$ whose expectation of the sample size is $N\pi$ is a Bernoulli design with inclusion probabilities π.*

Proof: The constraints are

$$\sum_{s \in \mathcal{S}} p(s) = 1 \quad \text{and} \quad \sum_{s \in \mathcal{S}} n_s p(s) = N\pi,$$

where n_s is the size of sample s. By canceling the derivative of the Lagrange function

$$\mathcal{L}(p, \lambda, \gamma) = -\sum_{s \in \mathcal{S}} p(s) \log p(s) + \lambda \left[\sum_{s \in \mathcal{S}} p(s) - 1 \right] + \gamma \left[\sum_{s \in \mathcal{S}} n_s p(s) - 1 \right],$$

with respect to $p(s)$, we obtain

$$\frac{\partial \mathcal{L}(p, \lambda, \gamma)}{\partial p(s)} = -\log p(s) - 1 + \lambda + \gamma n_s = 0.$$

Therefore, $p(s) = \exp(\lambda - 1 + n_s \gamma)$, for all $s \in \mathcal{S}$. Since

$$\sum_{s \in \mathcal{S}} p(s) = \sum_{s \in \mathcal{S}} \exp(\lambda - 1 + n_s \gamma) = \exp(\lambda - 1) \sum_{s \in \mathcal{S}} \exp(n_s \gamma) = 1,$$

$$\exp(\lambda - 1) = \frac{1}{\sum_{s \in \mathcal{S}} \exp(n_s \gamma)} = \frac{1}{\sum_{n_s = 0}^{N} \binom{N}{n_s} \exp(n_s \gamma)} = \frac{1}{[1 + \exp(\gamma)]^N},$$

and

$$p(s) = \frac{\exp(n_s \gamma)}{[1 + \exp(\gamma)]^N}.$$

Moreover,

$$\sum_{s \in \mathcal{S}} n_s p(s) = \sum_{s \in \mathcal{S}} \frac{n_s \exp(n_s \gamma)}{[1 + \exp(\gamma)]^N}$$

$$= \frac{N \exp(\gamma)[1 + \exp(\gamma)]^{N-1}}{[1 + \exp(\gamma)]^N} = \frac{N \exp(\gamma)}{1 + \exp(\gamma)} = N\pi.$$

Therefore,

$$\frac{\exp(\gamma)}{1 + \exp(\gamma)} = \pi,$$

and

$$p(s) = \frac{\exp(n_s \gamma)}{(1 + \exp(\gamma))^N} = \left(\frac{\exp(\gamma)}{1 + \exp(\gamma)} \right)^{n_s} \left(1 - \frac{\exp(\gamma)}{1 + \exp(\gamma)} \right)^{N - n_s}$$

$$= \pi^{n_s}(1 - \pi)^{N - n_s}. \qquad \blacksquare$$

The entropy of a Bernoulli design with any π is

$$I(p_{bern}) = -\sum_{s \in \mathcal{S}} p(s) \log p(s)$$

$$= -\sum_{s \in \mathcal{S}} \pi^{n_s}(1 - \pi)^{N - n_s} \log[\pi^{n_s}(1 - \pi)^{N - n_s}]$$

$$= -\sum_{n=0}^{N} \binom{N}{n} \pi^n (1 - \pi)^{N-n} \log[\pi^n (1 - \pi)^{N-n}]$$

$$= -\sum_{n=0}^{N} \binom{N}{n} \pi^n (1 - \pi)^{N-n} \left[n \log \frac{\pi}{1 - \pi} + N \log(1 - \pi) \right]$$

$$= -N\pi \log \frac{\pi}{1 - \pi} - N \log(1 - \pi)$$

$$= -N(1 - \pi) \log(1 - \pi) - N\pi \log \pi,$$

which generalizes Expression (3.24).

3.11.2 Entropy and Simple Random Sampling

Simple random sampling without replacement with fixed sample size is also a maximum entropy design.

Result 3.16 *The maximum entropy design with fixed sample size n is the simple random sampling design without replacement with fixed sample size.*

Proof: Let $\mathcal{S}_n$ the set of sample of size n. By canceling the derivative of the Lagrange function

$$\mathcal{L}(p, \lambda) = -\sum_{s \in \mathcal{S}_n} p(s) \log p(s) + \lambda \left(\sum_{s \in \mathcal{S}_n} p(s) - 1 \right),$$

with respect to $p(s)$, we obtain

$$\frac{\partial \mathcal{L}(p, \lambda)}{\partial p(s)} = -1 - \log p(s) + \lambda = 0.$$

Therefore, $p(s) = \exp(\lambda - 1)$, for all $s \in \mathcal{S}_n$. Since

$$\sum_{s \in \mathcal{S}_n} p(s) = \sum_{s \in \mathcal{S}_n} \exp(\lambda - 1) = \binom{N}{n} \exp(\lambda - 1) = 1$$

we obtain

$$p(s) = \binom{N}{n}^{-1},$$

for all $s \in \mathcal{S}_n$. ∎

The entropy of a simple random sampling design is

$$I(p) = -\sum_{s \in \mathcal{S}_n} \binom{N}{n}^{-1} \log \binom{N}{n}^{-1} = \log \binom{N}{n}.$$

Maximizing the entropy consists of choosing the most random sampling design possible on all the samples of size n. This design involves assigning the same probability to each sample of size n.

3.11.3   General Remarks

The entropy measures the randomness of the design. The larger the entropy, the more randomly the sample is selected. The mathematical developments given in Sections 3.11.1 and 3.11.2 show that the Bernoulli design with inclusion probabilities equal to 1/2 is the design with the largest entropy among all possible sampling designs. If we fix equal inclusion probabilities for all units, the design that maximizes the entropy is always a Bernoulli one. If we fix the sample size, then the maximum entropy design is a simple random sampling design without replacement. Instead, the systematic design has a very small entropy; only G samples have a nonzero probability of being selected. Each of these samples therefore has a $1/G$ probability of being selected. Therefore, the entropy of systematic sampling is

$$I(p_{bern}) = -\sum_{s \in \mathcal{S}} p(s) \log p(s) = -G \frac{1}{G} \log \frac{1}{G} = \log G.$$

Pea et al. (2007) have shown that the systematic designs are minimum entropy ones. They are in a way the antithesis of simple and Bernoulli designs which maximize the entropy.

Exercises

3.1   The sample variance can be written in three different ways:

$$s_y^2 = \frac{1}{n-1} \sum_{k \in S} (y_k - \widehat{\bar{Y}})^2$$

$$= \frac{1}{2n(n-1)} \sum_{k \in U} \sum_{\substack{\ell \in U \\ \ell \neq k}} (y_k - y_\ell)^2 a_k a_\ell \qquad (3.25)$$

$$= \frac{1}{n-1} \sum_{k \in U} y_k^2 a_k - \frac{n}{n-1} \widehat{\bar{Y}}^2. \qquad (3.26)$$

1. Under simple random sampling without replacement with a fixed sample size, show that s_y^2 is an unbiased estimator of s_y^2 by calculating the expectation of s_y^2 written in the form (3.25).

2. Show that s_y^2 is an unbiased estimator of s_y^2 by calculating the expectation of s_y^2 written in the form (3.26).

3.2 For simple random sampling design without replacement with a fixed sample size n from a population of size N:

1. calculate the probability that unit k is selected at least once in the sample;

2. show that $\Pr(k \in S) = \frac{n}{N} + O\left(\frac{1}{N^2}\right)$.

3.3 We are interested in estimating the proportion of men suffering from an occupational disease in a company of 1 500 workers. It is also known that three out of ten workers are usually affected by this disease in companies of the same type. It is proposed to select a sample by simple random sampling. Which sample size should be selected so that the total length of a confidence interval with a confidence level of 0.95 is less than 0.02 for simple designs with replacement and without replacement?

3.4 Your child would like to collect pictures of football players sold in sealed packages. The complete collection consists of 350 distinct pictures. Each package contains one picture "at random" in a totally independent manner from one package to another. Purchasing X packages is similar to taking X samples with replacement and with equal probability in the population of size $N = 350$. To simplify, your child does not trade any pictures.

1. What is the probability distribution of the number of pictures required to be purchased to obtain exactly r different players?

2. How many photos must be purchased on average in order to obtain the complete collection?

3.5 The objective is to estimate the number of rats present on an island. A trap is installed at a random location on the island. When a rat is trapped, it is marked and released. Another site is then randomly selected to install the trap and start again. If, for 50 captured rats, there are 42 distinct rats, estimate by the maximum likelihood method the number of rats living on the island assuming that the rats were captured at random and with replacement. Note: The maximum likelihood solution can only be obtained by trial and error using, for example, a spreadsheet.

3.6 A sample of 100 students is made up of a simple random sampling design without replacement in a population of 1 000 students. The results are presented the following table:

	Men	Women	Total
Success	$n_{11} = 35$	$n_{12} = 25$	$n_{1.} = 60$
Failure	$n_{21} = 20$	$n_{22} = 20$	$n_{2.} = 40$
Total	$n_{.1} = 55$	$n_{.2} = 45$	$n = 100$

1. Estimate the success rate of men and women.
2. Calculate the bias of the success rates.
3. Estimate the MSE of these success rates.

3.7 Select a sample of size 4 from a population of size 10 according to a simple random sampling design without replacement by means of the selection-rejection algorithm. Use the following realizations of a uniform random variable in $[0, 1]$:

0.375489 0.624004 0.517951 0.0454450 0.632912
0.246090 0.927398 0.32595 0.645951 0.178048.

4

Stratification

Stratification is a method allowing the use of auxiliary information in a sampling design. In this chapter, notation is introduced (Section 4.1) for the population and the strata. The stratified design is then defined in a general way (Section 4.2). After that the special cases of proportional allocation (Section 4.4) and optimal allocation (Section 4.5) are studied. The notion of optimality is applied to estimators other than totals (Section 4.6). Next, we discuss the question of costs (Section 4.8) and determine the minimum sample size needed for a fixed cost (Section 4.9). Power allocation (Section 4.7) allows us to find a compromise between different objectives. Finally, we discuss the question of strata construction in Section 4.10 and the question of stratifying with multiple objectives in Section 4.11.

4.1 Population and Strata

Stratification is one of the best ways to introduce auxiliary information in a survey in order to improve the precision of the estimators. As a general rule, if the information is available, it is almost always a good idea to stratify. The resulting total estimator will be more precise. Suppose that population $U = \{1, \dots, k, \dots, N\}$ is partitioned in H subgroups, $U_h, h = 1, \dots, H$, called strata such that

$$\bigcup_{h=1}^{H} U_h = U \quad \text{and} \quad U_h \bigcap U_i = \emptyset, h \neq i.$$

The number of elements in stratum N_h is called the size of the stratum. We have

$$\sum_{h=1}^{H} N_h = N,$$

where N is the population size U. We suppose the N_h are known.

Stratification is based on a qualitative variable known for all the units in the population. Geographic stratifications are often used based on areas of a country. In business surveys, strata are often constructed from classes formed by the number of workers by sectors of activity.

The goal is always to estimate a function of interest of the values taken by the variable of interest y, such as the total of values taken by the characteristic of interest in

Sampling and Estimation from Finite Populations, First Edition. Yves Tillé.
© 2020 John Wiley & Sons Ltd. Published 2020 by John Wiley & Sons Ltd.

the population:

$$Y = \sum_{k \in U} y_k = \sum_{h=1}^{H} \sum_{k \in U_h} y_k = \sum_{h=1}^{H} Y_h,$$

where Y_h is the total of the values taken by y in the units from U_h:

$$Y_h = \sum_{k \in U_h} y_k.$$

The general population total is simply the sum of the totals of the strata.

The population mean can be broken down as a weighted sum of the strata means:

$$\bar{Y} = \frac{1}{N} \sum_{k \in U} y_k = \frac{1}{N} \sum_{h=1}^{H} \sum_{k \in U_h} y_k = \frac{1}{N} \sum_{h=1}^{H} N_h \bar{Y}_h,$$

where $\bar{Y}_h$ denotes the mean calculated within stratum h:

$$\bar{Y}_h = \frac{1}{N_h} \sum_{k \in U_h} y_k.$$

Moreover, S_{yh}^2 denotes the corrected variance in stratum h:

$$S_{yh}^2 = \frac{1}{N_h - 1} \sum_{k \in U_h} (y_k - \bar{Y}_h)^2.$$

The variance S_y^2 can be broken down into two parts according to the law of total variance:

$$S_y^2 = \frac{1}{N-1} \sum_{k \in U} (y_k - \bar{Y})^2 = \frac{1}{N-1} \sum_{h=1}^{H} (N_h - 1) S_{yh}^2 + \frac{1}{N-1} \sum_{h=1}^{H} N_h (\bar{Y}_h - \bar{Y})^2.$$

This variance can be written as

$$S_y^2 = S_{y(within)}^2 + S_{y(between)}^2,$$

where $S_{y(within)}^2$ denotes the within-strata variance:

$$S_{y(within)}^2 = \frac{1}{N-1} \sum_{h=1}^{H} (N_h - 1) S_{yh}^2 \tag{4.1}$$

and $S_{y(between)}^2$ denotes the between-strata variance:

$$S_{y(between)}^2 = \frac{1}{N-1} \sum_{h=1}^{H} N_h (\bar{Y}_h - \bar{Y})^2.$$

4.2 Sample, Inclusion Probabilities, and Estimation

We will refer to the following definition, which is illustrated in Figure 4.1:

Definition 4.1 *A survey is said to be stratified if, in each stratum, a random sample of fixed size n_h is selected and if the sample selection within a stratum is independent of the sample selection in all other strata.*

Figure 4.1 Stratified design: the samples are selected independently from one stratum to another.

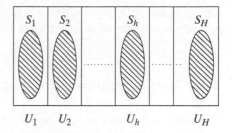

Let S_h denote the random sample selected in stratum h with sampling design $p_h(\cdot)$, where $p_h(s_h) = \Pr(S_h = s_h)$, and let s_h denote a possible value of S_h. The total random sample is made of the H samples selected within all strata:

$$S = \bigcup_{h=1}^{H} S_h.$$

Also, s generally denotes a possible value of S, where

$$s = \bigcup_{h=1}^{H} s_h.$$

The total sampling design is denoted by $p(\cdot)$, where $p(s) = \Pr(S = s)$. Because of the independence of the sampling within each stratum, we have

$$p(s) = \prod_{h=1}^{H} p_h(s_h), s = \bigcup_{h=1}^{H} s_h.$$

In a stratified design, we can determine the inclusion probabilities of the sampling design. If $k, \ell \in U_h$, then

$$\pi_k = \sum_{s_h \ni k} p_h(s_h) \quad \text{and} \quad \pi_{k\ell} = \sum_{s_h \supset \{k,\ell\}} p_h(s_h).$$

If two units are in two distinct strata, selecting unit k is independent from selecting unit ℓ. Therefore,

$$\pi_{k\ell} = \pi_k \pi_\ell, k \in U_h, \ell \in U_i, \text{ with } h \neq i.$$

In order to estimate the total, the expansion estimator under a stratified design can be used (STRAT):

$$\widehat{Y}_{\text{STRAT}} = \sum_{k \in S} \frac{y_k}{\pi_k} = \sum_{h=1}^{H} \sum_{k \in S_h} \frac{y_k}{\pi_k} = \sum_{h=1}^{H} \widehat{Y}_h,$$

where $\widehat{Y}_h$ is the unbiased estimator of Y_h:

$$\widehat{Y}_h = \sum_{k \in S_h} \frac{y_k}{\pi_k}.$$

As the samples are independent from stratum to stratum, the $\widehat{Y}_h$ are independent. Therefore,

$$\text{var}_p(\widehat{Y}_{\text{STRAT}}) = \text{var}_p\left(\sum_{h=1}^{H} \widehat{Y}_h\right) = \sum_{h=1}^{H} \text{var}_p(\widehat{Y}_h).$$

The independence of the samples greatly facilitates the form of the variance which can be written as a sum of the variances calculated within each stratum.

4.3 Simple Stratified Designs

The most common case corresponds to selecting a simple random sample without replacement of fixed size within each stratum. If n_h is the size of the selected sample in stratum h, we have

$$\sum_{h=1}^{H} n_h = n,$$

where n is the sample size.

If unit k is in stratum h, its inclusion probability can easily be calculated since, in stratum h, a simple random sampling design of size n_h is applied in a subpopulation of size N_h. Then,

$$\pi_k = \frac{n_h}{N_h}, k \in U_h.$$

In order to calculate the second-order inclusion probabilities, two cases must be distinguished:

- In the case where units k and ℓ are in the same stratum, we have

$$\pi_{k\ell} = \frac{n_h(n_h - 1)}{N_h(N_h - 1)}, k \text{ and } \ell \in U_h.$$

- If the two units k and ℓ belong to two different strata U_h and U_i, then from the independence of the two samples we obtain

$$\pi_{k\ell} = \frac{n_h n_i}{N_h N_i}, k \in U_h \text{ and } \ell \in U_i, h \neq i.$$

We therefore obtain

$$\Delta_{k\ell} = \begin{cases} \dfrac{n_h}{N_h} \dfrac{N_h - n_h}{N_h} & \text{if } \ell = k, k \in U_h \\[2mm] -\dfrac{n_h(N_h - n_h)}{N_h^2(N_h - 1)} & \text{if } k \text{ and } \ell \in U_h, k \neq \ell \\[2mm] 0 & \text{if } k \in U_h \text{ and } \ell \in U_i, h \neq i. \end{cases}$$

Matrix $\mathbf{\Delta}$ is block diagonal:

$$\mathbf{\Delta} = \begin{pmatrix} \mathbf{\Delta}_1 & \cdots & \mathbf{0} & \cdots & \mathbf{0} \\ \vdots & \ddots & \vdots & & \vdots \\ \mathbf{0} & \cdots & \mathbf{\Delta}_h & \cdots & \mathbf{0} \\ \vdots & & \vdots & \ddots & \vdots \\ \mathbf{0} & \cdots & \mathbf{0} & \cdots & \mathbf{\Delta}_H \end{pmatrix},$$

where

$$\Delta_h = \frac{n_h}{N_h - 1} \frac{N_h - n_h}{N_h} \mathbf{P}_h$$

and

$$\mathbf{P}_h = \begin{pmatrix} \frac{N_h-1}{N_h} & \cdots & \frac{-1}{N_h} & \cdots & \frac{-1}{N_h} \\ \vdots & \ddots & \vdots & & \vdots \\ \frac{-1}{N_h} & \cdots & \frac{N_h-1}{N_h} & \cdots & \frac{-1}{N_h} \\ \vdots & & \vdots & \ddots & \vdots \\ \frac{-1}{N_h} & \cdots & \frac{-1}{N_h} & \cdots & \frac{N_h-1}{N_h} \end{pmatrix}.$$

The $\mathbf{P}_h$ are idempotent matrices of dimension $N_h \times N_h$ and of rank $N_h - 1$. Matrix Δ is then of rank $N - H$.

The total and mean expansion estimators are equal, respectively, to

$$\hat{Y}_{\text{STRAT}} = \sum_{k \in S} \frac{y_k}{\pi_k} = \sum_{h=1}^{H} \frac{N_h}{n_h} \sum_{k \in S_h} y_k = \sum_{h=1}^{H} \hat{Y}_h,$$

and

$$\hat{\bar{Y}}_{\text{STRAT}} = \frac{1}{N} \sum_{k \in S} \frac{y_k}{\pi_k} = \frac{1}{N} \sum_{h=1}^{H} \frac{N_h}{n_h} \sum_{k \in S_h} y_k = \frac{1}{N} \sum_{h=1}^{H} N_h \hat{\bar{Y}}_h,$$

where $\hat{Y}_h$ is the stratum total estimator h

$$\hat{Y}_h = \frac{N_h}{n_h} \sum_{k \in S_h} y_k,$$

and $\hat{\bar{Y}}_h$ denotes the sample mean calculated within stratum h:

$$\hat{\bar{Y}}_h = \frac{1}{n_h} \sum_{k \in S_h} y_k.$$

We can determine the variance of this estimator from expression (2.4), page 21, but this variance can be directly calculated knowing the samples are independent from stratum to stratum and that the designs are simple in each stratum. Based on expression (3.2):

$$\text{var}_p(\hat{Y}_{\text{STRAT}}) = \text{var}_p \left(\sum_{h=1}^{H} \hat{Y}_h \right) = \sum_{h=1}^{H} \text{var}_p(\hat{Y}_h) = \sum_{h=1}^{H} N_h^2 \frac{N_h - n_h}{N_h} \frac{S_{yh}^2}{n_h}. \tag{4.2}$$

The variance of this estimator can be unbiasedly estimated by

$$v(\hat{Y}_{\text{STRAT}}) = \sum_{h=1}^{H} N_h^2 \frac{N_h - n_h}{N_h} \frac{s_{yh}^2}{n_h},$$

where

$$s_{yh}^2 = \frac{1}{n_h - 1} \sum_{k \in S_h} (y_k - \hat{\bar{Y}}_h)^2, h = 1, \dots, H.$$

4.4   Stratified Design with Proportional Allocation

When we want the sizes of the strata in the sample to be proportional to the sizes of the strata in the population, we are referring to stratification with proportional allocation. This design is sometimes referred to as representative stratification, but one should avoid using this expression as the word "representative" is not clearly defined. In Section 4.5 we will see that proportional allocation is not usually the best strategy. More specifically, we refer to the following definition:

Definition 4.2   *A stratified design is said to be with proportional allocation if, in all strata, the proportion of units in the sample is the same as the proportion in the population, in other words, if*

$$\frac{n_h}{n} = \frac{N_h}{N}.$$

We can derive that

$$n_h = \frac{nN_h}{N}, h = 1, \dots, N.$$

We assume that in the sample it is possible to calculate strata sizes $n_h = nN_h/N$ that are integers, which is rarely the case in practice. The expansion estimator of the total in a stratified design with proportional allocation (PROP) is

$$\hat{Y}_{\text{PROP}} = \sum_{h=1}^{H} \hat{Y}_h = \frac{N}{n} \sum_{k \in S} y_k,$$

and the estimator for the mean is

$$\hat{\bar{Y}}_{\text{PROP}} = \frac{1}{N} \sum_{h=1}^{H} N_h \hat{\bar{Y}}_h = \frac{1}{n} \sum_{k \in S} y_k,$$

where $\hat{\bar{Y}}_h$ is the sample mean calculated within stratum h and $\hat{Y}_h$ is the estimator of the total of stratum h:

$$\hat{\bar{Y}}_h = \frac{1}{n_h} \sum_{k \in S_h} y_k.$$

The variance of the estimator of the total can be simplified because of the proportional allocation:

$$\text{var}_p(\hat{Y}_{\text{PROP}}) = N^2 \frac{N-n}{Nn} \sum_{h=1}^{H} \frac{N_h}{N} S_{yh}^2, \tag{4.3}$$

and the variance of the mean estimator is

$$\text{var}_p(\hat{\bar{Y}}_{\text{PROP}}) = \frac{N-n}{Nn} \sum_{h=1}^{H} \frac{N_h}{N} S_{yh}^2.$$

If N_h is large, we have

$$\frac{N_h - 1}{N - 1} \approx \frac{N_h}{N}.$$

We can then write

$$\mathrm{var}_p(\widehat{\overline{Y}}_{\mathrm{PROP}}) = \frac{N-n}{nN^2} \sum_{h=1}^{H} N_h S_{yh}^2 \approx \frac{N-n}{N} \frac{S_{y(within)}^2}{n}, \tag{4.4}$$

where $S_{y(within)}^2$ is defined as (4.1).

Let us compare a stratified design with proportional allocation to a simple random sampling without replacement design. Indeed, those two designs have the same first-order inclusion probabilities and therefore the same expansion estimator. The variance of the expansion estimator for a simple random sampling design without replacement (SRS) is

$$\mathrm{var}_p(\widehat{\overline{Y}}_{\mathrm{SRS}}) = \frac{N-n}{N} \frac{S_y^2}{n}. \tag{4.5}$$

For a stratified design with proportional allocation, Expression (4.4) is practically identical to (4.5), except that the total variance is replaced by the within-stratum variance. By using a stratified design with proportional allocation, we almost always get better results than with a simple design without replacement. Indeed, we remove from the variance of the total estimator, the part due to the between-strata variance.

By stratifying, the results become better when the between-strata variance increases. The between-strata variance is large when the stratification variable is strongly dependent on the variable of interest. This is why, while developing the sampling design, one should always try to stratify with a variable dependent on the variable of interest.

The variance of the mean estimator can be unbiasedly estimated by

$$v(\widehat{\overline{Y}}_{\mathrm{PROP}}) = \frac{N-n}{nN^2} \sum_{h=1}^{H} N_h s_{yh}^2,$$

where

$$s_{yh}^2 = \frac{1}{n_h - 1} \sum_{k \in S_h} (y_k - \widehat{\overline{Y}}_h)^2, h = 1, \dots, H.$$

4.5 Optimal Stratified Design for the Total

Proportional allocation is not the best stratified sampling design. If we want to estimate a mean or a total, Neyman (1934) demonstrated that an optimal strata allocation exists for a sample with a fixed sample size. The problem is as follows: we are looking for the allocation of the strata sizes in the sample $n_1, \dots, n_h, \dots, n_H$ which minimizes the variance of the expansion estimator for a fixed sample size.

Result 4.1 *The allocation $n_1, \dots, n_h, \dots, n_H$ minimizing the variance of the expansion total estimator under a fixed size n constraint is*

$$n_h = \frac{nN_h S_{yh}}{\sum_{\ell=1}^{H} N_\ell S_{y\ell}}, h = 1, \dots, H. \tag{4.6}$$

Proof: So we have to minimize

$$\mathrm{var}_p(\widehat{Y}_{\mathrm{STRAT}}) = \sum_{h=1}^{H} N_h^2 \frac{N_h - n_h}{N_h} \frac{S_{yh}^2}{n_h} = \sum_{h=1}^{H} \frac{N_h^2 S_{yh}^2}{n_h} - \sum_{h=1}^{H} N_h S_{yh}^2, \tag{4.7}$$

in $n_1, \ldots, n_h, \ldots, n_H$ subject to

$$\sum_{h=1}^{H} n_h = n. \tag{4.8}$$

We can write the Lagrange equation:

$$\mathcal{L}(n_1, \ldots, n_H, \lambda) = \sum_{h=1}^{H} \frac{N_h^2 S_{yh}^2}{n_h} - \sum_{h=1}^{H} N_h S_{yh}^2 + \lambda \left(\sum_{h=1}^{H} n_h - n \right).$$

By canceling the partial derivative with respect to n_h and λ, we obtain

$$\frac{\partial \mathcal{L}}{\partial n_h} = -\frac{N_h^2 S_{yh}^2}{n_h^2} + \lambda = 0, h = 1, \ldots, H, \tag{4.9}$$

and

$$\frac{\partial \mathcal{L}}{\partial \lambda} = \sum_{h=1}^{H} n_h - n = 0. \tag{4.10}$$

By Expression (4.9), it follows that

$$n_h = \frac{N_h S_{yh}}{\sqrt{\lambda}}, h = 1, \ldots, H. \tag{4.11}$$

Hence, by (4.10) and (4.11), we have

$$\sum_{h=1}^{H} n_h = n = \frac{\sum_{h=1}^{H} N_h S_{yh}}{\sqrt{\lambda}},$$

which gives

$$\sqrt{\lambda} = \frac{\sum_{h=1}^{H} N_h S_{yh}}{n}. \tag{4.12}$$

By (4.11) and (4.12), we obtain Expression (4.6). ∎

The size of the stratum in the sample has to be proportional to the product of the population stratum size and the standard deviation of the variable of interest in the stratum. Obviously, to be able to apply Expression (4.6), the variance of the variable of interest has to be known for each stratum.

This result is enlightening as it shows that the individuals belonging to the strata with the highest spread have to be over-represented. For example, if a survey on consumption habits is being conducted, it is interesting to over-represent the socio-professional category of managers. Indeed, we can suppose, even without any auxiliary information, that managers have more varied consumption habits than retirees. This idea goes against the common notion of a representative sample which should be a small image of the population.

Expression (4.6) often leads to determining $n_h > N_h$. This comes from the fact that we should have introduced the inequalities $0 \le n_h \le N_h, h = 1, \dots, H$, on top of constraint (4.8). In this case, we can define the optimal allocation:

$$n_h = \min\left(N_h, \frac{CN_h S_{yh}}{\sum_{\ell=1}^{H} N_\ell S_{y\ell}} \right),$$

where C is such that

$$\sum_{h=1}^{H} \min\left(N_h, \frac{CN_h S_{yh}}{\sum_{\ell=1}^{H} N_\ell S_{y\ell}} \right) = n.$$

In practice, we proceed as follows: we calculate the n_h with Expression (4.6). If for a stratum $\ell, n_\ell > N_\ell$, we select all the individuals of the stratum. Then, we recalculate the n_h with (4.6) on the remaining strata. Of course, the problem can also occur at the next step. So we start this process over again until we have $n_h \le N_h, h = 1, \dots, H$.

Example 4.1 Let us define a population of size $N = 11\,100$ divided into three strata for which sizes are $N_1 = 10\,000, N_2 = 1000, N_3 = 100$ and the standard deviations are $S_{y1} = 5, S_{y2} = 50, S_{y2} = 500$. We want to select a sample of size $n = 450$. By calculating the optimal allocation with Expression (4.6), we obtain the results shown in Table 4.1. In the third stratum, the calculated size $n_3 = 150$ is larger than the size of the equivalent population stratum $N_3 = 100$. As such, we decide to sample the 100 units from stratum 3 ($N_3 = n_3 = 100$). Hence, a census is conducted in the third stratum. Therefore, $n - n_3 = 450 - 100 = 350$ units remain to be sampled in strata 1 and 2. By calculating the optimal allocation with Expression (4.6), we obtain the results shown in Table 4.2. We obtain $n_1 = 175, n_2 = 175, n_3 = 100$.

Table 4.1 Application of optimal allocation: the sample size is larger than the population size in the third stratum.

N_h	S_h	$N_h S_h$	n_h
10000	5	50000	150
1000	50	50000	150
100	500	50000	150
11100		150000	450

Table 4.2 Second application of optimal allocation in strata 1 and 2.

N_h	S_h	$N_h S_h$	n_h
10000	5	50000	175
1000	50	50000	175
11000		100000	350

Expression (4.6) does not necessarily lead to integer n_h. Therefore, the results will have to be rounded. If the n_h are integers and if $n_h < N_h$ for all h, then the variance of the expansion estimator in a stratified design with optimal allocation (OPT) can be derived from (4.6) and (4.2):

$$\text{var}_p(\widehat{Y}_{\text{OPT}}) = \sum_{h=1}^{H} N_h^2 \left(\frac{1}{n_h} - \frac{1}{N_h} \right) S_{yh}^2$$

$$= \sum_{h=1}^{H} N_h^2 \left(\frac{\sum_{\ell=1}^{H} N_\ell S_{y\ell}}{n N_h S_{yh}} - \frac{1}{N_h} \right) S_{yh}^2$$

$$= \sum_{h=1}^{H} N_h^2 \left(\frac{\sum_{\ell=1}^{H} N_\ell S_{y\ell}}{n N_h S_{yh}} \right) S_{yh}^2 - \sum_{h=1}^{H} N_h S_{yh}^2$$

$$= \frac{1}{n} \left(\sum_{h=1}^{H} N_h S_{yh} \right)^2 - \sum_{h=1}^{H} N_h S_{yh}^2.$$

We can then determine the gain in precision obtained by applying an optimal stratified design instead of a stratified design with proportional allocation:

$$\text{var}_p(\widehat{Y}_{\text{PROP}}) - \text{var}_p(\widehat{Y}_{\text{OPT}})$$

$$= \frac{N-n}{n} \sum_{h=1}^{H} N_h S_{yh}^2 - \frac{1}{n} \left(\sum_{h=1}^{H} N_h S_{yh} \right)^2 + \sum_{h=1}^{H} N_h S_{yh}^2$$

$$= \frac{N}{n} \sum_{h=1}^{H} N_h S_{yh}^2 - \frac{1}{n} \left(\sum_{h=1}^{H} N_h S_{yh} \right)^2$$

$$= \frac{N^2}{n} \sum_{h=1}^{H} \frac{N_h}{N} \left[S_{yh} - \left(\sum_{h=1}^{H} \frac{N_h}{N} S_{yh} \right) \right]^2 = \frac{N^2}{n} \text{var}(S_{yh}).$$

The gain in precision between both allocation types directly depends on the variance of the strata standard deviations. The more the breath of the spreads changes from one stratum to another, the larger the gain of precision achieved from optimal stratification.

4.6 Notes About Optimality in Stratification

Optimal allocation depends on the function of interest to estimate. Indeed, Expression (4.6) requires that the standard deviations S_{yh} of the variable of interest be known. Also, Expression (4.6) is only valid for total or mean estimation. For another function of interest, a different optimal allocation will be obtained. For example, let us take a population divided into two strata $H = 2$ for which we want to estimate the difference $D = \overline{Y}_1 - \overline{Y}_2$. The difference can be unbiasedly estimated by

$$\widehat{D} = \widehat{\overline{Y}}_1 - \widehat{\overline{Y}}_2.$$

Since the samples are independent from one stratum to another, we obtain

$$\text{var}_p(\widehat{D}) = \text{var}_p(\widehat{\overline{Y}}_1) + \text{var}_p(\widehat{\overline{Y}}_2) = \frac{N_1 - n_1}{n_1 N_1} S_{y1}^2 + \frac{N_2 - n_2}{n_2 N_2} S_{y2}^2. \tag{4.13}$$

Hence, we minimize (4.13) subject to $n_1 + n_2 = n$ and by following the same expansion as in the previous section we obtain

$$n_h = \frac{S_{yh}}{\sqrt{\lambda}}, h = 1, 2,$$

where λ is the Lagrange multiplier. As $n_1 + n_2 = n$, we obtain

$$n_h = \frac{nS_{yh}}{S_{y1} + S_{y2}}, h = 1, 2. \tag{4.14}$$

In this case, optimal allocation is different from the one obtained for estimating the total. Indeed, the n_h are no longer dependent on the sizes of the population strata.

This example clearly illustrates that the choice of allocation has to be based on the survey objective. However, a survey is rarely conducted to estimate a single function of interest, but to estimate multiple ones at the same time. For instance, national estimates and comparisons between regions can be of interest at the same time. In this type of situation, a compromise between various optimal allocations will have to be determined by considering the relative importance of each survey objective.

4.7 Power Allocation

Power allocation, as discussed by Kish (1976), Cochran (1977), Yates (1979), Fellegi (1981), Bankier (1988) and Lednicki & Wieczorkowski (2003), allows a compromise between global and local estimations to be found. To do this, we can minimize the following quantity:

$$F = \sum_{k=1}^{H} \frac{X_h^q \text{var}_p(\hat{Y}_h)}{Y_h^2},$$

where $0 \le q \le 1$ is a tuning parameter and X_h is a measure of the size of stratum h. By minimizing F under a fixed size constraint, we obtain

$$n_h = \frac{S_h N_h X_h^q / Y_h}{\sum_{i=1}^{H} S_i N_h X_h^q / Y_i}.$$

If $X_h = Y_h$ and $q = 1$, Neyman optimality for the total obtained in Expression (4.6) is found. If $q = 1$ and $X_h = Y_h/N_h$, we obtain the optimal allocation to estimate local means, with special case $H = 2$ given by Expression (4.14). If $q = 0$, we obtain the allocation

$$n_h = \frac{nS_h/\overline{Y}_h}{\sum_{i=1}^{H} S_i/\overline{Y}_i}.$$

This allocation tends to equalize the coefficients of variation

$$\frac{[\text{var}_p(\hat{Y}_h)]^{-1/2}}{Y_h}$$

in each stratum. By taking an intermediate value of q, $q = 1/2$, an allocation which compromises between global and local estimation objectives is obtained.

4.8 Optimality and Cost

Survey cost is a criterion that can be considered when determining the optimal allocation. The problem consists of estimating a total Y for a fixed cost C. Hence, we minimize Expression (4.7) subject to

$$\sum_{h=1}^{H} n_h C_h = C,$$

where C_h denotes the interview cost for a unit in stratum h. In fact, costs can be very different from one stratum to another. For example, an interview can be more costly in a rural area than in an urban one because of traveling costs.

By expanding in the same way as in Section 4.5, we obtain the following system of equations:

$$\begin{cases} n_h = \dfrac{N_h S_{yh}}{\sqrt{\lambda C_h}}, h = 1, \dots, H, \\ \sum_{h=1}^{H} n_h C_h = C, \end{cases}$$

where λ is the Lagrange multiplier. We directly obtain

$$n_h = \frac{C N_h S_{yh}}{\sqrt{C_h} \sum_{\ell=1}^{H} N_\ell S_{y\ell} \sqrt{C_\ell}}.$$

This result is very logical when compared to the usual optimal allocation. We can see that fewer units are selected in strata where the cost is larger.

4.9 Smallest Sample Size

Another way to state the optimality problem is by looking for the allocation leading to the smallest sample size for a given precision. Let us define $P_h = n_h/n, h = 1, \dots, H$. We then obtain $\sum_{h=1}^{H} P_h = 1$. From (4.7), we can write

$$\text{var}_p(\widehat{Y}_{\text{STRAT}}) = \sum_{h=1}^{H} N_h \frac{N_h - n P_h}{n P_h} S_{yh}^2. \tag{4.15}$$

We are looking for the smallest value of n in $P_1, \dots, P_H$, for a fixed value of $\text{var}_p(\widehat{Y}_{\text{STRAT}})$ labeled V. By replacing in (4.15) $\text{var}_p(\widehat{Y}_{\text{STRAT}})$ by V, we obtain

$$V = \frac{1}{n} \sum_{h=1}^{H} \frac{N_h^2}{P_h} S_{yh}^2 - \sum_{h=1}^{H} N_h S_{yh}^2,$$

which can be written as

$$n = \frac{\sum_{h=1}^{H} \frac{N_h^2}{P_h} S_{yh}^2}{V + \sum_{h=1}^{H} N_h S_{yh}^2}. \tag{4.16}$$

So we minimize (4.16) in $P_1, \ldots, P_H$, subject to

$$\sum_{h=1}^{H} P_h = 1,$$

and, after some calculations, we obtain

$$P_h = \frac{N_h S_{yh}}{\sum_{\ell=1}^{H} N_\ell S_{y\ell}}. \tag{4.17}$$

We obtain the same type of allocation as in Section 4.5. By (4.16) and (4.17), we can finally fix the sample size:

$$n^* = \frac{\left(\sum_{h=1}^{H} N_h S_{yh}\right)^2}{V + \sum_{h=1}^{H} N_h S_{yh}^2}.$$

4.10 Construction of the Strata

4.10.1 General Comments

When designing a stratified design, a delicate stage is the construction of the strata. Sometimes the auxiliary information is in the form of a quantitative variable which needs to be divided into strata. In other cases, a very detailed nomenclature allows us to form classes in various ways. Many aspects of the problem have to be considered when forming strata, as no unique method exists.

Generally speaking, the more strata are formed, the more precise the total estimator is. However, selecting one unit per stratum is not a very adequate design. Indeed, if only one unit is sampled per stratum, the second-order inclusion probabilities of units in the same stratum are zero, making it impossible to unbiasedly estimate the variance of the expansion estimator. Usually, the estimator of the variance becomes more unstable as the number of strata increases because of the loss in degrees of freedom coming from the strata mean estimation. The problem of nonresponse provides an additional reason not to use this design. A survey is always tarnished by nonresponse and, in such a design, some strata will be empty, which will make the estimation within these strata very difficult. Hence, it is highly recommended to anticipate this nonresponse by planning for strata that are large enough so that the probability of encountering this problem is small.

4.10.2 Dividing a Quantitative Variable in Strata

If a quantitative variable is available and one wishes to divide it into strata, Dalenius (1950, 1957) and Dalenius & Hodges Jr. (1959) suggest a division method minimizing the variance for a stratified design with optimal allocation (also see on this topic, Cochran, 1977, pp. 127–131, and Horgan, 2006). In Dalenius' method, we consider an infinite population and that the distribution of that the stratification variable can be represented by a density function $f(x)$. Then,

$$\mu = \int_{-\infty}^{\infty} x f(x) dx, \qquad \sigma^2 = \int_{-\infty}^{\infty} (x - \mu)^2 f(x) dx$$

and if k_{h-1} and k_h, respectively, denote the upper and lower bounds of the hth class,

$$W_h = \int_{k_{h-1}}^{k_h} f(x)dx, \quad \mu_h = \int_{k_{h-1}}^{k_h} xf(x)dx,$$

and

$$\sigma_h^2 = \int_{k_{h-1}}^{k_h} (x - \mu_h)^2 f(x)dx,$$

for $h = 1, \ldots, H$, with $k_0 = -\infty$ and $k_H = \infty$.

If we now select a sample with a simple design in each stratum, we have

$$\widehat{\mu}_h = \frac{1}{n_h} \sum_{k \in S_h} x_k, \quad \text{and} \quad \widehat{\mu} = \sum_{h=1}^{H} W_h \widehat{\mu}_h.$$

Since the population is supposedly infinite, the finite population correction factor is equal to 1 and the variance of this estimator is

$$\text{var}_p(\widehat{\mu}) = \sum_{h=1}^{H} W_h^2 \frac{\sigma_h^2}{n_n}. \tag{4.18}$$

If the Neyman optimal stratification is used, then

$$n_h = \frac{n W_h \sigma_h}{\sum_{\ell=1}^{H} W_\ell \sigma_\ell}, \tag{4.19}$$

and, by replacing n_h with (4.19) in (4.18), the variance becomes

$$\text{var}_p(\widehat{\mu}_{\text{OPT}}) = \frac{1}{n} \left(\sum_{h=1}^{H} W_h \sigma_h \right)^2 = \frac{1}{n} \left(\sum_{h=1}^{H} \sigma_h \int_{k_{h-1}}^{k_h} f(x)dx \right)^2. \tag{4.20}$$

Dalenius (1950) showed the allocation minimizing (4.20) must satisfy equality

$$\frac{\sigma_h^2 (k_h - \mu_h)^2}{\sigma_h} = \frac{\sigma_{h+1}^2 (k_h - \mu_{h+1})^2}{\sigma_{h+1}}, \tag{4.21}$$

for all $h = 1, \ldots, H$. This result is the starting point for most methods automatically forming strata, although it is not easy to find bounds k_h satisfying this equality, since μ_h and σ_h depend directly on these bounds.

Dalenius & Hodges Jr. (1957, 1959) supposed that, in each stratum, the distribution is uniform and therefore

$$\sigma_h \approx \frac{k_h - k_{h-1}}{\sqrt{12}}.$$

They show that, under this hypothesis, the equality (4.21) becomes

$$\int_{k_{h-1}}^{k_h} \sqrt{f(x)} dx = C,$$

where C is a constraint given by

$$C = \frac{1}{H} \int_{-\infty}^{\infty} \sqrt{f(x)} dx.$$

This method was used by Cochran (1961), who approximated the density $f(x)$ by a histogram with a large number of classes.

Bülher & Deutler (1975) used dynamic programming to find the bounds of the optimal strata. Lavallée & Hidiroglou (1988) generalized the method from Sethi (1963) to calculate the bounds of the strata while dealing with the question of take-all stratum (where all the units are selected). The algorithm from Lavallée & Hidiroglou (1988) starts with arbitrary bounds k_h. Then it iterates two steps: the calculation of parameters W_h, μ_h and σ_h and the calculation of the optimal bounds k_h. This technique was adapted by Rivest (2002) to consider the fact that the stratification variable is not equal to the variable of interest.

Finally, Gunning & Horgan (2004), Gunning et al. (2004) and Gunning et al. (2006) suggested a very simple method based on the hypothesis that the coefficients of variation in the strata are approximatively constant. If x_1 denotes the smallest value and x_N the largest, then the bounds are determined by the geometric progression:

$$k_h = x_1 \left(\frac{x_N}{x_1} \right)^h, h = 0, \dots, H.$$

Simulations on a true and asymmetric population presented in Horgan (2006) show that this method can lead to very good results.

For all these methods, a purely practical aspect should be considered: it is always preferable that the strata bounds coincide with the domains of interest. For instance, if we stratify according to the age, we can fix a stratum boundary as the retirement age. By making the strata and domains of interest coincide, we will increase the precision of the estimators in those domains.

4.11   Stratification Under Many Objectives

Surveys are never conducted to estimate the parameters of a single variable; they usually have many objectives. Hence, we have to look for an optimal solution according to a set of variables of interest. For example, we may want to minimize the weighted sum of the variances of the variables. The weighting coefficients determine the importance of each variable in this optimization process.

Unfortunately, it becomes difficult to solve the system of equations obtained by differentiating the function to minimize. We typically use numeric methods to solve these problems by considering the various constraints. For example, the sample size cannot be larger than the population size in each stratum. Various publications have been dedicated to this question (see Kish & Anderson, 1978; Dahmstroem & Hagnell, 1978; Jewett & Judkins, 1988; Shaw et al., 1995; Fitch, 2002; Lednicki & Wieczorkowski, 2003; Benedetti et al., 2008; Díaz-García & Cortez, 2008).

However, we have to keep in mind that the variances in the strata are rarely known. Theoretical developments on stratification give methodological directions rather than a method applicable in a straightforward way. In business surveys, the size of the strata is generally determined based on the number of workers and we usually consider that the standard deviations of the variables of interest are proportional to the upper bounds of the strata, which allows the calculation of the strata sizes in the sample.

Exercises

4.1 This exercise is inspired by an example in Deville (1988, p. 102). Let $U = \{1, 2, 3, 4\}$ be a population and $y_1 = y_2 = 0, y_3 = 1, y_4 = -1$ be the values taken by variable y.
   1. Calculate the variance of the estimator of the mean for simple random sampling without replacement with fixed sample size $n = 2$.
   2. Calculate the variance of the estimator of the mean for a stratified design for which only one unit is sampled per stratum, where the strata are $U_1 = \{1, 2\}$ and $U_2 = \{3, 4\}$.

4.2 A circus director has 100 elephants classified into two categories: "male" and "female". The director wants to estimate the total weight of his herd because he wants to cross a river by boat. However, the previous year, the same circus director weighed all the elephants in his herd and got the following results (in tons):

	Counts N_h	Means $\bar{Y}_h$	Variances S^2_{yh}
Males	60	6	4
Female	40	4	2.25

   1. Calculate the population variance of the "weight of the elephants" for the previous year.
   2. If, the previous year, the director had carried out a simple random sampling without replacement of ten elephants, what would have been the variance of the weight estimator of the herd?
   3. If the director had used a stratified sampling design with proportional allocation of ten elephants, what would have been the variance of the weight estimator of the herd?
   4. If the director had used an optimal stratified design of ten elephants, what would have been the strata numbers for the sample and what would have been the variance of the estimator of the total?

4.3 We want to estimate means for companies in a department. Companies are classified according to their turnover in three classes. Census data are as follows:

Turnover in millions of euros	Number of companies
From 0 to 1	1 000
From 1 to 10	100
From 10 to 100	10

We want to select a sample of 111 companies. Assuming that the distribution of companies is uniform within each stratum, give the variance of the estimator of the mean of the turnover for a stratified design with proportional allocation and for an optimal stratified design.

4.4 For a stratified design with proportional allocation:
1. Give an unbiased estimator for S_y^2.
2. Show that s_y^2 is a biased estimator of s_y^2 but this bias tends to zero when n goes to infinity.

4.5 In the population $U = \{1, 2, 3, 4, 5\}$, consider the following sampling design:

$$p(\{1, 2, 4\}) = 1/6, p(\{1, 2, 5\}) = 1/6, p(\{1, 4, 5\}) = 1/6,$$
$$p(\{2, 3, 4\}) = 1/6, p(\{2, 3, 5\}) = 1/6, p(\{3, 4, 5\}) = 1/6.$$

Calculate the first- and second-order inclusion probabilities as well as the $\Delta_{k\ell}$. Show that the design is stratified.

4.6 Consider a population U of size N partitioned in H strata denoted by $U_1, \ldots, U_h, \ldots, U_H$, of size $N_1, \ldots, N_h, \ldots, N_H$. Also, let $\overline{Y}_1, \ldots, \overline{Y}_h, \ldots, \overline{Y}_H$, denote the H means calculated within the strata. Obviously, we have

$$\overline{Y} = \frac{1}{N} \sum_{h=1}^{H} N_h \overline{Y}_h.$$

The objective of the survey is to compare a particular stratum U_i to the entire population. More specifically, we want to estimate $D_i = \overline{Y}_i - \overline{Y}$.
1. Calculate $\widehat{D}_{i\pi}$, the expansion estimator of D for any stratified design.
2. Give the variance of $\widehat{D}_{i\pi}$.
3. Give the optimal allocation that minimizes the variance of $\widehat{D}_{y\pi}$ for a fixed sample size n.
4. Briefly comment on this result (three lines maximum). How does it differ from the "classical" optimal allocation?

4.7 Suppose the purpose of a survey is to compare a metropolitan population with an overseas population. Each of these populations will be considered as a stratum. It is assumed that the variance of variable y is known in each stratum where a simple random sampling design without replacement with fixed sample size is realized. The objective is to estimate the difference between the two means of the strata:

$$D = \overline{Y}_1 - \overline{Y}_2,$$

where $\overline{Y}_1$ and $\overline{Y}_2$ are, respectively, the means of variable y in the metropolitan population and in the overseas population. We also know that an overseas interview costs twice as much as in the metropolitan population.
1. Define your notation and give an unbiased estimator of D.
2. Give the variance of the estimator of D.
3. Which criterion should be optimized to obtain the optimal allocation (giving the minimum variance) allowing D to be estimated for a fixed cost C?
4. Give the variance of the optimal estimator obtained.

4.8 A stratified survey of companies in a country is conducted. The strata are the regions and we study the "turnover" variable denoted by y. The objective is to compare the average turnover of each region with that of the other regions. We use the following criterion:

$$W = \sum_{h=1}^{H} \sum_{\substack{\ell=1 \\ \ell \neq h}}^{H} \text{var}_p(\widehat{\overline{Y}}_h - \widehat{\overline{Y}}_\ell),$$

where $\widehat{\overline{Y}}_h$ is the estimator of the mean of y in stratum h.

1. Show that W can also be written as

$$W = C \sum_{h=1}^{H} \frac{N_h - n_h}{N_h n_h} S_h^2,$$

where C is a constant that does not depend on h. Give the value of C.

2. Give the optimal allocation of the sizes of the strata in the sample which minimizes W.

5

Sampling with Unequal Probabilities

In most cases, simple random sampling designs are only used when no information is available on the population. If the values of an auxiliary variable are known, one can almost always use a more judicious design than simple random sampling. One way to introduce auxiliary information into the sampling design is to select units with unequal probabilities. However, the selection of a sample with unequal probabilities without replacement is a complex problem that has prompted the publication of several hundred articles (see among others Berger & Tillé, 2009) and several books (Brewer & Hanif, 1983; Gabler, 1990; Tillé, 2006).

After having posed the problem in all generality, we present in this chapter the most classic methods: sampling with replacement, successive sampling of the units, systematic sampling, Poisson's design, maximum entropy designs, and the splitting method. The algorithms are precisely described so that they can be directly implemented. Efficient variance estimators based on approximations are also given.

5.1 Auxiliary Variables and Inclusion Probabilities

In order to select a sample with unequal probabilities, it is necessary to have auxiliary information. Assume that the values of an auxiliary variable $x_k > 0, k \in U$, are known on all the units of U. If we can reasonably assume that variable x is approximately proportional to the variable of interest y, it may be very interesting to select the units with inclusion probabilities proportional to the values taken by x. In fact, if the design has a fixed sample size, the variance of the expansion estimator of Y is

$$\text{var}_p(\widehat{Y}) = -\frac{1}{2} \sum_{k \in U} \sum_{\substack{\ell \in U \\ \ell \neq k}} \left(\frac{y_k}{\pi_k} - \frac{y_\ell}{\pi_\ell} \right)^2 \Delta_{k\ell}. \tag{5.1}$$

If the π_k are proportional to the x_k and the x_k are approximately proportional to the y_k, then the ratios $y_k/\pi_k, k \in U$, are approximately constant and (5.1) is close to zero. The use of unequal probability designs is particularly interesting when most of the variables are linked by a size effect. If, for example, the observation units are companies, it is expected that variables such as turnover, profits, and the number of workers are more or less proportional. In this case, if one of the variables is known, one has benefits in sampling with unequal probabilities. We will also see the utility of selecting clusters or primary units with probabilities proportional to their size in Section 7.2.4. However, it

Sampling and Estimation from Finite Populations, First Edition. Yves Tillé.
© 2020 John Wiley & Sons Ltd. Published 2020 by John Wiley & Sons Ltd.

is recommended to be careful. Indeed, if the auxiliary variable x is not related to y or, even worse, correlated negatively, a sampling design with unequal probabilities can be catastrophic, with a significant increase in the variance of the estimators compared to simple random sampling.

5.2 Calculation of the Inclusion Probabilities

If we want to select exactly n units with probabilities proportional to a positive variable x, we search for a constant C such that

$$\sum_{k \in U} \min(C\, x_k, 1) = n.$$

Next we calculate $\pi_k = \min(Cx_k, 1)$, for all $k \in U$.

In practice, to solve this problem, we proceed as follows. We start by calculating the quantities

$$\pi_k = \frac{x_k n}{\sum_{\ell \in U} x_\ell}, \quad \text{for all } k \in U. \tag{5.2}$$

However, the π_k calculated according to Expression (5.2) can take values larger than one. In order to overcome this problem, units with π_k larger than 1 are automatically selected. The inclusion probabilities are then recalculated in the same way on the units that are not automatically selected. Of course, the size of the sample is reduced by the number of individuals automatically selected. This operation is repeated until all the first-order inclusion probabilities are either equal to one or strictly proportional to the $x_k > 0$.

Therefore, it is possible to separate the population into two groups: a group of automatically selected units having first-order inclusion probabilities equal to one and a second group having first-order inclusion probabilities strictly less than one and proportional to x_k. The selection of the units of the first group is not difficult. It is supposed that the algorithms to sample with unequal probabilities only apply to the second group, that is to say that the problem consists in selecting n units with fixed first-order inclusion probabilities such as

$$0 < \pi_k < 1, \quad \text{for all } k \in U, \text{ such that } \sum_{k \in U} \pi_k = n.$$

Example 5.1 If $N = 6$, $n = 3$, $x_1 = 1$, $x_2 = 9$, $x_3 = 10$, $x_4 = 70$, $x_5 = 90$, $x_6 = 120$, we have

$$X = \sum_{k \in U} x_k = 300,$$

and therefore

$$\frac{nx_1}{X} = \frac{1}{100}, \frac{nx_2}{X} = \frac{9}{100}, \frac{nx_3}{X} = \frac{1}{10}, \frac{nx_4}{X} = \frac{7}{10}, \frac{nx_5}{X} = \frac{9}{10}, \frac{nx_6}{X} = \frac{6}{5} > 1.$$

We automatically select unit 6 in the sample (so $\pi_6 = 1$). Then, we repeat the same procedure:

$$\sum_{k \in U \setminus \{6\}} x_k = 180,$$

and therefore

$$\frac{(n-1)x_1}{\sum_{\ell\in U\setminus\{6\}}x_\ell} = \frac{1}{90}, \frac{(n-1)x_2}{\sum_{\ell\in U\setminus\{6\}}x_\ell} = \frac{1}{10}, \frac{(n-1)x_3}{\sum_{\ell\in U\setminus\{6\}}x_\ell} = \frac{1}{9},$$

$$\frac{(n-1)x_4}{\sum_{\ell\in U\setminus\{6\}}x_\ell} = \frac{7}{9}, \frac{(n-1)x_5}{\sum_{\ell\in U\setminus\{6\}}x_\ell} = 1.$$

Therefore, the first-order inclusion probabilities are

$$\pi_1 = \frac{1}{90}, \pi_2 = \frac{1}{10}, \pi_3 = \frac{1}{9}, \pi_4 = \frac{7}{9}, \pi_5 = 1, \pi_6 = 1.$$

Units 5 and 6 are selected automatically and the problem is reduced to the selection of a unit in a subpopulation of size 4.

5.3 General Remarks

The calculation of the π_k represents the first step that precedes the selection of the sample. We have already seen that a sampling design is defined by the probabilities $p(s)$ and not by the first-order inclusion probabilities. Therefore, the problem can be formulated as follows: we search for a sampling design $p(\cdot)$ such that

$$\sum_{\substack{s\ni k \\ s\in\mathscr{S}_n}} p(s) = \pi_k, \quad \text{for all } k \text{ with } 0 < \pi_k < 1, \tag{5.3}$$

where $\mathscr{S}_n = \{s \subset U | \#s = n\}$. The search for a design with unequal probabilities is a very open problem. Indeed, there is in general an infinity of solutions satisfying the constraints (5.3).

However, the problem is quite complicated, which is why it has been written about extensively. It is easy to select a sample with unequal probabilities with replacement (multinomial sampling) or with a random sample size (Poisson design), but the fact of imposing simultaneously sampling without replacement and fixed sample size makes the problem much more difficult to solve.

Since there is an infinite number of sampling designs for fixed first-order inclusion probabilities, we can search for a method which has interesting second-order inclusion probabilities. We consider the following properties:

1. The second-order inclusion probabilities should be strictly positive.
2. The inclusion probabilities should satisfy the Yates–Grundy conditions: $\pi_{k\ell} \le \pi_k\pi_\ell$ for all $k \ne \ell$ (see Section 2.7).
3. The method should provide second-order inclusion probabilities such that the variance of the estimator of total is always lower than the design with replacement.

We have seen in Chapter 3 that simple random sampling and Bernoulli sampling maximize the entropy defined by

$$I(p) = -\sum_{\substack{s\ni k \\ s\in\mathscr{S}_n}} p(s)\log p(s).$$

Therefore, we could look for an unequal probability design that maximizes this same criterion. We could maximize $I(p)$ subject to constraints (5.3). This would give the most random design for given inclusion probabilities. This question, which is covered in Section 5.9, proves unfortunately quite complex and has been solved only recently (see Chen et al., 1994; Deville, 2000b). It is presented in detail in Tillé (2006).

The identification of a sampling design is not sufficient to have an applicable method. This design must also be implementable by means of an algorithm that avoids the enumeration of all possible samples. Finally, a good algorithm should have the following properties:

1. The method should be fast, the sample should be selected without calculating $p(s)$ for the $N!/\{n!(N-n)!\}$ possible samples of size n.
2. The implementation should be sequential, that is, we should be able to apply it to a data file in a single reading, examining units according to their order in the file.

Moreover, it would be interesting if the $p(s)$ as well as the second-order inclusion probabilities obtained by the sampling method do not depend on the order of the units of the data file. Several dozens methods have already been proposed. A fairly complete inventory of these is found in Hanif & Brewer (1980), Brewer & Hanif (1983), and Tillé (2006). We give below a description of the main methods.

5.4 Sampling with Replacement with Unequal Inclusion Probabilities

Before tackling the designs without replacement, we present the sampling design with replacement, also known as multinomial sampling, which can be interesting when the sample size m is small enough compared to the population size N in such a way that the probability of obtaining the same unit more than once is small. This method, proposed by Hansen & Hurwitz (1943), is very easy to implement and is also used as a reference for designs without replacement. Indeed, it is useless to use a method without replacement if it provides a less precise estimator than the design with replacement.

We define

$$p_k = \frac{x_k}{\sum_{\ell \in U} x_\ell}, k \in U,$$

and

$$v_k = \sum_{\ell=1}^{k} p_\ell, \text{ with } v_0 = 0.$$

The p_k are drawing probabilities and not inclusion probabilities. Indeed, the sum of p_k is not equal to the sample size but to 1. We generate a uniform continuous random variable u in $[0, 1[$ and we select the unit k such that $v_{k-1} \le u < v_k$. This operation is repeated m times independently to obtain the sample.

Let $a_k, k \in U$ be the number of times that unit k is selected in the sample. The random variable a_k follows a binomial distribution $a_k \sim B(m, p)$, with expectation $E_p(a_k) = mp_k$

and of variance $\text{var}_p(a_k) = mp_k(1 - p_k)$. Moreover, the vector $\mathbf{a} = (a_1, \ldots, a_k, \ldots, a_N)$ follows a multinomial distribution and therefore $\text{cov}_p(a_k, a_\ell) = -mp_kp_\ell$ for $k \neq \ell$. We can estimate Y by the Hansen–Hurwitz estimator:

$$\widehat{Y}_{\text{HH}} = \frac{1}{m} \sum_{k \in U} \frac{y_k a_k}{p_k}.$$

Since

$$E_p(\widehat{Y}_{\text{HH}}) = \frac{1}{m} \sum_{k \in U} \frac{y_k E_p(a_k)}{p_k} = \sum_{k \in U} y_k,$$

estimator $\widehat{Y}_{\text{HH}}$ is unbiased for Y.

Result 5.1 *The variance of the Hansen–Hurwitz estimator is*

$$\text{var}_p(\widehat{Y}_{\text{HH}}) = \frac{1}{m} \sum_{k \in U} p_k \left(\frac{y_k}{p_k} - Y\right)^2.$$

Proof:

$$\text{var}_p(\widehat{Y}_{\text{HH}}) = \frac{1}{m^2} \sum_{k \in U} \frac{y_k^2}{p_k^2} \text{var}_p(a_k) + \frac{1}{m^2} \sum_{k \in U} \sum_{\substack{\ell \in U \\ \ell \neq k}} \frac{y_k y_\ell}{p_k p_\ell} \text{cov}_p(a_k, a_\ell)$$

$$= \frac{1}{m^2} \sum_{k \in U} \frac{y_k^2}{p_k^2} mp_k(1 - p_k) - \frac{1}{m^2} \sum_{k \in U} \sum_{\substack{\ell \in U \\ \ell \neq k}} \frac{y_k y_\ell}{p_k p_\ell} mp_k p_\ell$$

$$= \frac{1}{m} \sum_{k \in U} \frac{y_k^2}{p_k}(1 - p_k) - \frac{1}{m} \sum_{k \in U} \sum_{\substack{\ell \in U \\ \ell \neq k}} y_k y_\ell = \frac{1}{m} \sum_{k \in U} \frac{y_k^2}{p_k} - \frac{1}{m} \sum_{k \in U} \sum_{\ell \in U} y_k y_\ell$$

$$= \frac{1}{m} \sum_{k \in U} \frac{y_k^2}{p_k} - \frac{Y^2}{m} = \frac{1}{m} \sum_{k \in U} p_k \left(\frac{y_k}{p_k} - Y\right)^2.$$
∎

In order to estimate $\text{var}_p(\widehat{Y}_{\text{HH}})$, we refer to the following result:

Result 5.2 *Under a multinomial design, the variance* $\text{var}_p(\widehat{Y}_{\text{HH}})$ *can be unbiasedly estimated by*

$$v(\widehat{Y}_{\text{HH}}) = \frac{1}{m(m-1)} \sum_{k \in U} \left(\frac{y_k}{p_k} - \widehat{Y}_{\text{HH}}\right)^2 a_k.$$

Proof: Since $v(\widehat{Y}_{\text{HH}})$ can be written as

$$v(\widehat{Y}_{\text{HH}}) = \frac{1}{m(m-1)} \sum_{k \in U} \left(\frac{y_k}{p_k} - Y\right)^2 a_k - \frac{1}{m-1}(\widehat{Y}_{\text{HH}} - Y)^2,$$

one can calculate the expectation:

$$
E_p[v(\hat{Y}_{HH})] = \frac{1}{m(m-1)} \sum_{k\in U} \left(\frac{y_k}{p_k} - Y\right)^2 E_p(a_k) - \frac{1}{m-1} \mathrm{var}_p(\hat{Y}_{HH})
$$

$$
= \frac{1}{m-1} \sum_{k\in U} p_k \left(\frac{y_k}{p_k} - Y\right)^2 - \frac{1}{m-1} \mathrm{var}_p(\hat{Y}_{HH})
$$

$$
= \frac{m}{m-1} \mathrm{var}_p(\hat{Y}_{HH}) - \frac{1}{m-1} \mathrm{var}_p(\hat{Y}_{HH}) = \mathrm{var}_p(\hat{Y}_{HH}).
$$ ∎

While this method is easy to implement, it is not admissible in the sense that it is always possible to define a more accurate estimator. The Hansen–Hurwitz estimator takes into account information about the multiplicity of units. Therefore, there is a Rao–Blackwellized estimator of $\hat{Y}_{HH}$ with a smaller mean squared error than $\hat{Y}_{HH}$ (see Expression (2.9), page 24). This Rao–Blackwellization is unfortunately so complex that it seems practically impossible to apply it for a sample size larger than 3. However, we know that the minimal sufficient statistic is obtained by removing from the observed data the information about the multiplicity of units and therefore that there is an estimator better than $\hat{Y}_{HH}$. For this reason, designs without replacement have attracted much interest.

A variant of the design with replacement with unequal probability consists of rejecting the sample if a unit is selected more than once. The procedure is then repeated until a sample consisting of distinct units is obtained.

This procedure is presented in Brewer & Hanif (1983, *Carroll–Hartley rejective procedure 14*, page 31, procedure 28-31, page 40) and is also discussed in Rao (1963), Carroll & Hartley (1964), Hájek (1964), Rao & Bayless (1969), Bayless & Rao (1970), and Cassel et al. (1977, 1993). However, its implementation is tricky. We can check on a sample size of 2 (see Brewer & Hanif, 1983, p. 24, procedure 5) that in order to obtain fixed order inclusion probabilities a priori π_k, the drawing probabilities p_k cannot be proportional to π_k. A correct algorithm is presented in Tillé (2006, pp. 90–91), but this method can be slow when n/N is large.

5.5 Nonvalidity of the Generalization of the Successive Drawing without Replacement

The method of successive drawing of units was used to implement simple random sampling without replacement in Section 3.9.3. It is unfortunately not directly generalizable to unequal probabilities. It is interesting to examine this method, which one would have intuitively wanted to apply, but which is in fact false, because it helps us to understand why the problem of drawing with unequal probabilities without replacement of fixed size is so complex.

The following reasoning aims at highlighting the problem posed by this generalization. First,

$$
p_k = \frac{x_k}{\sum_{\ell\in U} x_\ell}, k \in U.
$$

The generalization of the method is the following: select in the first step a unit according to the probabilities $p_k, k \in U$. The selected individual is denoted by j. Then, unit j is removed from U and we recalculate

$$p_k^j = \frac{p_k}{1 - p_j}, k \in U\backslash\{j\}.$$

We select a second unit with unequal probabilities $p_k^j, k \in U$, among the remaining $N - 1$ units and so on. However, this method does not provide first-order inclusion probabilities proportional to x_k. We will show it by taking $n = 2$. In this case,

$$\Pr(k \in S) = \Pr(k \text{ is selected at the first draw})$$
$$+ \Pr(k \text{ is selected at the second draw})$$

$$= p_k + \sum_{\substack{j \in U \\ j \neq k}} p_j p_k^j = p_k \left(1 + \sum_{\substack{j \in U \\ j \neq k}} \frac{p_j}{1 - p_j} \right). \tag{5.4}$$

This probability, given by (5.4), is not equal to the fixed inclusion probability which in this case should be

$$\pi_k = 2p_k, k \in U. \tag{5.5}$$

Instead of p_k, we could obviously use modified values p_k^* such that the inclusion probabilities (5.5) are respected. In the case where $n = 2$, we should search for p_k^* such that

$$p_k^* \left(1 + \sum_{\substack{j \in U \\ j \neq k}} \frac{p_j^*}{1 - p_j^*} \right) = \pi_k, k \in U.$$

This method is known as the procedure of Narain (1951) (also see Horvitz & Thompson, 1952; Yates & Grundy, 1953; Brewer & Hanif, 1983, p. 25). However, the determination of p_k^* is complex and becomes in practice impossible when n is large.

5.6 Systematic Sampling with Unequal Probabilities

Systematic sampling was proposed by Madow (1949) and remains one of the best solutions proposed for its simplicity of use and its accuracy. The sample size is fixed and the method exactly satisfies the inclusion probabilities π_k. We suppose we know the $0 < \pi_k < 1, k \in U$ with

$$\sum_{k \in U} \pi_k = n.$$

We want to select a sample of fixed size n with inclusion probabilities π_k. The cumulative inclusion probabilities are defined as

$$V_k = \sum_{\ell=1}^{k} \pi_\ell, \text{ for all } k \in U, \text{ with } V_0 = 0. \tag{5.6}$$

Next, we generate a uniform random variable u in $[0, 1[$, this number will give the "random start". The first selected unit k_1 is such that $V_{k_1-1} \leq u < V_{k_1}$, the second unit is such that $V_{k_2-1} \leq u + 1 < V_{k_2}$, and the jth unit is such that $V_{k_j-1} \leq u + j - 1 < V_{k_j}$.

Example 5.2 If $N = 6$, $n = 3$ and $\pi_1 = 0.2$, $\pi_2 = 0.7$, $\pi_3 = 0.8$, $\pi_4 = 0.5$, $\pi_5 = \pi_6 = 0.4$, and u takes the value 0.3658, we obtain $V_1 = 0.2$, $V_2 = 0.9$, $V_3 = 1.7$, $V_4 = 2.2$, $V_5 = 2.6$, $V_6 = 3$. We can represent graphically systematic sampling with Figure 5.1. The values $u, u + 1$ and $u + 2$ are in the intervals corresponding to units 2, 3, and 5 that are selected.

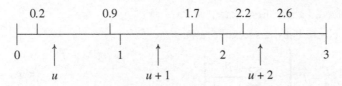

Figure 5.1 Systematic sampling: example with inclusion probabilities $\pi_1 = 0.2$, $\pi_2 = 0.7$, $\pi_3 = 0.8$, $\pi_4 = 0.5$, $\pi_5 = \pi_6 = 0.4$, and $u = 0.3658$.

Systematic sampling can be defined more formally in the following way: select the units k such that the intervals $[V_{k-1} - u, V_k - u[$ contain an integer. This definition makes it possible to define Algorithm 6, which is particularly simple. In this algorithm, $\lfloor x \rfloor$ represents the largest integer less than x.

Algorithm 6 Systematic sampling with unequal probabilities

Definition a, b, u reals; k integer;
$u = $ a uniform random variable in $[0,1[$;
$a = -u$;

Repeat for $k = 1, \ldots, N$ $\begin{array}{l} b = a; \\ a = a + \pi_k; \\ \text{if } \lfloor a \rfloor \neq \lfloor b \rfloor \text{ select } k. \end{array}$

However, the method has a drawback: many joint inclusion probabilities are zero. In the example shown in Figure 5.1, if unit 1 is selected, it is impossible to select units 2, 5, and 6, and units 3 and 4 are selected automatically. From Figure 5.1, one can easily calculate the matrix of the second-order inclusion probabilities:

$$
\begin{pmatrix}
- & 0 & 0.2 & 0.2 & 0 & 0 \\
0 & - & 0.5 & 0.2 & 0.4 & 0.3 \\
0.2 & 0.5 & - & 0.3 & 0.4 & 0.2 \\
0.2 & 0.2 & 0.3 & - & 0 & 0.3 \\
0 & 0.4 & 0.4 & 0 & - & 0 \\
0 & 0.3 & 0.2 & 0.3 & 0 & -
\end{pmatrix}.
$$

More formally, we can construct a general expression of the second-order inclusion probabilities for two individuals k and ℓ. These probabilities only depend on the first-order inclusion probabilities and quantity

$$V_{k\ell} = \begin{cases} \sum_{i=k}^{\ell-1} \pi_i, & \text{if } k < \ell \\ \sum_{i=k}^{N} \pi_i + \sum_{i=1}^{\ell-1} \pi_i = n - \sum_{i=\ell}^{k-1} \pi_i, & \text{if } k > \ell. \end{cases}$$

The joint inclusion probabilities are

$$\pi_{k\ell} = \min[\max(0, \pi_k - \delta_{k\ell}), \pi_\ell] + \min[\pi_k, \max(0, \delta_{k\ell} + \pi_\ell - 1)], k < \ell,$$

where $\delta_{k\ell} = V_{k\ell} - \lfloor V_{k\ell} \rfloor$ (see, for instance, Connor, 1966).

Pea et al. (2007) showed that systematic designs only assign nonzero probabilities to at most N samples. Systematic designs have a minimal support and therefore have very low entropy. The main problem with systematic sampling is the large number of null $\pi_{k\ell}$. To overcome this difficulty, one can apply systematic sampling after randomly sorting the file. However, random sorting does not completely solve the problem of null $\pi_{k\ell}$. Consider the following example, where $N = 5$ and $n = 2$, with $\pi_1 = \pi_2 = 0.25, \pi_3 = \pi_4 = \pi_5 = 0.5$. It can be checked that, whatever the order of the file, the probability of selecting together the two units that have an first-order inclusion probability of 0.25 is zero.

After a random sorting of the file, the determination of the second-order inclusion probabilities is then much more complex. Approximations of the second-order inclusion probabilities have been proposed by Hartley & Rao (1962) and Deville (Undated). However, if the file is sorted, it is better to directly estimate the variance using the estimators presented in Section 5.15. Connor (1966) and Hidiroglou & Gray (1980) give exact but relatively long methods for calculating $\pi_{k\ell}$.

5.7 Deville's Systematic Sampling

Deville (1998c) has proposed a particular systematic sampling method. The V_k are built as in Expression (5.6). The technique consists of selecting only one unit in each interval $[i - 1, i[$ with $i = 1, \dots, n$. To do this, we generate n uniform variables $u_i, i = 1, \dots, n$, in other words $u_i \sim \mathcal{U}[0, 1[$. For each variable u_i, unit k is selected if

$$V_{k-1} \leq u_i + (i - 1) < V_k.$$

Figure 5.2 illustrates the Deville method in the case where we want to select four units in a population of size $N = 7$.

Figure 5.2 shows that after generating u_1, u_2, u_3, and u_4, units 1, 2, 5, and 6 are selected. However, it is necessary to introduce a dependency between u_i in order not to select twice the boundary units. A unit is called a boundary if $[V_{\ell-1}, V_\ell[$ contains an integer.

Figure 5.2 Method of Deville.

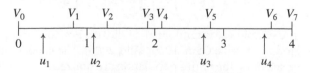

In Figure 5.2, units 2, 4, and 6 are boundary units. If the u_i were selected independently, the boundary units could be selected twice in the sample. Therefore, Deville proposed to introduce a dependency between the u_i so that the design remains without replacement.

Let ℓ be the boundary unit such that $[V_{\ell-1} \leq i - 1 < V_\ell]$. In this case, unit ℓ can be selected at step $i - 1$. The density function of u_i is set so that the unit ℓ is not selected twice.

- If ℓ is selected at step $i - 1$, u_i has density function

$$f_1(x) = \begin{cases} \dfrac{1}{i - V_\ell} & \text{if } x \geq V_\ell - (i - 1) \\ 0 & \text{if } x < V_\ell - (i - 1) \end{cases}, x \in [0, 1[.$$

- If ℓ is not selected at step $i - 1$, u_i has density function

$$f_2(x) = \begin{cases} 1 - \dfrac{(i - 1 - V_{\ell-1})(V_\ell - i + 1)}{[1 - (i - 1 - V_{\ell-1})][1 - (V_\ell - i + 1)]} & \text{if } x \geq V_\ell - i + 1 \\ \dfrac{1}{1 - (i - 1 - V_{\ell-1})} & \text{if } x < V_\ell - i + 1. \end{cases}$$

Knowing that the probability of selecting unit ℓ at step $i - 1$ equals $i - 1 - V_{\ell-1}$, u_i has a uniform distribution in $[0, 1[$. Indeed,

$$(i - 1 - V_{\ell-1})f_1(x) + [1 - (i - 1 - V_{\ell-1})]f_2(x) = 1, x \in [0, 1[.$$

Therefore, the method is without replacement and satisfies the fixed inclusion probabilities. Fuller (1970) proposed a very similar procedure. In Fuller's method, the units are placed on a circle and the starting point is determined randomly.

5.8 Poisson Sampling

Poisson sampling design is the generalization with unequal probabilities of the Bernoulli design described in Section 3.9.2. Like the latter, Poisson sampling is not of fixed sample size. It serves as a reference for many sampling designs, such as the conditional Poisson design (also called maximum entropy design). It also has an important theoretical interest in the treatment of nonresponse. Indeed, the response mechanism is often associated with a Poisson design, which consists of assuming that the units decide to answer or not independently, possibly with unequal probabilities.

In a Poisson design, each unit of U is selected independently with an inclusion probability π_k. The method can be easily implemented using Algorithm 7.

Since the units are selected independently of each other, we have $\pi_{k\ell} = \pi_k \pi_\ell$, and thus $\Delta_{k\ell} = \pi_{k\ell} - \pi_k \pi_\ell = 0$, for all $k \neq \ell$. The sampling design is

$$p(s) = \left(\prod_{k \in s} \pi_k\right) \times \left[\prod_{k \in U \backslash s}(1 - \pi_k)\right] \text{ for all } s \subset U.$$

The sample size is random and follows a Poisson binomial distribution (see, for example, Hodges Jr. & Le Cam, 1960; Tillé, 2006). The probabilities of selecting an empty sample or selecting the entire population are nonzero.

Algorithm 7 Algorithm for Poisson sampling

Definition u : real ; k : integer ;
$\qquad\qquad\qquad\qquad u =$ uniform random variable$[0, 1[$;
Repeat for $k = 1, \dots, N$ | if $u < \pi_k$ select unit k;
$\qquad\qquad\qquad\qquad$ otherwise skip unit k.

Since $\Delta_{k\ell} = \pi_{k\ell} - \pi_k \pi_\ell = 0$ when $k \neq \ell$, the variance of the expansion estimator is directly deduced from Expression (2.3), page 20:

$$\text{var}_p(\hat{Y}) = \sum_{k \in U} \frac{\pi_k(1 - \pi_k)y_k^2}{\pi_k^2}. \tag{5.7}$$

This variance can be estimated by

$$v(\hat{Y}) = \sum_{k \in S} \frac{(1 - \pi_k)y_k^2}{\pi_k^2}.$$

The following result shows that the Poisson sampling is the most random design, in the sense of the entropy, with fixed inclusion probabilities π_k.

Result 5.3 *The design that maximizes entropy*

$$I(p) = - \sum_{s \subset U} p(s) \log p(s),$$

on $\mathcal{S} = \{s | s \subset U\}$ with fixed inclusion probabilities $\pi_k, k \in U$, is the Poisson design.

Proof: Let us look for the design that maximizes $I(p)$ subject to

$$\sum_{\substack{s \subset U \\ s \ni k}} p(s) = \pi_k \quad \text{and} \quad \sum_{s \subset U} p(s) = 1.$$

We define the Lagrange function:

$$\mathcal{L}(p(s), \lambda_k, v)$$

$$= - \sum_{s \subset U} p(s) \log p(s) + \sum_{k \in U} \lambda_k \left[\sum_{s \ni k} p(s) - \pi_k \right] + v \left[\sum_{s \subset U} p(s) - 1 \right]$$

$$= - \sum_{s \subset U} p(s) \log p(s) + \sum_{s \subset U} p(s) \sum_{k \in s} \lambda_k - \sum_{k \in U} \lambda_k \pi_k + v \left[\sum_{s \subset U} p(s) - 1 \right].$$

By canceling the partial derivatives of $\mathcal{L}$ with respect to $p(s)$, we obtain

$$p(s) = \exp \left(\sum_{k \in s} \lambda_k + v - 1 \right), s \subset U.$$

If we let $\mu = 1 - v$, we have

$$p(s) = \exp \left(\sum_{k \in s} \lambda_k - \mu \right) = \frac{\prod_{k \in s} \exp \lambda_k}{\exp \mu}, s \subset U.$$

Since

$$\sum_{s \subset U} p(s) = \frac{\sum_{s \subset U} \prod_{k \in s} \exp \lambda_k}{\exp \mu} = \frac{\prod_{k \in U}(1 + \exp \lambda_k)}{\exp \mu} = 1,$$

we can determine μ:

$$\exp \mu = \prod_{k \in U}(1 + \exp \lambda_k).$$

Moreover, since

$$\pi_k = \sum_{\substack{s \ni k \\ s \subset U}} p(s) = \frac{\sum_{\substack{s \ni k \\ s \subset U}} \prod_{\ell \in s} \exp \lambda_\ell}{\exp \mu} = \frac{\exp \lambda_k \sum_{s \subset U \setminus \{k\}} \prod_{\ell \in s} \exp \lambda_\ell}{\exp \mu}$$

$$= \frac{\exp \lambda_k \prod_{\ell \in U \setminus \{k\}}(1 + \exp \lambda_\ell)}{\exp \mu} = \frac{\exp \lambda_k}{1 + \exp \lambda_k} \frac{\prod_{\ell \in U}(1 + \exp \lambda_\ell)}{\exp \mu}$$

$$= \frac{\exp \lambda_k}{1 + \exp \lambda_k},$$

we obtain

$$\exp \lambda_k = \frac{\pi_k}{1 - \pi_k},$$

and

$$p(s) = \frac{\prod_{k \in s} \frac{\pi_k}{1 - \pi_k}}{\prod_{k \in U} \frac{1}{1 - \pi_k}} = \left(\prod_{k \in s} \pi_k\right)\left[\prod_{k \notin s}(1 - \pi_k)\right], s \subset U,$$

which defines the Poisson sampling design. ∎

Therefore, the Poisson design is the most random design possible (in the sense of the entropy) which respects *a priori* fixed first-order inclusion probabilities. The Poisson design is of relatively limited interest in selecting a sample because the sample size is random. It does not allow the cost of a survey to be controlled. However, it can be used when a survey is applied in a population whose size is unknown and where the units come sequentially. For example, when we want to select customers in a store as they arrive and the total number of customers is not known in advance, we can select customers according to a Poisson design.

However, the Poisson design is extremely important for modeling nonresponse. It will often be assumed that individuals respond or not to a survey independently of each other with each a particular probability ϕ_k. Nonresponse is then considered as a second phase of sampling by means of a Poisson design (see Chapter 16, page 333).

In repeated surveys, we often want to manage the overlap from one wave to another. For example, we sometimes try to select a nonoverlapping sample or a sample with a maximum of common units. Often, a partial rotation is organized. For example, 20% of units are replaced each year. Poisson designs are easy to coordinate (see Section 8.2.4, page 175). The Swiss sample selection and coordination method mainly uses Poisson designs (see Section 8.2.9, page 178).

5.9 Maximum Entropy Design

The Poisson design defined in Section 5.8 maximizes the measure of entropy, but it has the drawback of not being of fixed size. We can also look for designs with fixed size n that maximize the entropy. We then limit ourselves to looking for the optimal design on all the samples of size n: $\mathcal{S}_n = \{s | \#s = n\}$. This design, also called conditional Poisson sampling, was studied by Hájek (1981), Chen et al. (1994), Deville (2000b), and Tillé (2006).

Result 5.4 *The maximum entropy design with fixed sample size with inclusion probabilities π_k can be written as*

$$p(s) = \frac{\exp \sum_{k \in s} \lambda_k}{\sum_{s \in \mathcal{S}_n} \exp \sum_{\ell \in s} \lambda_\ell}, s \in \mathcal{S}_n,$$

where the λ_k are determined by

$$\sum_{\substack{s \ni k \\ s \in \mathcal{S}_n}} p(s) = \pi_k, k \in U.$$

Proof: We minimize

$$I(p) = - \sum_{s \in \mathcal{S}_n} p(s) \log p(s),$$

subject to

$$\sum_{\substack{s \ni k \\ s \in \mathcal{S}_n}} p(s) = \pi_k \quad \text{and} \quad \sum_{s \in \mathcal{S}_n} p(s) = 1.$$

By introducing the Lagrange function:

$$\mathcal{L}(p(s), \lambda_k, v)$$

$$= - \sum_{s \in \mathcal{S}_n} p(s) \log p(s) + \sum_{k \in U} \lambda_k \left[\sum_{\substack{s \ni k \\ s \in \mathcal{S}_n}} p(s) - \pi_k \right] + v \left[\sum_{s \in \mathcal{S}} p(s) - 1 \right]$$

$$= - \sum_{s \in \mathcal{S}_n} p(s) \log p(s) + \sum_{s \in \mathcal{S}_n} p(s) \sum_{k \in s} \lambda_k - \sum_{k \in U} \lambda_k \pi_k + v \left[\sum_{s \in \mathcal{S}} p(s) - 1 \right]$$

and by setting to zero the partial derivatives of $\mathcal{L}$ with respect to $p(s)$, we have

$$p(s) = \exp \left(\sum_{k \in s} \lambda_k + v - 1 \right) = \exp(v - 1) \exp \left(\sum_{k \in s} \lambda_k \right), s \in \mathcal{S}_n.$$

Since

$$\sum_{s \in \mathcal{S}_n} p(s) = \exp(v - 1) \sum_{s \in \mathcal{S}_n} \exp \left(\sum_{k \in s} \lambda_k \right) = 1,$$

$$\exp(\nu - 1) = \sum_{s \in \mathcal{S}_n} p(s) - \left(\sum_{k \in s} \lambda_k \right).$$

Therefore,

$$p(s) = \frac{\exp \sum_{k \in s} \lambda_k}{\sum_{s \in \mathcal{S}_n} \exp \sum_{\ell \in s} \lambda_\ell}, s \in \mathcal{S}_n. \qquad \blacksquare$$

The λ_k can in principle be determined by solving the system of N equations:

$$\sum_{\substack{s \ni k \\ s \in \mathcal{S}_n}} p(s) = \sum_{\substack{s \ni k \\ s \in \mathcal{S}_n}} \frac{\exp \sum_{k \in s} \lambda_k}{\sum_{s \in \mathcal{S}_n} \exp \sum_{\ell \in s} \lambda_\ell} = \pi_k.$$

Unfortunately, the size of $\mathcal{S}_n$ is so large that it is practically impossible to solve this system of equations. We will see at the end of this section that it is possible to find shortcuts to identify λ_k.

Chen et al. (1994) showed that the λ_k exist and are unique subject to a change of origin whatever the inclusion probabilities $0 < \pi_k < 1, k \in U$.

Result 5.5 *Any maximum entropy design of parameter λ_k can be written as a conditional design with respect to the fixed size of all Poisson designs of inclusion probabilities:*

$$\tilde{\pi}_k = \frac{\exp(\lambda_k + C)}{1 + \exp(\lambda_k + C)}, \qquad (5.8)$$

for all $C \in \mathbb{R}$.

Proof: Expression (5.8) is equivalent to

$$\lambda_k = -C + \log \frac{\tilde{\pi}_k}{1 - \tilde{\pi}_k}, k \in U.$$

The maximum entropy design of fixed size can be written in the form

$$p(s) = \frac{\exp \sum_{k \in s} \lambda_k}{\sum_{s \in \mathcal{S}_n} \exp \sum_{\ell \in s} \lambda_\ell} = \frac{\prod_{k \in s} \dfrac{\tilde{\pi}_k}{1 - \tilde{\pi}_k}}{\sum_{s \in \mathcal{S}_n} \prod_{\ell \in s} \dfrac{\tilde{\pi}_\ell}{1 - \tilde{\pi}_\ell}}, s \in \mathcal{S}_n.$$

The $\tilde{\pi}_k$ of the Poisson design are obviously not equal to the π_k of the maximum entropy design with fixed sample size. Consider now a Poisson design $\tilde{p}(s)$ of inclusion probabilities $\tilde{\pi}_k$. Referring to Section 5.8, this design can be written as

$$\tilde{p}(s) = \frac{\prod_{k \in s} \dfrac{\tilde{\pi}_k}{1 - \tilde{\pi}_k}}{\sum_{s \in \mathcal{S}} \prod_{\ell \in s} \dfrac{\tilde{\pi}_\ell}{1 - \tilde{\pi}_\ell}}, s \in \mathcal{S}.$$

Design $\tilde{p}(s)$ differs from $p(s)$ because the sum of the denominator is $\mathcal{S}$ and not $\mathcal{S}_n$. The Poisson design has a random sample size denoted by $\tilde{n}_S$. The maximum entropy design of fixed size can be deduced from the Poisson design by conditioning with respect to the fixed size n. Indeed, we can directly check that

$$p(s) = \tilde{p}(s|\tilde{n}_S = n) = \frac{\tilde{p}(s)}{\sum_{s \in \mathcal{S}_n} \tilde{p}(s)}, s \in \mathcal{S}_n.$$

∎

One way to implement a maximum entropy design is to apply a rejective technique. Samples are generated using a Poisson design of inclusion probabilities $\tilde{\pi}_k$ until a sample of size n is obtained. Choosing C as

$$\sum_{k \in U} \tilde{\pi}_k = n,$$

we increase the chance of getting a sample of size n.

By conditioning with respect to the fixed sample size, the inclusion probabilities are modified. The inclusion probabilities π_k are *a priori* fixed. In order to construct an algorithm to implement the maximum entropy design, it is necessary to be able to calculate the vector $\tilde{\pi} = (\tilde{\pi}_1, \dots, \tilde{\pi}_k, \dots \tilde{\pi}_N)^\mathsf{T}$ from vector $\pi = (\pi_1, \dots, \pi_k, \dots, \pi_N)^\mathsf{T}$ and vice versa.

Result 5.6 *(Deville, 2000b) Consider a maximum entropy sampling design with parameter $\lambda_k, k \in U$. If $\pi(n) = (\pi_1(n), \dots, \pi_k(n), \dots, \pi_N(n))^\mathsf{T}$ represents the vector of the inclusion probabilities of the design given by $\tilde{p}(s|\tilde{n}_S = n)$, then*

$$\pi_k(n) = \frac{n \exp \lambda_k [1 - \pi_k(n-1)]}{\sum_{\ell \in U} \exp \lambda_\ell [1 - \pi_\ell(n-1)]}, k \in U. \tag{5.9}$$

and

$$\pi_k(1) = \frac{\exp \lambda_k}{\sum_{\ell \in U} \exp \lambda_\ell}.$$

Proof: We have

$$\pi_k(n) = \sum_{\substack{s \ni k \\ s \in \mathcal{S}_n}} \exp \left[\sum_{\ell \in s} \lambda_\ell - \mu(n) \right]$$

$$= \exp \lambda_k \frac{\exp \mu(n-1)}{\exp \mu(n)} \sum_{\substack{s \subset U \setminus \{k\} \\ s \in \mathcal{S}_{n-1}}} \exp \left[\sum_{\ell \in s} \lambda_\ell - \mu(n-1) \right]$$

$$= \exp \lambda_k \frac{\exp \mu(n-1)}{\exp \mu(n)} [1 - \pi_k(n-1)], \tag{5.10}$$

where

$$\mu(n) = \log \sum_{s \in \mathcal{S}_n} \sum_{k \in s} \exp \lambda_k.$$

By summing Expression (5.10) on the population, we obtain

$$\sum_{k \in U} \pi_k(n) = n = \sum_{k \in U} \exp \lambda_k \frac{\exp \mu(n-1)}{\exp \mu(n)} [1 - \pi_k(n-1)],$$

and therefore

$$\frac{\exp \mu(n-1)}{\exp \mu(n)} = \frac{n}{\sum_{k \in U} \exp \lambda_k [1 - \pi_k(n-1)]}. \tag{5.11}$$

By introducing (5.11) into (5.10), we obtain the recurrence equation:

$$\pi_k(n) = \frac{n \exp \lambda_k [1 - \pi_k(n-1)]}{\sum_{\ell \in U} \exp \lambda_\ell [1 - \pi_\ell(n-1)]}, k \in U.$$

Besides, $\boldsymbol{\pi}(1)$ is

$$\pi_k(1) = \sum_{\substack{s \ni k \\ s \in \mathcal{S}_1}} \exp \left(\sum_{\ell \in s} \lambda_\ell - \mu(1) \right) = \frac{\exp \lambda_k}{\sum_{\ell \in U} \exp \lambda_\ell}.$$

∎

This result shows that $\pi_k(n)$ can easily be derived from λ_k (and therefore $\widetilde{\pi}_k$) by successively calculating $\boldsymbol{\pi}(2), \boldsymbol{\pi}(3), \dots, \boldsymbol{\pi}(n-1), \boldsymbol{\pi}(n)$, using the recurrence equation (5.9). Therefore, it is possible to pass from $\widetilde{\pi}_k$ to π_k without having to enumerate all the samples of size n. We define the function $\boldsymbol{\phi}(\widetilde{\boldsymbol{\pi}}) = \boldsymbol{\pi}(n)$ from $\mathbb{R}^N$ into $\mathbb{R}^N$. Deville (2000b) has shown that it is possible to invert this function using the following variant of the Newton algorithm:

$$\widetilde{\boldsymbol{\pi}}^{(i+1)} = \widetilde{\boldsymbol{\pi}}^{(i)} + \boldsymbol{\pi}(n) - \boldsymbol{\phi}(\widetilde{\boldsymbol{\pi}}^{(i)}),$$

for $i = 0, 1, 2, 3, \dots$ and by taking $\widetilde{\boldsymbol{\pi}}^{(0)} = \boldsymbol{\pi}$. Therefore, we can also pass from π_k to $\widetilde{\pi}_k$ and therefore to λ_k without having to enumerate all the samples of size n.

Knowing the $\widetilde{\pi}_k$, it is then easy to implement the maximum entropy design according to a rejective principle: we select samples with a Poisson design of inclusion probabilities $\widetilde{\pi}_k$ until we obtain a fixed-size sample n. The selected sample then has the inclusion probabilities π_k. Several other faster algorithms are presented in Chen et al. (1994), Deville (2000b), and Tillé (2006). An implementation is available in the function UPmaxentropy of sampling package R developed by Tillé & Matei (2016).

5.10 Rao–Sampford Rejective Procedure

Rao (1965) and Sampford (1967) proposed a very ingenious procedure for selecting a sample with unequal inclusion probabilities π_k and a fixed sample size (see also Bayless & Rao, 1970; Asok & Sukhatme, 1976; Gabler, 1981). The implementation is very simple and the joint inclusion probabilities can be calculated relatively easily. Let $\pi_1, \dots, \pi_k, \dots, \pi_N$ be the inclusion probabilities such that

$$\sum_{k \in U} \pi_k = n$$

and

$$\omega_k = \frac{\pi_k}{1 - \pi_k}.$$

The Rao–Sampford design is defined as follows:

$$p_{RS}(s) = C_n \sum_{\ell \in S} \pi_\ell \prod_{\substack{k \in S \\ k \neq \ell}} \omega_k, \qquad (5.12)$$

or equivalently

$$p_{RS}(s) = C_n \left(\prod_{k \in S} \omega_k \right) \left(n - \sum_{\ell \in S}^{N} \pi_\ell \right),$$

where

$$C_n = \left(\sum_{t=1}^{n} tD_{n-t} \right)^{-1}$$

and

$$D_z = \sum_{s \in \mathcal{S}_z} \prod_{k \in s} \omega_k,$$

for $z = 1, \ldots, N$, and $D_0 = 1$, where $\mathcal{S}_z = \{ s \in \mathcal{S} \,|\, \#s = z \}$.

The validity of the Rao–Sampford method is based on the following result:

Result 5.7 *For the Rao–Sampford design defined in Expression (5.12),*

(i) $\sum_{s \in \mathcal{S}_n} p_{RS}(s) = 1.$

(ii) $\sum_{\substack{s \in \mathcal{S}_n \\ s \ni k}} p_{RS}(s) = \pi_k.$

(iii) $\pi_{k\ell} = C_n \omega_k \omega_\ell \sum_{t=2}^{n} (t - \pi_k + \pi_\ell) D_{n-t}(\overline{k}, \overline{\ell}),$

where

$$D_z(\overline{k}, \overline{\ell}) = \sum_{s \in \mathcal{S}_z(U \setminus \{k,\ell\})} \prod_{j \in s} \omega_j.$$

The demonstration is given in Sampford (1967) (see also Tillé, 2006, pp. 133–134). In order to calculate the joint inclusion probabilities, it is not necessary to enumerate all the possible samples. Sampford proved the following relation:

$$D_m = \frac{1}{m} \sum_{r=1}^{m} (-1)^{r-1} \left(\sum_{i=1}^{N} \omega^r \right) D_{m-r},$$

with the initial condition $D_0 = 1$. Then, a recursive relationship can be used to calculate $D_m(\overline{k}, \overline{\ell})$:

$$D_m(\overline{k}, \overline{\ell}) = D_m - (\omega_k + \omega_\ell) D_{m-1}(\overline{k}, \overline{\ell}) - \omega_k \omega_\ell D_{m-2}(\overline{k}, \overline{\ell}).$$

Gabler (1981) showed that the Rao–Sampford procedure has a smaller variance than sampling with unequal probabilities with replacement.

Algorithm 8 Sampford procedure

---

1. We select n units with replacement, the first draw being made with probabilities $q_k = \pi_k/n$ and all the following ones with probabilities proportional to $\pi_k/(1 - \pi_k)$.
2. If a unit is selected more than once, the sample is discarded and the procedure is restarted.

Sampford also proposed a very simple implementation by means of the multinomial rejection procedure presented in Algorithm 8.

However, this procedure can be slow when the sample size is large relative to the size of the population. More efficient implementations of the Rao–Sampford design are presented in Traat et al. (2004) and Tillé (2006).

5.11   Order Sampling

Order sampling methods are a class of sampling designs developed by Rosén (1997a, b). This sampling design is without replacement. The sample size is fixed but the inclusion probabilities are not exactly satisfied. Let $X_1, \dots, X_k, \dots, X_N$ be N continuous independent random variables. Each variable is associated with a unit. These variables are called order variables or rank variables. Each of these variables has a distribution function $F_k(\cdot)$ defined on $[0, \infty)$ and a density function $f_k(\cdot)$, $k = 1, \dots, N$. Ordered sampling is obtained by sorting the X_k in ascending order and selecting the first n units in that order.

The variables $X_1, \dots, X_k, \dots, X_N$ do not necessarily have the same distribution. When the X_k are identically distributed, we obtain a simple random sampling design.

If the X_k variables have different distributions, then the inclusion probabilities are unequal. The quantities λ_k are called the "target inclusion probabilities" and are calculated according to the method given in Section 5.2 from the auxiliary variable $X_k > 0, k \in U$, which is known for all the units of the population. We assume that $0 < \lambda_k < 1$, for all $k \in U$. Rosén (1997a) defines the order design with unequal probabilities with a fixed distribution $H(.)$ and a vector of target probabilities $\lambda = (\lambda_1, \dots, \lambda_N)^\top$ by defining

$$F_k(t) = H[tH^{-1}(\lambda_k)] \quad \text{and} \quad X_k = \frac{H^{-1}(u_k)}{H^{-1}(\lambda_k)},$$

where $H(.)$ is a cumulative distribution function defined on $[0, \infty)$, and where the u_k are independent continuous uniform random variables in $[0, 1]$.

Different distributions for $H(\cdot)$ lead to several types of order sampling. In particular, we have:

1. Order uniform sampling or sequential Poisson sampling algorithm (Ohlsson, 1990, 1998) which uses uniform distributions. In that case,

$$X_k = \frac{u_k}{\lambda_k}, \quad F_k(t) = \min(t\lambda_k, 1), \text{for all } k \in U.$$

2. Order exponential sampling or successive sampling (Hájek, 1964), which uses exponential distributions. In that case,

$$X_k = \frac{\log(1 - u_k)}{\log(1 - \lambda_k)}, \quad F_k(t) = 1 - (1 - \lambda_k)^t, \quad \text{for all } k \in U.$$

3. Pareto order sampling (see Rosén, 1997a,b; Saavedra, 1995), which uses Pareto distributions. In that case,

$$X_k = \frac{u_k/(1 - u_k)}{\lambda_k/(1 - \lambda_k)}, \quad F_k(t) = \frac{t\lambda_k}{1 - \lambda_k + t\lambda_k}, \quad \text{for all } k \in U.$$

Pareto sampling is the one that has been studied the most. It has the advantage of minimizing the variance estimator in the class of order designs with fixed shape parameters (see Rosén, 1997b). The Pareto order design has also been studied by Aires (1999, 2000), Holmberg & Swensson (2001), Aires & Rosén (2005), Traat et al. (2004), and Bondesson et al. (2006).

More generally, the $H(.)$ distribution can be written as a generalized Pareto distribution:

$$GPD(t, a, b) = \begin{cases} 1 - (1 - \frac{bt}{a})^{1/b}, & b \neq 0, \\ 1 - \exp\left(-\frac{t}{a}\right), & b = 0, \end{cases}$$

where with $b = 1, a = 1$, we obtain the uniform order sampling, with $b = 0, a = 1$, the exponential order sampling, and, with $b = -1, a = 1$, the Pareto order sampling.

For an order sample S, the values λ_k are not equal to the actually obtained probabilities π_k. However, under general conditions, Rosén (2000) showed that

$$\frac{\pi_k}{\lambda_k} \to 1, \quad \text{when } n, N \to \infty.$$

Using λ_k instead of π_k produces a small bias in the estimators. Therefore, it is important to calculate the true inclusion probabilities π_k when n is relatively small. Unfortunately, there are no analytical solutions for passing from λ_k to π_k. However, numeric integration techniques can be used to calculate the inclusion probabilities π_k.

In the case of the Pareto order design, Aires (1999) proposed an algorithm for calculating the first- and second-order inclusion probabilities. A second method has been proposed by Ng & Donadio (2006). Matei & Tillé (2007) proposed a simple way to calculate π_k based on the formula of Cao & West (1997). The simulations show that the true inclusion probabilities are very close to the target probabilities.

Order sampling does not exactly satisfy the target inclusion probabilities. However, these methods are interesting for positively coordinating two samples selected at different waves in the same population (see Section 8.2, page 172). To do this, we generate a permanent random number u_k for each unit that does not change from one wave to another. The method is then applied with probabilities λ_k that can change from one wave to another. Hence, the samples drawn have a very important overlap, since the units having received small u_k are more likely to be selected at all the waves.

5.12 Splitting Method

5.12.1 General Principles

The splitting method, proposed by Deville & Tillé (1998), is not, strictly speaking, a special method of sampling with unequal probabilities but a class of methods, or rather a

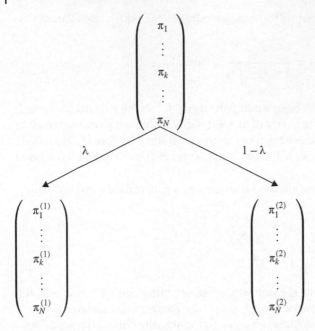

Figure 5.3 Splitting into two parts.

means to represent in a standardized way a set of methods of sampling with unequal probabilities. Indeed, all the methods (except the rejective and order methods) can benefit from being put into the splitting form. The splitting method also makes it possible to define several new sampling procedures. It consists of presenting a technique of sampling in the shape of a tree. As can be seen in Figure 5.3, at each step we choose at random a branch of this tree to satisfy the given fixed inclusion probabilities.

The basic idea is to split each π_k into two new inclusion probabilities $\pi_k^{(1)}$ and $\pi_k^{(2)}$ which satisfy the relations:

$$\pi_k = \lambda \pi_k^{(1)} + (1 - \lambda)\pi_k^{(2)},$$

$$0 \le \pi_k^{(1)} \le 1 \quad \text{and} \quad 0 \le \pi_k^{(2)} \le 1,$$

$$\sum_{k \in U} \pi_k^{(1)} = \sum_{k \in U} \pi_k^{(2)} = n,$$

where λ can be freely chosen provided that $0 < \lambda < 1$. Then one of these two vectors is chosen randomly. The method consists of selecting

$$\begin{cases} \pi_k^{(1)}, k \in U, & \text{with probability } \lambda \\ \pi_k^{(2)}, k \in U, & \text{with probability } 1 - \lambda. \end{cases}$$

Therefore, after one stage, the problem is reduced to another unequal probability sampling problem. If the splitting method is such that one or more of the $\pi_k^{(1)}$ and $\pi_k^{(2)}$ are equal to 0 or 1, the sampling problem is simpler in the next step because the splitting must then be applied to a smaller population. If this elementary splitting step is repeated on the $\pi_k^{(1)}$ and the $\pi_k^{(2)}$, then the problem can be reduced to obtaining the sample. The splitting can also be organized so that either the vector of $\pi_k^{(1)}$ or the vector of $\pi_k^{(2)}$ has equal probabilities. In this case, a simple random sampling design can be applied directly (see Section 3, page 31).

Figure 5.4 Splitting in *M* parts.

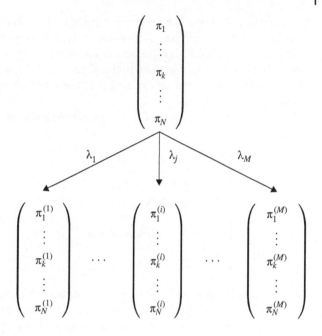

This approach can be generalized to the splitting technique in M vectors of inclusion probabilities, as can be seen in Figure 5.4. We first construct $\pi_k^{(j)}$ and λ_j such that

$$\sum_{j=1}^{M} \lambda_j = 1,$$

$$0 \leq \lambda_j \leq 1, \text{ for all } j = 1, \ldots, M,$$

$$\sum_{j=1}^{M} \lambda_j \pi_k^{(j)} = \pi_k, \text{ for all } k \in U,$$

$$0 \leq \pi_k^{(j)} \leq 1, \text{ for all } k \in U, j = 1, \ldots, M,$$

$$\sum_{k \in U} \pi_k^{(j)} = n, \text{ for all } j = 1, \ldots, M.$$

We select one of the vectors of $\pi_k^{(j)}$ with the probabilities $\lambda_j, j = 1, \ldots, M$. Again, the $\pi_k^{(j)}$ are constructed to obtain a simpler configuration in the next step.

5.12.2 Minimum Support Design

A minimal support design is defined as follows:

Definition 5.1 *Let $p(\cdot)$ be a sampling design on U with inclusion probabilities π_k and $B = \{s | p(s) > 0\}$. A sampling design $p(\cdot)$ is defined on a minimal support if and only if there does not exist a set $B_0 \subset B$ such that $B_0 \neq B$, and*

$$\sum_{s \in B_0} p_0(s) = \pi_k, k \in U, \tag{5.13}$$

has a solution in $p_0(s)$.

Wynn (1977, Thorem 1) showed that it is always possible to define a sampling design that satisfies any vector of given first-order inclusion probabilities with at most N samples with nonzero probabilities. Therefore, a minimal support design uses at most N samples s such that $p(s) > 0$. Unfortunately, the Wynn result does not allow this design to be built. However, the minimal support design is the most obvious application of the splitting method into two parts.

Let $\pi_{(1)}, \dots, \pi_{(k)}, \dots, \pi_{(N)}$ be the inclusion probabilities sorted in increasing order. Next, we define

$$\lambda = \min[1 - \pi_{(N-n)}, \pi_{(N-n+1)}],$$

$$\pi_{(k)}^{(1)} = \begin{cases} 0 & \text{if } k \leq N - n \\ 1 & \text{if } k > N - n, \end{cases}$$

$$\pi_{(k)}^{(2)} = \begin{cases} \dfrac{\pi_{(k)}}{1 - \lambda} & \text{if } k \leq N - n \\ \dfrac{\pi_{(k)} - \lambda}{1 - \lambda} & \text{if } k > N - n. \end{cases}$$

In other words, with probability λ, the sample with the largest n inclusion probabilities is selected and, with probability $1 - \lambda$, another sample of size n must be selected using the inclusion probabilities $\pi_k^{(2)}$. However, one of the $\pi_k^{(2)}$ is equal to 0 or 1. Indeed, if $1 - \pi_{(Nn)} < \pi_{(N-n+1)}$ then $\lambda = 1 - \pi_{(Nn)}$ and $\pi_{(Nn)}^{(2)} = 1$. On the other hand, if $\pi_{(N-n+1)} < 1 - \pi_{(Nn)}$ then $\lambda = \pi_{(N-n+1)}$ and $\pi_{(N-n+1)}^{(2)} = 0$. If the $\pi_{(k)}^{(1)}$ are drawn, the sample is selected immediately. On the other hand, if $\pi_{(k)}^{(2)}$ are drawn, the problem is reduced to the selection of a sample from a population size $N - 1$. The same procedure is then applied to $\pi_{(k)}^{(2)}$ and, in N steps at most, the sample is selected.

Example 5.3 Let $N = 6, n = 3, \pi_1 = 0.07, \pi_2 = 0.17, \pi_3 = 0.41, \pi_4 = 0.61, \pi_5 = 0.83, \pi_6 = 0.91$. As can be seen in Figure 5.5, the tree has only five leaves. In the first step, the vector of the inclusion probabilities is split into two parts given in columns two and three of Table 5.1. With probability $\lambda = 0.59$, the sample $\{4, 5, 6\}$ is selected and with probability $1 - \lambda = 0.41$, another splitting is applied with the unequal probabilities given by $(0.171, 0.415, 1, 0.049, 0.585, 0.780)^\top$. In four steps, the sample is selected. Therefore, the sampling design is $p(\{4, 5, 6\}) = 0.59$, $p(\{3, 5, 6\}) = (1 - 0.59) \times 0.585 = 0.24$, $p(\{2, 3, 6\}) = (1 - 0.59 - 0.24) \times 0.471 = 0.08$, $p(\{1, 2, 3\}) = (1 - 0.59 - 0.24 - 0.08) \times 0.778 = 0.07$, $p(\{2, 3, 4\}) = 1 - 0.59 - 0.24 - 0.08 - 0.7 = 0.02$.

5.12.3 Decomposition into Simple Random Sampling Designs

This method is based on a splitting of the inclusion probabilities into two parts. The value of λ is

$$\lambda = \min \left\{ \pi_{(1)} \frac{N}{n}, \frac{N}{N - n} [1 - \pi_{(N)}] \right\}.$$

Next we define for all $k \in U$,

$$\pi_{(k)}^{(1)} = \frac{n}{N}, \pi_{(k)}^{(2)} = \frac{\pi_k - \lambda \frac{n}{N}}{1 - \lambda}.$$

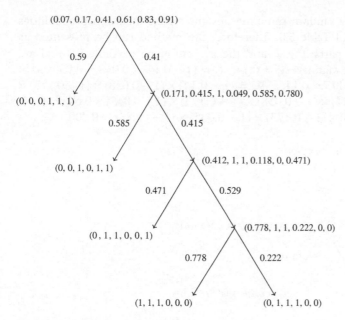

Figure 5.5 Minimum support design.

Table 5.1 Minimum support design.

π_k	Step 1 $\lambda = 0.59$		Step 2 $\lambda = 0.585$		Step 3 $\lambda = 0.471$		Step 4 $\lambda = 0.778$	
0.07	0	0.171	0	0.412	0	0.778	1	0
0.17	0	0.415	0	1	1	1	1	1
0.41	0	1	1	1	1	1	1	1
0.61	1	0.049	0	0.118	0	0.222	0	1
0.83	1	0.585	1	0	0	0	0	0
0.91	1	0.780	1	0.471	1	0	0	0

At each step, a random choice is made between a vector of equal inclusion probabilities and a vector of unequal inclusion probabilities. If the vector of equal inclusion probabilities is selected, then simple random sampling without replacement is applied. If $\lambda = \pi_{(1)}N/n$, then $\pi_{(1)}^{(2)} = 0$. If $\lambda = (1 - \pi_{(N)})N/(Nn)$, then $\pi_{(N)}^{(2)} = 1$. The probability of vector $\pi_{(k)}^{(2)}$ is set to 0 or 1. In the next step, the problem is reduced to selecting a sample of size n or $n - 1$ in a population size $N - 1$. In $N - 1$ steps at most, the problem is solved.

Example 5.4 With the same π_k as in Section 5.12.2, the tree is described in Figure 5.6. The result of the method is given in Table 5.2. The problem is finally to select one of the six simple random sampling designs whose inclusion probabilities are given in the columns of Table 5.3. Therefore, the method can be presented as a splitting method into N parts: the λ_j and the $\pi_k^{(j)}$ are given in Table 5.3 and are calculated in the following manner: $\lambda_1 = 0.14$; $\lambda_2 = (1 - 0.14) \times 0.058 = 0.050$; $\lambda_3 = (1 - 0.14) \times (1 - 0.058) \times 0.173 = 0.14$; $\lambda_4 = (1 - 0.14) \times (1 - 0.058) \times (1 - 0.173) \times 0.045 = 0.03$; $\lambda_5 = (1 - 0.14) \times (1 - 0.058) \times (1 - 0.173) \times (1 - 0.045) \times 0.688 = 0.44$; $\lambda_6 = (1 - 0.14) \times (1 - 0.058) \times (1 - 0.173) \times (1 - 0.045) \times (1 - 0.688) = 0.200$.

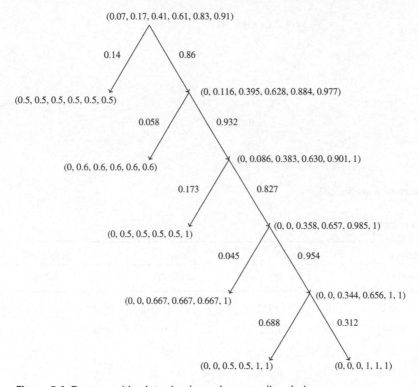

Figure 5.6 Decomposition into simple random sampling designs.

Table 5.2 Decomposition into simple random sampling designs.

π_k	Step 1 $\lambda = 0.14$		Step 2 $\lambda = 0.058$		Step 3 $\lambda = 0.173$		Step 4 $\lambda = 0.045$		Step 5 $\lambda = 0.688$	
0.07	0.5	0	0	0	0	0	0	0	0	0
0.17	0.5	0.116	0.6	0.086	0.5	0	0	0	0	0
0.41	0.5	0.395	0.6	0.383	0.5	0.358	0.667	0.344	0.5	0
0.61	0.5	0.628	0.6	0.630	0.5	0.657	0.667	0.656	0.5	1
0.83	0.5	0.884	0.6	0.901	0.5	0.985	0.667	1	1	1
0.91	0.5	0.977	0.6	1	1	1	1	1	1	1

Table 5.3 Decomposition into N simple random sampling designs.

k	$\lambda_1 = 0.14$	$\lambda_2 = 0.050$	$\lambda_3 = 0.14$	$\lambda_4 = 0.03$	$\lambda_5 = 0.44$	$\lambda_6 = 0.200$
1	0.5	0	0	0	0	0
2	0.5	0.6	0.5	0	0	0
3	0.5	0.6	0.5	0.667	0.5	0
4	0.5	0.6	0.5	0.667	0.5	1
5	0.5	0.6	0.5	0.667	1	1
6	0.5	0.6	1	1	1	1

5.12.4 Pivotal Method

The pivotal method, proposed by Deville & Tillé (1998), is also a special case of the splitting method. In a later publication, Srinivasan (2001) proposed the same procedure for computer applications. The pivotal method is based on splitting the vector of inclusion probabilities into two parts. At each step, only two inclusion probabilities are modified. Two units denoted by i and j are selected. Their inclusion probabilities are randomly modified. On the one hand, if $\pi_i + \pi_j > 1$, then

$$\lambda = \frac{1 - \pi_j}{2 - \pi_i - \pi_j},$$

$$\pi_k^{(1)} = \begin{cases} \pi_k & k \in U \backslash \{i, j\} \\ 1 & k = i \\ \pi_i + \pi_j - 1 & k = j, \end{cases}$$

$$\pi_k^{(2)} = \begin{cases} \pi_k & k \in U \backslash \{i, j\} \\ \pi_i + \pi_j - 1 & k = i \\ 1 & k = j. \end{cases}$$

On the other hand, if $\pi_i + \pi_j < 1$, then

$$\lambda = \frac{\pi_i}{\pi_i + \pi_j},$$

$$\pi_k^{(1)} = \begin{cases} \pi_k & k \in U \backslash \{i, j\} \\ \pi_i + \pi_j & k = i \\ 0 & k = j, \end{cases}$$

and

$$\pi_k^{(2)} = \begin{cases} \pi_k & k \in U \backslash \{i, j\} \\ 0 & k = i \\ \pi_i + \pi_j & k = j. \end{cases}$$

In the first case, an inclusion probability is set to 1, in the second, an inclusion probability is set to 0. The problem is reduced to the selection of a sample in a population size $N - 1$. By repeating this operation, in N steps at most, we obtain a sample.

Figure 5.7 contains an example of application of the pivotal method with the complete development of the tree until the selection of the sample. In practice, it is obviously not necessary to calculate the whole tree, it is enough to calculate the modification of the vector at each stage. A sequential implementation is possible. The algorithm is extremely fast.

There are several special cases of the pivotal method that differ only in how to select the two units at each step.

- In the *order pivotal method* (or sequential pivotal method), the units are taken in their original order. Unit 1 is first compared to unit 2. The one that does not have its probability rounded to 0 or 1 is then confronted with unit 3. Again, the unit that remains with a noninteger value is facing unit 4 and so on. Figure 5.7 is an application of the order pivotal method. Chauvet (2012) showed that this method gives exactly the same sampling design as the systematic sampling of Deville (1998c) (see Section 5.7, page 91). The method of Fuller (1970), which was proposed much earlier, is equivalent to the order pivotal method but with a starting point chosen randomly after ordering the units on a circle (see also Tillé, 2018). Figure 5.7 also shows that some inclusion

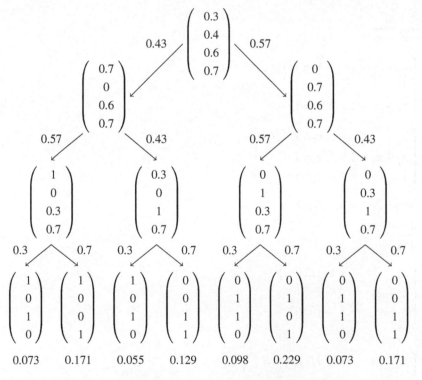

Figure 5.7 Pivotal method applied on vector $\pi = (0.3, 0.4, 0.6, 0.7)^{\mathsf{T}}$.

probabilities can be zero; this is the case for units 1 and 2 that are never selected together.

- The *random pivotal method* consists of randomly sorting the population. Then the sequential pivotal method is applied to this random order.
- The *local pivotal method* has been proposed by Grafström et al. (2012) for spatial sampling. At each stage, two units are selected that are geographically close. This method produces well-spread out samples in space and is described in Section 8.1.4, page 169.

5.12.5 Brewer Method

Brewer's method was first proposed for the special case $n = 2$ (see Brewer & Hanif, 1983, procedure 8, p. 26) but it was then generalized (Brewer, 1975). This method consists of selecting the units one after the other. In n steps, the sample is selected. Unlike Brewer's presentation, the method is presented here in the form of a splitting technique in N parts. We first define:

$$\lambda_j = \left[\sum_{z=1}^{N} \frac{\pi_z(n - \pi_z)}{1 - \pi_z} \right]^{-1} \frac{\pi_j(n - \pi_j)}{1 - \pi_j}.$$

Next, we calculate

$$\pi_k^{(j)} = \begin{cases} \dfrac{\pi_k(n - 1)}{n - \pi_j} & \text{if } k \neq j \\ 1 & \text{if } k = j. \end{cases}$$

The validity of Brewer's method comes from the following result:

Result 5.8 *For Brewer's method,*

$$\sum_{j=1}^{N} \lambda_j \pi_k^{(j)} = \pi_k,$$

for all $k = 1, \ldots, N$,

Proof: If

$$C = \left[\sum_{z=1}^{N} \frac{\pi_z(n - \pi_z)}{1 - \pi_z} \right]^{-1},$$

we have

$$\sum_{j=1}^{N} \lambda_j \pi_k^{(j)} = \sum_{\substack{j=1 \\ j \neq k}}^{N} C \frac{\pi_k \pi_j(n - 1)}{1 - \pi_j} + C \frac{\pi_k(n - \pi_k)}{1 - \pi_k}$$

$$= C\pi_k \left[\sum_{j=1}^{N} \frac{\pi_j(n - 1)}{1 - \pi_j} + n \right] = C\pi_k \sum_{j=1}^{N} \left[\frac{\pi_j(n - 1)}{1 - \pi_j} + \frac{\pi_j(1 - \pi_j)}{1 - \pi_j} \right] = \pi_k. \quad \blacksquare$$

At each iteration of the method, a unit is selected in the sample. Moreover, at each step, the $\pi_k^{(j)}$ must not all be calculated. Only the probabilities of the selected vector must be calculated. Practically, the procedure can be directly implemented in the following way: for the $i = 1, \ldots, n$ steps, a unit is selected in the sample with probability

$$
p_k^{S_{i-1}} = \left\{ \sum_{z \in U \backslash S_{i-1}} \frac{\pi_z \left(n - \sum_{\ell \in S_{i-1}} \pi_\ell - \pi_z \right)}{n - \sum_{\ell \in S_{i-1}} \pi_\ell - \pi_z [n - (i-1)]} \right\}^{-1}
$$

$$
\times \frac{\pi_k \left(n - \sum_{\ell \in S_{i-1}} \pi_\ell - \pi_k \right)}{n - \sum_{\ell \in S_{i-1}} \pi_\ell - \pi_k [n - (i-1)]},
$$

for $k \in U \backslash S_{i-1}$, where S_i represents all the units selected after step i and $S_0 = \emptyset$. We obviously have $\#S_i = i, i = 0, \ldots, n$.

The main problem of the method is that the second-order inclusion probabilities are very difficult to calculate. Brewer (1975) proposed a recursive formula, but it is inapplicable in practice because it involves a complete exploration of the splitting tree. Nevertheless, the second-order inclusion probabilities are strictly positive. This property results from the fact that at each split, all split inclusion probability vectors are composed of strictly positive probabilities. Despite the difficulty with the joint inclusion probabilities, Brewer's method is one of the few methods whose $p(s)$ do not depend on the order of the file. This method produces strictly positive joint inclusion probabilities and is simple to implement. This method is the correct way to successfully draw the units and corrects the problem of the method proposed in Section 5.5.

5.13 Choice of Method

There are many other methods of sampling with unequal probabilities, such as the method of Chao (1982) and the eliminatory method (Tillé, 1996a). A synthesis of the methods is presented in Brewer & Hanif (1983) and Tillé (2006). Unfortunately, there is no ideal method, as evidenced by the summary given in Table 5.4.

There is no one method that would be more precise than the others. Indeed, under a design with unequal probabilities the variance of the expansion estimator can be written as

$$
\mathrm{var}_p(\widehat{Y}) = \sum_{k \in U} \sum_{\ell \in U} \frac{y_k}{\pi_k} \frac{y_\ell}{\pi_\ell} \Delta_{k\ell} = \check{\mathbf{y}}^\mathsf{T} \Delta \check{\mathbf{y}},
$$

where $\check{\mathbf{y}}^\mathsf{T} = (y_1/\pi_1, \ldots, y_k/\pi_k, \ldots, y_N/\pi_N)^\mathsf{T}$ and Δ is the matrix of $\Delta_{k\ell} = \pi_{k\ell} - \pi_k \pi_\ell$. Therefore, the variance can be written as a quadratic form. Matrix Δ is symmetrical and has a null eigenvalue associated with the eigenvector $(1, \ldots, 1, \ldots, 1)^\mathsf{T}$. Indeed, as the design is of fixed size,

$$
\sum_{k \in U} \Delta_{k\ell} = 0, \text{ for all } \ell \in U.
$$

The sum of the eigenvalues $\lambda_1, \ldots, \lambda_k, \ldots, \lambda_N$ is equal to the trace of matrix Δ and is

$$
\sum_{k=1}^N \lambda_i = \sum_{k \in U} \Delta_{kk} = \sum_{k \in U} (\pi_k - \pi_k^2) = n - \sum_{k \in U} \pi_k^2.
$$

Table 5.4 Properties of the methods.

Method	Without replacement	π_k exact	Fixed size	Sequential	$\pi_{k\ell} > 0$
Multinomial	No	No	Yes	Yes	Yes
Systematic	Yes	Yes	Yes	Yes	No
Poisson	Yes	Yes	No	Yes	Yes
Rao–Sampford	Yes	Yes	Yes	No	Yes
Maximum entropy	Yes	Yes	Yes	No	Yes
Minimum support	Yes	Yes	Yes	No	No
Decomposition into simple	Yes	Yes	Yes	No	Yes
Pivotal method	Yes	Yes	Yes	No	No
Brewer's method	Yes	Yes	Yes	No	Yes

This sum depends only on the first-order inclusion probabilities.

Consider two designs $p_1(s)$ and $p_2(s)$ with the same inclusion probabilities π_k and whose variance operators are, respectively, Δ_1 and Δ_2. A design $p_1(s)$ is said to be better than a design $p_2(s)$ if $\mathbf{u}^\top \Delta_1 \mathbf{u} \le \mathbf{u}^\top \Delta_2 \mathbf{u}$ for all $\mathbf{u} \in \mathbb{R}^N$ and there is at least one vector $\mathbf{x}$ such that $\mathbf{x}^\top \Delta_1 \mathbf{x} < \mathbf{x}^\top \Delta_2 \mathbf{x}$. We refer to the following result:

Result 5.9 *For any design p(s), there is no better design than p(s) with the same first-order inclusion probability.*

Proof: Suppose there is a design $p_1(s)$ better than a design $p_2(s)$ with the same first-order inclusion probabilities. In this case, $\mathbf{u}^\top (\Delta_2 - \Delta_1)\mathbf{u} \ge 0$, for all $\mathbf{u} \in \mathbb{R}^N$, which implies that the array $(\Delta_2 - \Delta_1)$ is semi-definite positive. Since the trace of $(\Delta_2 - \Delta_1)$ is zero, all eigenvalues of $(\Delta_2 - \Delta_1)$ are zero, which contradicts the fact that there is at least one vector $\mathbf{x}$ such as $\mathbf{x}^\top \Delta_1 \mathbf{x} < \mathbf{x}^\top \Delta_2 \mathbf{x}$. ∎

This result shows that there is no design which is better than all the others. Therefore, we cannot use a precision criterion linked to the variance to find the optimal design. If additional auxiliary information is not available, it is reasonable to take the most random design possible, that is to say, the design that maximizes the entropy. However, several authors have shown that the Rao–Sampford designs and random systematic design are very close to the maximum entropy design (see Berger, 1998a, c, 2005; Tillé, 2006). Therefore, the variance of the expansion estimators of these designs is very similar. However, low entropy designs, such as order systematic sampling or minimal support design, behave very differently. They should only be used under very special conditions. For example, if the file is sorted according to an auxiliary variable related to the variable of interest, order systematic sampling can provide a stratification effect which will improve the accuracy of the expansion estimator.

5.14 Variance Approximation

Each of the methods presented above has special second-order inclusion probabilities. Their variances can theoretically be estimated by means of Expression (5.1) and can be unbiasedly estimated either by the Horvitz–Thompson estimator:

$$v_{HT}(\widehat{Y}) = \sum_{k \in S} \sum_{\substack{\ell \in S \\ \ell \neq k}} \frac{y_k}{\pi_k} \frac{y_\ell}{\pi_\ell} \frac{\Delta_{k\ell}}{\pi_{k\ell}},$$

or by the Sen–Yates–Grundy estimator:

$$v_{YG}(\widehat{Y}) = -\frac{1}{2} \sum_{k \in S} \sum_{\substack{\ell \in S \\ \ell \neq k}} \left(\frac{y_k}{\pi_k} - \frac{y_\ell}{\pi_\ell} \right)^2 \frac{\Delta_{k\ell}}{\pi_{k\ell}}.$$

However, these estimators are not recommended because the second-order inclusion probabilities are often difficult to calculate and it is necessary to sum n^2 terms to calculate the estimator. Moreover, all the simulations show that these two estimators are particularly unstable (see Brewer, 2000; Matei & Tillé, 2005a; Henderson, 2006). These authors have indeed shown that it is possible to validly approach the variance by a relatively simple estimator which depends only on the first-order inclusion probabilities. This approximate variance is suitable for all sampling designs that do not deviate too much from the maximum entropy design (see Berger, 1998a).

The construction of an approximation is based on the fact that the design has maximum entropy. Therefore, this design can be written as a Poisson design conditioned by a fixed sample size. Referring to Section 5.9 for the maximum entropy design of fixed size and using the same notation, it has been shown that the maximum entropy design $p(s)$ of inclusion probabilities π_k can be interpreted as the conditional of a Poisson design $\tilde{p}(s)$ with respect to $\tilde{n}_S = n$. If $\widehat{Y}$ denotes the expansion estimator, $\text{var}_{\text{POISS}}(\cdot)$ the variance under the Poisson design $\tilde{p}(s)$, and $\text{var}_p(\cdot)$ the variance under the design $p(\cdot)$, we can write that

$$\text{var}_p(\widehat{Y}) = \text{var}_{\text{POISS}}(\widehat{Y}|\tilde{n}_S = n).$$

If we assume that vector $(\widehat{Y}, \tilde{n}_S)^{\mathsf{T}}$ has a bivariate normal distribution (which can be proved under certain conditions), we have

$$\text{var}_{\text{POISS}}(\widehat{Y}|\tilde{n}_S = n) = \text{var}_{\text{POISS}}[\widehat{Y} + (n - \tilde{n}_S)\beta],$$

where

$$\beta = \frac{\text{cov}_{\text{POISS}}(\tilde{n}_S, \widehat{Y})}{\text{var}_{\text{POISS}}(\tilde{n}_S)},$$

$$\text{var}_{\text{POISS}}(\tilde{n}_S) = \sum_{k \in U} \tilde{\pi}_k(1 - \tilde{\pi}_k) \quad \text{and} \quad \text{cov}_{\text{POISS}}(\tilde{n}_S, \widehat{Y}) = \sum_{k \in U} \tilde{\pi}_k(1 - \tilde{\pi}_k)\frac{y_k}{\pi_k}.$$

The inclusion probabilities $\tilde{\pi}_k$ are different from π_k. The $\tilde{\pi}_k$ are the probabilities of the Poisson design whose fixed-size design with probabilities π_k is assumed to be the conditional. By setting $b_k = \tilde{\pi}_k(1 - \tilde{\pi}_k)$, we obtain a general approximation of the variance for a design with fixed sample size, without replacement and with large entropy:

$$\text{var}_{approx}(\widehat{Y}) = \sum_{k \in U} \frac{b_k}{\pi_k^2}(y_k - y_k^*)^2, \tag{5.14}$$

where
$$y_k^* = \pi_k \beta = \pi_k \frac{\sum_{\ell \in U} b_\ell y_\ell / \pi_\ell}{\sum_{\ell \in U} b_\ell}.$$

This variance is very similar to that of the Poisson design given in Expression (5.7), except that the y_k values are centered on the "predicted values" y_k^*. We could determine b_k by calculating $\tilde{\pi}_k$ from π_k according to the method described in Section 5.9. Several simpler approximations for b_k have also been proposed:

1. The approximation of Hájek is obtained by taking
$$b_k = \frac{\pi_k(1 - \pi_k)N}{N - 1}.$$

   This approximation is accurate for simple random sampling with replacement with fixed sample size.

2. A tighter approximation can be found at the cost of a slightly more sophisticated calculation. The general approximation given in Expression (5.14) can also be written as
$$\text{var}_{approx}(\hat{Y}) = \sum_{k \in U} \frac{y_k^2}{\pi_k^2}\left(b_k - \frac{b_k^2}{\sum_{\ell \in U} b_\ell}\right) - \frac{1}{\sum_{\ell \in U} b_\ell}\sum_{k \in U}\sum_{\substack{\ell \in U \\ \ell \neq k}} \frac{y_k y_\ell b_k b_\ell}{\pi_k \pi_\ell}.$$

   However, the exact variance is written as
$$\text{var}_p(\hat{Y}) = \sum_{k \in U} \frac{y_k^2}{\pi_k^2}\pi_k(1 - \pi_k) + \sum_{k \in U}\sum_{\substack{\ell \in U \\ \ell \neq k}} \frac{y_k y_\ell}{\pi_k \pi_\ell}(\pi_{k\ell} - \pi_k \pi_\ell).$$

   Therefore, we can look for b_k which are such that the coefficients of y_k^2 are exact, which amounts to looking for b_k that satisfy equality
$$b_k - \frac{b_k^2}{\sum_{\ell \in U} b_\ell} = \pi_k(1 - \pi_k).$$

   Such coefficients can be obtained by a fixed point technique by applying the recurrence equation up to convergence:
$$b_k^{(i)} = \frac{[b_k^{(i-1)}]^2}{\sum_{\ell \in U} b_\ell^{(i-1)}} + \pi_k(1 - \pi_k),$$

   for $i = 0, 1, 2, 3, \dots$, and initializing by
$$b_k^{(0)} = \pi_k(1 - \pi_k)\frac{N}{N - 1}, k \in U.$$

   Unfortunately, the convergence of this algorithm is only assured if
$$\frac{\pi_k(1 - \pi_k)}{\sum_{\ell \in U} \pi_\ell(1 - \pi_\ell)} \leq \frac{1}{2}, \text{for all } k \in U$$

   (see Deville & Tillé, 2005).

3. A variant of the previous technique (useful if the algorithm does not converge) consists of using only one iteration. We then obtain
$$b_k^{(1)} = \pi_k(1 - \pi_k)\left[\frac{N\pi_k(1 - \pi_k)}{(N - 1)\sum_{\ell \in U} \pi_\ell(1 - \pi_\ell)} + 1\right].$$

5.15 Variance Estimation

The general variance estimator from the approximation is built on the same principle and is

$$v(\widehat{Y}) = \sum_{k \in S} \frac{c_k}{\pi_k^2} (y_k - \hat{y}_k^*)^2, \tag{5.15}$$

where

$$\hat{y}_k^* = \pi_k \frac{\sum_{\ell \in S} c_\ell y_\ell / \pi_\ell}{\sum_{\ell \in S} c_\ell}.$$

Again, there are several possibilities for calculating the c_k.

1. The simplest approximation is to take

$$c_k = (1 - \pi_k) \frac{n}{n - 1}.$$

2. Deville (1999) (see also Ardilly, 1994, p. 338) proposes using

$$c_k = (1 - \pi_k) \left\{ 1 - \sum_{k \in S} \left[\frac{1 - \pi_k}{\sum_{\ell \in S} (1 - \pi_\ell)} \right]^2 \right\}^{-1}.$$

3. Berger (1998a) proposes taking

$$c_k = (1 - \pi_k) \frac{n}{n - 1} \frac{\sum_{k \in S} (1 - \pi_k)}{\sum_{k \in U} \pi_k (1 - \pi_k)}.$$

   This expression seems to have good asymptotic properties but its calculation requires knowledge of the inclusion probabilities on all the units of the population.

4. As for the variance, it is possible to construct a more interesting variance estimator at the cost of a somewhat more tedious calculation. The estimator given in Expression (5.15) can also be written as

$$v(\widehat{Y}) = \sum_{k \in S} \frac{y_k^2}{\pi_k^2} \left(c_k - \frac{c_k^2}{\sum_{\ell \in U} c_\ell} \right) - \frac{1}{\sum_{\ell \in S} c_\ell} \sum_{k \in S} \sum_{\substack{\ell \in S \\ \ell \neq k}} \frac{y_k y_\ell c_k c_\ell}{\pi_k \pi_\ell}.$$

However, the unbiased Horvitz–Thompson estimator of variance can be written as

$$v(\widehat{Y}) = \sum_{k \in S} \frac{y_k^2}{\pi_k^2} (1 - \pi_k) + \sum_{k \in S} \sum_{\substack{\ell \in S \\ \ell \neq k}} \frac{y_k y_\ell}{\pi_k \pi_\ell \pi_{k\ell}} (\pi_{k\ell} - \pi_k \pi_\ell).$$

Therefore, we can look for c_k which are such that the coefficients of y_k^2 are exact, which amounts to looking for c_k which satisfy equality

$$c_k - \frac{c_k^2}{\sum_{\ell \in S} c_\ell} = (1 - \pi_k).$$

Again, these coefficients can be obtained by a fixed point technique by applying the recurrence equation up to convergence:

$$c_k^{(i)} = \frac{[c_k^{(i-1)}]^2}{\sum_{\ell \in S} c_\ell^{(i-1)}} + (1 - \pi_k),$$

for $i = 0, 1, 2, 3 \dots$, and initializing by

$$c_k^{(0)} = (1 - \pi_k)\frac{n}{n-1}, k \in S.$$

Unfortunately, the convergence of this algorithm is only assured if

$$\frac{1 - \pi_k}{\sum_{\ell \in S}(1 - \pi_\ell)} \leq \frac{1}{2}, \text{ for all } k \in S$$

(see Deville & Tillé, 2005).

5. If the method does not converge, we can always use the variant of the previous technique which consists of using only one iteration. We then obtain

$$c_k^{(1)} = (1 - \pi_k)\left[\frac{n(1 - \pi_k)}{(n-1)\sum_{\ell \in S}(1 - \pi_\ell)} + 1\right].$$

Tillé (1996b), Brewer (2000), Matei & Tillé (2005a), and Henderson (2006) have also proposed other approximations. Using numerous simulations, these authors have shown that estimators based on approximations give much more accurate estimations than the Horvitz–Thompson and Sen–Yates–Grundy estimators.

Exercises

5.1 Consider a population of five units. We want to select a sample of two units with first-order inclusion probabilities proportional to the following values:

$$1, 1, 6, 6, 6.$$

Therefore, the first-order inclusion probabilities are

$$\pi_1 = \pi_2 = 0.1, \pi_3 = \pi_4 = \pi_5 = 0.6.$$

For both units whose values are 1, calculate the second-order inclusion probabilities for each of the possible orders.

5.2 In a finite population $U = \{1, \dots, k, \dots N\}$, we select three units with replacement with unequal drawing probabilities $p_k, k \in U$, where

$$\sum_{k \in U} p_k = 1.$$

Let a_k be the random variable that indicates the number of times unit k is selected in the sample with replacement.
1. Give the probability distribution of vector $(a_1, \dots, a_k, \dots, a_N)^\top$.
2. Derive $E_p(a_k)$, $\text{var}_p(a_k)$ and $\text{cov}_p(a_k, a_\ell), k \neq \ell$.
3. If S denotes the random sample of the distinct units, calculate $\Pr(k \notin S)$, $\Pr(S = \{k\})$, $\Pr(S = \{k, \ell\}), k \neq \ell, \Pr(S = \{k, \ell, m\}), k \neq \ell \neq m$.
4. What is the inclusion probability $\pi_k = \Pr(k \in S)$?
5. Determine $E_p[a_k \mid S = \{k\}]$, $E_p[a_k \mid S = \{k, \ell\}], k \neq \ell$, and

$$E_p[a_k \mid S = \{k, \ell, m\}], k \neq \ell \neq m.$$

6. Define the Hansen–Hurwitz estimator of the total by means of a sum on S and of the a_k.
7. Define the estimator obtained by applying the Rao–Blackwell theorem on the Hansen–Hurwitz estimator by considering separately the cases $S = \{k\}$, $S = \{k, \ell\}$, and $S = \{k, \ell, m\}$.

5.3 In a small municipality, there are six companies with sales of 40, 10, 8, 1, 0.5, and 0.5 million dollars, respectively. Select three companies at random and without replacement with unequal probabilities according to the turnover by systematic sampling (justify your approach). To do this, we use the following realization of a uniform random variable [0, 1]: 0.83021.

5.4 Let U be a population consisting of *six* units. We know the values taken by an auxiliary variable x on all the units of the population: $x_1 = 200, x_2 = 80, x_3 = 50, x_4 = 50, x_5 = 10, x_6 = 10$.
1. Calculate the first-order inclusion probabilities proportional to x_k for a sample size $n = 4$.
2. Let 0.23444 be a realization of a uniform random variable in [0, 1], select a sample with unequal probabilities using systematic sampling by conserving the initial order of the units.
3. Give the matrix of second-order inclusion probabilities.

5.5 The objective is to select a sample of size four from a population of eight establishments whose size is known (in terms of number of workers):

Establishment	1	2	3	4	5	6	7	8
Number of workers	300	300	150	100	50	50	25	25

We want to select the sample with unequal inclusion probabilities proportional to the number of workers.
1. Calculate the inclusion probabilities from the establishment size.
2. Select a sample using systematic sampling using the following random number: 0.278.
3. Give the second-order inclusion probability between units 3 and 5.

5.6 Consider the sampling design with unequal probabilities as follows. In a population U of size $N \geq 3$, we select a unit with unequal probabilities $a_k, k \in U$ (of course we have $\sum_{k \in U} a_k = 1$). This unit is definitively removed from the population and is not included in the sample. From the remaining $N - 1$ units, select n units from the sample according to simple random sampling without replacement.
1. Calculate the first- and second-order inclusion probabilities of this sampling design.
2. Do the second-order inclusion probabilities satisfy the Yates–Grundy conditions?
3. Determine the a_k in order to select the units according to the given inclusion probabilities $\pi_k, k \in U$,
4. Is this method (a) fixed size, (b) without replacement or (c) applicable to any fixed vector of inclusion probabilities. Explain.

5.7 Consider the following sampling design with unequal probabilities from a population U of size $N \geq 3$. We select a unit with unequal probabilities $\alpha_k, k \in U$, where

$$\sum_{k \in U} \alpha_k = 1.$$

Then, among the $N - 1$ remaining units, we select $n - 1$ units in the sample according to simple random sampling without replacement. Therefore, the final sample is of fixed sample size n.

1. Calculate the first- and second-order inclusion probabilities of this sampling design. Write the second-order inclusion probabilities according to the first-order inclusion probabilities.
2. Determine the α_k in order to select the units according to given $\pi_k, k \in U$.
3. Do the second-order inclusion probabilities satisfy the Yates–Grundy conditions?
4. Is this method (a) without replacement or (b) applicable to any vector of given inclusion probabilities? Explain.

5.8 Consider a sampling design $p(s)$ with fixed sample size n with unequal inclusion probabilities from a population U of size N whose probabilities of the first two orders are denoted by π_k and $\pi_{k\ell}$. A sampling design $p^*(\cdot)$ of size $n^* = N - n$ is said to be the complementary of $p(s)$ if $p^*(U \backslash s) = p(s)$, for all $s \subset U$.

1. Give the first- and second-order inclusion probabilities of design $p^*(\cdot)$ as a function of π_k and $\pi_{k\ell}$.
2. Show that, if the Yates–Grundy conditions are satisfied for a sampling design, they are also satisfied for the complementary design.
3. Brewer's method can be described by a succession of splitting steps. Give the splitting step of the complementary design of Brewer's method. Express this step as a function of the sample size and the inclusion probabilities of the complementary design.
4. Brewer's method consists of selecting a unit at each step. What about the complementary design?
5. The decomposition method into simple random sampling designs can also be described by a succession of splitting steps. Give the splitting step of the complementary designs of the decomposition method into simple random sampling designs.

5.9 Consider the variable "number of inhabitants in an urban island" of an area of a small town whose values are $28, 32, 45, 48, 52, 54, 61, 78, 125, 147, 165, 258, 356, 438, 462, 685, 729, 739, 749, 749$. Select an unequal probability sample of size eight using the following designs by means of R package `sampling`:

1. systematic sampling
2. eliminatory method
3. Brewer's method
4. maximum entropy design
5. minimal support design.

6

Balanced Sampling

6.1 Introduction

A balanced sampling design only selects samples whose expansion estimators are equal to or at least very close to the totals for one or more auxiliary variables. The values of these auxiliary variables must be known for every unit of the population. Therefore, under balanced sampling, the expansion estimator returns the true population totals of the auxiliary variables.

The idea of recovering totals of known auxiliary variables from the sample appears in an underlying way with the representative method proposed by the Norwegian Kiær (1896, 1899, 1903, 1905). In Italy, Gini (1928) and Gini & Galvani (1929) selected a sample of 29 administrative districts from the 214 Italian districts (see Langel & Tillé, 2011a). This sample was selected by a reasoned choice method in order to recover the means of seven auxiliary variables. This method was severely criticized by Neyman (1934), mainly because the selected sample was not random. Indeed, at that time, there was no method to select a sample that was both random and balanced. The selection of such samples is quite delicate. The first methods to select a balanced random sample were proposed by Yates (1949) and Thionet (1953, pp. 203–207). These consist of first selecting a sample according to simple random sampling. Then, units of the sample are randomly exchanged if this exchange allows the distance between the sample mean and the population mean to be reduced.

Hájek (1981) has discussed at length the interest of the rejective method, which consists of selecting a large number of samples using a classical technique (e.g. simple random sampling without replacement and fixed sample size) and then keeping only a well-balanced sample in the sense that the expansion estimators of the known variables are very close to the totals to be estimated. However, the rejective-type methods suffer from a significant defect: the inclusion probabilities of the rejective design are not the same as those of the original one and are generally impossible to calculate.

Ardilly (1991) proposed an enumerative method that consists of taking an inventory of all possible samples and then assigning nonzero probabilities only to sufficiently well-balanced samples. However, enumerative methods have the disadvantage of a prohibitive calculation time, which makes them of limited interest as soon as the population size exceeds a few dozen units.

In this chapter, the notion of balancing is defined in Section 6.2. Balanced sampling can be written as a linear programming problem defined in Section 6.3. However, it is

Sampling and Estimation from Finite Populations, First Edition. Yves Tillé.
© 2020 John Wiley & Sons Ltd. Published 2020 by John Wiley & Sons Ltd.

not possible to solve this program once the population exceeds a few dozen units. Therefore, a shortcut must be found to obtain a truly operational method. A simple method is
systematic sampling and this is examined in Section 6.4. However, systematic sampling
only balances on the rank of the variable on which the data file is sorted. In Section 6.5,
the method of Deville et al. (1988) is presented. The *cube method* proposed by Deville
& Tillé (2004) is described in Section 6.6. This method provides a general solution for
balanced sampling on several auxiliary variables using equal or unequal inclusion probabilities. A simple example is developed in Section 6.6.10. Variance approximations are
then proposed for this method (Section 6.7). These approximations allow simple variance estimators to be constructed (Section 6.8). Finally, the cube method is compared
to the other sampling methods (Section 6.9) and the way to choose auxiliary variables is
discussed in Section 6.10.

6.2 Balanced Sampling: Definition

Assume that J auxiliary variables $z_1, \ldots, z_J$ are available; the vectors of the values $\mathbf{z}_k = (z_{k1}, \ldots, z_{kj}, \ldots, z_{kJ})^\top$ taken from J auxiliary variables are known on all the units of the
population. The knowledge of the values of the auxiliary variables $\mathbf{z}_k$ allows us to calcu-
late the J total:

$$Z_j = \sum_{k \in U} z_{kj}, j = 1, \ldots, J,$$

or in vector form:

$$\mathbf{Z} = \sum_{k \in U} \mathbf{z}_k.$$

When the sample is selected, one can calculate the expansion estimator of the J auxiliary
variables:

$$\widehat{Z}_j = \sum_{k \in S} \frac{z_{kj}}{\pi_k},$$

which can also be written in vector form:

$$\widehat{\mathbf{Z}} = \sum_{k \in S} \frac{\mathbf{z}_k}{\pi_k}.$$

The objective is always to estimate the total Y of the variable of interest whose values
are known only on the units selected in the sample. The expansion estimator is used:

$$\widehat{Y} = \sum_{k \in S} \frac{y_k}{\pi_k}.$$

The goal of balanced sampling is to use all of the auxiliary information to select the
sample. We refer to the following definition:

Definition 6.1 *A sampling design $p(s)$ is said to be balanced on the auxiliary variable
vector $\mathbf{z}_k^\top = (z_1, \ldots, z_J)$ if and only if it satisfies the balancing equations given by*

$$\sum_{k \in S} \frac{\mathbf{z}_k}{\pi_k} = \sum_{k \in U} \mathbf{z}_k, \tag{6.1}$$

which can also be written as

$$\operatorname{var}_p(\widehat{\mathbf{Z}}) = \mathbf{0}.$$

The fixed size and stratified designs are special cases of balanced sampling.

Example 6.1   A sampling design with fixed sample size n is balanced on a single variable $z_k = \pi_k, k \in U$. Indeed, the left-hand term of Expression (6.1) is equal to the sample size:

$$\sum_{k \in S} \frac{z_k}{\pi_k} = \sum_{k \in S} \frac{\pi_k}{\pi_k} = \sum_{k \in S} 1 = n_S.$$

The right-hand term of Expression (6.1) is not random:

$$\sum_{k \in U} z_k = \sum_{k \in U} \pi_k.$$

Balancing equations simply mean that the sample size is not random.

Example 6.2   Assume that the sampling design is stratified. Let $U_h, h = 1, \dots, H$, denote the strata of size N_h, and suppose that

$$n_h = \sum_{k \in U_h} \pi_k$$

is an integer. In each of the strata U_h, a fixed-size sample n_h is selected with equal or unequal probabilities. In this case, the design is balanced on the H variables $z_{kh} = \delta_{kh} \pi_k$, where

$$\delta_{kh} = \begin{cases} 1 & \text{if } k \in U_h \\ 0 & \text{if } k \notin U_h. \end{cases}$$

Indeed, the left-hand term of the balancing equations given in Expression (6.1) becomes

$$\sum_{k \in S} \frac{z_{kh}}{\pi_k} = \sum_{k \in S} \delta_{kh} = \sum_{k \in S_h} 1,$$

and the right-hand term is

$$\sum_{k \in U} z_{kh} = \sum_{k \in U} \delta_{kh} \pi_k = \sum_{k \in U_h} \pi_k.$$

Therefore, the balancing equations mean that in each stratum the sample size is not random. Hence, a stratified design is a balanced design on the variables $z_{kh}, h =, 1, \dots, H$.

However, in many cases it is not possible to exactly satisfy the balancing equations given in Expression (6.1). Consider the following example.

Example 6.3   Suppose that $N = 10, n = 7, \pi_k = 7/10, k \in U$, and that the only auxiliary variable is $z_k = k, k \in U$. In other words, $z_1 = 1, z_2 = 2, z_3 = 3, \dots, z_{10} = 10$. So, a balanced design is such that

$$\sum_{k \in S} \frac{k}{\pi_k} = \sum_{k \in U} k.$$

However, the left-hand term of the balancing equations given in Expression (6.1) becomes

$$\sum_{k\in S} \frac{k}{\pi_k} = \frac{10}{7} \sum_{k\in S} k. \tag{6.2}$$

and the right-hand term is

$$\sum_{k\in U} k = 55. \tag{6.3}$$

The equalization of Expressions (6.2) and (6.3) implies that $\sum_{k\in S}k$ must be equal to $55 \times 7/10 = 38.5$, which is impossible because 38.5 is not an integer.

Example 6.3 emphasizes the difficulties of balanced sampling: the selection of a balanced sample is an integer problem that cannot always satisfy noninteger constraints. There may indeed be a rounding problem that prevents balancing constraints from being exactly satisfied. For this reason, the goal is to build a sampling design that exactly satisfies the constraints if it is possible and approximately if it is not. It will be seen later that the rounding problem becomes negligible when the sample size is large.

6.3 Balanced Sampling and Linear Programming

Theoretically, a balanced sampling problem can be solved by a linear program. To do this, a cost $C(s)$ is assigned to each sample s. This cost is 0 if the sample is perfectly balanced and increases when the sample is poorly balanced. For example, the following cost can be used:

$$C(s) = \sum_{j=1}^{J} \left(\frac{\hat{Z}_j(s) - Z_j}{Z_j} \right)^2,$$

where $\hat{Z}_j(s)$ is the value taken by the expansion estimator on sample s.

We then define the linear program. We search for the sampling design $p(s)$ which minimizes the average cost:

$$\sum_{s\subset U} C(s)p(s) \tag{6.4}$$

subject to satisfying the inclusion probabilities

$$\sum_{s\subset U} p(s) = 1 \quad \text{and} \quad \sum_{s\subset U, s\ni k} p(s) = \pi_k, k \in U.$$

Deville & Tillé (2004) have shown that the resolution of this linear program results in the selection of a minimal support design (see the definition in Section 5.12.2).

Result 6.1 *The linear program (6.4) has at least one solution defined on minimal support.*

Proof: Suppose the sampling design $p(\cdot)$ is not defined on a minimal support. So, the linear system (5.13) has a finite number of solutions defined on a minimal support denoted by $p_1(\cdot), \ldots, p_J(\cdot)$. As $p(\cdot)$ can be written as $p(s) = \sum_{j=1}^{J} \lambda_j p_j(s)$, with positive coefficients λ_j, and $\sum_{j=1}^{J} \lambda_j = 1$. The minimal support design with the lowest average cost $\sum_{s \in \mathscr{S}} p_j(s) C(s)$ has a lower average cost than $p(\cdot)$. ∎

This result implies that this design only assigns positive probabilities to at most N samples.

Unfortunately, the application of a linear program is in most cases impossible, as it is necessary to enumerate all possible samples. Beyond a population size of $N = 15$, the number of samples 2^N is much too large for this method to be applicable. Therefore, a truly balanced sampling method should bypass this enumeration problem.

6.4 Balanced Sampling by Systematic Sampling

Assume that the population is sorted in ascending order according to an auxiliary variable z_k. Let r_k denote the rank of this variable. For simplicity, assume that $N/n = G$ is an integer. If we apply systematic sampling, described in Section 3.10, page 57, we can only select the following G samples:

$$\{1, 1 + G, 1 + 2G, \ldots, 1 + (n-1)G\},$$
$$\{2, 2 + G, 2 + 2G, \ldots, 2 + (n-1)G\}, \ldots,$$
$$\{h, h + G, h + 2G, \ldots, h + (n-1)G\}, \ldots,$$
$$\{G, 2G, 3G, \ldots, (n-1)G\}.$$

Consider now the rank average in the population:

$$\overline{R} = \frac{1}{N} \sum_{k \in u} r_k = \frac{1}{N} \sum_{k=1}^{N} k = \frac{1}{N} \times \frac{N(N+1)}{2} = \frac{N+1}{2}.$$

The expansion estimator of $\overline{R}$ for the sample

$$\{h, h + G, h + 2G, \ldots, h + (n-1)G\}, h = 1, \ldots, G$$

is

$$\widehat{\overline{R}}(h) = \frac{1}{N} \sum_{k \in S} \frac{r_k}{\pi_k} = \frac{1}{n} \sum_{k \in S} r_k = \frac{1}{n} \sum_{j=0}^{n-1} (h + jG) = \frac{1}{n} \left(nh + G \sum_{j=0}^{n-1} j \right)$$

$$= \frac{1}{n} \left[nh + G \frac{(n-1)n}{2} \right] = \left[h + G \frac{(n-1)}{2} \right] = \frac{1}{2}(2h + Gn - G).$$

Since $G = N/n$,

$$\widehat{\overline{R}}(h) = \frac{1}{2} \left(2h + N - \frac{N}{n} \right) = \frac{N+1}{2} + h - \frac{N+n}{2n}.$$

So we can calculate the relative error:

$$\frac{\widehat{\overline{R}}(h) - \overline{R}}{\overline{R}} = \frac{2h - \frac{N}{n} - 1}{N+1} = O\left(\frac{1}{n} \right),$$

where $O(1/n)$ is a quantity that remains bounded when multiplied by n for any value of n. This simple result shows that systematic sampling is a relatively well-balanced design on the rank of the variable according to which the file is sorted. The sample is even perfectly balanced if

$$h = \frac{N+n}{2n} = \frac{G+1}{2}.$$

Unfortunately, this method is limited to univariate balancing with equal probabilities on the rank of a variable.

6.5 Methode of Deville, Grosbras, and Roth

The method of Deville et al. (1988) is also restricted to sampling with equal probabilities but is multivariate. We suppose that J auxiliary variables $z_j, j = 1, \ldots, J$ are available and

$$\overline{Z}_j = \frac{1}{N} \sum_{k \in U} z_{kj}.$$

We start by dividing the population into G groups so that the averages $\overline{Z}_j$ are very different from one group to another. If J is not too big we can divide the population into $G = 2^J$ parts, where the units are sorted according to the values of the sign vectors:

$$\text{sign } (z_{k1} - \overline{Z}_1, \ldots, z_{kj} - \overline{Z}_j, \ldots, z_{kJ} - \overline{Z}_J).$$

Then, let N_g denote the number of units of the group $g = 1, \ldots, G$. The algorithm consists of repeating the following four steps:

1. *Initialisation*

   We begin with a stratified design of size n with proportional allocation on the G groups. Also, let $\widehat{\overline{Z}}_j^g(0)$ denote the mean of variable z_j obtained within the sample selected in group g. The sizes n_g of the groups-strata in the sample must be the subject to the rounding procedure detailed in step 3. We then denote by

$$\widehat{\overline{Z}}_j(t) = \frac{1}{n} \sum_{g=1}^{G} n_g \widehat{\overline{Z}}_j^g(t)$$

   (within the first step $t = 0$) the mean obtained in the sample for variable j in group g.

2. *Modification of the sample*

   The sample is then modified to be closer to the population means. Therefore, we look for a set of integers $v_g, g = 1, \ldots, G$. The number v_g represents, according to the sign, the number of units to be added or removed to group g. In order for the sample size to remain unchanged, it is necessary that

$$\sum_{g=1}^{G} v_g = 0. \tag{6.4}$$

   After the modification, we would like to be closer in expectation to the population mean of each group, that is to say that

$$\sum_{g=1}^{G} (n_g + v_g) \widehat{\overline{Z}}_j^g(t) = \overline{Z}_j, j = 1, \ldots, J. \tag{6.5}$$

We look for the numbers v_g that minimize

$$\sum_{g=1}^{G} v_g^2$$

subject to constraints (6.4) and (6.5). We then obtain values of v_g.

3. *Rounding*

Since the v_g are not integers, we respect the sizes v_g in expectation. We denote by $\lfloor v_g \rfloor$ the largest integer lower to v_g and $r_g = v_g - \lfloor v_g \rfloor$. We select a sample with unequal probabilities r_g of size $\sum_g r_g$ in $\{1, \ldots, G\}$. Rounded values are then $\lfloor v_g \rfloor$ for unselected units and $\lfloor v_g \rfloor + 1$ for selected units.

4. *Stop condition*

If $\sum_g |v_g| > A$, we then keep the sample and the algorithm stops. Otherwise, we increment t and calculate $A = \sum_g |v_g|$. For all the values of g we remove $|v_g|$ units from groups g for which the v_g are negative, these units are drawn with simple random sampling. We add v_g new units in g groups for which the v_g are positive. Then we recalculate the $\widetilde{\widehat{Z}}_j^g(t)$ and $\widehat{\widetilde{Z}}_j(t)$, and return to step 2.

This multivariate method is relatively fast because it avoids the enumeration of all possible samples and is not rejective. However, it is limited to the use of equal inclusion probabilities.

6.6 Cube Method

6.6.1 Representation of a Sampling Design in the form of a Cube

The cube method (see Deville & Tillé, 2004) enables a balanced sample to be selected while exactly satisfying fixed inclusion probabilities that can be equal or unequal. This method is based on a geometric representation of the sampling design. The 2^N possible samples (or subsets) of U (if we consider that the empty set $\emptyset$ is a sample) can be represented by 2^N vectors of $\mathbb{R}^N$ as follows. Let

$$\mathbf{a} = (a_1, \ldots, a_k, \ldots, a_N)^{\top}$$

denote the random sample whose components a_k are Bernoulli variables that are equal to 1 if $k \in S$ and 0 otherwise. Also,

$$\mathbf{a}_s = (a_{s1}, \ldots, a_{sk}, \ldots, a_{sN})^{\top}$$

is the value taken by $\mathbf{a}$ on sample $s \subset U$. Therefore, $\mathbf{a}_s \in \{0, 1\}^N$.

We can then interpret each vector $\mathbf{a}_s$ as a vertex of an N-cube (hypercube of $\mathbb{R}^N$). Figure 6.1 shows the geometric representation of a sampling design in a population of size $N = 3$.

The number of possible samples is the number of vertices of the N-cube and equals 2^N. A sampling design with inclusion probabilities π_k can then be defined as a distribution of probabilities $p(s)$ on all the vertices of the N-cube such that

$$E_p(\mathbf{a}) = \sum_{s \in \mathcal{S}} p(s)\mathbf{a}_s = \boldsymbol{\pi},$$

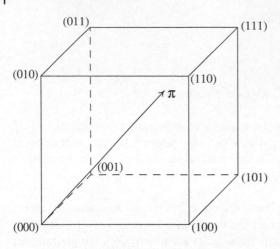

Figure 6.1 Possible samples in a population of size $N = 3$.

where $\boldsymbol{\pi} = (\pi_1, \dots, \pi_k, \dots, \pi_N)^\top$ is the vector of inclusion probabilities. If $0 < \pi_k < 1, k \in U$, then vector $\boldsymbol{\pi}$ is inside the cube. Therefore, a sampling design consists of expressing the vector of probabilities $\boldsymbol{\pi}$ as a convex linear combination (i.e. with positive or null coefficients and sum equal to 1) vertices of an N-cube.

6.6.2 Constraint Subspace

The cube method is based on the fact that the balancing equations given in Expression (6.1) can also be represented geometrically. These balancing equations define an affine subspace. Let us take the simple example of a sample with fixed sample size $n = 2$ in a population of size $N = 3$. Figure 6.2 shows that the three fixed-size samples can be connected by the affine subspace Q defined by

$$Q = \left\{ \mathbf{v} \in \mathbb{R}^3 \mid \sum_{k=1}^{3} v_k = 2 \right\}.$$

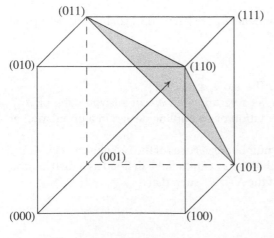

Figure 6.2 Fixed size constraint: the three samples of size $n = 2$ are connected by an affine subspace.

More generally, a sample s is balanced if

$$\sum_{k\in U}\frac{\mathbf{z}_k a_{sk}}{\pi_k} = \sum_{k\in U}\mathbf{z}_k.$$

This system of equations can also be written as

$$\begin{cases}\displaystyle\sum_{k\in U}\frac{\mathbf{z}_k a_{sk}}{\pi_k} = \sum_{k\in U}\frac{\mathbf{z}_k}{\pi_k}\pi_k \\ a_{sk}\in\{0,1\}, k\in U\end{cases}. \tag{6.6}$$

If

$$\check{\mathbf{Z}}^{\mathsf{T}} = \left(\frac{\mathbf{z}_1}{\pi_1}\cdots\frac{\mathbf{z}_k}{\pi_k}\cdots\frac{\mathbf{z}_N}{\pi_N}\right)$$

is a matrix of dimension $J\times N$, we can write system (6.6) in matrix form:

$$\check{\mathbf{Z}}^{\mathsf{T}}\mathbf{a}_s = \check{\mathbf{Z}}^{\mathsf{T}}\boldsymbol{\pi}.$$

The system of equations (6.6) with given $\mathbf{z}_k/\pi_k$ and unknown a_{sk} defines an affine subspace Q in $\mathbb{R}^N$ of dimension $N-J$ which is written as

$$Q = \{\mathbf{a}_s\in\mathbb{R}^N\mid\check{\mathbf{Z}}^{\mathsf{T}}\mathbf{a}_s = \check{\mathbf{Z}}^{\mathsf{T}}\boldsymbol{\pi}\}.$$

We can also define the affine subspace Q using the notion of kernel:

$$Q = \boldsymbol{\pi} + \operatorname{Ker}\check{\mathbf{Z}}^{\mathsf{T}},$$

where $\operatorname{Ker}\check{\mathbf{Z}}^{\mathsf{T}}$ is the kernel of $\check{\mathbf{Z}}^{\mathsf{T}}$, in other words,

$$\operatorname{Ker}\check{\mathbf{Z}}^{\mathsf{T}} = \{\mathbf{u}\in\mathbb{R}^N\mid\check{\mathbf{Z}}^{\mathsf{T}}\mathbf{u} = \mathbf{0}\}.$$

Example 6.4 If we consider the case where a single balancing variable $z_k = \pi_k$ is used, the balancing equations are reduced to a fixed size constraint. In this case, matrix $\check{\mathbf{Z}}^{\mathsf{T}}$ is reduced to a vector of ones:

$$\check{\mathbf{Z}}^{\mathsf{T}} = (1,\dots,1,\dots,1) = \mathbf{1}^{\mathsf{T}}.$$

Therefore, the affine subspace Q is

$$Q = \{\mathbf{a}_s\in\mathbb{R}^N\mid\check{\mathbf{Z}}^{\mathsf{T}}\mathbf{a}_s = \check{\mathbf{Z}}^{\mathsf{T}}\boldsymbol{\pi}\} = \left\{\mathbf{a}_s\in\mathbb{R}^N\,\middle|\,\sum_{k\in U}a_{sk} = \sum_{k\in U}\pi_k\right\}.$$

The subspace Q is represented in Figure 6.2 for the case where $N=3$ and $n=2$.

6.6.3 Representation of the Rounding Problem

The rounding problem can also be geometrically represented. Let us look at some cases. In the case of Example 6.4 where the only constraint is the fixed size of the sample, there is no rounding problem because it is always possible to select a sample of fixed size. We can see in Figure 6.2 that the vertices of the intersection of the cube and the subspace are vertices of the cube and therefore samples.

The rounding problem appears when the vertices of the intersection of the subspace and the cube are not vertices of the cube. Consider the following example.

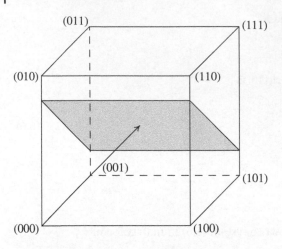

Figure 6.3 None of the vertices of K is a vertex of the cube.

Example 6.5 Suppose that the only constraint ($J = 1$) is given by the auxiliary variable $z_1 = 0, z_2 = 3$ and $z_3 = 2$ and that the inclusion probabilities are $\pi_1 = \pi_2 = \pi_3 = 1/2$. Matrix $\check{\mathbf{Z}}^\mathsf{T}$ is reduced to the vector

$$\check{\mathbf{Z}}^\mathsf{T} = \left(\frac{z_1}{\pi_1}, \frac{z_2}{\pi_2}, \frac{z_3}{\pi_3} \right) = (0, 6, 4),$$

and the balancing equation is

$$0 \times a_{s1} + 6 \times a_{s2} + 4 \times a_{s3} = 0 + 3 + 2 = 5.$$

The subspace of constraints $Q = \{\mathbf{a}_s \in \mathbb{R}^3 | 6a_{s2} + 4a_{s3} = 5\}$ of dimension 2 is represented in Figure 6.3. This subspace does not go through any of the vertices of the cube. Therefore, the vertices of the intersection of the cube and the subspace of the constraints are not vertices of the cube (samples). This example shows the geometric representation of the rounding problem. Here, there is no sample that exactly satisfies the constraints because no vertex of the cube belongs to the constraint subspace. We then say that the system of balancing equations can only be approximately satisfied.

Finally, there are mixed cases, where some vertices of the intersection of the cube and the subspace are samples and some are not. Consider the following example.

Example 6.6 Suppose the only constraint ($J = 1$) is given by the auxiliary variable $z_1 = 0.8, z_2 = 2.4$, and $z_3 = 0.8$ and that $\pi_1 = \pi_2 = \pi_3 = 0.8$. Matrix $\check{\mathbf{Z}}^\mathsf{T}$ is reduced to vector

$$\check{\mathbf{Z}}^\mathsf{T} = \left(\frac{z_1}{\pi_1}, \frac{z_2}{\pi_2}, \frac{z_3}{\pi_3} \right) = (1, 3, 1),$$

and the balancing equation is

$$1 \times a_{s1} + 3 \times a_{s2} + 1 \times a_{s3} = 0.8 + 2.4 + 0.8 = 4.$$

The subspace of the constraints $Q = \{\mathbf{a}_s \in \mathbb{R}^3 | a_{s1} + 3a_{s2} + a_{s3} = 4\}$ of dimension 2 is represented in Figure 6.4. In this case, the subspace of the constraints passes through two vertices of the cube but a vertex of the intersection is not a vertex of the cube. The balancing equation can only be satisfied in some cases. It is said that the system of balancing equations can sometimes be satisfied.

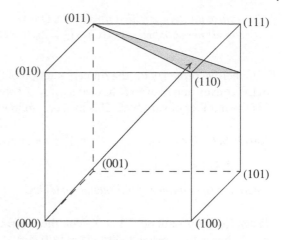

Figure 6.4 Two vertices of *K* are vertices of the cube, but the third is not.

When the balancing equation system can sometimes be satisfied, some samples are exactly balanced. However, samples that are not exactly balanced need to be considered in the sampling design in order to satisfy inclusion probabilities.

These examples can be formalized. We denote by:

- $C = [0, 1]^N$ the *N*-cube in $\mathbb{R}^N$ whose vertices are the samples of U,
- $Q = \pi + \text{Ker } \check{\mathbf{Z}}^\top$ the constraint subspace of dimension $N - J$, and
- $K = Q \cap C = \{[0, 1]^N \cap (\pi + \text{Ker } \check{\mathbf{Z}}^\top)\}$, the intersection of the constraint subspace and the cube.

The intersection *K* between *C* and *Q* is not empty because π is inside *C* and belongs to *Q*. The intersection between an *N*-cube and a subspace defines a convex polyhedron *K* of dimension $N - J$.

The study of the properties of the convex polyhedron *K* allows whether or not it is possible to select an exactly balanced sample to be determined.

Definition 6.2 *Let D be a convex polyhedron, a vertex of D is a point that cannot be written as a convex linear combination of the other points of D. The set of all vertices of D is written as Ext(D).*

For example, we have $\#\text{Ext}(C) = 2^N$. Since *K* is a convex polyhedron, the number of vertices Ext(*K*) of *K* is finite. In addition, each inner point of *K* can be written as at least one convex linear combination with positive vertex coefficients.

Definition 6.3 *A sample* $\mathbf{a}_s$ *is exactly balanced if and only if*

$$\mathbf{a}_s \in \text{Ext}(C) \cap Q.$$

A necessary condition for the existence of an exactly balanced sample is that $\text{Ext}(C) \cap Q \neq \emptyset$.

Definition 6.4 *A system of balancing equations can be*

i) exactly satisfied if $\text{Ext}(C) \cap Q = \text{Ext}(C \cap Q)$

ii) *approximately satisfied if* $\mathrm{Ext}(C) \cap Q = \emptyset$

iii) *sometimes satisfied if* $\mathrm{Ext}(C) \cap Q \neq \mathrm{Ext}(C \cap Q)$, *and* $\mathrm{Ext}(C) \cap Q \neq \emptyset$.

A decisive property for the implementation of the cube method is that all the vertices of the convex polyhedron K have at least $N - J$ components equal to either 0 or 1, where J is the number of variables. This property follows from the following result.

Result 6.2 *If* $\mathbf{r} = (r_1, \ldots, r_k, \ldots, r_N)^\top$ *is an extreme point of* K *then*

$$\#\{k | 0 < r_k < 1\} \leq J,$$

where J *is the number of balancing variables.*

Proof: (by contradiction) Let $\check{\mathbf{Z}}^{*\top}$ be the matrix $\check{\mathbf{Z}}^\top$ restricted to noninteger units of $\mathbf{r}$, in other words, restricted to $U^* = \{k | 0 < r_k < 1\}$. If $q = \#U^* > J$, then $\mathrm{Ker}\,\check{\mathbf{Z}}^{*\top}$ is of dimension $q - J > 0$ and $\mathbf{r}$ is not an extreme point of K. ∎

6.6.4 Principle of the Cube Method

The main idea for obtaining a balanced sample is to choose a vertex of the N-cube (a sample) that is on subspace Q when it is possible or that is in the neighbourhood of Q when it is not possible. The cube method is divided into two phases called the *flight phase* and the *landing phase*.

- In the flight phase described in Section 6.6.5, the objective is to randomly round off to 0 or 1 a maximum of inclusion probabilities by satisfying exactly the balancing equations. The flight phase is a random walk that starts from the inclusion probability vector and ends on a vertex of the intersection between the cube and the constraint subspace K. At the end of the flight phase, a vertex of K is chosen at random in such a way that the inclusion probabilities $\pi_k, k \in U$, and the balancing equations given in Expression (6.1) are exactly satisfied.
- The *landing phase*, described in Section 6.6.6, consists of managing at best the fact that the balancing equations (6.1) cannot be exactly satisfied. This phase starts with the vertex of K obtained at the end of the flight phase and ends with a sample that is close to that vertex.

The *flight phase* ends on a vertex of K, that is, a vertex that is almost a sample because, by virtue of Result 6.2, the vertices of K contain all at least $N - J$ components equal to 0 or 1. The landing phase applies only in the case where there is a rounding problem, i.e. if the vertex of K reached is not a vertex of C. This phase consists of managing at best the relaxation of the constraints (6.1) so as to select a sample, that is to say, a vertex of C. It consists of randomly rounding the noninteger components to 0 or 1 while moving as little as possible away from the balancing equations.

6.6.5 The Flight Phase

Before giving a rigorous definition of the flight phase algorithm, consider the example of Figure 6.5. In this example, we consider a population of size $N = 3$ in which we want

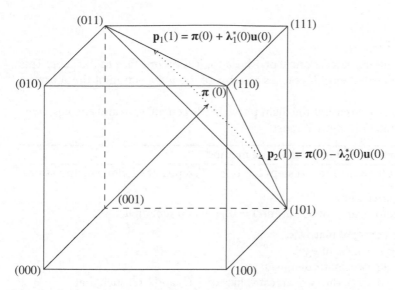

Figure 6.5 Flight phase in a population of size $N = 3$ with a constraint of fixed size $n = 2$.

to select a sample of fixed size $n = 2$. In other words, only one auxiliary variable is used, $z_k = \pi_k, k \in U$. In this example, the balancing constraint can be exactly satisfied.

The algorithm starts with the initialization $\pi(0) = \pi$ and will define a sequence of random vectors, $\pi(0)$, $\pi(1)$, $\pi(2)$, ..., $\pi(T)$, that satisfy the martingale property, in other words, such that

$$E_p[\pi(t) \mid \pi(t-1)] = \pi(t-1), t =, 1 \ldots, T. \tag{6.7}$$

This martingale property ensures that the inclusion probabilities are exactly satisfied. Indeed, it ensures that by recurrence for all t, we have

$$E_p[\pi(t)] = E_p[\pi(t-1)] = \cdots = E_p[\pi(1)] = E_p[\pi(0)] = \pi.$$

In the first step of the flight phase algorithm, a vector $\mathbf{u}(0)$ is generated in the kernel of the matrix $\check{\mathbf{Z}}^{\top}$. This vector can be defined freely, randomly or not. Each of the possible directions of $\mathbf{u}(0)$ defines a particular variant of the method. Following the direction given by vector $\mathbf{u}(0)$ from the point π, we necessarily cross a face of the cube at a point denoted by $\mathbf{p}_1(1)$ on Figure 6.5. Since $\mathbf{u}(0)$ is in the kernel of $\check{\mathbf{Z}}^{\top}$, the point $\mathbf{p}_1(1)$ is necessarily in the constraint subspace.

Then, following the opposite direction, that is to say the direction of vector $-\mathbf{u}(0)$ from point π, we cross another face of the cube at a point denoted by $\mathbf{p}_2(1)$ on Figure 6.5. Since $\mathbf{u}(0)$ is in the kernel of $\check{\mathbf{Z}}^{\top}$, the point $\mathbf{p}_2(1)$ is necessarily in the constraint subspace.

We then define the random vector $\pi(1)$:

$$\pi(1) = \begin{cases} \mathbf{p}_1(1) & \text{with probability } q_1(1) \\ \mathbf{p}_2(1) & \text{with probability } q_2(1), \end{cases}$$

where the probabilities $q_1(1)$ and $q_2(1)$ are defined to satisfy the martingale property given in Expression (6.7), i.e. $q_1(1)\mathbf{p}_1(1) + q_2(1)\mathbf{p}_2(1) = \pi(0)$. At the end of this phase, vector $\pi_1(1)$ is on one face of the cube and satisfies the balancing constraints in the sense that

$$\check{\mathbf{Z}}^{\mathsf{T}}\boldsymbol{\pi}(1) = \check{\mathbf{Z}}^{\mathsf{T}}\boldsymbol{\pi}.$$

The vector $\boldsymbol{\pi}_1(1)$ contains at least one 0 or 1 because it is on one face of the cube. This basic step is then repeated until it reaches a vertex of the intersection of the subspace and the cube.

Algorithm 9 precisely describes the flight phase in the general case and enables a vertex of K to be reached in N steps at most.

Algorithm 9 General algorithm for the cube method

First initialize with $\boldsymbol{\pi}(0) = \boldsymbol{\pi}$. Then, at steps $t = 1, \dots, T$, repeat the following three steps:

1. *Generation of a direction*
   Generate any vector (random or not) $\mathbf{u}(t) = [u_k(t)] \neq 0$ such that
   (i) $\mathbf{u}(t)$ is in the kernel of matrix $\check{\mathbf{Z}}^{\mathsf{T}}$
   (ii) $u_k(t) = 0$ if $\pi_k(t-1)$ is integer.
2. *Identification of the two points on the faces*
   Calculate $\lambda_1^*(t)$ and $\lambda_2^*(t)$, the two largest values of $\lambda_1(t)$ and $\lambda_2(t)$ such that

$$0 \leq \pi_k(t-1) + \lambda_1(t)u_k(t) \leq 1$$

and

$$0 \leq \pi_k(t-1) - \lambda_2(t)u_k(t) \leq 1,$$

for all $k \in U$. Note that $\lambda_1(t) > 0$ and $\lambda_2(t) > 0$.
3. *Random choice of one of the two points*
   Select

$$\boldsymbol{\pi}(t) = \begin{cases} \boldsymbol{\pi}(t-1) + \lambda_1^*(t)\mathbf{u}(t) & \text{with probability } q_1(t) \\ \boldsymbol{\pi}(t-1) - \lambda_2^*(t)\mathbf{u}(t) & \text{with probability } q_2(t), \end{cases}$$

where $q_1(t) = \lambda_2^*(t)/\{\lambda_1^*(t) + \lambda_2^*(t)\}$ and $q_2(t) = \lambda_1^*(t)/\{\lambda_1^*(t) + \lambda_2^*(t)\}$.

These three steps are repeated until it is no longer possible to find a vector $\mathbf{u}(t) = [u_{(t)}]$ (random or not) such as (i) $\mathbf{u}(t)$ in the kernel of $\check{\mathbf{Z}}^{\mathsf{T}}$, (ii) $u_k(t) = 0$ if $\pi_k(t-1)$ is an integer.

At each step of Algorithm 9, we see that $E_p[\boldsymbol{\pi}(t)] = \boldsymbol{\pi}(0)$. Indeed, since $q_1(t)\lambda_1^*(t) - \lambda_2^*(t)q_2(t) = 0$, we have

$$E_p[\boldsymbol{\pi}(t)|\boldsymbol{\pi}(t-1)] = \boldsymbol{\pi}(t-1).$$

From this martingale property, we can deduce recursively that $E_p[\boldsymbol{\pi}(t)] = \boldsymbol{\pi}(0)$. At each stage of the algorithm, the inclusion probabilities are perfectly satisfied.

Since the $\mathbf{u}(t)$ vectors are in the kernel of matrix $\check{\mathbf{Z}}^{\mathsf{T}}$,

$$\check{\mathbf{Z}}^{\mathsf{T}}\mathbf{u}(t) = \sum_{k \in U} \frac{\mathbf{z}_k u_k(t)}{\pi_k} = \mathbf{0}$$

for all t. It follows that

$$\sum_{k \in U} \frac{\mathbf{z}_k \pi_k(t)}{\pi_k} = \sum_{k \in U} \frac{\mathbf{z}_k \pi_k(t-1)}{\pi_k} = \cdots = \sum_{k \in U} \frac{\mathbf{z}_k \pi_k(0)}{\pi_k} = \sum_{k \in U} \mathbf{z}_k.$$

Therefore, the balancing equations remain always satisfied.

At each stage, at least one component of process $\boldsymbol{\pi}(t)$ is definitely rounded to 0 or 1, in other words $\boldsymbol{\pi}(t)$ has at least one more integer component than $\boldsymbol{\pi}(t-1)$. So, $\boldsymbol{\pi}(1)$ is on one face of the N-cube (on a cube of dimension at most $N-1$), $\boldsymbol{\pi}(2)$ is on a cube of dimension at most $N-2$ and so on.

Let T be the moment the flight phase is over. The fact of no longer finding a vector $\mathbf{u}(T+1)$ such that (i) $\mathbf{u}(T+1)$ is in the kernel of the matrix $\check{\mathbf{Z}}^{\mathsf{T}}$ and (ii) $u_k(T+1) = 0$ if $\pi_k(T)$ is an integer shows that the sequence $\boldsymbol{\pi}(t)$ has reached a vertex of K, and so by Result 6.2, page 130, that $\#\{0 < \pi_k(T) < 1\} \leq J$.

There are several variants of Algorithm 9 that depend on the method of construction of vectors $\mathbf{u}(t)$. The variant proposed by Chauvet & Tillé (2006) allows a particularly fast execution. This variant is implemented in the SAS® program developed by Chauvet & Tillé (2005a,b, 2007), in the package R sampling developed by Tillé & Matei (2016), and in the R BalancedSampling package developed by Grafström & Lisic (2016). Chauvet (2009) describes how to select balanced stratified designs by compensating for rounding problems between strata. Hasler & Tillé (2014) proposed an implementation for highly stratified populations and use this algorithm to perform balanced imputations to handle item nonresponse (Hasler & Tillé, 2016).

6.6.6 Landing Phase by Linear Programming

At the end of the flight phase, the sequence $\boldsymbol{\pi}(t)$ has reached a vertex of K but not necessarily a vertex of the cube C. Let q be the number of noninteger components of this vertex. If $q = 0$, the algorithm is complete. If $q > 0$, some constraints cannot be reached exactly. The constraints must then be relaxed to reach a vertex of the N-cube close in some sense to the extreme point already reached. Let T be the last step of the flight phase, for simplicity we write $\boldsymbol{\pi}^* = [\pi_k^*] = \boldsymbol{\pi}(T)$.

Definition 6.5 *A sample $\mathbf{a}_s$ is said to be compatible with a vector $\boldsymbol{\pi}^*$ if $\pi_k^* - \mathbf{a}_s = 0$ for any k such that π_k^* is an integer. We write $C(\boldsymbol{\pi}^*)$ for all the samples compatible with $\boldsymbol{\pi}^*$.*

Since the number of noninteger components of vector $\boldsymbol{\pi}^*$ is smaller than the number of balancing variables (see Result 6.2) when q is not too large, it is again possible to solve this problem by linear programming as described in Section 6.3. For each sample s compatible with $\boldsymbol{\pi}^*$, we assign a cost $C(s)$ which is zero if the sample is perfectly balanced and which increases as the sample moves away from the balancing constraints. We then solve the following program:

$$\min_{p(.|\boldsymbol{\pi}^*)} \sum_{s\in C(\boldsymbol{\pi}^*)} C(s)p(s|\boldsymbol{\pi}^*),$$

subject to

$$\sum_{s\in C(\boldsymbol{\pi}^*)} p(s|\boldsymbol{\pi}^*) = 1, \tag{6.8}$$

$$\sum_{s\in C(\boldsymbol{\pi}^*)|s\ni k} p(s|\boldsymbol{\pi}^*) = \pi_k^*, k \in U,$$

$$0 \leq p(s|\boldsymbol{\pi}^*) \leq 1, s \in C(\boldsymbol{\pi}^*).$$

If $U^* = \{k \in U | 0 < \pi_k^* < 1\}$, $q = \#U^*$, and $s^* = s \cap U^*$, the linear program (6.8) can also be written as

$$
\begin{vmatrix}
\min_{p^*(\cdot)} \sum_{s^* \subset U^*} C(s^*)p^*(s^*), \\
\text{subject to} \\
\sum_{s^* \subset U^*} p^*(s^*) = 1, \\
\sum_{s^* \subset U^* | s^* \ni k} p^*(s^*) = \pi_k^*, k \in U^*, \\
0 \le p^*(s^*) \le 1, s \subset U^*.
\end{vmatrix}
\tag{6.9}
$$

The linear program (6.9) no longer depends on the population size but only the number of balancing variables, because $q \le J$. Therefore, it restricts itself to 2^q possible samples and can be generally solved with a dozen balancing variables. In the case where the design is of fixed size and the sum of the inclusion probabilities is an integer, we can restrict ourselves to the

$$
\begin{pmatrix} q \\ \sum_{k \in U^*} \pi_k^* \end{pmatrix}
$$

samples with fixed sample sizes.

6.6.7 Choice of the Cost Function

A tricky problem is to choose the cost $C(s)$. The choice of $C(\cdot)$ is a decision that depends on the survey manager's priorities. A simple cost can be built using the sum of the squared relative deviations:

$$
C_1(s) = \sum_{j=1}^{J} \frac{[\hat{Z}_j(s) - Z_j]^2}{Z_j^2},
$$

where $\hat{Z}_j(s)$ is the value taken by $\hat{Z}_j$ on sample s.

Another possibility is to use

$$
C_2(s) = (\mathbf{a}_s - \boldsymbol{\pi}^*)^\top \check{\mathbf{Z}} (\check{\mathbf{Z}}^\top \check{\mathbf{Z}})^{-1} \check{\mathbf{Z}}^\top (\mathbf{a}_s - \boldsymbol{\pi}^*).
$$

The choice of $C_2(\cdot)$ has a natural interpretation given by the following result.

Result 6.3 *The distance between a sample* $\mathbf{a}_s$ *and its orthogonal projection on the subspace of the constraints is*

$$
C_2(s) = (\mathbf{a}_s - \boldsymbol{\pi}^*)^\top \check{\mathbf{Z}} (\check{\mathbf{Z}}^\top \check{\mathbf{Z}})^{-1} \check{\mathbf{Z}}^\top (\mathbf{a}_s - \boldsymbol{\pi}^*).
\tag{6.10}
$$

Proof: The projection of a sample $\mathbf{a}_s$ on the constraint subspace is

$$
\mathbf{a}_s - \check{\mathbf{Z}} (\check{\mathbf{Z}}^\top \check{\mathbf{Z}})^{-1} \check{\mathbf{Z}}^\top (\mathbf{a}_s - \boldsymbol{\pi}).
$$

Therefore, the Euclidian distance between $\mathbf{a}_s$ and its projection is

$$(\mathbf{a}_s - \boldsymbol{\pi})^\top \check{\mathbf{Z}} (\check{\mathbf{Z}}^\top \check{\mathbf{Z}})^{-1} \check{\mathbf{Z}}^\top (\mathbf{a}_s - \boldsymbol{\pi})$$

$$= (\mathbf{a}_s - \boldsymbol{\pi}^* + \boldsymbol{\pi}^* - \boldsymbol{\pi})^\top \check{\mathbf{Z}} (\check{\mathbf{Z}}^\top \check{\mathbf{Z}})^{-1} \check{\mathbf{Z}}^\top (\mathbf{a}_s - \boldsymbol{\pi}^* + \boldsymbol{\pi}^* - \boldsymbol{\pi})$$

and since $\check{\mathbf{Z}}^\top (\boldsymbol{\pi} - \boldsymbol{\pi}^*) = \mathbf{0}$, Expression (6.10) is directly derived. ∎

The cost $C_2(\cdot)$ can be interpreted as a simple distance in $\mathbb{R}^N$ between the sample and the constraint subspace.

6.6.8 Landing Phase by Relaxing Variables

If the number of auxiliary variables is too large to solve the linear program by the simplex algorithm, then, at the end of the flight phase, an auxiliary variable can be directly deleted. A constraint is then relaxed and it is possible to return to the flight phase until it is again no longer possible to move in the subspace of constraints. The constraints are relaxed successively. For this reason, it is necessary to sort the variables in order of importance so that the least important constraints are relaxed first.

6.6.9 Quality of Balancing

Whatever variant is used for the landing phase, the gap between the deviation estimator and the population total can be bounded.

Result 6.4 *For any application of the cube method,*

$$\left| \hat{Z}_j - Z_j \right| \leq J \times \max_{k \in U} \left| \frac{z_{kj}}{\pi_k} \right|.$$

Proof:

$$\left| \hat{Z}_j - Z_j \right| = \left| \sum_{k \in S} \frac{z_{kj}}{\pi_k} - \sum_{k \in U} \frac{z_{kj}}{\pi_k} \pi_k \right| = \left| \sum_{k \in S} \frac{z_{kj}}{\pi_k} - \sum_{k \in U} \frac{z_{kj}}{\pi_k} \pi_k^* \right|$$

$$= \left| \sum_{k \in S^*} \frac{z_{kj}}{\pi_k} - \sum_{k \in U^*} \frac{z_{kj}}{\pi_k} \pi_k^* \right| = \left| \sum_{k \in U^*} \frac{z_{kj}}{\pi_k} (s_k^* - \pi_k^*) \right| \leq \sup_{F | \#F = q} \sum_{k \in F} \left| \frac{z_{kj}}{\pi_k} \right|$$

$$\leq J \times \max_{k \in U} \left| \frac{z_{kj}}{\pi_k} \right|.$$

∎

Under the realistic assumptions that $z_k/N = O(1/N)$ and $\pi_k = O(n/N)$, the bound becomes

$$\frac{\left| \hat{Z}_j - Z_j \right|}{Z_j} \leq O\left(\frac{J}{n} \right).$$

The rounding problem becomes negligible when the sample size is large. It must of course be mentioned that this bound is built for the worst case, while the landing phase aims to find the best one.

6.6.10   An Example

Balanced sampling also allows an important gain of precision for the selection of small samples in small populations. In the dummy example presented in Table 6.1, we consider $N = 20$ students. Four variables are used, a constant variable that is equal to 1 for all the units of the population, gender that takes values 1 for men and 2 for women, age in years, and a mark of 20 in a statistics exam.

The objective is to select a balanced sample of 10 students with equal inclusion probabilities. The inclusion probabilities are all $\pi_k = 1/2, k \in U$. Therefore, there is $\binom{20}{10} =$ 184 756 possible samples of size 10 in this population. At the end of the flight phase, vector $\boldsymbol{\pi}^*$ contains at most four noninteger components because the number of balancing variables is four. The landing phase should consider at most only

$$\binom{4}{2} = 6$$

possible samples. Using the package R `sampling` (Tillé & Matei, 2016) and by applying a linear program for the landing phase with the C_2 cost function, the cube algorithm

Table 6.1 Population of 20 students with variables, constant, gender (1, male, 2 female), age, and a mark of 20 in a statistics exam.

Constant	Gender	Age	Grade
1	1	18	5
1	1	18	13
1	1	19	10
1	1	19	12
1	1	20	14
1	1	20	6
1	1	20	12
1	1	21	16
1	1	22	15
1	1	25	3
1	2	18	8
1	2	19	19
1	2	19	14
1	2	19	9
1	2	20	7
1	2	22	19
1	2	24	18
1	2	25	15
1	2	24	11
1	2	28	14

Table 6.2 Totals and expansion estimators for balancing variables.

Variable	Z_j	$\hat{Z}_j$	$100 \times (\hat{Z}_j - Z_j)/Z_j$
Constant	20	20	0
Gender	30	30	0
Age	420	420	0
Grade	240	242	0.8333

Table 6.3 Variances of the expansion estimators of the means under simple random sampling and balanced sampling.

Variables	Simple random sampling $\text{var}_p(\hat{\bar{Z}}_{\text{SIMPLE}})$	Balanced sampling $\text{var}_p(\hat{\bar{Z}}_{\text{BAL}})$	Gain of variance $\dfrac{\text{var}_p(\hat{\bar{Z}}_{\text{BAL}})}{\text{var}_p(\hat{\bar{Z}}_{\text{SIMPLE}})}$
Gender	0.01332	0.00028	0.02095
Age	0.41396	0.07545	0.18226
Grade	1.03371	0.15901	0.15383

selects the following sample:

$$\mathbf{a}_s = (1, 1, 1, 0, 0, 1, 0, 0, 1, 0, 0, 0, 1, 0, 1, 1, 1, 0, 0, 1)^\top,$$

which can also be written as $s = \{1, 2, 3, 6, 9, 13, 15, 16, 17, 20\}$.

Table 6.2 contains the totals and their expansion estimators for the four variables as well as the relative deviation of the estimator. For this sample, only the "Grade" variable is not exactly balanced.

The balanced design is then compared to simple random sampling with fixed sample size. Using 1 000 simulations, the variance of the expansion estimators of the means of the four balancing variables is approximated for these two designs. The results are shown in Table 6.3.

It can be seen that even for a very small sample size, the variance reduction is very important because the cube method avoids the selection of poorly balanced samples. Other examples for much larger populations are presented in Chauvet & Tillé (2005a), Tillé (2006), and Chauvet & Tillé (2007).

6.7 Variance Approximation

The variance of $\hat{Y}$ can theoretically be expressed using second-order inclusion probabilities. Unfortunately, even in very simple cases, the calculation seems to be impossible. However, in the case of the Poisson design, the variance of $\hat{Y}$ can be easily calculated because it only depends on first-order inclusion probabilities.

If $\tilde{S}$ is the random sample selected by means of a Poisson design and $\tilde{\pi}_k, k \in U$, the first-order inclusion probabilities of this Poisson design, then

$$\text{var}_{p|poiss}(\widehat{Y}) = \text{var}_{p|poiss}\left(\sum_{k \in \tilde{S}} \frac{y_k}{\pi_k}\right) = \sum_{k \in U} \frac{y_k^2}{\pi_k^2} \tilde{\pi}_k(1 - \tilde{\pi}_k).$$

The Poisson design maximizes the entropy for the given first-order inclusion probabilities (see Section 5.8). If the balanced design has maximum entropy or is close to maximum entropy, Deville & Tillé (2005) have shown that the variance can be calculated as the conditional variance with respect to the balancing equations of a particular Poisson design of inclusion probabilities $\tilde{\pi}_k$. In other words,

$$\text{var}_{p|bal}(\widehat{Y}) = \text{var}_{p|poiss}(\widehat{Y}|\widehat{\mathbf{Z}} = \mathbf{Z}),$$

where $\text{var}_{p|bal}$ is the variance under the balanced design with maximum entropy of inclusion probabilities π_k and $\text{var}_{p|poiss}$ is the variance under the Poisson design of inclusion probabilities $\tilde{\pi}_k$. The probabilities $\tilde{\pi}_k$ are unknown and must be approximated.

If we assume that, for a Poisson design, vector $(\widehat{Y} \ \widehat{\mathbf{Z}}^\top)^\top$ has in first approximation a multivariate normal distribution, we obtain

$$\text{var}_{p|poiss}(\widehat{Y}|\widehat{\mathbf{Z}} = \mathbf{Z}) \approx \text{var}_{p|poiss}[\widehat{Y} + (\mathbf{Z} - \widehat{\mathbf{Z}})^\top \boldsymbol{\beta}], \tag{6.11}$$

where

$$\boldsymbol{\beta} = \text{var}_{p|poiss}(\widehat{\mathbf{Z}})^{-1}\text{cov}_{POISS}(\widehat{\mathbf{Z}}, \widehat{Y}),$$

$$\text{var}_{p|poiss}(\widehat{\mathbf{Z}}) = \sum_{k \in U} \frac{\mathbf{z}_k \mathbf{z}_k^\top}{\pi_k^2} \tilde{\pi}_k(1 - \tilde{\pi}_k),$$

and

$$\text{cov}_{POISS}(\widehat{\mathbf{Z}}, \widehat{Y}) = \sum_{k \in U} \frac{\mathbf{z}_k y_k}{\pi_k^2} \tilde{\pi}_k(1 - \tilde{\pi}_k).$$

Based on (6.11), we obtain the variance approximation:

$$\text{var}_{p,app}(\widehat{Y}) = \sum_{k \in U} \frac{b_k}{\pi_k^2}(y_k - y_k^*)^2, \tag{6.12}$$

where

$$y_k^* = \mathbf{z}_k^\top \left(\sum_{\ell \in U} b_\ell \frac{\mathbf{z}_\ell \mathbf{z}_\ell}{\pi_\ell^2}\right)^{-1} \sum_{\ell \in U} b_\ell \frac{\mathbf{z}_\ell y_\ell}{\pi_\ell^2}$$

is a regression prediction of y_k. The weights $b_k = \tilde{\pi}_k(1 - \tilde{\pi}_k)$ are used because the inclusion probabilities of the Poisson design $\tilde{\pi}_k$ are not exactly equal to the inclusion probabilities of the balanced design. Since the exact values of the $\tilde{\pi}_k$ are unknown, the exact values of the b_k are also unknown and are approximated.

Expression (6.12) can also be written as

$$\text{var}_{p,app}(\widehat{Y}) = \sum_{k \in U}\sum_{\ell \in U} \frac{y_k}{\pi_k}\frac{y_\ell}{\pi_\ell}\Delta_{appk\ell},$$

where $\mathbf{\Delta}_{app} = (\mathbf{\Delta}_{appk\ell})_{N \times N}$ is the approximation of the variance operator whose general element is

$$
\mathbf{\Delta}_{appk\ell} = \begin{cases} b_k - b_k \dfrac{\mathbf{z}_k^{\mathsf{T}}}{\pi_k} \left(\displaystyle\sum_{i \in U} b_i \dfrac{\mathbf{z}_i \mathbf{z}_i^{\mathsf{T}}}{\pi_i^2} \right)^{-1} \dfrac{\mathbf{z}_k}{\pi_k} b_k & k = \ell \\[4mm] b_k \dfrac{\mathbf{z}_k^{\mathsf{T}}}{\pi_k} \left(\displaystyle\sum_{k \in U} b_i \dfrac{\mathbf{z}_i \mathbf{z}_i^{\mathsf{T}}}{\pi_i^2} \right)^{-1} \dfrac{\mathbf{z}_\ell b_\ell}{\pi_\ell} & k \neq \ell. \end{cases}
\tag{6.13}
$$

Hence, the main problem is to find a good approximation of b_k. Four values were tested in Deville & Tillé (2004) for b_k. These values are denoted respectively by $b_{k\alpha}, \alpha = 1, 2, 3, 4$, and allow us to define four variance estimators denoted by $V_\alpha, \alpha = 1, 2, 3, 4$ and four variance operators denoted by $\mathbf{\Delta}_\alpha, \alpha = 1, 2, 3, 4$ replacing b_k in (6.12) and (6.13) by, respectively, b_{k1}, b_{k2}, b_{k3}, and b_{k4}.

1. The first approximation is obtained by considering that when n is large, $\pi_k \approx \tilde{\pi}_k, k \in U$. We then take $b_{k1} = \pi_k$.
2. The second approximation is obtained by applying a correction for the loss of degrees of freedom

$$
b_{k2} = \pi_k (1 - \pi_k) \frac{N}{N - J}.
$$

This correction allows the variance of a simple random sampling design to be exactly obtained without replacement with fixed sample size.
3. The third correction results from an adjustment on the diagonal elements of the variance operator $\mathbf{\Delta}$ of the given variance. Indeed, these diagonal elements are always known and equal to $\pi_k (1 - \pi_k)$. Then, we use

$$
b_{k3} = \pi_k (1 - \pi_k) \frac{\text{trace } \mathbf{\Delta}}{\text{trace } \mathbf{\Delta}_1}.
$$

Therefore, we can define an approximate variance operator $\mathbf{\Delta}_3$ which has the same trace as $\mathbf{\Delta}$.
4. Finally, the fourth approximation comes from the fact that the diagonal elements of $\mathbf{\Delta}_{app}$ are given in Expression (6.13). However, the diagonal elements of $\mathbf{\Delta}$ only depend on the first-order inclusion probabilities and are equal to $\pi_k (1 - \pi_k)$. The b_{k4} are constructed so that $\mathbf{\Delta}$ and $\mathbf{\Delta}_{app}$ have the same diagonal, in other words,

$$
\pi_k (1 - \pi_k) = b_k - b_k \frac{\mathbf{z}_k^{\mathsf{T}}}{\pi_k} \left(\sum_{k \in U} b_k \frac{\mathbf{z}_k \mathbf{z}_k^{\mathsf{T}}}{\pi_k^2} \right)^{-1} \frac{\mathbf{z}_k}{\pi_k} b_k, k \in U.
\tag{6.14}
$$

The determination of b_{k4} requires the resolution of a system of nonlinear equations. However, this fourth approximation is the only one that provides the exact expression of the variance for a stratified design with equal inclusion probabilities.

A set of simulations described by Deville & Tillé (2005) shows that b_{k4} gives the best approximation. Unfortunately the solution of the system of equations (6.14) does not always exist. In this case, we can use b_{k3}.

6.8 Variance Estimation

The variance can be estimated using the same principle. The general estimator is

$$v(\hat{Y}) = \sum_{k \in S} \frac{c_k}{\pi_k^2}(y_k - \hat{y}_k^*)^2, \tag{6.15}$$

where

$$\hat{y}_k^* = \mathbf{z}_k^\top \left(\sum_{\ell \in S} c_\ell \frac{\mathbf{z}_\ell \mathbf{z}_\ell'}{\pi_\ell^2} \right)^{-1} \sum_{\ell \in S} c_\ell \frac{\mathbf{z}_\ell y_\ell}{\pi_\ell^2}$$

is the estimation of the predictor by the regression of y_k. Again, four estimators are derived from four definitions of c_k. Expression (6.15) can also be written as

$$\sum_{k \in S} \sum_{\ell \in S} \frac{y_k}{\pi_k} \frac{y_\ell}{\pi_\ell} D_{k\ell},$$

where

$$D_{k\ell} = \begin{cases} c_k - c_k \dfrac{\mathbf{z}_k^\top}{\pi_k} \left(\displaystyle\sum_{i \in S} c_i \dfrac{\mathbf{z}_i \mathbf{z}_i^\top}{\pi_i^2} \right)^{-1} \dfrac{\mathbf{z}_k}{\pi_k} c_k & k = \ell \\[3ex] c_k \dfrac{\mathbf{z}_k^\top}{\pi_k} \left(\displaystyle\sum_{k \in S} c_i \dfrac{\mathbf{z}_i \mathbf{z}_i^\top}{\pi_i^2} \right)^{-1} \dfrac{\mathbf{z}_\ell c_\ell}{\pi_\ell} & k \neq \ell. \end{cases}$$

The four definitions of c_k are denoted by c_{k1}, c_{k2}, c_{k3}, and c_{k4}, and allow four approximations of the variance to be defined by replacing c_k in Expression (6.15) by, respectively, c_{k1}, c_{k2}, c_{k3}, and c_{k4}.

1. The first estimator is obtained by taking $c_{k1} = (1 - \pi_k)$.
2. The second is obtained by applying a correction for the loss of degrees of freedom:

$$c_{k2} = (1 - \pi_k)\frac{n}{n - J}.$$

   This correction gives an unbiased estimator for simple random sampling without replacement.
3. The third estimator is derived from the fact that the diagonal elements of the true matrix of $\Delta_{k\ell}/\pi_{k\ell}$ are always known and are equal to $1 - \pi_k$. So, we can write

$$c_{k3} = (1 - \pi_k)\frac{\sum_{k \in S}(1 - \pi_k)}{\sum_{k \in S} D_{kk}}.$$

4. Finally, the fourth estimator is derived from the fact that the diagonal elements D_{kk} are known. The c_{k4} are constructed in such a way that

$$1 - \pi_k = D_{kk}, k \in U, \tag{6.16}$$

   or

$$1 - \pi_k = c_k - c_k \frac{\mathbf{z}_k^\top}{\pi_k} \left(\sum_{i \in S} c_i \frac{\mathbf{z}_i \mathbf{z}_i^\top}{\pi_i^2} \right)^{-1} \frac{\mathbf{z}_k}{\pi_k} c_k, k \in U.$$

This fourth approximation is the only one which provides an exactly unbiased estimator for a stratified design.

These four estimators are derived directly from the four approximations of variance. In some cases, the system of equations (6.16) has no solution, so we use c_{k3}. If the sample size is large relative to the number of auxiliary variables, the four estimators give very similar results.

6.9 Special Cases of Balanced Sampling

Unequal probability sampling is a special case of the cube method. Indeed, when the only auxiliary variable is the inclusion probability, the sample is of fixed size. The cube method is a generalization of the splitting method of Deville & Tillé (1998), which itself generalizes several algorithms with unequal probabilities (Brewer method, pivot method, minimal support design). Stratification is also a special case of the cube method. In this case, the balancing variables are the indicators of the strata. The advantage of the cube method is that it allows us to balance on overlapping strata. For example, the sample can be balanced on the marginal totals of a contingency table. In addition, qualitative and quantitative variables can be used together.

Almost all other sampling techniques are special cases of balanced sampling (except multistage sampling). In fact, balanced sampling is simply more general, in the sense that many sampling methods can be implemented with the cube method. Balanced sampling allows us to use any balancing variable. With the more general concept of balancing, strata can overlap, quantitative and qualitative variables can be used together, and inclusion probabilities can be chosen freely.

6.10 Practical Aspects of Balanced Sampling

The main recommendation is that the balancing variables should be correlated to the variables of interest. As in a regression problem, the balancing variables must be chosen by observing a parsimony principle. The selection of too many balancing variables must be avoided because, for each additional variable, a degree of freedom is lost for the variance estimation. In practice, one wants to estimate several variables of interest. Therefore, the auxiliary variables must be globally correlated to the variables of interest without being too correlated with each other. Lesage (2008) proposed a method for balancing on complex statistics. The main idea is to balance on the linearized variables of the complex statistics.

In many cases, the balancing variables contain measurement errors. In most registers, data may be flawed or simply not up to date. Missing values may have been imputed. As for the calibration methods described in Chapter 12, the errors in the balancing variables are not very important. Indeed, with balanced sampling, the expansion estimator is used and is unbiased, even if the balancing variables contain errors. The gain in efficiency depends only on the correlation between the balancing variables and the variables of interest. This correlation is rarely affected by some errors in the auxiliary variables. Tillé (2011) gives a list of applications of the cube method in official statistics.

Exercise

6.1 Select a balanced sample of size eight using the data of Table 6.3 and R software.

7

Cluster and Two-stage Sampling

Auxiliary information is not always used to increase the precision of estimators. It can also be used to improve the organization of the survey. Cluster sampling designs (Section 7.1) target a reduction in costs. Two-stage sampling designs permit more complex survey strategies (Section 7.2) by using information available on the primary sampling units. We introduce multi-stage sampling in Section 7.3. Finally, we present two-phase sampling designs in Section 7.5 and the case where the sample is presented as the intersection of two independent samples in Section 7.6.

7.1 Cluster Sampling

Cluster sampling designs are very useful when the totality of the sampling frame is not known while some information is known about the primary units. For example, if we want to select a sample of individuals from a particular country, we usually do not have a complete list of all its inhabitants (unless a national registry exists). However, a list of municipalities (primary units) is usually easy to find. If these municipalities have an exhaustive list of inhabitants or residences, the use of cluster sampling allows a survey to be conducted without constructing a complete survey frame or even knowing the size of the population. Cluster sampling also allows the use of auxiliary information. However, it is used in a different way than when used for stratification. The objective of constructing clusters is to reduce costs by simplifying the organization of the survey. This can be quite notable when clusters are geographic entities. However, the variance of estimators of totals is generally larger than for simple designs.

7.1.1 Notation and Definitions

Suppose the population $U = \{1, \dots, k, \dots, N\}$ is partitioned in M subsets, U_i, $i = 1, \dots, M$, called clusters such that

$$\bigcup_{i=1}^{M} U_i = U \quad \text{and} \quad U_i \cap U_j = \emptyset, i \neq j.$$

The number of elements N_i in cluster i is called the size of the cluster. We have

$$\sum_{i=1}^{M} N_i = N,$$

where N is the size of population U.

Sampling and Estimation from Finite Populations, First Edition. Yves Tillé.
© 2020 John Wiley & Sons Ltd. Published 2020 by John Wiley & Sons Ltd.

The total can be written as

$$Y = \sum_{k \in U} y_k = \sum_{i=1}^{M} \sum_{k \in U_i} y_k = \sum_{i=1}^{M} Y_i$$

and the mean is

$$\overline{Y} = \frac{1}{N} \sum_{k \in U} y_k = \frac{1}{N} \sum_{i=1}^{M} \sum_{k \in U_i} y_k = \frac{1}{N} \sum_{i=1}^{M} N_i \overline{Y}_i,$$

where Y_i is the total within the cluster i and $\overline{Y}_i$ is the mean of the cluster i:

$$Y_i = \sum_{k \in U_i} y_k, i = 1, \dots, M,$$

$$\overline{Y}_i = \frac{1}{N_i} \sum_{k \in U_i} y_k, i = 1, \dots, M.$$

Additionally, let S_{yi}^2 denote the corrected variance within cluster i:

$$S_{yi}^2 = \frac{1}{N_i - 1} \sum_{k \in U_i} (y_k - \overline{Y}_i)^2.$$

A sampling design is called cluster sampling if:

- A sample of clusters s_I is randomly selected given some sampling method $p_I(s_I)$, let S_I denote the random sample such that $\Pr(S_I = s_I) = p_I(s_I)$ and $m = \#S_I$, the number of selected clusters.
- All the sampling units from the selected clusters are observed.

The design is illustrated in Figure 7.1.

The complete random sample is

$$S = \bigcup_{i \in S_I} U_i.$$

The size of S is

$$n_S = \sum_{i \in S_I} N_i.$$

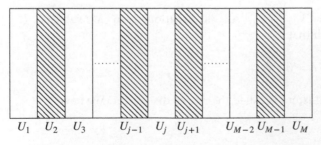

$$U_1 \quad U_2 \quad U_3 \qquad U_{j-1} \quad U_j \quad U_{j+1} \qquad U_{M-2} \, U_{M-1} \, U_M$$

Figure 7.1 Cluster sampling: the population is divided into clusters. Clusters are randomly selected. All units from the selected clusters are included in the sample.

The size of the sample is generally random, even if the number of selected clusters is not. To start, we consider a general cluster sampling design. We will then look at two specific examples: cluster sampling with equal probabilities and cluster sampling with probabilities proportional to size.

In the general case, the probability of selecting a cluster comes from the sampling method and is

$$\pi_{Ii} = \sum_{s_I \ni i} p_I(s_I), i = 1, \dots, M,$$

and the probability of selecting two distinct clusters is

$$\pi_{Iij} = \sum_{s_I \ni i,j} p_I(s_I), i = 1, \dots, M, j = 1, \dots, M, i \neq j.$$

The inclusion probabilities of the sampling units come from the selection probabilities of the clusters. The first-order inclusion probability of a unit is equal to the selection probability of its cluster. If unit k belongs to cluster i, we have

$$\pi_k = \pi_{Ii}, k \in U_i.$$

For second-order inclusion probabilities, we must consider two cases:

- If k and ℓ belong to the same cluster i, their joint selection probability is equal to the selection probability of their cluster:

$$\pi_{k\ell} = \pi_{Ii}, k \text{ and } \ell \in U_i.$$

- If k and ℓ belong to two different clusters i and j, respectively, their joint selection probability is equal to the joint probability of selecting both clusters:

$$\pi_{k\ell} = \pi_{Iij}, k \in U_i \text{ and } \ell \in U_j, i \neq j.$$

Estimating the mean has some specific hurdles. First, because the sample size is generally random. Then, because the Yates–Grundy condition (see Section 2.7, page 20) on the second-order probabilities is not satisfied when two units come from the same cluster. Indeed, if k and $\ell \in U_i$ then $\pi_k \pi_\ell - \pi_{k\ell} = \pi_{Ii}^2 - \pi_{Ii} = -\pi_{Ii}(1 - \pi_{Ii})$.

The expansion estimators of the total and the mean are

$$\hat{Y} = \sum_{i \in S_I} \frac{Y_i}{\pi_{Ii}} \quad \text{and} \quad \hat{\bar{Y}} = \frac{1}{N} \sum_{i \in S_I} \frac{N_i \bar{Y}_i}{\pi_{Ii}}.$$

However, in cluster sampling, N is rarely known. In order to estimate the mean, we can use the Hájek estimator (see Expression (2.2), page 20).

The variance of the expansion estimator of the total is

$$\text{var}_p(\hat{Y}) = \sum_{i=1}^{M} \frac{Y_i^2}{\pi_{Ii}}(1 - \pi_{Ii}) + \sum_{i=1}^{M} \sum_{\substack{j=1 \\ j \neq i}}^{M} \frac{Y_i Y_j}{\pi_{Ii} \pi_{Ij}}(\pi_{Iij} - \pi_{Ii}\pi_{Ij}). \tag{7.1}$$

If the number of selected clusters is fixed, we can write it as

$$\text{var}_p(\hat{Y}) = \frac{1}{2} \sum_{i=1}^{M} \sum_{\substack{j=1 \\ j \neq i}}^{M} \left(\frac{Y_i}{\pi_{Ii}} - \frac{Y_j}{\pi_{Ij}} \right)^2 (\pi_{Ii}\pi_{Ij} - \pi_{Iij}). \tag{7.2}$$

The variance can be unbiasedly estimated. Given (7.1), we obtain

$$v_1(\widehat{Y}) = \sum_{i \in S_I} \frac{Y_i^2}{\pi_{Ii}^2}(1 - \pi_{Ii}) + \sum_{i \in S_I} \sum_{\substack{j \in S_I \\ j \neq i}} \frac{Y_i Y_j}{\pi_{Ii} \pi_{Ij}} \frac{\pi_{Iij} - \pi_{Ii}\pi_{Ij}}{\pi_{Iij}}.$$

When the number of clusters is fixed, we can construct another estimator of this variance with Expression (7.2):

$$v_2(\widehat{Y}) = \frac{1}{2} \sum_{i \in S_I} \sum_{\substack{j \in S_I \\ j \neq i}} \left(\frac{Y_i}{\pi_{Ii}} - \frac{Y_j}{\pi_{Ij}}\right)^2 \frac{\pi_{Ii}\pi_{Ij} - \pi_{Iij}}{\pi_{Iij}}.$$

However, these two estimators require a sum of n^2 terms. If the cluster design has unequal probabilities with fixed size, we can construct a much more practical estimator using Expression (5.15):

$$v_3(\widehat{Y}) = \sum_{i \in S_I} \frac{c_{Ii}}{\pi_{Ii}^2}(Y_i - \widehat{Y}_i^*)^2, \tag{7.3}$$

where

$$\widehat{Y}_i^* = \pi_{Ii} \frac{\sum_{j \in S} c_{Ij} Y_j / \pi_{Ij}}{\sum_{j \in S} c_{Ij}},$$

where the c_{Ii} can be defined by

$$c_{Ii} = (1 - \pi_{Ii}) \frac{m}{m - 1}.$$

In cluster designs, the expansion estimator has the undesirable property raised in Expression (2.1), page 20. The expansion estimator of the mean is not linearly invariant. If the variable of interest is constant ($y_k = C$, for all $k \in U$), we obtain

$$\widehat{\bar{Y}} = C \frac{1}{N} \sum_{i \in S_I} \frac{N_i}{\pi_{Ii}}.$$

In this case, to estimate the mean, it is preferable to use the Hájek estimator defined in Expression (2.2), page 20:

$$\widehat{\bar{Y}}_{\mathrm{HAJ}} = \left(\sum_{i \in S_I} \frac{N_i}{\pi_{Ii}}\right)^{-1} \left(\sum_{i \in S_I} \frac{Y_i}{\pi_{Ii}}\right).$$

7.1.2 Cluster Sampling with Equal Probabilities

A classical method of selecting clusters consists of using simple random sampling without replacement of fixed size m. In this case, we have

$$\pi_{Ii} = \frac{m}{M}, i = 1, \dots, M,$$

and

$$\pi_{Iij} = \frac{m(m - 1)}{M(M - 1)}, i = 1, \dots, M.$$

Therefore, all units have equal inclusion probabilities. The sample size is random and its mean is

$$E_p(n_S) = E_p\left(\sum_{i\in S_I} N_i\right) = \sum_{i\in U_I} \frac{N_i m}{M} = \frac{Nm}{M}.$$

The expansion estimators of the total and the mean are

$$\hat{Y} = \frac{M}{m}\sum_{i\in S_I} Y_i \quad \text{and} \quad \hat{\overline{Y}} = \frac{M}{Nm}\sum_{i\in S_I} N_i\overline{Y}_i.$$

The expansion estimator of the mean is not linearly invariant. The undesirable property raised in Expression (2.1) is again present, except in cases where the clusters are the same size.

As the number of selected clusters is fixed, the variance is derived from (7.2)

$$\text{var}_p(\hat{Y}) = \frac{M^2}{m}\left(1 - \frac{m}{M}\right)\frac{1}{M-1}\sum_{i=1}^{M}\left(Y_i - \frac{Y}{M}\right)^2$$

and can be unbiasedly estimated by

$$v(\hat{Y}) = \frac{M^2}{m}\left(1 - \frac{m}{M}\right)\frac{1}{m-1}\sum_{i\in S_I}\left(Y_i - \frac{\hat{Y}}{M}\right)^2.$$

7.1.3 Sampling Proportional to Size

To carry out a sample selection proportional to the size of the clusters, we apply a without replacement design of fixed size m with the cluster selection probabilities proportional to the size of the clusters using the methods described in Chapter 5. Therefore, we select clusters without replacement with selection probabilities

$$\pi_{Ii} = \frac{mN_i}{N}, i = 1,\ldots,M,$$

supposing $mN_i/N \leq 1, i = 1,\ldots,M$. The sample size is still random and its mean is

$$E_p(n_S) = E_p\left(\sum_{i\in S_I} N_i\right) = \sum_{i\in U_I} N_i\frac{N_i m}{N} = \frac{m}{N}\sum_{i\in U_I} N_i^2.$$

The expansion estimator of the mean is

$$\hat{\overline{Y}} = \frac{1}{m}\sum_{i\in S_I} \overline{Y}_i$$

and is linearly invariant. The issues due to the undesirable property raised in Expression (2.1), page 20 disappear. Indeed, in this case, if the variable of interest is constant, the mean is equal to the parameter of interest to estimate, as the means within clusters are all equal. The expansion estimator of the total is

$$\hat{Y} = \frac{N}{m}\sum_{i\in S_I} \overline{Y}_i.$$

Since the number of selected clusters is fixed, the variance is derived from (7.2):

$$\text{var}_p(\hat{Y}) = \frac{N^2}{2m^2} \sum_{i=1}^{M} \sum_{\substack{j=1 \\ j \neq i}}^{M} (\overline{Y}_i - \overline{Y}_j)^2 \left(\frac{mN_i}{N} \frac{mN_j}{N} - \pi_{Iij} \right).$$

The variance estimator is straightforward:

$$v_1(\hat{Y}) = \frac{N^2}{2m^2} \sum_{i \in S_I} \sum_{\substack{j \in S_I \\ j \neq i}} \frac{(\overline{Y}_i - \overline{Y}_j)^2}{\pi_{Iij}} \left(\frac{mN_i}{N} \frac{mN_j}{N} - \pi_{Iij} \right).$$

Again, this estimator is not practical. If the cluster sampling design has unequal probabilities, we can construct a simpler estimator using the approximation given by (7.3) and obtain

$$v_2(\hat{Y}) = \frac{N^2}{m^2} \sum_{i \in S_I} c_{Ii} \left(\overline{Y}_i - \hat{\overline{Y}}^* \right)^2,$$

where

$$\hat{\overline{Y}}^* = \frac{\sum_{j \in S_I} c_{Ij} \overline{Y}_j}{\sum_{j \in S_I} c_{Ij}},$$

and where c_{Ii} is

$$c_{Ii} = \frac{m}{m-1}(1 - \pi_{Ii}).$$

This sampling design has good properties. However, to be able to apply it, the size of all clusters must be known. Furthermore, for the same number of selected clusters, it gives a design that has on average a larger number of sampling units than the equal probability design.

7.2 Two-stage Sampling

Two-stage sampling is based on a double sampling procedure: one on the primary units and another on the secondary units. The objective of two-stage sampling is to help organize data collection. The use of two-stage sampling is relevant in the following two cases:

- When no information exists on the units of interest, it is not possible to directly select these units. We can sample intermediary units (called primary units) which have well-known characteristics. We then enumerate the units of interest in the primary units to achieve a second sample. Therefore, a complete sampling frame is not required to implement the survey.
- When units are spread over a large area, interviewer travel costs can become very large. We can then group sampled units together to reduce costs.

For example, in a household survey we can first select municipalities. Then, in the selected municipalities, a sample of households can be selected. Interviewer travel is greatly reduced since the selected households are found in a restricted number of municipalities. Furthermore, listing is only required for households in the selected municipalities. Therefore, we achieve a significant reduction in the resources needed for data collection.

7.2.1 Population, Primary, and Secondary Units

Suppose that population $U = \{1, \ldots, k, \ldots, N\}$ is composed of M subpopulations $U_i, i = 1, \ldots, M$, called primary units. Each primary unit U_i is composed of N_i secondary units or individuals and we have

$$\sum_{i=1}^{M} N_i = N,$$

where N is the size of the population U.

Generally, a two-stage sampling design is defined in the following way:

- A sample of primary units is selected under design $p_I(s_I)$. Let S_I denote the random sample such that $\Pr(S_I = s_I) = p_I(s_I)$ and $m = \#S_I$.
- If a primary unit U_i is selected in the first sample, we select a secondary sample s_i of secondary units with design $p_i(s_i)$. Let S_i denote the random sample of secondary units such that $\Pr(S_i = s_i) = p_i(s_i)$ and $n_i = \#S_i$.
- Särndal et al. (1992) adds that two-stage designs need the properties of *invariance* and *independence*. Invariance means the second-stage sampling designs $p_i(s_i)$ do not depend on the first stage, in other words that $\Pr(S_i = s_i) = \Pr(S_i = s_i | S_I)$. Independence means the second-stage sampling designs are independent from one other (like in stratification).

noindentThe two-stage sampling design is illustrated in Figure 7.2.

The concept of invariance is discussed in Beaumont & Haziza (2016), who distinguish two types of invariance. Under their definition, strong invariance implies the first and second stage are selected independently. Weak invariance only implies that the first- and second-order inclusion probabilities of the second stage do not depend on the first-stage sample. In the following, we assume strong invariance.

In the case of strong invariance, we can define a two-stage sampling design as the intersection between a stratified design and a cluster design. The cluster design is defined by $S_1 = \bigcup_{i \in S_I}^{M} U_i$ and the stratified design by $S_2 = \bigcup_{i=1}^{M} S_i$. These two samples must be independent and the cluster definition must correspond to the stratums.

The random sample is

$$S = \bigcup_{i \in S_I} S_i = S_1 \cap S_2.$$

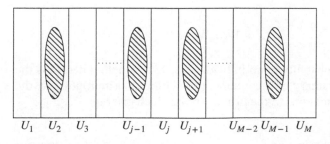

$$U_1 \quad U_2 \quad U_3 \qquad U_{j-1} \quad U_j \quad U_{j+1} \qquad U_{M-2} \; U_{M-1} \; U_M$$

Figure 7.2 Two-stage sampling design: we randomly select primary units in which we select a sample of secondary units.

For the variable of interest y, the total in the population can be written as

$$Y = \sum_{k \in U} y_k = \sum_{i=1}^{M} \sum_{k \in U_i} y_k = \sum_{i=1}^{M} Y_i,$$

where Y_i represents the total in the primary unit U_i:

$$Y_i = \sum_{k \in U_i} y_k, i = 1, \dots, M.$$

Similarly, the mean in the population can be written as

$$\overline{Y} = \frac{1}{N} \sum_{k \in U} y_k = \frac{1}{N} \sum_{i=1}^{M} \sum_{k \in U_i} y_k = \frac{1}{N} \sum_{i=1}^{M} N_i \overline{Y}_i,$$

where $\overline{Y}_i$ represents the mean within the primary unit U_i

$$\overline{Y}_i = \frac{1}{N_i} \sum_{k \in U_i} y_k, i = 1, \dots, M.$$

Let S_{yi}^2 be the corrected variance within the primary unit U_i:

$$S_{yi}^2 = \frac{1}{N_i - 1} \sum_{k \in U_i} (y_k - \overline{Y}_i)^2.$$

The size of S is

$$n_S = \sum_{i \in S_I} n_i.$$

The size of the sample is generally random.

Let π_{Ii} be the inclusion probabilities in the first sample, in other words the probability of selecting primary unit U_i. Let also π_{Iij} be the second-order inclusion probabilities for the first sample. This is the probability of selecting primary unit U_i and primary unit U_j. These probabilities result from design $p_I(s_I)$. We have

$$\Delta_{Iij} = \begin{cases} \pi_{Iij} - \pi_{Ii} \pi_{Ij} & \text{if } i \neq j \\ \pi_{Ii}(1 - \pi_{Ii}) & \text{if } j = i. \end{cases}$$

Let $\pi_{k|i}$ be the probability of selecting secondary unit k given primary unit i was selected and $\pi_{k\ell|i}$ the probability of jointly selecting secondary units k and ℓ given unit i was selected. Finally, we have

$$\Delta_{k\ell|i} = \begin{cases} \pi_{k\ell|i} - \pi_{k|i} \pi_{\ell|i} & \text{if } k \neq \ell \\ \pi_{k|i}(1 - \pi_{k|i}) & \text{if } k = \ell \end{cases}, i = 1, \dots, M.$$

The first-order selection probability of an individual for this sampling design is the probability of selecting the primary unit containing this individual multiplied by the probability of selecting this individual in the primary unit, which gives us

$$\pi_k = \pi_{Ii} \pi_{k|i}, k \in U_i.$$

For the second-order selection probabilities, we have to take into account the invariance and independence properties. There are two cases:

- If two individuals k and ℓ belong to the same primary unit U_i, then the second-order inclusion probability is

$$\pi_{k\ell} = \pi_{Ii}\pi_{k\ell|i}.$$

- If two individuals k and ℓ belong to two distinct primary units U_i and U_j, respectively, then the second-order inclusion probability is

$$\pi_{k\ell} = \pi_{Iij}\pi_{k|i}\pi_{\ell|j}.$$

7.2.2   The Expansion Estimator and its Variance

Generally, the expansion estimator of the total is

$$\widehat{Y} = \sum_{i\in S_I}\sum_{k\in S_i}\frac{y_k}{\pi_{Ii}\pi_{k|i}} = \sum_{i\in S_I}\frac{\widehat{Y}_i}{\pi_{Ii}},$$

where $\widehat{Y}_i$ is the expansion estimator of Y_i:

$$\widehat{Y}_i = \sum_{k\in S_i}\frac{y_k}{\pi_{k|i}}$$

and $\widehat{\overline{Y}}$ is the expansion estimator of the mean:

$$\widehat{\overline{Y}} = \frac{1}{N}\sum_{i\in S_I}\sum_{k\in S_i}\frac{y_k}{\pi_{Ii}\pi_{k|i}}.$$

Knowing the first- and second-order inclusion probabilities, we can in principle deduce the expression of the variance of this estimator using the general formulas for the expansion estimator. However, this process is tedious. We can calculate this variance using the law of total variance by conditioning on either the first or the second stage.

Result 7.1   *In a two-stage sampling design,*

$$\text{var}_p(\widehat{Y}) = V_\alpha + V_\beta,$$

where

$$V_\alpha = \sum_{i=1}^{M}\sum_{j=1}^{M}\frac{Y_iY_j}{\pi_{Ii}\pi_{Ij}}\Delta_{Iij}, \qquad V_\beta = \sum_{i=1}^{M}\frac{\text{var}_p(\widehat{Y}_i)}{\pi_{Ii}},$$

and

$$\text{var}_p(\widehat{Y}_i) = \sum_{k\in U_i}\sum_{\ell\in U_i}\frac{y_ky_\ell}{\pi_{k|i}\pi_{\ell|i}}\Delta_{k\ell|i}, i = 1,\dots,M. \tag{7.4}$$

Proof: The variance can be decomposed in two parts:

$$\text{var}_p(\widehat{Y}) = \text{var}_p\,\text{E}_p(\widehat{Y}|S_I) + \text{E}_p\,\text{var}_p(\widehat{Y}|S_I).$$

The variance of the conditional expectation becomes

$$\text{var}_p\,\text{E}_p(\widehat{Y}|S_I) = \text{var}_p\,\text{E}_p\left(\sum_{i\in S_I}\frac{\widehat{Y}_i}{\pi_{Ii}}\bigg|S_I\right).$$

From the invariance property we have

$$
E_p\left(\sum_{i\in S_I}\frac{\hat{Y}_i}{\pi_{Ii}}\bigg|S_I\right) = \sum_{i\in S_I}E_p\left(\frac{\hat{Y}_i}{\pi_{Ii}}\bigg|S_I\right) = \sum_{i\in S_I}E_p\left(\frac{\hat{Y}_i}{\pi_{Ii}}\right) = \sum_{i\in S_I}\frac{Y_i}{\pi_{Ii}}.
$$

Therefore,

$$
\mathrm{var}_p\;E_p(\hat{Y}|S_I) = \mathrm{var}_p\left(\sum_{i\in S_I}\frac{Y_i}{\pi_{Ii}}\right) = \sum_{i=1}^{M}\sum_{j=1}^{M}\frac{Y_iY_j}{\pi_{Ii}\pi_{Ij}}\Delta_{Iij}.
$$

The expectation of the conditional variance becomes

$$
E_p\;\mathrm{var}_p(\hat{Y}|S_I) = E_p\;\mathrm{var}_p\left(\sum_{i\in S_I}\frac{\hat{Y}_i}{\pi_{Ii}}\bigg|S_I\right).
$$

By the invariance and independence properties,

$$
\mathrm{var}_p\left(\sum_{i\in S_I}\frac{\hat{Y}_i}{\pi_{Ii}}\bigg|S_I\right) = \sum_{i\in S_I}\mathrm{var}_p\left(\frac{\hat{Y}_i}{\pi_{Ii}}\bigg|S_I\right) = \sum_{i\in S_I}\frac{\mathrm{var}_p(\hat{Y}_i)}{\pi_{Ii}^2}.
$$

Hence,

$$
E_p\;\mathrm{var}_p(\hat{Y}|S_I) = E_p\left[\sum_{i\in S_I}\frac{\mathrm{var}_p(\hat{Y}_i)}{\pi_{Ii}^2}\right] = \sum_{i=1}^{M}\frac{\mathrm{var}_p(\hat{Y}_i)}{\pi_{Ii}},
$$

where $\mathrm{var}_p(\hat{Y}_i)$ is given in Expression (7.4). ∎

We can see that V_α is exactly the variance of a cluster design. The two terms V_α and V_β are often interpreted as inter-cluster variance (V_α) and intra-cluster variance (V_β). However, this interpretation is abusive, as it is possible to obtain another decomposition by conditioning this time on the second stage. This second decomposition is often called the reverse approach.

Result 7.2 *In a two-stage design,*

$$
\mathrm{var}_p(\hat{Y}) = V_A + V_B,
$$

where

$$
V_A = E_p\left(\sum_{i=1}^{M}\sum_{j=1}^{M}\frac{\hat{Y}_i\hat{Y}_j}{\pi_{Ii}\pi_{Ij}}\Delta_{Iij}\right) \tag{7.5}
$$

and

$$
V_B = \sum_{i=1}^{M}\mathrm{var}_p(\hat{Y}_i).
$$

Proof: The variance is decomposed in two parts:

$$
\mathrm{var}_p(\hat{Y}) = E_p\;\mathrm{var}_p(\hat{Y}|S_2) + \mathrm{var}_p\;E_p(\hat{Y}|S_2).
$$

The expectation of the conditional variance becomes

$$E_p \, \text{var}_p(\widehat{Y}|S_2) = E_p \, \text{var}_p\left(\sum_{i \in S_I} \frac{\widehat{Y}_i}{\pi_{Ii}} \middle| S_2\right).$$

From the invariance property,

$$\text{var}_p\left(\sum_{i \in S_I} \frac{\widehat{Y}_i}{\pi_{Ii}} \middle| S_2\right) = \sum_{i=1}^{M}\sum_{j=1}^{M} \frac{\widehat{Y}_i\widehat{Y}_j}{\pi_{Ii}\pi_{Ij}}\Delta_{Iij}.$$

Therefore,

$$E_p \, \text{var}_p(\widehat{Y}|S_2) = E_p\left(\sum_{i=1}^{M}\sum_{j=1}^{M} \frac{\widehat{Y}_i\widehat{Y}_j}{\pi_{Ii}\pi_{Ij}}\Delta_{Iij}\right).$$

The variance of the conditional expectation becomes

$$\text{var}_p \, E_p(\widehat{Y}|S_2) = \text{var}_p\left(\sum_{i=1}^{M} \widehat{Y}_i\right).$$

By the independence property,

$$\text{var}_p \, E_p(\widehat{Y}|S_2) = \text{var}_p\left(\sum_{i=1}^{M} \widehat{Y}_i\right) = \sum_{i=1}^{M} \text{var}_p(\widehat{Y}_i).$$

∎

In this second decomposition, we notice the term V_B corresponds to the variance of a stratified design. Even if the decompositions of Results 7.1 and 7.2 may seem different, these two variances are equal. In order to prove this, we must consider that

$$E_p\left(\widehat{Y}_i\widehat{Y}_j|S_I\right) = \begin{cases} \text{var}_p(\widehat{Y}_i) + Y_i^2 & \text{if } i = j \\ Y_iY_j & \text{if } i \neq j, \end{cases}$$

Then, we introduce this expectation in Expression (7.5). We obtain

$$V_A = E_p\left(\sum_{i=1}^{M}\sum_{j=1}^{M} \frac{\widehat{Y}_i\widehat{Y}_j}{\pi_{Ii}\pi_{Ij}}\Delta_{Iij}\right) = \sum_{i=1}^{M}\sum_{j=1}^{M} \frac{E_p(\widehat{Y}_i\widehat{Y}_j)}{\pi_{Ii}\pi_{Ij}}\Delta_{Iij}$$

$$= \sum_{i=1}^{M}\sum_{j=1}^{M} \frac{Y_iY_j}{\pi_{Ii}\pi_{Ij}}\Delta_{Iij} + \sum_{i=1}^{M} \frac{\text{var}_p(\widehat{Y}_i)}{\pi_{Ii}^2}\pi_{Ii}(1-\pi_{Iij})$$

$$= \sum_{i=1}^{M}\sum_{j=1}^{M} \frac{Y_iY_j}{\pi_{Ii}\pi_{Ij}}\Delta_{Iij} + \sum_{i=1}^{M} \text{var}_p(\widehat{Y}_i)\left(\frac{1}{\pi_{Ii}}-1\right). \tag{7.6}$$

By introducing Expression (7.6) in Result 7.2, we obtain Result 7.1. It is now possible to unbiasedly estimate this variance.

Result 7.3 *In a two-stage sampling design, if all inclusion probabilities are strictly positive, then*

$$v_1(\widehat{Y}) = v_A + v_B$$

is an unbiased estimator of $\mathrm{var}_p(\hat{Y})$, *where* v_A *is the variance term calculated at the primary unit level:*

$$v_A = \sum_{i \in S_I} \sum_{j \in S_I} \frac{\hat{Y}_i \hat{Y}_j}{\pi_{Ii} \pi_{Ij}} \frac{\Delta_{Iij}}{\pi_{Iij}},$$

(with $\pi_{Iii} = \pi_{Ii}$*) and* v_B *is the variance term calculated at the secondary unit level:*

$$v_B = \sum_{i \in S_I} \frac{v(\hat{Y}_i)}{\pi_{Ii}},$$

and

$$v(\hat{Y}_i) = \sum_{k \in S_i} \sum_{\ell \in S_i} \frac{y_k y_\ell}{\pi_{k|i} \pi_{\ell|i}} \frac{\Delta_{k\ell|i}}{\pi_{k\ell|i}},$$

with $\pi_{kk|i} = \pi_{k|i}$.

Proof: First,

$$E_p(v_A) = E_p E_p \left(\sum_{i \in S_I} \sum_{j \in S_I} \frac{\hat{Y}_i \hat{Y}_j}{\pi_{Ii} \pi_{Ij}} \frac{\Delta_{Iij}}{\pi_{Iij}} \middle| S_2 \right)$$

$$= E_p \left(\sum_{i=1}^{M} \sum_{j=1}^{M} \frac{\hat{Y}_i \hat{Y}_j}{\pi_{Ii} \pi_{Ij}} \frac{\Delta_{Iij}}{\pi_{Iij}} \right) = V_A.$$

Then,

$$E_p(v_B) = E_p E_p \left[\sum_{i \in S_I} \frac{v(\hat{Y}_i)}{\pi_{Ii}} \middle| S_2 \right] = E_p \left[\sum_{i=1}^{M} v(\hat{Y}_i) \right]$$

$$= \sum_{i=1}^{M} E_p[v(\hat{Y}_i)] = \sum_{i=1}^{M} \mathrm{var}_p(\hat{Y}_i) = V_B.$$ ∎

This result shows that v_A and v_B are unbiased estimators of the two terms in the reverse approach V_A and V_B. Conversely, v_A is not an unbiased estimator of V_α and v_B is not an unbiased estimator of V_β. The estimator v_A overestimates V_α. In practice, v_A is much larger than v_B.

In the case where we select the primary and secondary units with unequal probabilities, we prefer a more practical estimator coming from Expression (5.15):

$$v_2(\hat{Y}) = \sum_{i \in S_I} \frac{c_{Ii}}{\pi_{Ii}^2} \left(\hat{Y}_i - \hat{\overline{Y}}_i^* \right)^2 + \sum_{i \in S_I} \frac{1}{\pi_{Ii}} \sum_{k \in S_i} \frac{c_{k|i}}{\pi_{k|i}^2} \left(y_k - \hat{y}_k^* \right)^2,$$

where

$$\hat{\overline{Y}}_i^* = \pi_{Ii} \frac{\sum_{j \in S_I} c_{Ij} \hat{Y}_j / \pi_{Ij}}{\sum_{j \in S_I} c_{Ij}},$$

$$c_{Ii} = (1 - \pi_{Ii}) \frac{m}{m-1},$$

$$\hat{y}_k^* = \pi_{k|i} \frac{\sum_{k \in S_i} c_{k|i} y_k / \pi_{k|i}}{\sum_{k \in S_i} c_{k|i}},$$

$$c_{k|i} = (1 - \pi_{k|i}) \frac{n_i}{n_i - 1}.$$

7.2.3 Sampling with Equal Probability

A classical design consists of selecting the primary and secondary units using simple random sampling without replacement. The first-order inclusion probabilities for the first sample are

$$\pi_{Ii} = \frac{m}{M}, i = 1, \dots, M,$$

and

$$\pi_{Iij} = \frac{m(m - 1)}{M(M - 1)}, i = 1, \dots, M, j = 1, \dots, M, i \neq j.$$

Therefore, for the second sample, the size of the samples within the primary units is n_i. The inclusion probability for the whole design is

$$\pi_k = \frac{m n_i}{M N_i}.$$

The expansion estimator becomes

$$\hat{Y} = \frac{M}{m} \sum_{i \in S_I} \sum_{k \in S_i} \frac{N_i y_k}{n_i},$$

and its variance is

$$\text{var}_p(\hat{Y}) = V_\alpha + V_\beta = M^2 \left(1 - \frac{m}{M}\right) \frac{S_I^2}{m} + \frac{M}{m} \sum_{i=1}^{M} N_i^2 \left(1 - \frac{n_i}{N_i}\right) \frac{S_i^2}{n_i},$$

where

$$S_I^2 = \frac{1}{M - 1} \sum_{i=1}^{M} \left(Y_i - \frac{Y}{M}\right)^2$$

and

$$S_i^2 = \frac{1}{N_i - 1} \sum_{k \in U_i} (y_k - \overline{Y}_i)^2.$$

Finally, the variance estimator simplifies to

$$v(\hat{Y}) = v_A + v_B = M^2 \left(1 - \frac{m}{M}\right) \frac{s_I^2}{m} + \frac{M}{m} \sum_{i \in S_I} N_i^2 \left(1 - \frac{n_i}{N_i}\right) \frac{s_i^2}{n_i},$$

where

$$s_I^2 = \frac{1}{m - 1} \sum_{i \in S_I} \left(\hat{Y}_i - \frac{\hat{Y}}{M}\right)^2$$

and

$$s_i^2 = \frac{1}{n_i - 1} \sum_{k \in S_i} \left(y_k - \frac{\widehat{Y}_i}{N_i} \right)^2.$$

As in the general case, the two terms in $v(\widehat{Y})$ do not separately estimate the two terms of $\mathrm{var}_p(\widehat{Y})$, even though $v(\widehat{Y})$ is unbiased for $\mathrm{var}_p(\widehat{Y})$. In particular, s_i^2 overestimates S_I^2.

In order to conserve equal inclusion probabilities, we can take sample size s in the primary units proportional to their sizes:

$$n_i = \frac{n \, M \, N_i}{m \, N},$$

which obviously can have some rounding problems. We obtain the probability of selecting a secondary unit in the primary unit as

$$\pi_{k|i} = \frac{M}{m} \frac{n}{N}, k \in U_i.$$

Finally, the inclusion probabilities for the entire design are

$$\pi_k = \pi_{Ii} \pi_{k|i} = \frac{n}{N}, k \in U_i.$$

This design has important drawbacks. The sample size n_S is random and its mean is

$$E_p(n_S) = E_p \left(\sum_{i \in S_I} n_i \right) = E_p \left(\sum_{i \in S_I} \frac{n \, M \, N_i}{m \, N} \right) = \sum_{k \in U_I} \frac{n \, N_i}{N} = n.$$

Therefore, the survey cost is random. Furthermore, it is difficult to divvy up the work between interviewers because in each primary unit (usually geographical), the number of people to survey is different, since it depends on the size of the primary unit.

7.2.4 Self-weighting Two-stage Design

Self-weighting designs solve the two problems described in the previous method. In the first step, the primary units are selected with inclusion probabilities proportional to their size. The selection probabilities of the primary units are

$$\pi_{Ii} = \frac{N_i \, m}{N},$$

and are assumed to be less than 1. In the second step, we select secondary units using simple random sampling without replacement with fixed sample size $n_i = n_0$ (whatever the size of the primary unit), which gives

$$\pi_{k|i} = \frac{n_0}{N_i}.$$

Then, the first-order inclusion probabilities are

$$\pi_k = \pi_{Ii} \pi_{k|i} = \frac{N_i \, mn_0}{N \, N_i} = \frac{mn_0}{N}, k \in U_i.$$

Therefore, the inclusion probabilities are constants for all individuals in the population. The self-weighting two-stage design has an important advantage: it has fixed size. In fact, the sample size is always $n = mn_0$. This sampling method is often used in official statistics. The primary units are geographical units, and the secondary units individuals or households.

Even though the primary units are selected with unequal probabilities, the secondary units are selected with equal probabilities, and the expansion estimator of the total becomes

$$\hat{Y} = \frac{N}{n} \sum_{k \in S} y_k.$$

The variance is

$$\text{var}_p(\hat{Y}) = \sum_{i=1}^{M} \sum_{j=1}^{M} \frac{Y_i Y_j}{\pi_{Ii} \pi_{Ij}} \Delta_{Iij} + \frac{N}{n} \sum_{i \in U_I} (N_i - n_0) S_i^2,$$

where

$$S_i^2 = \frac{1}{N_i - 1} \sum_{k \in U_i} \left(y_k - \frac{Y_i}{N_i} \right)^2.$$

The variance estimator simplifies as well:

$$v_1(\hat{Y}) = \sum_{i \in S_I} \sum_{j \in S_I} \frac{\hat{Y}_i \hat{Y}_j}{\pi_{Ii} \pi_{Ij}} \frac{\Delta_{Iij}}{\pi_{Iij}} + \frac{N}{n} \sum_{i \in S_I} (N_i - n_0) s_i^2,$$

where

$$s_i^2 = \frac{1}{n_i - 1} \sum_{k \in S_i} \left(y_k - \frac{\hat{Y}_i}{N_i} \right)^2.$$

Basing ourselves on the estimator given in Expression (5.15), we can construct the following more practical estimator:

$$v_2(\hat{Y}) = \sum_{i \in S_I} \frac{c_{Ii}}{\pi_{Ii}^2} (\hat{Y}_i - \hat{Y}_i^*)^2 + \frac{N}{n} \sum_{i \in S_I} (N_i - n_0) s_i^2,$$

where

$$\hat{Y}_i^* = \pi_{Ii} \frac{\sum_{j \in S_I} c_{Ij} \hat{Y}_j / \pi_{Ij}}{\sum_{j \in S_I} c_{Ij}},$$

with $c_{Ii} = (1 - \pi_{Ii}) m / (m - 1)$.

7.3 Multi-stage Designs

In practice, two stages are often not enough to properly organize a survey. Surveys on individuals or households require a larger number of stages. For example, we first select municipalities. In the selected municipalities, we select blocks of dwellings, from which we select households and then finally select individuals. We can then survey sampling units for which we do not possess *a priori* any information. Sometimes we can use a different sampling design according to geographical areas, for example choosing different designs for large cities. The variance estimators become very complicated. However, there is a general principle to calculate them.

The expansion estimator can be easily obtained if we know the inclusion probabilities at each stage. Suppose we have M primary units and the first-stage sampling design consists of selecting m primary units with inclusion probabilities π_{Ii} for $i = 1, \dots, M$. Let S_I be the random sample of selected primary units. Suppose also that within each primary unit, we can calculate the expansion estimator $\widehat{Y}_i$ of total Y_i within primary unit U_i for the m selected primary units. The expansion estimator of the total is

$$\widehat{Y} = \sum_{i \in S_I} \frac{\widehat{Y}_i}{\pi_{Ii}}.$$

Using the exact same reasoning as for the two-stage designs, the variance of the expansion estimator is

$$\operatorname{var}_p(\widehat{Y}) = \sum_{i=1}^{M} \sum_{j=1}^{M} \frac{Y_i Y_j}{\pi_{Ii} \pi_{Ij}} \Delta_{Iij} + \sum_{i=1}^{M} \frac{\operatorname{var}_p(\widehat{Y}_i)}{\pi_{Ii}},$$

and can be unbiasedly estimated by

$$v_1(\widehat{Y}) = \sum_{i \in S_I} \sum_{j \in S_I} \frac{\widehat{Y}_i \widehat{Y}_j}{\pi_{Ii} \pi_{Ij}} \frac{\Delta_{Iij}}{\pi_{Iij}} + \sum_{i \in S_I} \frac{v(\widehat{Y}_i)}{\pi_{Ii}},$$

or in a more operational manner by

$$v_2(\widehat{Y}) = \sum_{i \in S_I} \frac{c_{Ii}}{\pi_{Ii}^2} (\widehat{Y}_i - \widehat{\widehat{Y}}_i^*)^2 + \sum_{i \in S_I} \frac{v(\widehat{Y}_i)}{\pi_{Ii}},$$

where

$$\widehat{Y}_i^* = \pi_{Ii} \frac{\sum_{j \in S} c_{Ij} Y_j / \pi_{Ij}}{\sum_{j \in S} c_{Ij}},$$

and where a way to define the c_{Ii} is

$$c_{Ii} = (1 - \pi_{Ii}) \frac{m}{m - 1}.$$

The importance of this formulation resides in its recursiveness. The expansion estimator and its variance estimator can be written as functions of expansion and variance estimators calculated at a lower stage. Indeed, we can write

$$\widehat{Y} = T(\widehat{Y}_i, \pi_{Ii}, i \in S_I)$$

and

$$v(\widehat{Y}) = Q(\widehat{Y}_i, v(\widehat{Y}_i), \pi_{Ii}, i \in S_I).$$

Estimator $v(\widehat{Y}_i)$ can also be written as such a function. By using recursion, we can construct a variance estimator for any stage of the multi-stage design.

7.4 Selecting Primary Units with Replacement

For multi-stage designs, the selection of primary units with replacement using a multinomial design is of particular interest because of the simplicity of the variance estimator. We consider m primary units selected using a replacement design with unequal selection probabilities p_{Ii} as described in Section 5.4. In each primary unit, any single or multi-stage design can be used. Since the primary units are selected with replacement, they can be selected several times. If a primary unit is selected several times, we select independent samples from this primary unit.

The random variable a_{Ii} represents the number of times primary unit U_i is selected. We then have $E_p(a_{Ii}) = mp_{Ii}$. We can refer to the following estimator:

$$\hat{Y} = \frac{1}{m} \sum_{i=1}^{M} \sum_{j=1}^{a_{Ii}} \frac{\hat{Y}_{i(j)}}{p_{Ii}},$$

where $\hat{Y}_{i(j)}$ is the jth estimator of the total Y_i in primary unit i. If a primary unit is not selected, we have

$$\sum_{j=1}^{0} \frac{\hat{Y}_i(j)}{p_{Ii}} = 0.$$

If $E_p(\hat{Y}_{i(j)}) = Y_i$ for all $\hat{Y}_{i(j)}$, then estimator $\hat{Y}$ is unbiased. Indeed, since

$$E_p(\hat{Y}|a_{I1}, \dots, a_{IM}) = \frac{1}{m} \sum_{i=1}^{M} \sum_{j=1}^{a_{Ii}} \frac{E_p(\hat{Y}_{i(j)}|a_{I1}, \dots, a_{IM})}{p_{Ii}}$$

$$= \frac{1}{m} \sum_{i=1}^{M} \sum_{j=1}^{a_{Ii}} \frac{Y_i}{p_{Ii}} = \frac{1}{m} \sum_{i=1}^{M} \frac{Y_i}{p_{Ii}} a_{Ii},$$

we have

$$E_p(\hat{Y}) = E_p E_p(\hat{Y}|a_{I1}, \dots, a_{IM}) = E_p \left(\frac{1}{m} \sum_{i=1}^{M} \frac{Y_i}{p_{Ii}} a_{Ii} \right) = Y.$$

The variance is calculated in the same way as in Result 7.1 by conditioning with respect to the primary units:

Result 7.4

$$\text{var}_p(\hat{Y}) = \frac{1}{m} \sum_{i=1}^{M} p_{Ii} \left(\frac{Y_i}{p_{Ii}} - Y \right)^2 + \frac{1}{m} \sum_{i=1}^{M} \frac{\text{var}_p(\hat{Y}_i)}{p_{Ii}}.$$

Proof: By conditioning on the selection of the primary units, we obtain the following decomposition:

$$\text{var}_p(\hat{Y}) = \text{var}_p E_p(\hat{Y}|a_{I1}, \dots, a_{IM}) + E_p \text{var}_p(\hat{Y}|a_{I1}, \dots, a_{IM}).$$

The variance of the conditional expectation is

$$\text{var}_p E_p(\hat{Y}|a_{I1}, \dots, a_{IM}) = \frac{1}{m} \sum_{i=1}^{M} p_{Ii} \left(\frac{Y_i}{p_{Ii}} - Y \right)^2.$$

and the conditional variance is

$$\text{var}_p(\widehat{Y}|a_{I1}, \dots, a_{IM}) = \frac{1}{m^2} \sum_{i=1}^{M} \sum_{j=1}^{a_{Ii}} \frac{\text{var}_p(\widehat{Y}_{i(j)})}{p_{Ii}}.$$

Since the $\text{var}_p(\widehat{Y}_{i(j)}), j = 1, \dots, a_{Ii}$ are all equal, we can write $\text{var}_p(\widehat{Y}_{i(j)}) = \text{var}_p(\widehat{Y}_i)$. Therefore,

$$\text{var}_p(\widehat{Y}|a_{I1}, \dots, a_{IM}) = \frac{1}{m^2} \sum_{i=1}^{M} \frac{\text{var}_p(\widehat{Y}_i)}{p_{Ii}^2} a_{Ii}$$

and

$$E_p\text{var}_p(\widehat{Y}|a_{I1}, \dots, a_{IM}) = \frac{1}{m^2} \sum_{i=1}^{M} \frac{\text{var}_p(\widehat{Y}_i)}{p_{Ii}^2} E_p(a_{Ii}) = \frac{1}{m} \sum_{i=1}^{M} \frac{\text{var}_p(\widehat{Y}_i)}{p_{Ii}}.$$

By summing both terms, we obtain

$$\text{var}_p(\widehat{Y}) = \frac{1}{m} \sum_{i=1}^{M} p_{Ii} \left(\frac{Y_i}{p_{Ii}} - Y \right)^2 + \frac{1}{m} \sum_{i=1}^{M} \frac{\text{var}_p(\widehat{Y}_i)}{p_{Ii}}.$$

∎

When the primary units are selected with replacement, the variance estimator contains only a single term.

Result 7.5 *The estimator*

$$v(\widehat{Y}) = \frac{1}{m(m-1)} \sum_{i=1}^{M} \sum_{j=1}^{a_{Ii}} \left(\frac{\widehat{Y}_{i(j)}}{p_{Ii}} - \widehat{Y} \right)^2$$

is unbiased for $\text{var}_p(\widehat{Y})$.

Proof: The estimator can be written as

$$v(\widehat{Y}) = \frac{1}{m(m-1)} \sum_{i=1}^{M} \sum_{j=1}^{a_{Ii}} \frac{\widehat{Y}_{i(j)}^2}{p_{Ii}^2} - \frac{\widehat{Y}^2}{m-1}.$$

Its conditional expectation is

$$E_p[v(\widehat{Y})|a_{I1}, \dots, a_{IM}]$$

$$= \frac{1}{m(m-1)} \sum_{i=1}^{M} \sum_{j=1}^{a_{Ii}} \frac{E_p(\widehat{Y}_{i(j)}^2|a_{I1}, \dots, a_{IM})}{p_{Ii}^2} - \frac{E_p(\widehat{Y}^2|a_{I1}, \dots, a_{IM})}{m-1}.$$

Again, the $E_p(\widehat{Y}_{i(j)}^2|a_{I1}, \dots, a_{IM}), j = 1, \dots, a_{Ii}$ are all equal and given the invariance property, we can write $E_p[\widehat{Y}_{i(j)}^2|a_{I1}, \dots, a_{IM}] = E_p(\widehat{Y}_i^2)$. Therefore,

$$E_p(v(\widehat{Y})|a_{I1}, \dots, a_{IM}) = \frac{1}{m(m-1)} \sum_{i=1}^{M} \frac{E_p(\widehat{Y}_{i(j)}^2)}{p_{Ii}^2} a_{Ii} - \frac{E_p(\widehat{Y}^2|a_{I1}, \dots, a_{IM})}{m-1}.$$

Finally, we can calculate the unconditional expectation:

$$E_p[v(\hat{Y})] = \frac{1}{(m-1)} \sum_{i=1}^{M} \frac{E_p(\hat{Y}_i^2)}{p_{Ii}} - \frac{E_p(\hat{Y}^2)}{m-1}$$

$$= \frac{1}{(m-1)} \sum_{i=1}^{M} \frac{\text{var}_p(\hat{Y}_i^2) + Y_i^2}{p_{Ii}} - \frac{\text{var}_p(\hat{Y}^2) + Y^2}{m-1}$$

$$= \frac{1}{m-1} \sum_{i=1}^{M} p_{Ii} \left(\frac{Y_i}{p_{Ii}} - Y \right)^2 + \frac{1}{(m-1)} \sum_{i=1}^{M} \frac{\text{var}_p(\hat{Y}_i^2)}{p_{Ii}} - \frac{\text{var}_p(\hat{Y}^2)}{m-1}.$$

By replacing $\text{var}_p(\hat{Y})$ by its value given in Result 7.4, we obtain

$$E_p[v(\hat{Y})] = \frac{1}{m-1} \sum_{i=1}^{M} p_{Ii} \left(\frac{Y_i}{p_{Ii}} - Y \right)^2 + \frac{1}{(m-1)} \sum_{i=1}^{M} \frac{\text{var}_p(\hat{Y}_i^2)}{p_{Ii}}$$

$$- \frac{1}{m(m-1)} \sum_{i=1}^{M} p_{Ii} \left(\frac{Y_i}{p_{Ii}} - Y \right)^2 - \frac{1}{m(m-1)} \sum_{i=1}^{M} \frac{\text{var}_p(\hat{Y}_i)}{p_{Ii}}$$

$$= \frac{1}{m} \sum_{i=1}^{M} p_{Ii} \left(\frac{Y_i}{p_{Ii}} - Y \right)^2 + \frac{1}{m} \sum_{i=1}^{M} \frac{\text{var}_p(\hat{Y}_i^2)}{p_{Ii}} = \text{var}_p(\hat{Y}). \quad \blacksquare$$

If the primary units are selected with replacement, the variance estimator is extremely simple, as it is not necessary to calculate the variance within the primary units. It is sufficient to know the estimated totals of each sampled primary unit. If the sample is selected using a without replacement design, the variance of the with replacement design is easy to calculate and will overestimate the variance by a small amount. This overestimation is negligible when the number of selected primary units is small with respect to the total number of primary units. For this reason, some statistical software uses the with replacement variance estimator by default (see Chauvet & Vallée, 2018).

7.5 Two-phase Designs

7.5.1 Design and Estimation

Two-phase sampling designs give a larger framework than two-stage surveys (for more on this topic, see Särndal & Swensson, 1987). This idea is illustrated in Figure 7.3. In the first phase, we select a sample given some design $p_I(S_a)$ of size n_a (eventually a two-stage design). In the second phase, we select a random sample S_b of size n_b given

Figure 7.3 Two-phase design: a sample S_b is selected in sample S_a.

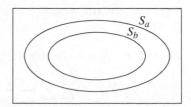

another sampling design within S_a using design $p(s_b|S_a) = \Pr(S_b = s_b|S_a)$. The second phase design can depend on the sample obtained in the first phase.

Example 7.1 We first use a cluster sampling design. Municipalities are selected and all individuals from these municipalities form sample S_a. We then select in S_a a stratified sample S_b of people where the strata are defined by age groups. In this case, the second-phase design depends on the first-phase design as in S_a the number of people in each age group is random.

Let

$$\pi_{ak} = \Pr(k \in S_a),$$

$$\pi_{ak\ell} = \Pr(k \text{ and } \ell \in S_a), k \neq \ell, \text{with } \pi_{akk} = \pi_{ak},$$

$$\Delta_{ak\ell} = \begin{cases} \pi_{ak\ell} - \pi_{ak}\pi_{a\ell}, & k \neq \ell \\ \pi_{ak}(1 - \pi_{ak}), & k = \ell, \end{cases}$$

$$\pi_{bk} = \Pr(k \in S_b|S_a),$$

$$\pi_{bk\ell} = \Pr(k \text{ and } \ell \in S_b|S_a), k \neq \ell, \text{with } \pi_{bkk} = \pi_{bk},$$

$$\Delta_{bk\ell} = \begin{cases} \pi_{bk\ell} - \pi_{bk}\pi_{b\ell}, & k \neq \ell \\ \pi_{bk}(1 - \pi_{bk}), & k = \ell. \end{cases}$$

In the following derivations, we must not forget that π_{bk}, $\pi_{bk\ell}$ and $\Delta_{bk\ell}$ are random variable functions of S_a.

The inclusion probability of unit k is

$$\pi_k = \pi_{ak}E_p(\pi_{bk}).$$

However, this probability will never be calculated. We will estimate the total by the two-phase estimator (2P):

$$\hat{Y}_{2P} = \sum_{k \in S_b} \frac{y_k}{\pi_{ak}\pi_{bk}},$$

which is not an expansion estimator, as we do not divide the y_k by the inclusion probabilities $\pi_{ak}E_p(\pi_{bk})$. However, it is unbiased. Indeed,

$$E_p(\hat{Y}_{2P}) = E_p E_p \left(\sum_{k \in S_b} \frac{y_k}{\pi_{ak}\pi_{bk}} \middle| S_a \right) = E_p \left(\sum_{k \in S_a} \frac{y_k}{\pi_{ak}} \right) = Y.$$

7.5.2 Variance and Variance Estimation

The variance and variance estimator have been calculated by Särndal & Swensson (1987).

Result 7.6 *In a two-phase design, the variance estimator is*

$$\text{var}_p(\hat{Y}_{2P}) = \sum_{k \in U}\sum_{\ell \in U} \frac{y_k y_\ell}{\pi_{ak}\pi_{a\ell}}\Delta_{ak\ell} + E_p\left(\sum_{k \in S_a}\sum_{\ell \in S_a} \frac{y_k y_\ell}{\pi_{ak}\pi_{bk}\pi_{a\ell}\pi_{b\ell}}\Delta_{bk\ell} \right),$$

and can be estimated by

$$
v(\hat{Y}_{2P}) = \sum_{k \in S_b} \sum_{\ell \in S_b} \frac{y_k y_\ell}{\pi_{ak} \pi_{a\ell}} \frac{\Delta_{ak\ell}}{\pi_{ak\ell} \pi_{bk\ell}} + \sum_{k \in S_b} \sum_{\ell \in S_b} \frac{y_k y_\ell}{\pi_{ak} \pi_{bk} \pi_{a\ell} \pi_{b\ell}} \frac{\Delta_{bk\ell}}{\pi_{bk\ell}}.
\tag{7.7}
$$

Proof: If, for this proof, we define

$$
\hat{Y}_a = E_p\left(\sum_{k \in S_b} \frac{y_k}{\pi_{ak} \pi_{bk}} \,\bigg|\, S_a \right) = \sum_{k \in S_a} \frac{y_k}{\pi_{ak}},
$$

then

$$
\begin{aligned}
\mathrm{var}_p(\hat{Y}_{2P}) &= \mathrm{var}_p \, E_p(\hat{Y}_{2P}|S_a) + E_p \, \mathrm{var}_p(\hat{Y}_{2P}|S_a) \\
&= \mathrm{var}_p(\hat{Y}_a) + E_p \, \mathrm{var}_p(\hat{Y}_{2P}|S_a) \\
&= \sum_{k \in U} \sum_{\ell \in U} \frac{y_k y_\ell}{\pi_{ak} \pi_{a\ell}} \Delta_{ak\ell} + E_p\left(\sum_{k \in S_a} \sum_{\ell \in S_a} \frac{y_k y_\ell}{\pi_{ak} \pi_{bk} \pi_{a\ell} \pi_{b\ell}} \Delta_{bk\ell} \right).
\end{aligned}
\tag{7.8}
$$

If S_a was known, we could estimate (7.8) by

$$
v_a(\hat{Y}_{2P}) = \sum_{k \in S_a} \sum_{\ell \in S_a} \frac{y_k y_\ell}{\pi_{ak} \pi_{a\ell}} \frac{\Delta_{ak\ell}}{\pi_{ak\ell}} + \sum_{k \in S_a} \sum_{\ell \in S_a} \frac{y_k y_\ell}{\pi_{ak} \pi_{bk} \pi_{a\ell} \pi_{b\ell}} \Delta_{bk\ell}.
$$

Finally, $v(\hat{Y}_{2P})$ can be unbiasedly estimated on S_b by (7.7). ∎

The main problem resides in the practical calculation of (7.7). Caron (1999) developed an explicit formulation in the case where the second-phase design uses Poisson sampling, simple random sampling without replacement, or stratified sampling. For example, for Poisson sampling in the second phase, the variance estimator can be simplified:

$$
\begin{aligned}
v(\hat{Y}_{2P}) &= \sum_{k \in S_b} \sum_{\ell \in S_b} \frac{y_k y_\ell}{\pi_{ak} \pi_{a\ell}} \frac{\Delta_{ak\ell}}{\pi_{ak\ell} \pi_{bk} \pi_{b\ell}} \\
&\quad - \sum_{k \in S_b} \frac{y_k^2}{\pi_{ak}^2} (1 - \pi_{ak}) \frac{1 - \pi_{bk}}{\pi_{bk}^2} + \sum_{k \in S_b} \frac{y_k^2}{\pi_{ak}^2} \frac{1 - \pi_{bk}}{\pi_{bk}^2} \\
&= \sum_{k \in S_b} \sum_{\ell \in S_b} \frac{y_k y_\ell}{\pi_{ak} \pi_{a\ell}} \frac{\Delta_{ak\ell}}{\pi_{ak\ell} \pi_{bk} \pi_{b\ell}} + \sum_{k \in S_b} \frac{y_k^2}{\pi_{ak}^2} \pi_{ak} \frac{1 - \pi_{bk}}{\pi_{bk}^2}.
\end{aligned}
\tag{7.9}
$$

7.6   Intersection of Two Independent Samples

Let S_A and S_B be two independent samples. We are interested in the intersection $S = S_A \cap S_B$ of these two samples as shown in Figure 7.4. This case does not include Example 7.1, but encapsulates two-stage designs (see Section 7.2). In fact, two-stage designs can be seen as the intersection of two independent samples. The first sample S_A is selected using cluster sampling, the second sample S_B using stratified sampling. The

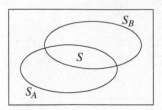

Figure 7.4 The sample S is the intersection of samples S_A and S_B.

clusters in S_A must correspond to the strata in S_B. The case where the two samples are independent is called strong invariance by Beaumont & Haziza (2016).

Let $\pi_{Ak}, \pi_{Ak\ell}, \pi_{Bk}, \pi_{Bk\ell}$ be the first- and second-order inclusion probabilities of samples S_A and S_B. Furthermore, we define

$$\Delta_{Ak\ell} = \begin{cases} \pi_{Ak\ell} - \pi_{Ak}\pi_{A\ell}, & k \neq \ell \\ \pi_{Ak}(1 - \pi_{Ak}), & k = \ell, \end{cases} \quad \text{and} \quad \Delta_{Bk\ell} = \begin{cases} \pi_{Bk\ell} - \pi_{Bk}\pi_{B\ell}, & k \neq \ell \\ \pi_{Bk}(1 - \pi_{Bk}), & k = \ell, \end{cases}$$

Since these two samples are independent we have

$$\pi_k = \pi_{Ak}\pi_{Bk}, \pi_{k\ell} = \pi_{Ak\ell}\pi_{Bk\ell}.$$

Furthermore,

$$\Delta_{k\ell} = \pi_{k\ell} - \pi_k\pi_\ell = \pi_{Ak\ell}\pi_{Bk\ell} - \pi_{Ak}\pi_{Bk}\pi_{A\ell}\pi_{B\ell},$$

which can be written in two different ways:

$$\Delta_{k\ell} = \pi_{Ak\ell}(\Delta_{Bk\ell} + \pi_{Bk}\pi_{B\ell}) - \pi_{Ak}\pi_{Bk}\pi_{A\ell}\pi_{B\ell} = \pi_{Ak\ell}\Delta_{Bk\ell} + \Delta_{Ak\ell}\pi_{Bk}\pi_{B\ell} \quad (7.10)$$

and

$$\Delta_{k\ell} = (\Delta_{Ak\ell} + \pi_{Ak}\pi_{A\ell})\pi_{Bk\ell} - \pi_{Ak}\pi_{Bk}\pi_{A\ell}\pi_{B\ell} = \pi_{Bk\ell}\Delta_{Ak\ell} + \Delta_{Bk\ell}\pi_{Ak}\pi_{A\ell}. \quad (7.11)$$

We define the expansion estimator:

$$\hat{Y} = \sum_{k \in S} \frac{y_k}{\pi_{Ak}\pi_{Bk}}.$$

Using Expressions (7.10) and (7.11), we give two expressions of variance that correspond to the two ways of conditioning:

$$\text{var}_p(\hat{Y}) = E_p\text{var}_p(\hat{Y}|S_A) + \text{var}_pE_p(\hat{Y}|S_A)$$

$$= \sum_{k \in U}\sum_{\ell \in U} \frac{y_k y_\ell}{\pi_{Ak}\pi_{Bk}\pi_{A\ell}\pi_{B\ell}}\pi_{Ak\ell}\Delta_{Bk\ell}$$

$$+ \sum_{k \in U}\sum_{\ell \in U} \frac{y_k y_\ell}{\pi_{Ak}\pi_{Bk}\pi_{A\ell}\pi_{B\ell}}\Delta_{Ak\ell}\pi_{Bk}\pi_{B\ell}$$

and

$$\text{var}_p(\hat{Y}) = E_p\text{var}_p(\hat{Y}|S_B) + \text{var}_pE_p(\hat{Y}|S_B)$$

$$= \sum_{k \in U}\sum_{\ell \in U} \frac{y_k y_\ell}{\pi_{Ak}\pi_{Bk}\pi_{A\ell}\pi_{B\ell}}\pi_{Bk\ell}\Delta_{Ak\ell}$$

$$+ \sum_{k \in U}\sum_{\ell \in U} \frac{y_k y_\ell}{\pi_{Ak}\pi_{Bk}\pi_{A\ell}\pi_{B\ell}}\Delta_{Bk\ell}\pi_{Ak}\pi_{A\ell}.$$

We can then construct two variance estimators:

$$v_1(\widehat{Y}) = \sum_{k \in U} \sum_{\ell \in U} \frac{y_k y_\ell}{\pi_{Ak} \pi_{Bk} \pi_{A\ell} \pi_{B\ell}} \frac{\Delta_{Bk\ell}}{\pi_{Bk\ell}} + \sum_{k \in U} \sum_{\ell \in U} \frac{y_k y_\ell}{\pi_{Ak} \pi_{A\ell}} \frac{\Delta_{Ak\ell}}{\pi_{Ak\ell} \pi_{Bk\ell}}$$

and

$$v_2(\widehat{Y}) = \sum_{k \in S} \sum_{\ell \in S} \frac{y_k y_\ell}{\pi_{Ak} \pi_{Bk} \pi_{A\ell} \pi_{B\ell}} \frac{\Delta_{Ak\ell}}{\pi_{Ak\ell}} + \sum_{k \in U} \sum_{\ell \in U} \frac{y_k y_\ell}{\pi_{Bk} \pi_{B\ell}} \frac{\Delta_{Bk\ell}}{\pi_{Ak\ell} \pi_{Bk\ell}}.$$

Tillé & Vallée (2017b) discussed the importance of these two estimators and suggested using the conditioning with respect to the design that has the simplest second-order inclusion probabilities. For example, if S_B is selected using Poisson sampling, then it is easier to use $v_2(\widehat{Y})$, which becomes

$$v_2(\widehat{Y}) = \sum_{k \in S} \sum_{\ell \in S} \frac{y_k y_\ell}{\pi_{Ak} \pi_{Bk} \pi_{A\ell} \pi_{B\ell}} \frac{\Delta_{Ak\ell}}{\pi_{Ak\ell}} + \sum_{k \in U} \frac{y_k^2 (1 - \pi_{Bk})}{\pi_{Ak} \pi_{Bk}^2}.$$

The case where S_B uses Poisson sampling is often used to model total nonresponse. Nonresponse is then considered as a second-phase sampling design. In this case, $v_1(\widehat{Y})$ is called the two-phase approach estimator and $v_2(\widehat{Y})$ the reverse approach estimator (see Fay, 1991; Shao & Steel, 1999).

Another application of this result is the two-stage design. In order to estimate the variance, we can condition with respect to the first- or second-stage sampling design. Result 7.3, page 151, shows that it is better to condition with respect to the second stage to estimate the variance.

Exercises

7.1 Consider a survey with a cluster design with equal probabilities. Suppose all clusters have the same size. Give an expression of the variance of the estimator of the mean as a function of the inter-cluster variance.

7.2 The objective is to estimate the mean household income. In a city neighbourhood composed of 60 residential blocks, we select three blocks with equal probabilities and without replacement. In addition, we know that 5 000 households are found in this neighbourhood. The result is given in Table 7.1.

Table 7.1 Block number, number of households, and total household income.

Block number	Number of households in the block	Total household income in the block
1	120	1 500
2	100	2 000
3	80	2 100

1. Estimate the mean household income in the neighbourhood and the total household income of the neighbourhood using the expansion estimator.
2. Estimate the variance of the expansion estimator of the mean.
3. Estimate the mean household income of the neighbourhood using the Hájek estimator.

7.3 Consider the population $\{1, 2, 3, 4, 5, 6, 7, 8, 9\}$ and the following sampling design:

$$p(\{1,2\}) = 1/6, \quad p(\{1,3\}) = 1/6, \quad p(\{2,3\}) = 1/6,$$
$$p(\{4,5\}) = 1/12, \ p(\{4,6\}) = 1/12, \ p(\{5,6\}) = 1/12,$$
$$p(\{7,8\}) = 1/12, \ p(\{7,9\}) = 1/12, \ p(\{8,9\}) = 1/12.$$

1. Give the first-order inclusion probabilities.
2. Is this design simple, stratified, cluster, two-stage or none of these? Justify your answer.

8

Other Topics on Sampling

8.1 Spatial Sampling

8.1.1 The Problem

Spatial data is often autocorrelated. If two nearby points are selected in a sample, we will probably obtain very similar measurements. Therefore, more information is collected in the sample if the data points are spread out in space. When we select a sample in space, it is useful to spread the observations. A group of methods have been developed to obtain points that are spread out in space while controlling inclusion probabilities.

Grafström & Lundström (2013) even argue for the use of a spread out sample for non-spatial data with distances calculated on auxiliary variables such as business income or the number of workers in a company. The spreading in the space of variables creates a kind of multivariate stratification.

There exist some simple techniques to spread points in space. The most obvious one, used in environmental monitoring, consists of selecting a systematic sample. An alternative to systematic sampling is stratification. The space is split in strata and a small number of units is selected in each stratum. In the example presented in Figure 8.1, 64 units are selected in a grid of $40 \times 40 = 1\,600$ points using a systematic design and a stratified design with a single unit per stratum.

However, these two methods cannot be applied in all situations. For example, if the units are dispersed irregularly in the space, it is not possible to do a systematic sampling. It is also difficult to construct strata of the same size. Similarly, systematic sampling with unequal probabilities is not generalizable in two dimensions. Therefore, it is necessary to consider other methods for more general cases.

8.1.2 Generalized Random Tessellation Stratified Sampling

The *Generalized Random Tessellation Stratified* (GRTS) method was proposed by Stevens Jr. & Olsen (2004). It allows a sample spread in space to be selected with equal or unequal inclusion probabilities (see also Stevens Jr. & Olsen, 1999, 2003; Theobald et al., 2007; McDonald, 2016; Pebesma & Bivand, 2005).

The objective of the method consists of sorting points given some order (in one dimension) that satisfies spatial proximity in two dimensions. To achieve this, a recursive quadrant function is used as shown in Figure 8.2. In this example, the space is divided

Sampling and Estimation from Finite Populations, First Edition. Yves Tillé.
© 2020 John Wiley & Sons Ltd. Published 2020 by John Wiley & Sons Ltd.

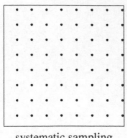

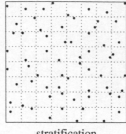

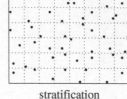

systematic sampling   stratification

Figure 8.1 In a 40×40 grid, a systematic sample and a stratified sample with one unit per stratum are selected.

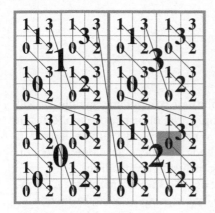

Figure 8.2 Recursive quadrant function used for the GRTS method with three subdivisions.

into four squares. Each square is further divided by four squares and so on, until there is at most one unit per square. Therefore, each square has a label. For example, the grey cell in Figure 8.2 has label $(2, 3, 0)$.

Stevens Jr. & Olsen (2004) then performed a random permutation inside each square at all levels. Figure 8.3 shows four examples of permutations inside the squares.

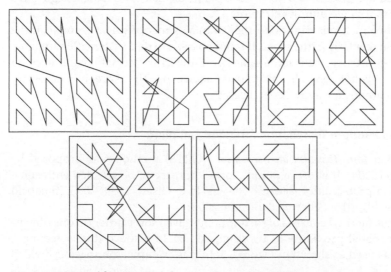

Figure 8.3 Original function with four random permutations.

After randomly selecting a permutation, it is possible to order the squares and consequently the units. There is at most one unit per square. Two units that follow each other on one of these curves are most often close to each other. Then, the sample is selected by using a systematic sampling on the ordered units (see Section 3.10). The sample is then relatively well spaced out.

8.1.3   Using the Traveling Salesman Method

Dickson & Tillé (2016) proposed simply using the shortest path between points. This calculation is known as the "traveling salesman problem". The traveling salesman must pass through a set of cities and come back to his original city while minimizing the traveled distance. The traveling salesman is a hard problem for which there is no algorithm to find the optimal solution without enumerating all possible paths. In fact, the number of possible paths is equal to $(V-1)!/2$, where V is the number of cities. Therefore, it is impossible to enumerate all paths when V is large. However, there exist many algorithms to obtain a local minimum. When a short path is identified, a systematic sampling design can then be applied on the sorted units using the method presented in Section 3.10.

8.1.4   The Local Pivotal Method

An equally simple approach was proposed by Grafström et al. (2012). They proposed using the pivotal method presented in Section 5.12.4 to perform a spatial sampling. At each stage, two units are selected which are very close to each other. Then, the pivotal method is applied on these two units. This method is called the *local pivotal method*. If the probability of one of these two units is increased, the probability of the other is decreased and vice versa. This induces a repulsion of nearby units and the resulting sample is then well spread out.

Grafström et al. (2012) proposed many variants of this method that give very similar results. These variants only differ in how to select two nearby units in the population and are implemented in the `BalancedSampling` R package (Grafström & Lisic, 2016).

8.1.5   The Local Cube Method

The local cube method, proposed by Grafström & Tillé (2013), is an extension of the local pivotal method. This algorithm allows us to obtain a sample that is both spread out geographically and balanced on auxiliary variables in the sense of Definition 6.1, page 121. This method, called the *local cube method*, consists of running the flight phase of the cube method (see Section 6.6.5, page 130) on a subset of $J+1$ neighbouring units, where J is the number of auxiliary variables on which we want to balance the sample. After this stage, the inclusion probabilities are updated such that one of the $J+1$ units has its inclusion probability updated to 0 or 1 and the balancing equations are still satisfied.

In this group of J units, when a unit is selected it decreases the inclusion probabilities of the other J other units in the group. When a unit is definitely excluded from the sample, it increases the inclusion probabilities of the other J units of the group. Consequently, this induces a repulsion in neighbouring units, which spreads out the sample.

Figure 8.4 contains an example of a sample of 64 points in a grid of $40 \times 40 = 1\,600$ points using simple random sampling, the local pivotal method, and the local cube

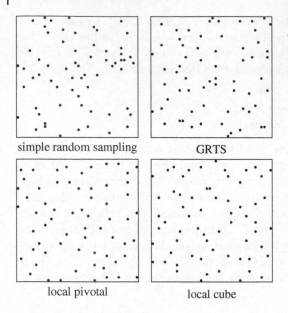

Figure 8.4 Samples of 64 points in a grid of $40 \times 40 = 1\,600$ points using simple designs, GRTS, the local pivotal method, and the local cube method.

simple random sampling   GRTS

local pivotal   local cube

method. The samples were selected using package `BalancedSampling` and `SDraw` of R software (see Grafström & Lisic, 2016; McDonald, 2016).

8.1.6 Measures of Spread

A Voronoï polygon centered on a point k is the set of points that are closer to k than all other points. Figure 8.5 shows the Voronoï polygons around selected points in the six samples presented in Figures 8.1 and 8.4. For this example, the area of Voronoï polygons is less dispersed for well-spread samples.

From this idea, Stevens Jr. & Olsen (2004) proposed using the Voronoï polygons to measure the spread of the sample. For each selected point k in the sample, one calculates v_k the sum of inclusion probabilities of all the points in the population that are closer to k then all other points in the sample. Therefore, the measure v_k is the sum of the inclusion probabilities of the units found in the Voronoï polygons centered on selected unit k. If the sample has fixed size n,

$$\frac{1}{n}\sum_{k\in U} v_k a_{sk} = \frac{1}{n}\sum_{k\in U} \pi_k = 1.$$

To measure the spread of the sample $\mathbf{a}_s$, one simply calculates the variance of the v_k:

$$B(\mathbf{a}_s) = \frac{1}{n}\sum_{k\in U} a_{sk}(v_k - 1)^2.$$

More recently, Tillé et al. (2018) proposed using a measure based on the Moran (1950) index of spatial autocorrelation. Tillé et al. (2018) modified the Moran index in order to make it strictly between −1 and 1. It can then be interpreted as a correlation coefficient. Consider the vector of indicators in sample $\mathbf{a}_s$. We calculate its correlation with the vector of local means. The local mean of unit k is the mean of the $1/\pi_k - 1$ closest values

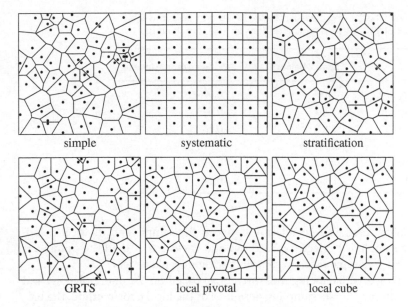

Figure 8.5 Sample of 64 points in a grid of $40 \times 40 = 1\,600$ points and Voronoï polygons. Applications to simple, systematic, and stratified designs, the local pivotal method, and the local cube method.

of k. More precisely, we first construct an $N \times N$ contiguity matrix $\mathbf{W} = (w_{k\ell})$. For each unit k, we calculate $q_k = 1/\pi_k - 1$. Then, we define

$$
w_{k\ell} = \begin{cases}
0 & \text{if } \ell = k \\
1 & \text{if } \ell \text{ belongs to the } \lfloor q_k \rfloor \text{ closest values of } k \\
q_k - \lfloor q_k \rfloor & \text{if } \ell \text{ is the } \lfloor q_k \rfloor \text{th closest neighbour of } k \\
0 & \text{otherwise.}
\end{cases}
$$

Then, the vector containing N times the weighted mean is

$$
\bar{\mathbf{a}}_w = \mathbf{1} \frac{\mathbf{a}^\top \mathbf{W} \mathbf{1}}{\mathbf{1}^\top \mathbf{W} \mathbf{1}},
$$

where $\mathbf{1}$ is a vector of N ones. Finally, we define

$$
I_B = \frac{(\mathbf{a} - \bar{\mathbf{a}}_w)^\top \mathbf{W} (\mathbf{a} - \bar{\mathbf{a}}_w)}{\sqrt{(\mathbf{a} - \bar{\mathbf{a}}_w)^\top \mathbf{D}(\mathbf{a} - \bar{\mathbf{a}}_w) \ (\mathbf{a} - \bar{\mathbf{a}}_w)^\top \mathbf{B}(\mathbf{a} - \bar{\mathbf{a}}_w)}},
$$

where $\mathbf{D}$ is a diagonal matrix containing the

$$
w_{k.} = \sum_{\ell \in U} w_{k\ell}
$$

on its diagonal,

$$
\mathbf{A} = \mathbf{D}^{-1} \mathbf{W} - \frac{\mathbf{1}\mathbf{1}^\top \mathbf{W}}{\mathbf{1}^\top \mathbf{W} \mathbf{1}}
$$

and

$$
\mathbf{B} = \mathbf{A} \mathbf{D} \mathbf{A}^\top = \mathbf{W}^\top \mathbf{D}^{-1} \mathbf{W} - \frac{\mathbf{W}^\top \mathbf{1}\mathbf{1}^\top \mathbf{W}}{\mathbf{1}^\top \mathbf{W} \mathbf{1}}.
$$

Table 8.1 Means of spatial balancing measures based on Voronoï polygons $B(\mathbf{a}_s)$ and modified Moran indices I_B for six sampling designs on 1000 simulations.

Plan	Spatial balance	Modified Moran index I_B
Simple random sampling	0.297	−0.006
Systematic sampling	0.041	−0.649
Stratification	0.063	−0.167
GRTS	0.063	−0.218
Local pivotal method	0.058	−0.210
Local cube method	0.058	−0.210

Index I_B can be interpreted as a correlation between the sample and its local mean. It equals −1 if the sample is perfectly spread out and is close to 1 if all the points are clumped. It is near 0 for simple random sampling without replacement.

Table 8.1 contains spatial balancing measures $B(\mathbf{a}_s)$ based on Voronoï polygons and modified Moran indices I_B for six sampling designs. For the local cube method, the balancing variables are the coordinates of the points and the squares of these coordinates. The table confirms that the systematic design is the most spread out. The local pivotal method spreads as well as the GRTS method.

8.2 Coordination in Repeated Surveys

8.2.1 The Problem

The question of coordination is of particular importance for business surveys. Negative coordination can be differentiated from positive coordination. With negative coordination, the number of common units between successive samples through time in a population is minimized, while with positive coordination, this number is maximized. The first articles on coordination were written by Patterson (1950) and Keyfitz (1951), but these methods are limited to two successive samples and small sample sizes.

Brewer et al. (1972) proposed using Permanent Random Numbers (PRNs) to coordinate Poisson designs. PRNs are outcomes of random variables that are assigned to the units on the first wave and that are then used for all samples. PRNs can also be used to perform the order sampling presented in Section 5.11, page 100 (see Rosén, 1997a,b). Many statistical institutes use methods based on PRNs. Ohlsson (1995) gives a list of the methods used by various countries.

Coordination in stratified samples is a complex problem. The main reason is that, through time, units often change from one stratum to another. Many methods have been developed to control the coverage between stratified samples selected at different occasions. These methods are the Kish and Scott method (Kish & Scott, 1971), the Cotton and Hesse method (Cotton & Hesse, 1992), the Netherlands method (EDS) described in De Ree (1983), Van Huis et al. (1994a,b), and Koeijers & Willeboordse (1995), the method of Rivière (1998, 1999, 2001a,b), and the family of methods presented in Nedyalkova et al. (2008). In Switzerland, at the Federal Statistical Office, the coordination method is based on Poisson designs. This method is described in Qualité (2009) and Graf & Qualité (2014).

8.2.2 Population, Sample, and Sample Design

In coordination problems, the population evolves through time. Suppose that we are interested in a population at time $t = 1, 2, \ldots, T-1, T$. Let U^t be the population at time t. The set $U^t \setminus U^{t-1}$ contains births between times $t-1$ and t. The set $U^{t-1} \setminus U^t$ contains deaths between times $t-1$ and t. The population U contains all the units from time 1 to T. Therefore,

$$U = \bigcup_{t=1}^{T} U^t.$$

In wave t, a sample without replacement is a subset s^t of population U^t. Since $U^t \subset U$, a sample s_t of U_t is also a subset of U. The joint sample design $p(s^1, \ldots, s^t, \ldots, s^T)$ is the probability distribution for all waves. The random samples $S^1, \ldots, S^t, \ldots, S^T$ are defined in the following way:

$$\Pr(S^1 = s^1, \ldots, S^t = s^t, \ldots, S^T = s^T) = p(s^1, \ldots, s^t, \ldots, s^T).$$

The size of S^t is denoted by n^t.

From the joint sample design, we can deduce the marginal surveys for a given time t:

$$\sum_{s^1, \ldots, s^{t-1}, s^{t+1}, \ldots, s^T} p(s^1, \ldots, s^t, \ldots, s^T) = p_t(s^t).$$

At wave t, the first- and second-order inclusion probabilities are denoted, respectively, by

$$\pi_k^t = \Pr(k \in S^t) \quad \text{and} \quad \pi_{k\ell}^t = \Pr(k \in S^t \text{ and } \ell \in S^t),$$

where $k, \ell \in U^t, t = 1, \ldots, T$. The longitudinal inclusion probability, for waves t and u, is

$$\pi_k^{tu} = \Pr(k \in S^t \text{ and } k \in S^u), k \in U^t \cap U^u, t, u = 1, \ldots, T.$$

Finally, the joint longitudinal probability has the form

$$\pi_{k\ell}^{tu} = \Pr(k \in S^t \text{ and } \ell \in S^u), k, \ell \in U^t \cup U^u, t, u = 1, \ldots, T.$$

This probability is not symmetric. In fact, $\pi_{k\ell}^{tu} \neq \pi_{\ell k}^{tu}$ and $\pi_{k\ell}^{tu} \neq \pi_{\ell k}^{ut}$.

The following result gives bounds for the longitudinal inclusion probabilities.

Result 8.1 *For waves t and u, we have*

$$\max(0, \pi_k^t + \pi_k^u - 1) \leq \pi_k^{tu} \leq \min(\pi_k^t, \pi_k^u).$$

Proof: By definition, we have

$$\pi_k^{tu} = \Pr(k \in S^t \text{ and } k \in S^u) \leq \min[\Pr(k \in S^t), \Pr(k \in S^u)] = \min(\pi_k^t, \pi_k^u).$$

In addition,

$$\pi_k^t - \pi_k^{tu} = \Pr(k \in S^t) - \Pr(k \in S^t \text{ and } k \in S^u) = \Pr(k \in S^t \text{ and } k \notin S^u)$$
$$\leq \min[\Pr(k \in S^t), \Pr(k \notin S^u)] = \min(\pi_k^t, 1 - \pi_k^u).$$

Hence,

$$\pi_k^{tu} \geq \pi_k^t - \min(\pi_k^t, 1 - \pi_k^u) = \max(0, \pi_k^t + \pi_k^u - 1). \qquad \blacksquare$$

8.2.3 Sample Coordination and Response Burden

Definition 8.1 *The overlap is the number of common units at times t and u:*

$$n^{tu} = \#(S^t \cap S^u).$$

Overlap can be random. Its expectation equals

$$E_p(n^{tu}) = E_p \left[\sum_{k \in U} \mathbb{1}(k \in S^t \text{ and } k \in S^u) \right]$$

$$= \sum_{k \in U} \Pr(k \in S^t \text{ and } k \in S^u) = \sum_{k \in U} \pi_k^{tu}.$$

Definition 8.2 *The overlap rate is defined by*

$$v^{tu} = \frac{2n^{tu}}{n^t + n^u}.$$

If n^t and n^u are not random, then the expectation of the overlap rate is

$$\tau^{tu} = \frac{2E_p(n^{tu})}{n^t + n^u}.$$

Let ALB $= \sum_{k \in U} \max(0, \pi_k^t + \pi_k^u - 1)$ be the *Absolute Lower Bound* (ALB) and AUB $= \sum_{k \in U} \min(\pi_k^t, \pi_k^u)$ be the *Absolute Upper Bound* (AUB) (see Matei & Tillé, 2005b). Then, from Result 8.1, we can directly deduce the bounds for the expectation of the overlap

$$\text{ALB} \leq E_p(n^{tu}) \leq \text{AUB}.$$

Except in some very special cases, like simple random sampling, the bounds ALB and AUB cannot be reached.

If, in waves 1 and 2, two samples are independently selected without coordination, then, for all $k \in U$,

$$\pi_k^1 \pi_k^2 = \pi_k^{12}.$$

In the case of positive coordination, the longitudinal inclusion probabilities must satisfy conditions

$$\pi_k^1 \pi_k^2 \leq \pi_k^{12} \leq \min(\pi_k^1, \pi_k^2).$$

In the case of negative coordination, for all $k \in U$, the inclusion probability must satisfy conditions

$$\max(0, \pi_k^1 + \pi_k^2 - 1) \leq \pi_k^{12} \leq \pi_k^1 \pi_k^2.$$

In this last case, the longitudinal inclusion probability is only zero if

$$\pi_k^1 + \pi_k^2 \leq 1.$$

The *response burden* of a survey is generally quantified in terms of the required time to complete the questionnaire. However, other aspects of response burden exist, for example how difficult it is to provide information or how sensitive the question is to the respondent. Consequently, response burden can vary from one survey to another. At time t, a survey has a burden denoted by b^t, which can be proportional to the time required to complete the form or can simply be equal to one. After wave T, the total burden of unit k is defined as the sum of the loads from all the surveys where unit k was included:

$$c_k^T = \sum_{t=1}^{T} b^t \mathbb{1}(k \in S^t).$$

The cumulative burden from wave m to wave T is defined as

$$c_k^{m,T} = \sum_{t=m}^{T} b^t \mathbb{1}(k \in S^t).$$

The quality of a coordination procedure can be measured by using the following four criteria:

1. the procedure provides a controllable overlap rate
2. the sample design is satisfied for each sample
3. for each unit, a fixed period out of a sample is satisfied
4. the procedure is simple to implement.

8.2.4 Poisson Method with Permanent Random Numbers

The use of PRNs was proposed by Brewer et al. (1972) to coordinate Poisson designs. The idea is as clever as it is simple. To each statistical unit we attach the outcome of a uniform random variable on [0,1] denoted by u_k^1 that the unit keeps during its whole existence. These random numbers can be modified from one wave to another, but the modification is deterministic and not random.

To create negatively coordinated Poisson samples with inclusion probabilities π_k^t, we can proceed in the following way:

- In the first wave, if $u_k^1 < \pi_k^1$, then unit k is selected in S_1. The PRNs are then modified by calculating

$$u_k^2 = (u_k^1 - \pi_k^1) \mod 1.$$

  The random variable u_k^2 always has a uniform distribution on $[0, 1]$.
- At the second wave, if $u_k^2 < \pi_k^2$, then unit k is selected in S_2. The PRNs are again modified by calculating

$$u_k^3 = (u_k^2 - \pi_k^2) \mod 1.$$

- At wave t, if $u_k^t < \pi_k^t$, then unit k is selected in S_t. The PRNs are modified by calculating

$$u_k^{t+1} = (u_k^t - \pi_k^t) \mod 1.$$

Using this procedure, at each wave, the samples are selected using a Poisson design. The negative coordination is optimal. If a unit dies, it is just erased from the survey frame. When a unit is birthed, it is added to the survey frame with a permanent random number. The main drawback of this method is the Poisson design. The sample size is random (see Section 5.8, page 92).

Qualité (2009) generalized this method to positively or negatively coordinate a sample with respect to a set of waves given a priority order. This method was implemented at the Swiss Federal Statistical Office for sample selection in business surveys (see Section 8.2.9).

Another variant of this method, proposed by Rosén (1997a,b) and Ohlsson (1990, 1998), consists of applying to the PRN the ordered sampling method described in Section 5.11, page 100. This method allows a fixed sample size to be obtained, but the target inclusion probabilities are not exactly satisfied.

8.2.5 Kish and Scott Method for Stratified Samples

Kish & Scott (1971) proposed a substitution method to coordinate stratified samples that allows changes in the strata definitions. Even though they introduced this method as a way to do positive coordination, presented in Algorithm 10, the method also makes a negative coordination. At wave 1 and 2, the definition of the strata can change.

To be able to present this algorithm in a rigorous way, we must formalize the notation. Suppose there are no births or deaths in the population. Suppose that the design is stratified in wave 1 with H strata $U_1^1, \dots, U_h^1, \dots, U_H^1$, and in wave 2 with G strata $U_1^2, \dots, U_g^2, \dots, U_G^2$ as follows:

$$U = \bigcup_{h=1}^{H} U_h^1 = \bigcup_{g=1}^{G} U_g^2.$$

Let N_h^1 be the size of U_h^1, N_g^2 the size of U_g^2, and N_{hg}^{12} the size of $U_{hg}^{12} = U_h^1 \cap U_g^2$. Suppose two independent samples s^1 and s^2 are selected at waves 1 and 2. The following notation is used:

- s_g^i is the set of units of the stratum U_g^2 that are selected in s^i, for $i = 1, 2$, with $n_g^i = \#(s_g^i)$
- s_{hg}^i is the set of units of $U_h^1 \cap U_g^2$ that are selected in s^i, for $i = 1, 2$, with $n_{hg}^i = \#(s_{hg}^i)$
- $s_{hg}^{12} = s_{hg}^1 \cap s_{hg}^2$, with $n_{hg}^{12} = \#(s_{hg}^{12})$.

The Kish and Scott method is described in Algorithm 10. It can be shown that Algorithm 10 selects a stratified design at waves 1 and 2. However, coordinating more than two sample becomes very difficult.

8.2.6 The Cotton and Hesse Method

The Cotton and Hesse method from the National Institute of Statistics and Economic Studies (Insee) in France is described in Cotton & Hesse (1992). This method works when the strata change over time and can be used to obtain a negative coordination. It works in the following way: each unit receives a PRN u_k using a $U[0, 1]$ uniform distribution. At the first wave, the sample is defined, in each stratum, as the set of units having the smallest permanent random number. After sample selection, the PRN are permuted in each stratum in such a way that the units selected in the first wave receive the largest PRN and the unselected units receive the smallest PRN. In the two subsets of

---

Algorithm 10 Positive coordination using the Kish and Scott method

---

In wave 1, select a sample using a stratified design in U, denoted by s^1.
In wave 2, select a sample using a stratified design in U, denoted by s^2.
For each intersection of strata U_{hg}^{12}:

- If $n_{hg}^2 - n_{hg}^{12} \geq n_{hg}^1 - n_{hg}^{12}$, then
  replace $n_{hg}^1 - n_{hg}^{12}$ units of $s_{hg}^2 \backslash s_{hg}^{12}$ with units of $s_{hg}^1 \backslash s_{hg}^{12}$ using simple random sampling without replacement of fixed size.
- Otherwise,
  replace $n_{hg}^2 - n_{hg}^{12}$ units of $s_{hg}^2 \backslash s_{hg}^{12}$ with units of $s_{hg}^1 \backslash s_{hg}^{12}$ using simple random sampling without replacement of fixed size.

---

selected and unselected units, the order of permuted PRN must not change. Then, the first procedure is applied in the following waves. Dead units lose their PRN while births simply receive a new PRN.

The main advantage of this method is that the strata can change in time. This method is correct because after permutation the PRN are still independent uniform random numbers. Therefore, the method is very simple to apply. Another advantage of this method is that only the permuted PRN must be conserved from one wave to another. The drawback of this method is that it allows only one type of coordination. Once it is decided to make a negative coordination between two waves, it is not possible to do a positive coordination with another sample. In addition, the order of surveys is determined once and for all. Therefore, it is not possible anymore to prioritize coordination with an old survey.

A simple example, developed in Nedyalkova et al. (2006, 2008), shows that the ALB is not reached by the Cotton and Hesse method or the Kish and Scott method. However, the Cotton and Hesse method gives a slightly better solution than the Kish and Scott method. This result has also been confirmed by a simulation study. Consequently, it is preferable to use the Cotton and Hesse method over the Kish and Scott method, as the Kish and Scott method does not allow more then two waves to be coordinated.

8.2.7 The Rivière Method

This method, based on the use of microstrata, has been proposed in numerous publications (Rivière, 1998, 1999, 2001a,b). Two software applications have been developed: SALOMON in 1998 (see Mészáros, 1999) and MICROSTRAT in 2001. The method is based on four ideas.

- The use of PRNs that are assigned to each unit.
- The use of a burden measure, which can be the number of times a unit has already been selected for all the waves we want to coordinate.
- The use of microstrata constructed in each wave by crossing all the strata of the waves we want to coordinate.
- The permutation of the PRNs proportionally to the burden measure in the microstrata, in such a way that the units having the smallest burden measures obtain the smallest random numbers.

At the start of the algorithm, each unit receives a PRN from a uniform distribution on [0,1] and a response burden equal to 0. In the Rivière method, if t is the first wave to coordinate, the microstrata at wave T are defined as the intersection of the strata of waves t to $T - 1$. The permutations are done inside each microstrata as a function of the cumulative burden. In the subsets of equal burden, the order of the permuted PRN must not change. Then, Algorithm 11 can be applied to realize the coordination by using the Rivière method.

Algorithm 11 Negative coordination with the Rivière method

Assign PRN u_k^1 to each unit $k \in U$.
Assign a zero burden to each unit $k \in U$, in other words $c_k^1 = 0$.
For $T = 2, \dots$, number of waves, repeat:

- Calculate the burden $c_k^T = \sum_{t=1}^{T-1} b^t \mathbb{1}(k \in S^t)$.
- Construct microstrata by crossing the strata of waves 1 to $T - 1$.
- Permute u_k^1, in each microstrata, in such a way as to sort the units in increasing order of burden and the rank does not change in a subset of equal burden. Select the n_h^T units in each stratum.

A proof of the validity of the method was given by Rubin-Bleuer (2002). Nevertheless, if the algorithm is only applied once, the procedure has a major drawback: by crossing the strata from all the previous surveys, the microstrata become very small and the coordination is not very good. For this reason, and to obtain good coordination with the latest surveys, Rivière (1999, p. 5) recommends the use of only three crosses.

8.2.8 The Netherlands Method

The Centraal Bureau voor de Statistiek of the Netherlands has developed a system named EDS to sort stratified samples for business surveys (De Ree, 1983; Van Huis et al., 1994a,b; Koeijers & Willeboordse, 1995; De Ree, 1999).

The method is based on two principles:

- Response Burden Control (RBC) groups. RBC groups are basic strata that are defined once and for all. RBC groups cannot change in time and all strata in all waves must be constructed as unions of RBC groups.
- The use of a burden measure. The weight of each survey is measured with a positive coefficient. This coefficient can be proportional to the time required to complete the survey, or can be equal to 1. For each unit, a burden measure is calculated. It is equal to the total of the coefficient of charge for all samples in which the unit is included.

The method is described in Algorithm 12.

8.2.9 The Swiss Method

This coordination system can be applied to surveys on both households and establishments. It is described in Graf & Qualité (2014). This method is described below using an

---

Algorithm 12 Negative coordination with EDS method

---

1. The register is divided in RBC groups. The RBC groups are defined in a way such that the strata of all waves are defined as a union of RBC groups.
2. Initialize by giving a zero burden measure to each unit.
3. For $t = 1, \dots, T$, repeat:
   - In each RBC group, units are sorted in increasing order of burden. Then, in each group having equal charge, units are sorted by increasing order of their random numbers.
   - Select the n_h first units in each RBC group.
   - The response burden of the selected units is updated.

---

example and figures from Chapter 6 of Lionel Qualité's PhD thesis (Qualité, 2009), therefore all designs use Poisson sampling. Hence, the method can be described separately for each unit.

Suppose that we want to coordinate unit k for three successive surveys with longitudinal inclusion probabilities $\pi_k^1, \pi_k^2, \pi_k^3$.

- At the first wave, the unit has inclusion probability π_k^1 and receives a permanent random number u_k which is the outcome of a uniform random variable on $[0, 1]$. Unit k is selected if $u_k \leq \pi_k^1$. Figure 8.6 shows that the line segment $[0, 1]$ is divided in two subsets $[0, \pi_k^1]$ and $[\pi_k^1, 1]$. Unit k is selected if the permanent random number u_k falls inside the interval $[0, \pi_k^1]$.
- At the second wave, a new selection zone is defined for unit k. This zone may be made up of one or two intervals whose total length is π_k^2. The zones are defined depending on whether we want to apply negative or positive coordination.

  Figures 8.7 and 8.8 show the cases where we want to obtain a positive coordination between waves 1 and 2. A selection zone is defined at stage 2 that maximizes the overlap with the selection zone in wave 1. In the case where $\pi_k^2 \leq \pi_k^1$, the selection zone at the second wave is included in the selection zone at the first wave as shown in Figure 8.7. If π_k^2 is larger than π_k^1, we obtain the situation presented in Figure 8.8.

Figure 8.6 Interval corresponding to the first wave (extract from Qualité, 2009).

Figure 8.7 Positive coordination when $\pi_k^2 \leq \pi_k^1$ (extract from Qualité, 2009).

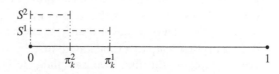

Figure 8.8 Positive coordination when $\pi_k^2 \geq \pi_k^1$ (extract from Qualité, 2009).

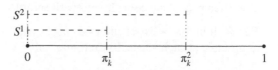

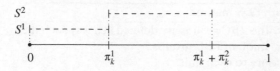

Figure 8.9 Negative coordination when $\pi_k^1 + \pi_k^2 \leq 1$ (extract from Qualité, 2009).

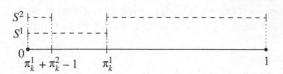

Figure 8.10 Negative coordination when $\pi_k^1 + \pi_k^2 \geq 1$ (extract from Qualité, 2009).

Table 8.2 Selection intervals for negative coordination and selection indicators in the case where the PRNs falls within the interval. On the left, the case where $\pi_k^1 + \pi_k^2 \leq 1$ (Figure 8.9). On the right, the case where $\pi_k^1 + \pi_k^2 \geq 1$ (Figure 8.10).

Interval	$k \in S_1$	$k \in S_2$	Interval	$k \in S_1$	$k \in S_2$
$h_k^1 = [0, \pi_k^1]$	1	0	$h_k^1 = [0, \pi_k^1 + \pi_k^2 - 1]$	1	1
$h_k^2 = [\pi_k^1, \pi_k^1 + \pi_k^2]$	0	1	$h_k^2 = [\pi_k^1 + \pi_k^2 - 1, \pi_k^1]$	1	0
$h_k^3 = [\pi_k^1 + \pi_k^2, 1]$	0	0	$h_k^3 = [\pi_k^1, 1]$	0	1

On the other hand, if we want a negative coordination, we define a selection zone that minimizes the overlap with the selection zone at wave 1. If $\pi_k^1 + \pi_k^2 \leq 1$, the selection zone is included in $[\pi_k^1, 1]$, as shown in Figure 8.9. If $\pi_k^1 + \pi_k^2 \geq 1$, the selection zone at the second wave is the union of $[\pi_k^1, 1]$ and of $[0, \pi_k^2 + \pi_k^1 - 1]$, as shown in Figure 8.10. In the cases presented in Figures 8.7, 8.8, 8.9 and 8.10, the interval $[0, 1]$ is split into three. Each interval corresponds to a longitudinal sample. For example, in the case of negative coordination (Figures 8.9 and 8.10), the longitudinal sample is given in Table 8.2.

• Therefore, an additional interval appears after each survey wave. After t survey waves, the $[0,1]$ interval is split into $t + 1$ intervals. At wave $t + 1$, we assign a score to the $t + 1$ intervals as a function of the coordination priority with previous surveys. If sample S^{t+1} must be positively coordinated with S^i, with maximum priority, each interval that corresponds to a longitudinal sample where unit k is selected in S^i receives a higher score than all the intervals where unit k is not selected in S^i. Then, inside each of these two groups, the intervals receive sorted scores as a function of the desired coordination with the second wave sample by order of priority, and so on. The selection zone for unit k in sample S^{t+1} is the union of the intervals with the highest scores until a total length equal to or lower than π_k^{t+1} and a subinterval of the following interval by decreasing score, up to a total length of π_k^{t+1}.

For example, if the first two sampling designs are positively coordinated, as in Figure 8.7, that sample S^3 must be, in priority, positively coordinated with sample S^2 and then negatively coordinated with sample S^1, so we obtain Figure 8.11.

For each unit k, a list of intervals $h_k^1, \ldots, h_k^{t+1}$ forming a partition of $[0, 1]$ is created and updated at each wave. To each interval h_k^t we associate an indicator of the selection

Figure 8.11 Coordination of a third sample (extract from Qualité, 2009).

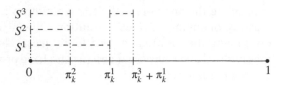

Table 8.3 Selection indicators for each selection interval for unit k.

Interval	$k \in S^1$	$k \in S^2$	...	$k \in S^3$
h_k^1	1	0	$\cdots$	0
h_k^2	1	1	$\cdots$	0
$\vdots$	$\vdots$	$\vdots$		$\vdots$
h_k^{t+1}	0	0	$\cdots$	1

of unit k as presented in Table 8.3. These data allow the efficient selection of a sample, which is determined by the permanent random number u_k. These indicators must be saved to calculate the sampling design during the next wave.

This method requires the saving of the PRN, the interval bounds, and the indicator variables belonging to each unit. Hence, the information that needs to be conserved is relatively reduced. It is possible to realize a positive or negative coordination with respect to multiple dozen previous waves. Managing the births and deaths is very easy as well.

8.2.10 Coordinating Unequal Probability Designs with Fixed Size

Poisson designs are relatively easy to coordinate. If we impose a fixed size, the problem becomes much more complicated. If all unequal probability samples are selected simultaneously, Deville & Tillé (2000) proposed a fast and efficient method. Ernst (1986), Causey et al. (1985), Ernst & Ikeda (1995), and Ernst & Paben (2002) proposed using transport theory to calculate the designs giving the best coordination, but this method requires a lot of computing time. Other solutions have been proposed by Matei & Tillé (2005b) and Matei & Skinner (2009). The ordered sample method described in Section 5.11 allows very good coordination. However, the inclusion probabilities are not exactly satisfied (see Rosén, 1995, 1997a,b, 1998, 2000; Ng & Donadio, 2006; Matei & Tillé, 2007). Tillé & Favre (2004) have also proposed approximate solutions to coordinate samples selected using a balanced design.

8.2.11 Remarks

Nedyalkova et al. (2009) have shown that if the longitudinal design is fixed so that coordination is optimal, after a certain number of waves the transversal design can no longer be chosen freely. Therefore, there exists a contradiction between wanting to coordinate the units optimally and wanting to impose a transversal design. For these reasons, obtaining a definitive solution allowing the choice of transversal design while optimizing the coordination is a fool's quest. Nedyalkova et al. (2009) describe some longitudinal and transversal designs and have proposed some compromise solutions in a general framework.

8.3 Multiple Survey Frames

8.3.1 Introduction

Sampling from multiple sources refers to surveys where independent samples are selected in multiple survey frames. Each survey frame covers at least partially the target population. Inference is then done by combining the various samples. Many reasons justify using multiple survey frames.

- If we have multiple samples selected independently in the same population, and at least a part of the variables are in common, then we can estimate by integrating all available information.
- In certain cases, we have multiple survey frames and we know that each one does not completely cover the target population. Selecting a sample in each frame then allows the coverage of the target population to be increased. The coverage rate is the proportion of the population that is selected in at least one sample.
- When we are interested in a rare population and we want to compare it to the rest of the total population, we can do a sampling design on the total population and complete it by a sample drawn on the rare population.

Example 8.1 In a health survey, we select individuals in a population register that we complete with individuals selected from a hospital register or a social security register. Therefore, we over-represent individuals who have been hospitalized, which can be interesting for a health survey.

Example 8.2 In a person survey, we select individuals in multiple phone books: cell phones, land lines, and work numbers. We can then obtain a larger coverage rate.

Formally, we are interested in a target population U and have multiple registers $U_1, \ldots, U_i, \ldots, U_M$ where each may be incomplete. To be able to construct an unbiased estimator, the union of the registers must cover the target population:

$$\bigcup_{i=1}^{M} U_i = U.$$

In each population, we usually select independent samples denoted by $S_1, \ldots, S_i, \ldots, S_M$ having inclusion probabilities $\pi_k^1, \ldots, \pi_k^2, \ldots, \pi_k^M$, respectively. The main question consists of knowing how to unbiasedly estimate a total Y by using the information available on all samples.

The simplest case is presented in Figure 8.12. Two samples S_A and S_B are selected in two survey frames, U_A and U_B, respectively, with inclusion probabilities π_{kA} and π_{kB}.

Figure 8.12 Two survey frames U_A and U_B cover the population. In each one, we select a sample.

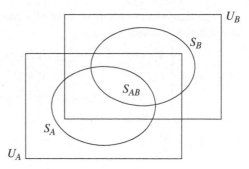

Let

$$U_a = U_A \backslash U_B, U_b = U_B \backslash U_A, U_{ab} = U_A \cap U_B,$$

and

$$Y_A = \sum_{k \in U_A} y_k, \quad Y_B = \sum_{k \in U_B} y_k, \quad Y_a = \sum_{k \in U_a} y_k, \quad Y_b = \sum_{k \in U_b} y_k, \quad Y_{ab} = \sum_{k \in U_{ab}} y_k.$$

We can then write

$$Y = Y_a + Y_b + Y_{ab} = Y_A + Y_B - Y_{ab}.$$

Totals Y_a, Y_b, Y_A, and Y_B can be estimated separately using the expansion estimator:

$$\hat{Y}_A = \sum_{k \in S_A} \frac{y_k}{\pi_{kA}}, \quad \hat{Y}_B = \sum_{k \in S_B} \frac{y_k}{\pi_{kB}}, \quad \hat{Y}_a = \sum_{k \in S_a} \frac{y_k}{\pi_{kA}}, \quad \hat{Y}_b = \sum_{k \in S_b} \frac{y_k}{\pi_{kB}}.$$

However, to estimate Y we must construct an estimator of the total Y_{ab} of the intersection U_{ab} that can be estimated in many ways.

8.3.2 Calculating Inclusion Probabilities

In order to estimate the total Y, we can try to calculate the probability that a unit is selected at least once in one of the samples. In the case where we have M survey frames, we first calculate

$$\pi_k = \Pr \left(k \in \bigcup_{i=1}^{M} S_i \right).$$

Then, we can use the expansion estimator:

$$\hat{Y} = \sum_{k \in \bigcup_{i=1}^{M} S_i} \frac{y_k}{\pi_k}.$$

This solution may at first seem simple, but it is in fact the hardest to put into practice. When a unit is selected in a sample S_i, is it necessary to know its inclusion probability in all other samples to be able to calculate its global inclusion probability. This can be particularly complicated when M is large. In addition, we must conserve in the estimator units that are selected in the multiple sample only once. To do this, we must have a unique and reliable identifier for each statistical unit of each survey.

In the case where only two survey frames U_A and U_B are used, and two independent samples are selected, we have

$$\pi_k = \begin{cases} \pi_{kA} & \text{if } k \in S_A \cap U_a \\ \pi_{kB} & \text{if } k \in S_B \cap U_b \\ \pi_{kA} + \pi_{kB} - \pi_{kA}\pi_{kB} & \text{if } k \in (S_A \cup S_B) \cap U_{ab}. \end{cases}$$

The expansion estimator is

$$\hat{Y} = \sum_{k \in S_A \cup S_B} \frac{y_k}{\pi_k}$$

$$= \sum_{k \in S_A \cap U_a} \frac{y_k}{\pi_{kA}} + \sum_{k \in S_B \cap U_b} \frac{y_k}{\pi_{kB}} + \sum_{k \in (S_A \cup S_B) \cap U_{ab}} \frac{y_k}{\pi_{kA} + \pi_{kB} - \pi_{kA}\pi_{kB}}. \tag{8.1}$$

8.3.3 Using Inclusion Probability Sums

A simple way to do this is to calculate the mathematical expectation μ_k of the number of times unit k is in the sample:

$$\mu_k = \sum_{i=1}^{M} \pi_k^i,$$

where $\pi_k^i = 0$ if $k \notin U_i$. We can then construct an estimator

$$\hat{Y}_\mu = \sum_{i=1}^{M} \sum_{k \in S_i} \frac{y_k}{\mu_k} = \sum_{k \in U} \frac{y_k}{\mu_k} \sum_{i=1}^{M} a_k^i,$$

where

$$a_k^i = \begin{cases} 1 & \text{if } k \in S_i \\ 0 & \text{if } k \notin S_i. \end{cases}$$

This estimator is unbiased. Indeed,

$$E_p(\hat{Y}_\mu) = \sum_{k \in U} \frac{y_k}{\mu_k} E_p\left(\sum_{i=1}^{M} a_k^i \right) = \sum_{k \in U} \frac{y_k}{\mu_k} \sum_{i=1}^{M} \pi_k^i = Y.$$

When a unit is selected in sample S_i, it is again necessary to known its inclusion probability in all samples to be able to calculate the expectation μ_k of the number of times unit k is in the sample. However, the tedious calculation of the global inclusion probability is avoided. Unfortunately, the estimator $\hat{Y}_\mu$ depends of the unit multiplicity and is not admissible. A Rao–Blackwellization would theoretically improve these estimations.

In the case where only two survey frames U_A and U_B are used, we have

$$\mu_k = \begin{cases} \pi_{kA} & \text{if } k \in S_A \cap U_a \\ \pi_{kB} & \text{if } k \in S_B \cap U_b \\ \pi_{kA} + \pi_{kB} & \text{if } k \in (S_A \cup S_B) \cap U_{ab}. \end{cases}$$

We then obtain the estimator proposed by Bankier (1986) and Kalton & Anderson (1986):

$$\hat{Y}_\mu = \sum_{k \in S_A} \frac{y_k}{\mu_k} + \sum_{k \in S_B} \frac{y_k}{\mu_k}$$

$$= \sum_{k \in S_A \cap U_a} \frac{y_k}{\pi_{kA}} + \sum_{k \in S_B \cap U_b} \frac{y_k}{\pi_{kB}}$$

$$+ \sum_{k \in S_A \cap U_{ab}} \frac{y_k}{\pi_{kA} + \pi_{kB}} + \sum_{k \in S_B \cap U_{ab}} \frac{y_k}{\pi_{kA} + \pi_{kB}}.$$

If the two samples are selected independently, we can then Rao–Blackwellize (see Section 2.8, page 22) the estimator $\hat{Y}_\mu$ based on the following result.

Result 8.2

$$E_p(a_{kA} + a_{kB}|a_{kA} + a_{kB} \geq 1) = \frac{\pi_{kA} + \pi_{kB}}{\pi_{kA} + \pi_{kB} - \pi_{kA}\pi_{kB}}, \text{ for all } k \in U_{ab},$$

where a_{kA} and a_{kB} are the indicator variables of unit k in samples S_A and S_B.

Proof: Since

$$\Pr(a_{kA} + a_{kB} = 1) = \pi_{kA}(1 - \pi_{kB}) + \pi_{kB}(1 - \pi_{kA}) = \pi_{kA} + \pi_{kB} - 2\pi_{kA}\pi_{kB},$$

and

$$\Pr(a_{kA} + a_{kB} = 2) = \pi_{kA}\pi_{kB},$$

we have

$$E_p(a_{kA} + a_{kB}|a_{kA} + a_{kB} \geq 1)$$
$$= \frac{1 \times \Pr(a_{kA} + a_{kB} = 1) + 2 \times \Pr(a_{kA} + a_{kB} = 2)}{\Pr(a_{kA} + a_{kB} = 1) + \Pr(a_{kA} + a_{kB} = 2)}$$
$$= \frac{(\pi_{kA} + \pi_{kB} - 2\pi_{kA}\pi_{kB}) + 2 \times \pi_{kA}\pi_{kB}}{\pi_{kA} + \pi_{kB} - \pi_{kA}\pi_{kB}} = \frac{\pi_{kA} + \pi_{kB}}{\pi_{kA} + \pi_{kB} - \pi_{kA}\pi_{kB}}. \quad \blacksquare$$

The Rao–Blackwellized estimator is none other then the expansion estimator $\hat{Y}$ given in Expression (8.1). The expansion estimator $\hat{Y}$ is always more accurate than $\hat{Y}_\mu$ but is harder to calculate.

8.3.4 Using a Multiplicity Variable

The indicator variable L_{ik} can also be used:

$$L_{ik} = \begin{cases} 1 & \text{if } k \in U_i \\ 0 & \text{if } k \notin U_i \end{cases}$$

of which the sum on the various survey frames,

$$L_{.k} = \sum_{i=1}^{M} L_{ik},$$

is the number of survey frames where unit k appears. To be able to unbiasedly estimate a total, each unit k must belong to at least one survey frame, in other words, $L_{.k} \geq 1$, for all $k \in U$. Using this variable, we can then rewrite the population total by

$$Y = \sum_{k \in U} y_k = \sum_{i=1}^{M} \sum_{k \in U_i} \frac{y_k}{L_{.k}}.$$

The totals can then separately be unbiasedly estimated in each population and we obtain

$$\hat{Y}_L = \sum_{i=1}^{M} \sum_{k \in S_i} \frac{y_k}{\pi_k^i L_{.k}}. \tag{8.2}$$

This estimator is much easier to calculate than $\hat{Y}$ and $\hat{Y}_\mu$, as it is sufficient to know $L_{.k}$ to be able to construct it. Estimator $\hat{Y}_L$ also depends on the multiplicity of the units.

In the case where only two survey frames U_A and U_B are used, we obtain the classical estimator proposed and generalized by Hartley (1962, 1974):

$$\hat{Y}_L = \sum_{k \in S_A \cap U_a} \frac{y_k}{\pi_{kA}} + \sum_{k \in S_B \cap U_b} \frac{y_k}{\pi_{kB}} + \sum_{k \in S_A \cap U_{ab}} \frac{y_k}{2\pi_{kA}} + \sum_{k \in S_B \cap U_{ab}} \frac{y_k}{2\pi_{kB}}.$$

If the two samples are selected independently, we have the following result.

Result 8.3

$$E_p \left(\frac{a_{kA}}{\pi_{kA}} + \frac{a_{kB}}{\pi_{kB}} \,\middle|\, a_{kA} + a_{kB} \ge 1 \right) = \frac{2}{\pi_{kA} + \pi_{kB} - \pi_{kA}\pi_{kB}}.$$

Proof:

$$E_p \left(\frac{a_{kA}}{\pi_{kA}} + \frac{a_{kB}}{\pi_{kB}} \,\middle|\, a_{kA} + a_{kB} \ge 1 \right)$$

$$= \left[\frac{1}{\pi_{kA}} \times \Pr(a_{kA} = 1 \text{ and } a_{kB} = 0) + \frac{1}{\pi_{kB}} \times \Pr(a_{kA} = 0 \text{ and } a_{kB} = 1) \right.$$

$$\left. + \left(\frac{1}{\pi_{kA}} + \frac{1}{\pi_{kB}} \right) \times \Pr(a_{kA} + a_{kB} = 2) \right]$$

$$\times \{ \Pr(a_{kA} + a_{kB} = 1) + \Pr(a_{kA} + a_{kB} = 2) \}^{-1}$$

$$= \frac{\frac{1}{\pi_{kA}}\pi_{kA}(1 - \pi_{kB}) + \frac{1}{\pi_{kB}}\pi_{kB}(1 - \pi_{kA}) + \left(\frac{1}{\pi_{kA}} + \frac{1}{\pi_{kB}} \right) \pi_{kA}\pi_{kB}}{\pi_{kA} + \pi_{kB} - \pi_{kA}\pi_{kB}}$$

$$= \frac{2}{\pi_{kA} + \pi_{kB} - \pi_{kA}\pi_{kB}}. \qquad \blacksquare$$

This results allows us to show that the Rao–Blackwellized estimator of $\hat{Y}_L$ is again the expansion estimator $\hat{Y}$.

8.3.5 Using a Weighted Multiplicity Variable

The use of a multiplicity variable can be generalized by weighting on each survey frame. We can always write the general total in the form

$$Y = \sum_{k \in U} y_k = \sum_{i=1}^{M} \sum_{k \in U_i} \frac{\theta_i}{\sum_{j=1}^{M} \theta_j L_{jk}} y_k = \sum_{k \in U} y_k \sum_{i=1}^{M} \frac{\theta_i}{\sum_{j=1}^{M} \theta_j L_{jk}} L_{ik},$$

where the θ_i are the positive weights for each survey frame.

We can then construct the estimator:

$$\hat{Y}_\theta = \sum_{i=1}^{M} \sum_{k \in S_i} \frac{\theta_i}{\sum_{j=1}^{M} \theta_j L_{jk}} \frac{y_k}{\pi_k^i}.$$

When $\theta_i = 1$, we recover Estimator (8.2). We can try to find weights θ_i that minimize the variance of the estimator.

In the case where only two survey frames U_A and U_B are used, we obtain an estimator proposed by Hartley (1962, 1974):

$$\hat{Y}_\theta = \sum_{k \in S_A \cap U_a} \frac{y_k}{\pi_{kA}} + \sum_{k \in S_B \cap U_b} \frac{y_k}{\pi_{kB}} + \theta \sum_{k \in S_A \cap U_{ab}} \frac{y_k}{\pi_{kA}} + (1-\theta) \sum_{k \in S_B \cap U_{ab}} \frac{y_k}{\pi_{kB}},$$

where $\theta \in [0, 1]$ is a weighting coefficient. Hartley (1962, 1974) calculated the coefficient θ that minimizes the variance of $\hat{Y}_\theta$. However, this estimator is not really optimal, as it still depends on the multiplicity of the units. It is again possible to show that the Rao–Blackwellization of $\hat{Y}_\theta$ consists of calculating the expansion estimator regardless of the value of coefficient θ.

8.3.6 Remarks

For all estimators, a lower weight is applied to units that are likely to be selected several times in the sample. Estimator $\hat{Y}_L$ is the easiest to calculate as it does not require knowledge of any information other than the multiplicity indicator $L_{.k}$ for selected units in the sample. In the case where two survey frames are available, all estimators that depends on the multiplicity are in principle Rao–Blackwellizable. We obtain, in all cases, the expansion estimator. However, the expansion estimator, even if it is preferable, requires more information to be calculated. That is why, in most cases, inadmissible estimators (see Section 2.8, page 22) are used because they have the benefit of being simple.

Sampling from multiple survey frames can be seen as a special case of indirect sampling. A large number of publications deal with integrattion of additional information and optimization of the estimators (see Fuller & Burmeister, 1972; Lepkowski & Groves, 1986; Iachan & Dennis, 1993; Skinner et al., 1994; Rao & Skinner, 1996; Haines & Pollock, 1998; Kott et al., 1998; Lohr & Rao, 2000, 2006; Mecatti, 2007; Clark et al., 2007). An overview of these methods is presented in Lohr (2007, 2009a).

8.4 Indirect Sampling

8.4.1 Introduction

The survey frame of a target population is often unknown, partially known, or not up to date. It is then not possible to directly select units in this population. Indirect sampling is a set of methods where we sample in a population different to the target population. Existing links between the two populations can then be used in order to indirectly select units from this target population.

Example 8.3 Suppose that a register of individuals is available and that we are interested in households. A sample of individuals is selected. Then, all households in which these individuals belong are selected. If several sampled individuals of the initial sample belong to the same household, it is possible to select the same households several times.

Example 8.4 Suppose that a register of dwellings is available. We are interested in the individuals of a population. We randomly select a sample of dwellings and we then select all individuals of this dwelling. The simplest case is in fact a cluster survey, which can be interpreted as a special case of indirect sampling.

Example 8.5 A sample of S_{t-1} businesses has been selected at time $t - 1$. The goal consists of selecting a sample of businesses at time t which contains the maximum number of common businesses with S_{t-1}. However, many businesses have merged, split, closed or started up. At time S_{t-1}, we select all businesses that were selected at time $t - 1$ and all the businesses that arose from mergers or splits involving at least one business selected at time $t - 1$. This sample is then completed with a sample of new businesses selected separately.

Example 8.6 We want to select a sample of high net worth individuals. To do this, we select an initial sample. If in this sample someone has high net worth, we interview all his neighbors as well.

All these examples can be grouped under the indirect sampling moniker. They show that indirect sampling covers a very large set of methods allowing sophisticated sampling strategies to be put in place to select units in a population to which we do not have direct access. Before approaching the subject more generally, we will first examine a few interesting special cases.

8.4.2 Adaptive Sampling

A survey is adaptive if, after selecting a set of units using a basic design, we select additional units as a function of the results obtained in the first sample. This method has been mainly developed by Thompson & Ramsey (1983), Thompson (1990, 1991a,b, 1992), and Thompson & Seber (1996). Adaptive sampling is often used to geographically select units. For example, suppose a forest is divided into squares. We are looking to estimate the prevalence of a parasite on trees. A set of squares is selected and if a square is contaminated, we select the eight adjacent squares. If in one of the adjacent squares we find a new contaminated tree, we again select the adjacent squares, as shown in Figure 8.13.

Adaptive sampling selects a cluster of contaminated squares for which it is difficult to calculate the inclusion probabilities. In addition, taking into account contaminated squares at the edge of the zone is a special problem that must be dealt with (see Thompson & Seber, 1996). However, it is possible to propose an unbiased estimator using the generalized weight sharing method.

8.4.3 Snowball Sampling

Snowball sampling is a method where sampled individuals recruit other new individuals themselves using their network of acquaintances. At each stage, each individuals can recruit k new individuals in such a way that the sample grows like a snowball.

Figure 8.13 In this example, the points represent contaminated trees. During the initial sampling, the shaded squares are selected. The borders in bold surround the final selected zones.

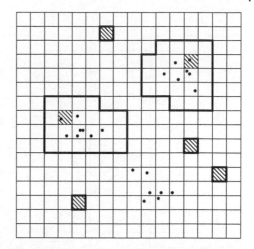

This technique is often used to select individuals that present a rare characteristic in the population or that are hidden or clandestine (e.g. homeless people, drug users). The method consists of selecting clusters of individuals linked by a social network and can be seen as a special case of indirect sampling.

8.4.4 Indirect Sampling

Indirect sampling generalizes all the methods described previously. Indirect sampling is a set of methods used to estimate functions of interest on a population U_B of which we know nothing using another survey frame U_A whose units have links to the units of U_B. In addition, we suppose that the units of U_B are grouped into clusters $U_{B1}, \dots, U_{Bi}, \dots, U_{BM}$, of size $N_{B1}, \dots, N_{Bi}, \dots, N_{BM}$.

Units $j \in U_A$ can have links with units $k \in U_B$. Let L_{jk} denote the general element of the link matrix $\mathbf{L}$, which is

$$L_{jk} = \begin{cases} 1 & \text{if unit } j \text{ of } U_A \text{ has a link with unit } k \text{ of } U_B \\ 0 & \text{otherwise.} \end{cases}$$

Let also $L_{.k}$ denote the number of links arriving on unit k of U_B:

$$L_{.k} = \sum_{j \in U_A} L_{jk}.$$

To be able to construct an estimator, only one condition is necessary. It must be that

$$\sum_{k \in U_{Bi}} L_{.k} > 0, \text{ for all } i = 1, \dots, M,$$

that is to say, any cluster U_B must contain at least one unit that has a link with a unit of U_A as seen in the example presented in Figure 8.14.

Indirect sampling consists of selecting a sample in U_A. Then, we select all the statistical units in clusters that have at least one link to a unit selected in U_A. In the sample presented in Figure 8.14, all the clusters of U_B contain at least one unit that has a link towards a unit of U_A. In U_A, the units surrounded by a circle are selected. By using the links of these units, we finally select all the units of clusters U_{B1} and U_{B3}. Only the units of cluster U_{B2} are not selected.

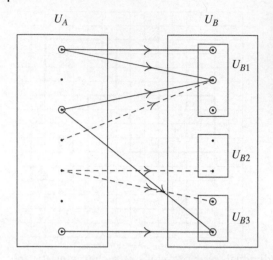

U_A U_B

Figure 8.14 Example of indirect sampling. In population U_A, the units surrounded by a circle are selected. Two clusters (U_{B1} and U_{B3}) of population U_B each contain at least one unit that has a link with a unit selected in population U_A. Units of U_B surrounded by a circle are selected at the end.

8.4.5 The Generalized Weight Sharing Method

The generalized weight sharing method developed by Lavallée (2002, 2007) constructs an estimator of the total for population U_B by using a sampling design on U_A. Suppose a sample S_A of U_A has been selected using design $p_A(\cdot)$ having inclusion probabilities π_{Ak}. Suppose as well that we have weights w_{Aj} that allow us to estimate a variable of interest y_A measured on units of S_A. The total estimator is then

$$\hat{Y}_A = \sum_{j \in S_A} w_{Aj} y_{Aj}. \tag{8.3}$$

The weights w_{Aj} can be equal to the inverse inclusion probabilities $1/\pi_{Ak}$ of design $p_A(\cdot)$, but can also be weights obtained by a regression estimator or a calibration estimator, as described in Chapters 9, 10, 11, and 12.

The weight sharing method is split into four stages:

1. At the first stage, we calculate preliminary weights w^*_{Bk}:

$$w^*_{Bk} = \begin{cases} \sum\limits_{j \in S_A} L_{jk} w_{Aj} & \text{if } L_{.k} > 0 \\ 0 & \text{if } L_{.k} = 0. \end{cases}$$

2. We then calculate the number of links between unit k of U_B and the units of U_A:

$$L_{.k} = \sum_{j \in U_A} L_{jk}.$$

3. The weights are shared inside clusters:

$$w_{Bi} = \frac{\sum_{k \in U_{Bi}} w^*_{Bk}}{\sum_{k \in U_{Bi}} L_{.k}}, i \in S_I.$$

4. We then attach the weights w_{Bi} of cluster U_{Bi} to all individuals belonging to this cluster:

$$w_{Bk} = w_{Bi}, \text{ for all } k \in U_i, i \in S_I$$

and so, inside a cluster, all the units have the same weight.

By denoting by S_{IB} the sample of selected clusters in U_B, we can define the estimator of the total of a variable of interest y_B in multiple ways:

$$\hat{Y}_B = \sum_{k \in S_B} w_{Bk} y_{Bk} = \sum_{i \in S_{IB}} w_{Bi} \sum_{k \in U_i} y_{Bk} = \sum_{i \in S_{IB}} \frac{\sum_{k \in U_{Bi}} w^*_{Bk}}{\sum_{k \in U_{Bi}} L_{.k}} \sum_{k \in U_i} y_{Bk}$$

$$= \sum_{i \in U_{IB}} \frac{\sum_{k \in U_{Bi}} w^*_{Bk}}{\sum_{k \in U_{Bi}} L_{.k}} \sum_{k \in U_i} y_{Bk} = \sum_{i \in U_{IB}} \frac{\sum_{k \in U_{Bi}} \sum_{j \in S_A} L_{jk} w_{Aj}}{\sum_{k \in U_{Bi}} L_{.k}} \sum_{k \in U_i} y_{Bk}. \tag{8.4}$$

Result 8.4 *If estimator $\hat{Y}_A$ on U_A given in Expression (8.3) is unbiased, then estimator $\hat{Y}_B$ by the generalized weight sharing method given in Expression (8.4) is also unbiased.*

Proof: By denoting by a_{Aj} the indicator variable which equals 1 if j is in S_A and 0 otherwise, the expectation of $\hat{Y}_B$ can be calculated by

$$E_p(\hat{Y}_B) = \sum_{i \in U_{IB}} \frac{\sum_{k \in U_{Bi}} \sum_{j \in U_A} L_{jk} E_p(a_{Aj} w_{Aj})}{\sum_{k \in U_{Bi}} L_{.k}} \sum_{k \in U_i} y_{Bk}.$$

Yet, a sufficient and necessary condition for $\hat{Y}_A$ to be unbiased is that

$$E_p(a_{Ak} w_{Aj}) = 1, \quad \text{for all } k \in U_A.$$

Hence,

$$E_p(\hat{Y}_B) = \sum_{i \in U_{IB}} \frac{\sum_{k \in U_{Bi}} \sum_{j \in U_A} L_{jk}}{\sum_{k \in U_{Bi}} L_{.k}} \sum_{k \in U_i} y_{Bk}$$

$$= \sum_{i \in U_{IB}} \frac{\sum_{k \in U_{Bi}} L_{.k}}{\sum_{k \in U_{Bi}} L_{.k}} \sum_{k \in U_i} y_{Bk} = \sum_{i \in U_{IB}} \sum_{k \in U_i} y_{Bk} = \sum_{k \in U_B} y_{Bk}. \qquad \blacksquare$$

The usefulness of the generalized weight sharing method is that it requires only knowledge of the weights of selected units in S_A and the number of links $L_{.k}$ arriving on unit k. The drawback of this estimator is that it can depend on the multiplicity of the units. A Rao–Blackwellization (see Section 2.8, page 22) is in general not a viable solution as it requires knowledge of much more information on the sampling design on population U_A.

8.5 Capture–Recapture

Capture–recapture methods (also called *mark and recapture*) are often used in biology to estimate the abundance of animal populations. Hundreds of publications deal with this particular field of applied survey theory. One can refer to overview articles by Seber (1986, 1992), Schwarz & Seber (1999), and Pollock (2000), and to the books of Pollock et al. (1990), Seber (2002), and Amstrup et al. (2005).

We will limit ourselves to outlining the simplest capture–recapture method, which consists of randomly selecting a sample of animals, marking them, and then releasing them. We then select a second independent sample and count the number of marked and unmarked animals. Under certain assumptions, we can then estimate the size of the population.

Formally, we are interested in a population U of size N. This size is unknown and the main objective is to estimate it.

- An initial sample of size n_1 is selected using simple random sampling without replacement of fixed size. The animals are marked and then released.
- We then select a second sample of size n_2 in which we count m_2 marked animals.

If the animals are actually selected randomly and independently in each wave, then m_2 is a random variable having a hypergeometric distribution $m_2 \sim \mathcal{H}(n_2, N, n_1/n)$. Its expectation and variance are

$$E_p(m_2) = \frac{n_2 n_1}{N} \quad \text{and} \quad \text{var}_p(m_2) = n_2 \frac{n_1}{N} \left(1 - \frac{n_1}{N}\right) \frac{N - n_2}{N - 1}.$$

Since

$$\frac{E_p(m_2)}{n_2} = \frac{n_1}{N},$$

m_2/n_2 is an unbiased estimator of n_1/N. From this result, we can empirically construct the so called Petersen–Lincoln estimator given by

$$\widehat{N}_{PL} = \frac{n_1 n_2}{m_2}.$$

This estimator is slightly biased and has a nonzero probability of taking an infinite value (when $m_2 = 0$).

Chapman (1951) proposed an alternative estimator that has the benefit of never taking an infinite value

$$\widehat{N}_C = \frac{(n_1 + 1)(n_2 + 1)}{m_2 + 1} - 1.$$

Seber (1970) proposed a variance estimator of $\widehat{N}_C$

$$v(\widehat{N}_{PL}) = \frac{(n_1 + 1)(n_2 + 1)(n_1 - m_2)(n_2 - m_2)}{(m_2 + 1)^2 (m_2 + 2)}.$$

Using the Chapman estimator and its variance estimator, it is possible to construct a confidence interval for the population size. We can use a normal distribution when $m_2 > 50$. If $m_2 < 50$, the probability distribution of $\widehat{N}_C$ is strongly asymmetric. Seber (2002) suggests using the properties of the hypergeometric distribution to construct this interval or to use an approximation by a Poisson variable.

Example 8.7 In a lake, we randomly and successively catch 400 fishes. After the first session of catching, we mark the fish. During the second catching, we see that 70 caught fishes are marked. We have the following results:

$$\widehat{N}_{PL} = \frac{400 \times 400}{700} \approx 2285.71,$$

$$\widehat{N}_C = \frac{(400 + 1)(400 + 1)}{70 + 1} - 1 \approx 2263.8,$$

and

$$v(\widehat{N}_{PL}) = \frac{(400+1)(400+1)(400-70)(400-70)}{(70+1)^2(70+2)} \approx 48246.7.$$

The confidence interval is

$$IC(95\%) = [\widehat{N}_C - 1.96 \times \sqrt{v(\widehat{N}_{PL})}, \widehat{N}_C + 1.96 \times \sqrt{v(\widehat{N}_{PL})}]$$
$$= [2263.8 - 1.96 \times 219.651, 2263.8 + 1.96 \times 219.651]$$
$$= [1833.29, 2694.32].$$

This interval is relatively large, as the variance of the Chapman estimator is large.

The capture–recapture method can be obviously generalized to surveys with more than two waves and for more complex sampling designs. Many other factors can be taken into consideration, such as the fact that populations evolve over time depending on births and deaths. The gregarious nature of some animals and their capacity to learn to avoid capture can also bias the results.

The application of capture–recapture methods is not limited to the estimation of population abundance. These methods can also apply to quality control techniques. There are more applications in epidemiology or estimation in hard to target populations such as the homeless. This is obviously not about capturing individuals, but asking them if they have already been interviewed in the survey.

9

Estimation with a Quantitative Auxiliary Variable

9.1 The Problem

Suppose that an auxiliary variable is a variable that is known partially or totally over the entire population. The knowledge of this variable can come from a census or an administrative file. To be able to use an auxiliary variable at the estimation stage, it is not necessary to know the value taken by this variable by all the units of the population. The knowledge of the total is sufficient to calculate the estimators considered in this chapter: the ratio estimator, the difference estimator, and the regression estimator.

In practice, auxiliary information is often not available at the level of the statistical units. A law often protects individuals and businesses from disclosure of their individual data. Access to administrative files can be very difficult.

In this chapter, we will consider the simple problem of using an auxiliary variable X whose total is known at the population level:

$$X = \sum_{k \in U} x_k,$$

where $x_1, \ldots, x_k, \ldots, x_N$ are the N values taken by variable x on U units. If we want to estimate

$$Y = \sum_{k \in U} y_k,$$

the total of the values taken by the variable of interest y, and that auxiliary variable x is suspected to be related to variable y, it may be interesting to use this auxiliary information to construct an estimator for variable y that takes into account knowledge of X. We will examine three basic estimators using such auxiliary information: the difference estimator, the ratio estimator, and the regression estimator.

In the selected sample, we measure the values taken by variables x and y on all the units of the sample. Therefore, we can calculate the expansion estimators of Y and X by

$$\widehat{Y} = \sum_{k \in S} \frac{y_k}{\pi_k} \quad \text{and} \quad \widehat{X} = \sum_{k \in S} \frac{x_k}{\pi_k}.$$

Since the population total X is known, we can then compare $\widehat{X}$ to X. Depending on the selected sample, $\widehat{X}$ is more or less close to X. The basic idea of using auxiliary information is to use the deviation between $\widehat{X}$ and X to improve the estimation of Y. Therefore, the estimator using auxiliary information is based on two sources of information: the data observed on the sample and the known population total X.

Sampling and Estimation from Finite Populations, First Edition. Yves Tillé.
© 2020 John Wiley & Sons Ltd. Published 2020 by John Wiley & Sons Ltd.

All the estimators presented below are linear estimators within the meaning of the following definition:

Definition 9.1 *An estimator of the total Y is said to be linear if it can be written as*

$$\hat{Y}_w = w_0(S) + \sum_{k \in S} w_k(S) y_k, \tag{9.1}$$

where the weights $w_k(S)$ may depend on sample S and one or more auxiliary variables x.

In a linear estimator, the weight assigned to a particular unit may vary according to the selected sample. The expansion estimator is obviously a linear estimator in the sense of this definition with $w_0(S) = 0$ and $w_k(S) = 1/\pi_k$.

9.2 Ratio Estimator

9.2.1 Motivation and Definition

The ratio estimator is the simplest way to integrate auxiliary information given by a total X to estimate an unknown total Y that can be estimated using an expansion estimator $\hat{Y}$. We will see, in Section 11.7.1, that the ratio estimator can be particularly interesting when variable x is proportional to variable y, in other words, when the observations are aligned along a line passing through the origin, as shown in Figure 9.1.

For example, if in municipalities we count the number of inhabitants and their total income, it is legitimate to suppose that the observations will be aligned along a line passing through the origin because if the municipality does not have inhabitants, the total income is necessarily zero.

The ratio estimator is defined as

$$\hat{Y}_{\text{RATIO}} = \frac{X \hat{Y}}{\hat{X}}. \tag{9.2}$$

This estimator is linear in the sense of the definition given in Expression (9.1), where $w_0(S) = 0$ and

$$w_k(S) = \frac{X}{\pi_k \hat{X}}.$$

The ratio estimator has the following property: if we replace y_k by x_k in Expression (9.2), we have $\hat{X}_{\text{RATIO}} = X$. In other words, the estimator of the total X by means of the ratio estimator is equal to the known total X. We can then say that the ratio estimator

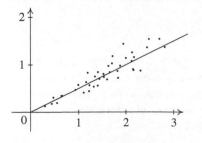

Figure 9.1 Ratio estimator: observations aligned along a line passing through the origin.

is calibrated on X (see Chapter 12). In this case, $\text{var}_p(\widehat{X}_{\text{RATIO}}) = 0$. It follows that if $y_k \propto x_k$, then $\widehat{Y}_{\text{RATIO}} = Y$. Therefore, totals of variables that are proportional to the x_k are calibrated, in other words, they are estimated exactly.

However, the ratio estimator can be a very poor estimator if the line along which the observations are aligned passes very far from the origin or if the slope of the regression line is negative. The ratio estimator is generally biased. It is not possible to calculate this bias exactly. It is necessary to resort to an approximation.

9.2.2 Approximate Bias of the Ratio Estimator

The ratio estimator is biased. In order to calculate the bias, we can calculate the expectation of

$$\widehat{Y}_{\text{RATIO}} - Y = \frac{X \widehat{Y}}{\widehat{X}} - Y = X \frac{\widehat{Y} - R \widehat{X}}{\widehat{X}},$$

where $R = Y/X$. It is not possible to calculate this bias exactly because the ratio estimator is a ratio of two random variables. If we pose

$$\varepsilon = \frac{X - \widehat{X}}{X},$$

we can write

$$\widehat{Y}_{\text{RATIO}} - Y = \frac{\widehat{Y} - R \widehat{X}}{1 - \varepsilon}.$$

If we do a series development of $\widehat{Y}_{\text{RATIO}} - Y$ with respect to ε, we obtain

$$\widehat{Y}_{\text{RATIO}} - Y = (\widehat{Y} - R \widehat{X})(1 + \varepsilon + \varepsilon^2 + \varepsilon^3 + \cdots)$$

$$\approx (\widehat{Y} - R \widehat{X})(1 + \varepsilon) = (\widehat{Y} - R \widehat{X}) \left(1 - \frac{\widehat{X} - X}{X} \right).$$

The expectation is $E_p(\varepsilon) = 0$ and

$$\text{var}_p(\varepsilon) = \frac{\text{var}_p(\widehat{X})}{X^2}.$$

It is reasonable to assume that ε is small when n is large, and so we obtain an approximation of the bias of this estimator:

$$E_p(\widehat{Y}_{\text{RATIO}} - Y) \approx E_p \left[(\widehat{Y} - R\widehat{X}) \left(1 - \frac{\widehat{X} - X}{X} \right) \right]$$

$$= E_p(\widehat{Y} - R \widehat{X}) - \frac{E_p[\widehat{Y}(\widehat{X} - X)] - R E_p[\widehat{X}(\widehat{X} - X)]}{X}$$

$$= \frac{R \, \text{var}_p(\widehat{X}) - \text{cov}_p(\widehat{X}, \widehat{Y})}{X}.$$

In particular, under a simple design, the bias becomes

$$E_p(\widehat{Y}_{\text{RATIO}} - Y) \approx N^2 \left(1 - \frac{n}{N} \right) \frac{1}{n} \frac{R \, S_x^2 - S_{xy}}{X},$$

where

$$S_y^2 = \frac{1}{N-1}\sum_{k \in U}(y_k - \overline{Y})^2 \text{ and } S_{xy} = \frac{1}{N-1}\sum_{k \in U}(x_k - \overline{X})(y_k - \overline{Y}).$$

The expectation of the ratio estimator is then

$$E_p(\widehat{Y}_{\text{RATIO}} - Y) + Y \approx Y + N^2\left(1 - \frac{n}{N}\right)\frac{1}{n}\frac{R\,S_x^2 - S_{xy}}{X}.$$

The bias is negligible when n is large or when $R\,S_x^2 - S_{xy}$ is close to zero, i.e. when the slope of the regression line $B = S_{xy}/S_x^2$ is close to R. Let the constant of the regression line be

$$A = \frac{Y}{N} - B\frac{X}{N} = \frac{X}{N}(R - B).$$

If $B = R$, then $A = 0$, which means that if the regression line passes through the origin, then the bias is approximately zero.

9.2.3 Approximate Variance of the Ratio Estimator

The variance of the ratio estimator cannot be calculated directly because this estimator is a ratio of two random variables. Using the linearization techniques described in Chapter 15, we can obtain an approximation of the variance presented in Expression 15.4, page 304:

$$\text{var}_p(\widehat{Y}_{\text{RATIO}}) \approx \text{var}_p(\widehat{E}),$$

where

$$\widehat{E} = \sum_{k \in S}\frac{E_k}{\pi_k}$$

and $E_k = y_k - Rx_k$. Therefore, the approximate variance can be written as

$$\text{var}_p(\widehat{Y}_{\text{RATIO}}) \approx \text{var}_p(\widehat{E}) = \sum_{k \in U}\sum_{\ell \in U}\frac{E_k E_\ell}{\pi_k \pi_\ell}\Delta_{k\ell}.$$

We can also write

$$\text{var}_p(\widehat{Y}_{\text{RATIO}}) \approx \text{var}_p(\widehat{Y} - R\widehat{X}) = \text{var}_p(\widehat{Y}) + R^2\,\text{var}_p(\widehat{X}) - 2R\text{var}_p(\widehat{Y}, \widehat{X}).$$

This expression of the variance allows cases where the ratio estimator is appropriate to be identified. The variance is small when E_k is close to zero, that is, when y_k is close to Rx_k. Therefore, in order to obtain a small variance, variable x must be proportional to variable y, which means that the observations are aligned along a line passing through the origin as in Figure 9.1. This variance can be estimated (see Expression 15.5) by

$$v(\widehat{Y}_{\text{RATIO}}) = \sum_{k \in S}\sum_{\ell \in S}\frac{e_k e_\ell}{\pi_k \pi_\ell}\frac{\Delta_{k\ell}}{\pi_{k\ell}},$$

where $e_k = y_k - \widehat{R}\,x_k$ and $\widehat{R} = \widehat{Y}/\widehat{X}$.

Under a simple design, the approximate variance becomes

$$\text{var}_p(\widehat{Y}_{\text{RATIO}}) \approx N^2\left(1 - \frac{n}{N}\right)\frac{1}{n}(S_y^2 + R^2\,S_x^2 - 2R\,S_{xy}). \tag{9.3}$$

This variance can be estimated by

$$v(\hat{Y}_{\text{RATIO}}) = N^2 \left(1 - \frac{n}{N}\right) \frac{1}{n} (s_y^2 + \hat{R}^2 \, s_x^2 - 2\hat{R} \, s_{xy}),$$

where $\hat{R} = \hat{Y}/\hat{X}, \widehat{\bar{X}} = \hat{X}/N, \widehat{\bar{Y}} = \hat{Y}/N,$

$$s_y^2 = \frac{1}{n-1} \sum_{k \in S} (y_k - \widehat{\bar{Y}})^2, s_x^2 = \frac{1}{n-1} \sum_{k \in S} (x_k - \widehat{\bar{X}})^2$$

and

$$s_{xy} = \frac{1}{n-1} \sum_{k \in S} (x_k - \widehat{\bar{X}})(y_k - \widehat{\bar{Y}}).$$

9.2.4 Bias Ratio

It is now possible to check that the bias of the ratio estimator is small with respect to its standard deviation. Indeed, under a simple design, it is possible to calculate the bias ratio by means of the approximate variance of the ratio estimator calculated in Expression (9.3):

$$
\begin{aligned}
\text{BR}(\hat{Y}_{\text{RATIO}}) &= \frac{E_p(\hat{Y}_{\text{RATIO}} - Y)}{\sqrt{\text{var}_p(\hat{Y}_{\text{RATIO}})}} \\
&\approx \frac{N^2 \left(1 - \frac{n}{N}\right) \frac{1}{n\bar{X}} (R \, S_x^2 - S_{xy})}{\sqrt{N^2 \left(1 - \frac{n}{N}\right) \frac{1}{n} (S_y^2 + R^2 \, S_x^2 - 2R \, S_{xy})}} \\
&= \frac{1}{\sqrt{n}} \sqrt{1 - \frac{n}{N}} \frac{R \, S_x^2 - S_{xy}}{\bar{X} \sqrt{S_y^2 + R^2 \, S_x^2 - 2R \, S_{xy}}} = O\left(\frac{1}{\sqrt{n}}\right).
\end{aligned}
$$

When the regression line passes through the origin, then the intercept of the regression line $A = \bar{Y} - \bar{X} S_{xy}/S_x^2$ is zero. In this case, $R \, S_x^2 - S_{xy}$ is zero and the estimator is unbiased. In addition, when the sample size is large, the bias is negligible with respect to the standard deviation of the estimator. Therefore, it is possible to make valid inference despite the existence of a bias.

9.2.5 Ratio and Stratified Designs

The ratio estimator is very frequently used, especially in business surveys. In this case, the auxiliary variable is often the number of workers in the enterprise. However, this estimator is rarely used at the population level, but is used in relatively small strata. In business surveys, strata are often formed by crossing the sector of activities and firm size classes measured by the number of workers. An estimator is called a separated ratio estimator if it is a sum of ratio estimators calculated separately in each stratum. A combined ratio estimator is an estimator that uses a ratio only at the population level.

Assume that the population is partitioned into H strata $U_1, \ldots, U_h, \ldots, U_H$. Let $\widehat{Y}_h$ and $\widehat{X}_h$ be the expansion estimators of the totals of the variable of interest and the auxiliary variable in each stratum:

$$Y_h = \sum_{k \in U_h} y_k \quad \text{and} \quad X_h = \sum_{k \in U_h} x_k.$$

If the design is stratified and the totals of the strata X_h are known, we can construct a ratio estimator in each stratum (RATIO-STRAT) and then sum the estimators obtained. Therefore, we obtain

$$\widehat{Y}_{\text{RATIO-STRAT}} = \sum_{h=1}^{H} \widehat{Y}_{\text{RATIO},h},$$

where

$$\widehat{Y}_{\text{RATIO},h} = \frac{X_h \, \widehat{Y}_h}{\widehat{X}_h}$$

is the ratio estimator in stratum h, for $h = 1, \ldots, H$,

$$\widehat{Y}_h = \sum_{k \in S_h} \frac{y_k}{\pi_k}, \qquad \widehat{X}_h = \sum_{k \in S_h} \frac{x_k}{\pi_k},$$

and S_h represents the random sample selected in stratum h.

In a stratified sampling design, the drawings are independent from one stratum to another. We can write

$$\text{var}_p(\widehat{Y}_{\text{RATIO-STRAT}}) = \sum_{h=1}^{H} \text{var}_p(\widehat{Y}_{\text{RATIO},h}).$$

Therefore, variances and variance estimators can be deduced directly from the previous sections since the problem can be decomposed within each stratum.

The combined ratio is simply a ratio calculated at the global level. We define

$$X = \sum_{h=1}^{H} X_h, \quad Y = \sum_{h=1}^{H} Y_h, \quad \widehat{X} = \sum_{h=1}^{H} \widehat{X}_h \quad \text{and} \quad \widehat{Y} = \sum_{h=1}^{H} \widehat{Y}_h.$$

The combined ratio estimator is a ratio calculated at the global level:

$$\widehat{Y}_{\text{RATIO-COMB}} = \frac{X \, \widehat{Y}}{\widehat{X}} = \sum_{i=1}^{H} X_i \frac{\sum_{h=1}^{H} \widehat{Y}_h}{\sum_{h=1}^{H} \widehat{X}_h}.$$

To calculate the separated ratio estimator, one needs to know more auxiliary information, since all X_h must be known. To calculate the combined estimator, we only need to know X. The separated estimator is almost always preferable to the combined estimator, except when the ratios Y_h/X_h are equal or nearly equal, which is a difficult assumption, for instance, in business surveys.

9.3   The Difference Estimator

The difference estimator is a second simple way to introduce auxiliary information into an estimator. This estimator is particularly prescribed when the variables x and y are expressed in the same units and measure comparable phenomena, i.e. when the slope of the regression line is equal to or close to 1, as presented on Figure 9.2.

For example, if we realize a survey to take inventory of a store, y_k can be the inventory value calculated on the sampled products and x_k a known book value for all products in the store. In this case, the values y_k and x_k are very close and the gain in precision is very large.

The difference estimator (D) is

$$\widehat{Y}_D = \widehat{Y} + X - \widehat{X}.$$

It is a linear estimator as defined in Expression (9.1), where $w_k(S) = 1/\pi_k, k \in U$, and $w_0(S) = X - \widehat{X}$. The difference estimator is unbiased. Indeed,

$$E_p(\widehat{Y}_D) = E_p(\widehat{Y}) + E_p(X) - E_p(\widehat{X}) = Y + X - X = Y.$$

Since

$$\widehat{Y}_D = X + \widehat{E},$$

where

$$\widehat{E} = \sum_{k \in S} \frac{E_k}{\pi_k}$$

and $E_k = y_k - x_k$, the variance is

$$\mathrm{var}_p(\widehat{Y}_D) = \mathrm{var}_p(\widehat{E}) = \mathrm{var}_p(\widehat{Y}) + \mathrm{var}_p(\widehat{X}) - 2\mathrm{cov}_p(\widehat{X}, \widehat{Y}).$$

Under a simple design, this variance becomes

$$\mathrm{var}_p(\widehat{Y}_D) = N^2 \left(1 - \frac{n}{N}\right) \frac{1}{n} S_E^2 = N^2 \left(1 - \frac{n}{N}\right) \frac{1}{n} (S_y^2 + S_x^2 - 2S_{xy}),$$

where

$$S_E^2 = \frac{1}{N-1} \sum_{k \in U} (E_k - \overline{E})^2$$

and

$$\overline{E} = \frac{1}{N} \sum_{k \in U} E_k.$$

Figure 9.2 Difference estimator: observations aligned along a line of slope equal to 1.

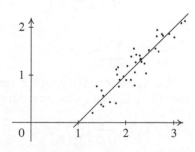

This variance can be unbiasedly estimated by

$$v(\widehat{Y}_D) = N^2 \left(1 - \frac{n}{N}\right) \frac{1}{n} s_E^2 = N^2 \left(1 - \frac{n}{N}\right) \frac{1}{n} (s_y^2 + s_x^2 - 2s_{xy}),$$

where

$$s_E^2 = \frac{1}{n-1} \sum_{k \in U} (E_k - \widehat{\overline{E}})^2$$

and

$$\widehat{\overline{E}} = \frac{1}{n} \sum_{k \in S} E_k.$$

The difference estimator is worth using if $\widehat{Y} - \widehat{X}$ has a small variance, i.e. if the $(y_k - x_k)/\pi_k$ are constant or almost constant, which is the case when the observations are aligned along a line of slope equal to 1, as in Figure 9.2. For example, if y_k and x_k are the same variable, but measured in another context or at another time, the difference estimator is entirely appropriate.

9.4 Estimation by Regression

The regression estimator, which generalizes the ratio, and the difference estimator, is appropriate when the variable of interest y depends linearly on the auxiliary variable x, the ordinate on the origin is different from zero, and the slope is different from 0 and 1. Indeed, if we know that the slope is equal to 1, then we will prefer the difference estimator. If we know that the intercept is equal to 0, we will prefer the ratio estimator. If the slope of the regression line is zero, then the regression estimator may have a larger variance than the expansion estimator.

The regression estimator (REG) is defined by

$$\widehat{Y}_{\text{REG}} = \widehat{Y} + (X - \widehat{X})\widehat{B},$$

where

$$\widehat{B} = \frac{\widehat{S}_{xy}}{\widehat{S}_x^2},$$

$$\widehat{S}_{xy} = \frac{1}{\widehat{N} - 1} \sum_{k \in S} \frac{1}{\pi_k} \left(x_k - \frac{\widehat{X}}{\widehat{N}}\right) \left(y_k - \frac{\widehat{Y}}{\widehat{N}}\right),$$

$$\widehat{S}_x^2 = \frac{1}{\widehat{N} - 1} \sum_{k \in S} \frac{1}{\pi_k} \left(x_k - \frac{\widehat{X}}{\widehat{N}}\right)^2,$$

and

$$\widehat{N} = \sum_{k \in S} \frac{1}{\pi_k}.$$

The regression estimator belongs to the class of linear estimators defined in Expression (9.1). Indeed, as the covariance estimated in the sample can also be written as

$$\widehat{S}_{xy} = \frac{1}{\widehat{N} - 1} \sum_{k \in S} \frac{1}{\pi_k} \left(x_k - \frac{\widehat{X}}{\widehat{N}}\right) y_k,$$

we have $w_0(S) = 0$ and

$$w_k(S) = \frac{1}{\pi_k}\left[1 + (X - \hat{X})\frac{1}{\hat{S}_x^2(\hat{N} - 1)}\left(x_k - \frac{\hat{X}}{\hat{N}}\right)\right], k \in S.$$

We can then write the regression estimator in the form

$$\hat{Y}_{\text{REG}} = \sum_{k \in S} w_k(S)y_k.$$

It is not possible to accurately calculate the expectation and variance of the regression estimator. It is nevertheless possible to show that the regression estimator is unbiased. Using the linearization techniques developed in Chapter 15, we can obtain an approximation of the variance (see Expression 15.9, page 309):

$$\text{var}_p(\hat{Y}_{\text{REG}}) \approx \text{var}_p\left(\sum_{k \in S} \frac{E_k}{\pi_k}\right) = \sum_{k \in U}\sum_{\ell \in U}\frac{E_k E_\ell}{\pi_k \pi_\ell}\Delta_{k\ell},$$

where $E_k = y_k - x_k B$ and $B = S_{xy}/S_x^2$. We can also write

$$\text{var}_p(\hat{Y}_{\text{REG}}) \approx \text{var}_p[\hat{Y} + (\overline{X} - \hat{\overline{X}})B] = \text{var}_p(\hat{Y}) - 2B\,\text{cov}_p(\hat{X}, \hat{Y}) + B^2\,\text{var}_p(\hat{X}).$$

We can then estimate this approximate variance by

$$v\left(\sum_{k \in S}\frac{e_k}{\pi_k}\right) = \sum_{k \in S}\sum_{\ell \in S}\frac{e_k e_\ell \Delta_{k\ell}}{\pi_k \pi_\ell \pi_{k\ell}},$$

where $e_k = y_k - x_k\hat{B}$ and $\hat{B} = \hat{S}_{xy}/\hat{S}_x^2$. We can also write

$$v(\hat{Y}_{\text{REG}}) = v(\hat{Y}) - 2\hat{B}\,\text{cov}(\hat{X}, \hat{Y}) + \hat{B}^2\,v(\hat{X}),$$

where $\text{cov}(\hat{X}, \hat{Y})$ is an estimator of $\text{cov}_p(\hat{X}, \hat{Y})$.

Under a simple design, the approximate variance becomes

$$\text{var}_p(\hat{Y}_{\text{REG}}) \approx N^2\left(1 - \frac{n}{N}\right)\frac{1}{n}(S_y^2 - 2BS_{xy} + B^2 S_x^2)$$
$$= N^2\left(1 - \frac{n}{N}\right)\frac{1}{n}S_y^2(1 - \rho_{xy}^2),$$

where

$$\rho_{xy} = \frac{S_{xy}}{S_x S_y}$$

is the correlation coefficient between x and y. This approximate variance can be estimated by

$$v(\hat{Y}_{\text{REG}}) = N^2\left(1 - \frac{n}{N}\right)\frac{1}{n}s_y^2(1 - \hat{\rho}_{xy}^2),$$

where

$$\hat{\rho}_{xy} = \frac{s_{xy}}{s_x s_y}.$$

It can be seen that the larger the correlation between x and y, the smaller the variance of the regression estimator.

The regression estimator generalizes both the difference estimator and the ratio estimator. It is particularly interesting when $y_k - x_k B$ is small, i.e. when the observations are aligned along a line. If this line passes through the origin, then the regression estimator is reduced to a ratio estimator. If this line has a slope equal to 1, then it is reduced to a difference estimator. The regression estimation can be generalized to the use of several auxiliary variables (see Chapter 11).

9.5   The Optimal Regression Estimator

Montanari (1987) proposed an alternative approach for estimating the regression coefficient of the regression estimator (see also Berger et al., 2001). In this approach, we are interested in expression

$$\tilde{Y}_B = \hat{Y} + (X - \hat{X})B.$$

Parameter B is not random. So, $\tilde{Y}_B$ is not really an estimator in the sense that it depends on the unknown parameter B. We will nevertheless look for the value of B which minimizes the variance $\text{var}_p(\tilde{Y}_B)$. This variance is

$$\text{var}_p(\tilde{Y}_B) = \text{var}_p(\hat{Y}) + B^2\text{var}_p(\hat{X}) - 2B\text{cov}_p(\hat{X}, \hat{Y}).$$

The minimum can be found by solving the equation obtained by canceling the derivative of $\text{var}_p(\hat{Y}_B)$ with respect to B. We obtain

$$\frac{d\text{var}_p(\tilde{Y}_B)}{dB} = 2B\text{var}_p(\hat{X}) - 2\text{cov}_p(\hat{X}, \hat{Y}) = 0.$$

Therefore,

$$B_{\text{OPT}} = \frac{\text{cov}_p(\hat{X}, \hat{Y})}{\text{var}_p(\hat{X})}.$$

Expression

$$\tilde{Y}_{\text{REG-OPT}} = \hat{Y} + (X - \hat{X})B_{\text{OPT}}$$

is not an estimator because $\text{cov}_p(\hat{X}, \hat{Y})$ and $\text{var}_p(\hat{X})$ are unknown. The variance of $\tilde{Y}_B$ is

$$
\begin{aligned}
\text{var}_p(\tilde{Y}_{\text{REG-OPT}}) &= \text{var}_p(\hat{Y}) + B_{\text{OPT}}^2\text{var}_p(\hat{X}) - 2B_{\text{OPT}}\text{cov}_p(\hat{X}, \hat{Y}) \\
&= \text{var}_p(\hat{Y}) + \frac{\text{cov}_p^2(\hat{X}, \hat{Y})}{\text{var}^2(\hat{X})}\text{var}_p(\hat{X}) - 2\frac{\text{cov}_p(\hat{X}, \hat{Y})}{\text{var}_p(\hat{X})}\text{cov}_p(\hat{X}, \hat{Y}) \\
&= \text{var}_p(\hat{Y}) - \frac{\text{cov}_p^2(\hat{X}, \hat{Y})}{\text{var}_p(\hat{X})} = \text{var}_p(\hat{Y})[1 - \text{cor}(\hat{X}, \hat{Y})],
\end{aligned}
$$

where

$$\text{cor}(\hat{X}, \hat{Y}) = \frac{\text{cov}_p(\hat{X}, \hat{Y})}{\sqrt{\text{var}_p(\hat{X})\text{var}_p(\hat{Y})}}.$$

In order to obtain an estimator, we must estimate $\text{var}_p(\hat{X})$ and $\text{cov}_p(\hat{X}, \hat{Y})$. We then obtain the optimal regression estimator (REG-OPT):

$$\hat{Y}_{\text{REG-OPT}} = \hat{Y} + (X - \hat{X})\hat{B}_{\text{OPT}},$$

where

$$\hat{B}_{\text{OPT}} = \frac{\text{cov}(\hat{X}, \hat{Y})}{v(\hat{X})}.$$

Of course, $\hat{Y}_{\text{REG-OPT}}$ is not quite optimal anymore because B_{OPT} is estimated. It is nevertheless close to the optimal estimator when the sample size is large. Therefore, we can use the variance of $\tilde{Y}_{\text{REG-OPT}}$ as an approximate variance of $\hat{Y}_{\text{REG-OPT}}$, in other words,

$$\text{var}_p(\hat{Y}_{\text{REG-OPT}}) \approx \text{var}_p(\tilde{Y}_{\text{REG-OPT}}) = \text{var}_p(\hat{Y})[1 - \text{cor}^2(\hat{X}, \hat{Y})].$$

Under a simple design, the regression estimator and the optimal estimator $\hat{Y}_{\text{REG-OPT}}$ are equal because

$$\hat{B}_{\text{OPT}} = \frac{\hat{S}_{xy}}{\hat{S}_x^2}.$$

Therefore, the regression estimator is asymptotically optimal in this particular case.

9.6 Discussion of the Three Estimation Methods

The choice of an estimation method depends on the relationship between the variables x and y. Since these estimators are asymptotically unbiased, we will choose the estimator with the smallest variance or approximate variance. It is possible to draw precise conclusions in the case where the design is simple random sampling without replacement and when n is large. Under this design, the estimators and their variances are presented in Table 9.1.

- If we compare the variances of the difference estimator and the expansion estimator, we obtain

$$\text{var}_p(\hat{Y}) - \text{var}_p(\hat{Y}_D)$$
$$= N^2 \left(1 - \frac{n}{N}\right)\frac{1}{n}S_y^2 - N^2 \left(1 - \frac{n}{N}\right)\frac{1}{n}(S_y^2 + S_x^2 - 2S_{xy})$$
$$= N^2 \left(1 - \frac{n}{N}\right)\frac{1}{n}(2S_{xy} - S_x^2).$$

The difference estimator is better than the expansion estimator when

$$2S_{xy} - S_x^2 > 0,$$

which can be written as

$$B > \frac{1}{2}.$$

Table 9.1 Estimation methods: summary table.

Estimator	Definition	$\left[N^2\left(1-\frac{n}{N}\right)\frac{1}{n}\right]^{-1} \times \text{var}_p$
Expansion	$\hat{Y} = \dfrac{N}{n}\displaystyle\sum_{k\in S} y_k$	S_y^2
Difference	$\hat{Y}_D = \hat{Y} + X - \hat{X}$	$S_y^2 + S_x^2 - 2S_{xy}$
Ratio	$\hat{Y}_{\text{RATIO}} = X\,(\hat{Y}/\hat{X})$	$S_y^2 + R^2\,S_x^2 - 2R\,S_{xy}$
Regression	$\hat{Y}_{\text{reg}} = \hat{Y} + (X - \hat{X})\hat{B}$	$S_y^2(1 - \rho_{xy}^2)$

- Now compare the ratio estimator and the expansion estimator:

$$\text{var}_p(\hat{Y}) - \text{var}_p(\hat{Y}_{\text{RATIO}})$$

$$\approx N^2\left(1-\frac{n}{N}\right)\frac{1}{n}S_y^2 - N^2\left(1-\frac{n}{N}\right)\frac{1}{n}(S_y^2 + R^2\,S_x^2 - 2R\,S_{xy})$$

$$= N^2\left(1-\frac{n}{N}\right)\frac{1}{n}(2R\,S_{xy} - R^2\,S_x^2).$$

While recalling that the expression of the variance for the ratio estimator is an approximation, we can say that the ratio estimator is, in general, better than the expansion estimator when

$$2R\,S_{xy} - R^2\,S_x^2 > 0,$$

which is when

$$B > \frac{R}{2} \ \text{ if } R > 0 \ \text{ and } \ B < \frac{R}{2} \ \text{ if } R < 0.$$

In the case where Y and X are positive, inequality $B > R/2$ implies that

$$A < \frac{Y}{2N},$$

where

$$A = \frac{Y}{N} - B\frac{X}{N}$$

is the ordinate at the origin of the regression line. Therefore, the ratio estimator is better than the expansion estimator if the intercept A is not too far from 0.

- Finally, compare the ratio estimator to the difference estimator:

$$\text{var}_p(\hat{Y}_D) - \text{var}_p(\hat{Y}_{\text{RATIO}})$$

$$\approx N^2\left(1-\frac{n}{N}\right)\frac{1}{n}(S_y^2 + S_x^2 - 2S_{xy})$$

$$- N^2\left(1-\frac{n}{N}\right)\frac{1}{n}(S_y^2 + R^2\,S_x^2 - 2R\,S_{xy})$$

$$= N^2\left(1-\frac{n}{N}\right)\frac{1}{n}[(1 - R^2)S_x^2 - 2(1 - R)S_{xy}].$$

The ratio estimator is usually better when

$$2(1 - R)S_{xy} - (1 - R^2)S_x^2 < 0,$$

which is when

$$2(1 - R)B < (1 - R^2).$$

If we divide this equality by $(1 - R)\, \text{sign}(Y - X) < 0$, we obtain

$$2B\, \text{sign}(Y - X) > (1 + R)\, \text{sign}(Y - X).$$

Since

$$B = R - \frac{AN}{X},$$

we obtain

$$2\left(R - \frac{AN}{X}\right)\, \text{sign}(Y - X) > (1 + R)\, \text{sign}(Y - X),$$

which gives, if $X > 0$,

$$A\, \text{sign}(Y - X) < \frac{|Y - X|}{N}.$$

The ratio estimator is preferable to the difference estimator if the intercept of the regression line A is not too far from 0.

- Approximately, the regression estimator is better than the other three estimators because under a simple design it is equal to the optimal estimator. Indeed, we have

$$\text{var}_p(\widehat{Y}) - \text{var}_p(\widehat{Y}_{\text{REG}}) \approx \rho_{xy}^2\, \text{var}_p(\widehat{Y}) \geq 0,$$

$$\text{var}_p(\widehat{Y}_D) - \text{var}_p(\widehat{Y}_{\text{REG}})$$

$$\approx N^2\left(1 - \frac{n}{N}\right)\frac{1}{n}(S_y^2 + S_x^2 - 2S_{xy}) - N^2\left(1 - \frac{n}{N}\right)\frac{1}{n}(1 - \rho_{xy}^2)S_y^2$$

$$= N^2\left(1 - \frac{n}{N}\right)\frac{1}{n}\left(S_x - \frac{S_{xy}}{S_x}\right)^2 \geq 0,$$

and

$$\text{var}_p(\widehat{Y}_{\text{RATIO}}) - \text{var}_p(\widehat{Y}_{\text{REG}})$$

$$\approx N^2\left(1 - \frac{n}{N}\right)\frac{1}{n}(S_y^2 + R^2\, S_x^2 - 2R\, S_{xy}) - N^2\left(1 - \frac{n}{N}\right)\frac{1}{n}(1 - \rho_{xy}^2)S_y^2$$

$$= N^2\left(1 - \frac{n}{N}\right)\frac{1}{n}\left(R\, S_x - \frac{S_{xy}}{S_x}\right)^2 \geq 0.$$

Therefore, the regression estimator is preferable to the other three estimators when the population size is large. However, it is advisable to be careful because some variances are approximated. The regression estimator requires the estimation of the regression coefficient B. This causes the loss of a degree of freedom that is not taken into account in the approximate variance. So, if the gain from using the regression estimator is small compared to the ratio estimator or the difference estimator, it may be better not to use it.

Exercises

9.1 We would like to estimate the mean $\overline{Y}$ of a variable y by means of a sample selected according to a simple random sampling design of size 1000 in a population of size 1 000 000. We know the mean $\overline{X} = 15$ of the auxiliary variable x. We have the following results:

$$s_y^2 = 20, s_x^2 = 25, s_{xy} = 15, \widehat{\overline{X}} = 14, \widehat{\overline{Y}} = 10.$$

1. Estimate $\overline{Y}$ using the expansion, difference, ratio, and regression estimators.
2. Estimate the variances of these estimators.
3. Which estimator would you choose to estimate $\overline{Y}$?

9.2 The director of a shoe-making company wants to estimate the average length of the right foot of adult men (18+) in a city. Let y be the variable "length of the right foot" and x the size. The director also knows from census results that the average height of adult men in this city is 168 cm. To estimate foot length, the director organizes a survey with a simple random sampling design without replacement of 100 adult men. The results are as follows:

$$\widehat{\overline{X}} = 169, \widehat{\overline{Y}} = 24, s_{xy} = 15, s_x = 10, s_y = 2.$$

Knowing that 400 000 adult men live in this city:
1. Calculate the expansion estimator, the ratio estimator and the difference estimator.
2. Calculate the variance estimators of these three estimators.
3. Would you advise the director to use the ratio estimator or the difference estimator?

9.3 In a city populated by one million households, we want to measure the household income. A survey of 400 households is conducted according to a simple random sampling without replacement. Denoting "age of the household head" by x and "household income" by y, the survey results are as follows:

$$\widehat{\overline{X}} = 40, \widehat{\overline{Y}} = 150, s_x^2 = 100, s_y^2 = 900, s_{xy} = 150.$$

A census also gives us the average age of the inhabitants for the whole city $\overline{X} = 38$.
1. Estimate the average household income by the ratio estimator, the difference estimator, and the expansion estimator.
2. Estimate the variances of these three estimators.

10

Post-Stratification and Calibration on Marginal Totals

10.1 Introduction

In this chapter, the objective is still to estimate a total Y using the results of the survey and totals known for the complete population. We first deal with the case where the available auxiliary information is the population total of some qualitative variable. In this case, the auxiliary variables are indicator variables of the presence of the units in this group or post-strata. We can then do a post-stratification. We will then examine the case of calibration which applies when the totals of two qualitative variables are known for a population.

10.2 Post-Stratification

10.2.1 Notation and Definitions

Post-stratification can be considered as the basic method of using auxiliary information during estimation. This intuitive technique simply consists of giving subgroups within the sample the same weight as these same subgroups in the population. Post-stratification has been studied, among others, by Holt & Smith (1979), Doss et al. (1979), Jagers (1986), Jagers et al. (1985), and Valliant (1993).

This issue is graphically represented in Figure 10.1. Suppose that the auxiliary variable is qualitative and takes J distinct values. Similarly to stratification, this variable can partition the population $U = \{1, \dots, k, \dots, N\}$ in J subsets, $U_j, j = 1, \dots, J$, called post-strata, such that

$$\bigcup_{j=1}^{J} U_j = U \text{ and } U_j \bigcap U_i = \emptyset, \text{ and } j \neq i.$$

The number of elements in the post-stratum N_j is called the size of the post-stratum. We have

$$\sum_{j=1}^{J} N_j = N.$$

Sampling and Estimation from Finite Populations, First Edition. Yves Tillé.
© 2020 John Wiley & Sons Ltd. Published 2020 by John Wiley & Sons Ltd.

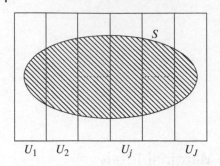

Figure 10.1 Post-stratification: the population is divided in post-strata, but the sample is selected without taking post-strata into account.

$U_1 \quad U_2 \qquad U_j \qquad\qquad U_J$

The functions of interest N_j are supposed to be known. These constitute the auxiliary information used to improve the estimator of the total of variable y.

The sample S is selected without taking the post-strata into account. Post-stratification is fundamentally different to stratification. Indeed, for post-stratification, the auxiliary information is used after sample selection, while for stratification, auxiliary information is integrated into the sampling design.

Here, the auxiliary variables are the indicators $\delta_1, \dots, \delta_j, \dots, \delta_J$. The vector δ_k contains J indicator variables for the presence of unit k in each of the post-strata, that is to say,

$$\delta_k = (\delta_{k1}, \dots, \delta_{kj}, \dots, \delta_{kJ})^{\mathsf{T}},$$

where

$$\delta_{kj} = \begin{cases} 1 & \text{if } k \in U_j \\ 0 & \text{if } k \notin U_j. \end{cases}$$

We then have

$$\sum_{k \in U} \delta_k = (N_1, \dots, N_j, \dots, N_J)^{\mathsf{T}},$$

which is the vector of totals in the population.

As for stratification, the population total can be written as

$$Y = \sum_{k \in U} y_k = \sum_{j=1}^{J} \sum_{k \in U_j} y_k,$$

where Y_j represents the total in post-stratum j:

$$Y_j = \sum_{k \in U} \delta_{kj} y_k = \sum_{k \in U_j} y_k.$$

Similarly, the mean can be written as

$$\overline{Y} = \frac{1}{N} \sum_{k \in U} y_k = \sum_{j=1}^{J} \frac{N_j}{N} \overline{Y}_j,$$

where $\overline{Y}_j$ is the mean in post-stratum j:

$$\overline{Y}_j = \frac{1}{N_j} \sum_{k \in U_j} y_k, j = 1, \dots, J.$$

Furthermore, let S_{yj}^2 denote the variance in post-stratum j:

$$S_{yj}^2 = \frac{1}{N_j - 1} \sum_{k \in U_j} (y_k - \overline{Y}_j)^2.$$

The total variance S_y^2 can be decomposed according to the law of total variance:

$$S_y^2 = \frac{1}{N-1} \sum_{k \in U} (y_k - \overline{Y})^2 = \frac{1}{N-1} \sum_{k \in U} (y_k - \overline{Y}_j + \overline{Y}_j - \overline{Y})^2$$

$$= \frac{1}{N-1} \sum_{j=1}^{J} (N_j - 1)S_{yj}^2 + \frac{1}{N-1} \sum_{j=1}^{J} N_j(\overline{Y}_j - \overline{Y})^2.$$

10.2.2 Post-Stratified Estimator

In the general case, suppose we select a sample S using some sampling design with inclusion probabilities π_k. The expansion estimator with weights $d_k = 1/\pi_k$ is defined by

$$\hat{Y} = \sum_{k \in S} \frac{y_k}{\pi_k} = \sum_{k \in S} d_k y_k,$$

the expansion estimator of the total in post-stratum U_j:

$$\hat{Y}_j = \sum_{k \in S_j} \frac{y_k}{\pi_k}, \tag{10.1}$$

where $S_j = U_j \cap S$, and the mean estimator in post-stratum U_j:

$$\hat{\overline{Y}}_j = \frac{\hat{Y}_j}{\hat{N}_j}, j = 1, \dots, J,$$

where

$$\hat{N}_j = \sum_{k \in S_j} \frac{1}{\pi_k}, j = 1, \dots, J,$$

is the expansion estimator of N_j.

If we refer to Section 9.2 relative to the ratio estimator, it is possible to estimate the total Y_j of post-stratum U_j by using either the expansion estimator given in Expression (10.1) or the known auxiliary information N_j using a ratio estimator:

$$\hat{Y}_{j\text{RATIO}} = N_j \hat{\overline{Y}}_j = \frac{N_j \hat{Y}_j}{\hat{N}_j}, j = 1, \dots, J.$$

The post-stratified estimator is the estimator of the total obtained by summing $\hat{Y}_{j\text{RATIO}}$ over the post-strata:

$$\hat{Y}_{\text{POST}} = \sum_{j=1}^{J} \hat{Y}_{j\text{RATIO}} = \sum_{\substack{j=1 \\ \hat{N}_j > 0}}^{J} \frac{N_j \hat{Y}_j}{\hat{N}_j}.$$

The knowledge of population totals N_j is necessary to calculate this estimator. We must be aware that some of the S_j may be empty. Some $\widehat{N}_j$ (which are random) can be equal to zero. In this case, one or multiple $\widehat{Y}_{j\text{RATIO}}$ may not exist. This small issue induces a small bias in the post-stratified estimator. To overcome this small problem, we try to make sure that the post-strata are sufficiently full by defining the strata such that

$$\sum_{k \in U_j} \pi_k \geq 30,$$

which makes zero $\widehat{N}_j$ quite unlikely. The estimator is linearly homogeneous, as it can be written using (9.1) with

$$w_0(S) = 0 \quad \text{and} \quad w_k(S) = \frac{N_j}{\pi_k \widehat{N}_j}, k \in U_j.$$

Therefore, the weights $w_k(S)$ are random variables, as the $\widehat{N}_j$ are random.

The post-stratified estimator is calibrated on the totals of post-strata N_j in the sense that, when estimating N_j using the post-stratified estimator, we recover exactly the post-stratum totals. Indeed,

$$N_{jpost} = \sum_{k \in S} w_k(S) \delta_{kj} = \sum_{k \in S} \frac{N_j}{\pi_k \widehat{N}_j} \delta_{kj} = \frac{N_j}{\widehat{N}_j} \sum_{k \in S} \frac{\delta_{kj}}{\pi_k} = \frac{N_j}{\widehat{N}_j} \widehat{N}_j = N_j,$$

for all $j = 1, \ldots, J$.

10.3 The Post-Stratified Estimator in Simple Designs

10.3.1 Estimator

Under simple designs, it is possible to give more precise results than in a general case. In particular, we can calculate the expectation and variance of the post-stratified estimator. If in the population we select sample S using a simple random sampling without replacement, the sample sizes of the post-strata n_j are random variables. Since $\pi_k = n/N, k \in U$, and $d_k = N/n$, the expansion estimator of Y becomes

$$\widehat{Y} = \frac{N}{n} \sum_{k \in S} y_k = \frac{N}{n} \sum_{\substack{j=1 \\ n_j > 0}}^{J} n_j \overline{\widehat{Y}}_j,$$

where $\overline{\widehat{Y}}_j$ is the mean of sample within post-stratum j:

$$\overline{\widehat{Y}}_j = \frac{1}{n_j} \sum_{k \in S_j} y_k.$$

The post-stratified estimator is defined by

$$\widehat{Y}_{\text{POST}} = \sum_{\substack{j=1 \\ n_j > 0}}^{J} N_j \overline{\widehat{Y}}_j. \tag{10.2}$$

10.3.2 Conditioning in a Simple Design

Under a simple design, one can study the properties of this estimator by conditioning the post-stratified estimator with respect to vector $(n_1, \ldots, n_j, \ldots, n_J)$ that has a multivariate hypergeometric distribution, i.e.

$$\Pr(n_j = n_{0j}, j = 1, \ldots, J) = \frac{\prod_{j=1}^{J} \binom{N_j}{n_{0j}}}{\binom{N}{n}}, \quad \text{with} \quad \sum_{j=0}^{J} n_{0j} = n.$$

Each n_j has a hypergeometric distribution:

$$\Pr(n_j = n_{0j}) = \frac{\binom{N_j}{n_{0j}} \binom{N - N_j}{n - n_{0j}}}{\binom{N}{n}}, n_{0j} = \max[0, n - (N - N_j)], \ldots, \min(n, N_j).$$

$$(10.3)$$

If we calculate the probability distribution of the random sample S conditioned on vector $(n_1, \ldots, n_j, \ldots, n_J)^{\mathsf{T}}$, we obtain the following fundamental result.

Result 10.1

$$\Pr(S = s | n_j, j = 1, \ldots, J) = \prod_{j=1}^{J} \binom{N_j}{n_j}^{-1}.$$

Proof: We have

$$\Pr(S = s | n_j = n_{0j}, j = 1, \ldots, J) = \frac{\Pr(S = s \text{ and } n_j = n_{0j}, j = 1, \ldots, J)}{\Pr(n_j = n_{0j}, j = 1, \ldots, J)}.$$

Since

$$\Pr(S = s \text{ and } n_j = n_{0j}, j = 1, \ldots, J)$$

$$= \begin{cases} \Pr(S = s) = \binom{N}{n}^{-1} & \text{if } \#s_j = n_{0j}, j = 1, \ldots, J, \\ 0 & \text{otherwise} \end{cases}$$

we obtain, for all s such that $\#s_j = n_{0j}, j = 1, \ldots, J,$

$$\Pr(S = s | n_j = n_{0j}, j = 1, \ldots, J)$$

$$= \frac{\Pr(S = s \text{ and } n_j = n_{0j}, j = 1, \ldots, J)}{\Pr(n_j = n_{0j}, j = 1, \ldots, J)} = \frac{\binom{N}{n}^{-1}}{\dfrac{\prod_{j=1}^{J} \binom{N_j}{n_{0j}}}{\binom{N}{n}}} = \prod_{j=1}^{J} \binom{N_j}{n_{0j}}^{-1}.$$

∎

Result 10.1 shows that, conditionally on n_j, the design is stratified with size n_j in each post-stratum. Furthermore, conditionally on n_j, we obtain simple random sampling without replacement of fixed size n_j in each post-stratum.

We can directly infer that

$$\Pr(k \in S | n_j, j = 1, \ldots, J) = \frac{n_j}{N_j}, \quad \text{for all } k \in U_j,$$

$$\Pr(k, \ell \in S | n_j, j = 1, \ldots, J) = \frac{n_j(n_j - 1)}{N_j(N_j - 1)}, \quad \text{for all } k, \ell \in U_j,$$

and

$$\Pr(k, \ell \in S | n_j, j = 1, \ldots, J) = \frac{n_j n_i}{N_j N_i}, \quad \text{for all } k \in U_j, \ell \in U_i, j \neq i.$$

10.3.3 Properties of the Estimator in a Simple Design

To determine the expectation of $\widehat{Y}_{\text{POST}}$, we start by calculating the conditional expectation:

$$E_p(\widehat{Y}_{\text{POST}} | n_j, j = 1, \ldots, J) = \sum_{\substack{j=1 \\ n_j > 0}}^{J} N_j E_p(\widehat{\overline{Y}}_j | n_j, j = 1, \ldots, J).$$

Conditionally to n_j, since there is simple design without replacement in each post-stratum, we have

$$E_p(\widehat{\overline{Y}}_j | n_j, j = 1, \ldots, J) = \overline{Y}_j, \quad \text{if } n_j > 0.$$

Therefore,

$$E_p(\widehat{Y}_{\text{POST}} | n_j, j = 1, \ldots, J) = \sum_{\substack{j=1 \\ n_j > 0}}^{J} N_j \overline{Y}_j = Y - \sum_{\substack{j=1 \\ n_j = 0}}^{J} N_j \overline{Y}_j. \tag{10.4}$$

The estimator is then approximately conditionally unbiased. The bias is only due to potentially empty post-strata. The unconditional expectation can be easily calculated by

$$E_p E_p(\widehat{Y}_{\text{POST}} | n_j, j = 1, \ldots, J) = E_p \left(Y - \sum_{\substack{j=1 \\ n_j = 0}}^{J} N_j \overline{Y}_j \right)$$

$$= Y - \sum_{j=1}^{J} N_j \overline{Y}_j \Pr(n_j = 0).$$

As we saw in Expression (10.3), each n_j has a hypergeometric distribution, so we directly obtain

$$\Pr(n_j = 0) = \frac{(N - N_j)^{[n]}}{N^{[n]}},$$

where

$$N^{[n]} = \frac{N!}{(N-n)!} = N \times (N-1) \times \cdots \times (N-n+2) \times (N-n+1),$$

which finally gives

$$E_p(\widehat{Y}_{\text{POST}}) = Y - \sum_{j=1}^{J} N_j \overline{Y}_j \frac{(N-N_j)^{[n]}}{N^{[n]}}.$$

When n is large,

$$\frac{(N-N_j)^{[n]}}{N^{[n]}} < \left(\frac{N-N_j}{N}\right)^n \approx 0.$$

The bias is then negligible. Therefore, we can write

$$E_p(\widehat{Y}_{\text{POST}}) \approx Y.$$

The estimator is also approximatively unconditionally unbiased.

The variance can be obtained using the classical decomposition formula:

$$\text{var}_p(\widehat{Y}_{\text{POST}}) = \text{var}_p\, E_p(\widehat{Y}_{\text{POST}}|n_j, j = 1, \ldots, J) + E_p\, \text{var}_p(\widehat{Y}_{\text{POST}}|n_j, j = 1, \ldots, J).$$

By (10.4), we obtain

$$\text{var}_p\, E_p(\widehat{Y}_{\text{POST}}|n_j, j = 1, \ldots, J) \approx 0,$$

and therefore

$$\text{var}_p(\widehat{Y}_{\text{POST}}) \approx E_p\, \text{var}_p(\widehat{Y}_{\text{POST}}|n_j, j = 1, \ldots, J). \tag{10.5}$$

However, conditionally to the n_j, Result 10.1 teaches us that the design is simple without replacement within each post-stratum. Therefore, the conditional variance is identical to that of a classical stratified design:

$$\text{var}_p(\widehat{Y}_{\text{POST}}|n_j, j = 1, \ldots, J) = \sum_{\substack{j=1 \\ n_j > 0}}^{J} N_j \frac{N_j - n_j}{n_j} S_{yj}^2. \tag{10.6}$$

The unconditional variance is a little trickier to obtain. By (10.5), we have

$$\text{var}_p(\widehat{Y}_{\text{POST}}) \approx E_p\left(\sum_{\substack{j=1 \\ n_j > 0}}^{J} N_j \frac{N_j - n_j}{n_j} S_{yj}^2\right) \approx \sum_{j=1}^{J} N_j \left[N_j E_p\left(\frac{1}{n_j}\right) - 1\right] S_{yj}^2.$$

$$\tag{10.7}$$

We must then calculate the expectation of n_j^{-1}. *Stricto sensu*, this expectation does not exist since n_j has a nonzero probability to be equal to zero. We will still look for an approximation by supposing that this probability is negligible.

By setting

$$\varepsilon = 1 - \frac{n_j}{E_p(n_j)} = 1 - \frac{N n_j}{n N_j},$$

we have

$$E_p\left(\frac{1}{n_j}\right) = \frac{1}{E_p(n_j)}E_p\left(\frac{1}{1-\varepsilon}\right).$$

When n is large, we consider that ε is close to zero and, using a Taylor series expansion, we obtain

$$E_p\left(\frac{1}{n_j}\right) \approx \frac{1}{E_p(n_j)}E_p(1 + \varepsilon + \varepsilon^2).$$

Since n_j has a hypergeometric distribution, we can calculate its expectation and variance. We have

$$E_p(n_j) = n\frac{N_j}{N} \text{ and } \mathrm{var}_p(n_j) = \frac{n\,N_j}{N}\frac{N-N_j}{N}\frac{N-n}{N-1},$$

and we obtain

$$E_p\left(\frac{1}{n_j}\right) \approx \frac{1}{E_p(n_j)}E_p\left[1 + \left(1 - \frac{n_j N}{n N_j}\right) + \left(1 - \frac{n_j N}{n N_j}\right)^2\right]$$

$$= \frac{N}{N_j n}\left[1 + 0 + \frac{N^2 \mathrm{var}_p(n_j)}{n^2 N_j^2}\right] = \frac{N}{N_j n} + \frac{(N-N_j)N}{N_j^2}\frac{N-n}{n^2(N-1)}. \qquad (10.8)$$

Using the result in Expression (10.8) in Expression (10.7), we obtain the variance

$$\mathrm{var}_p(\widehat{Y}_{\mathrm{POST}}) \approx \frac{N-n}{n}\sum_{j=1}^{J}N_j S_{yj}^2 + \frac{(N-n)N^2}{n^2(N-1)}\sum_{j=1}^{J}\frac{N-N_j}{N}S_{yj}^2, \qquad (10.9)$$

which is estimated by

$$v(\widehat{Y}_{\mathrm{POST}}) = \frac{N-n}{n}\sum_{j=1}^{J}N_j s_{yj}^2 + \frac{(N-n)N^2}{n^2(N-1)}\sum_{j=1}^{J}\frac{N-N_j}{N}s_{yj}^2,$$

where

$$s_{yj}^2 = \frac{1}{n_j - 1}\sum_{k \in S_j}(y_k - \widehat{\overline{Y}}_j)^2.$$

Variance (10.9) is composed of two terms. The first term is equal to the variance of the expansion estimator for the stratified design with proportional allocation given in Expression 4.3, page 70. Therefore, we can write

$$\frac{\mathrm{var}_p(\widehat{Y}_{\mathrm{POST}})}{\mathrm{var}_p(\widehat{Y}_{\mathrm{PROP}})} = \left(\frac{N-n}{n}\sum_{j=1}^{J}N_j S_{yj}^2\right)^{-1}$$

$$\times \left[\frac{N-n}{n}\sum_{j=1}^{J}N_j S_{yj}^2 + \frac{(N-n)N^2}{n^2(N-1)}\sum_{j=1}^{J}\frac{N-N_j}{N}S_{yj}^2\right]$$

$$= 1 + \frac{N}{n(N-1)}\left(\sum_{j=1}^{J}\frac{N_j}{N}S_{yj}^2\right)^{-1}\sum_{j=1}^{J}\frac{N-N_j}{N}S_{yj}^2 = 1 + O(n^{-1}),$$

where $n \times O(n^{-1})$ is bounded when n tends towards infinity. Therefore, the second term is a correction of order n^{-1} with respect to the proportional allocation. This term is always positive. This shows that it is always advantageous to use *a priori* stratification. However, this is not always possible and the rectification by post-stratification can be an interesting estimation strategy. As in for *a priori* stratification, the post-stratification variable must be strongly associated with the variable of interest for the variance reduction to be substantial. If the post-stratification variable is independent of the variable of interest, the variance of the post-stratified estimator can then be superior to the variance of the expansion estimator.

10.4 Estimation by Calibration on Marginal Totals

10.4.1 The Problem

Now suppose that two qualitative auxiliary variables are used. The first variable allows the partitioning of the population into J subsets $U_{1.}, \ldots, U_{j.}, \ldots, U_{J.}$, and the second into I subsets $U_{.1}, \ldots, U_{.i}, \ldots, U_{.I}$. These two auxiliary variables allow us to define the crossed partition presented in Table 10.1.

The totals coming from the crossing of these two variables are denoted by $N_{ji} = \#U_{ji}, j = 1, \ldots, J, i = 1, \ldots, I,$. If the N_{ji} were known, we could use them to do a post-stratification. Therefore, suppose that the N_{ji} are unknown or that they are unusable in stratification. Indeed, the N_{ji} coming from the crossing of two variables may be too small and it may be dangerous to use them in a post-stratification as too many post-strata could be empty.

However, the marginal totals denoted by $N_{j.} = \#U_{j.}, j = 1, \ldots, J$, and $N_{.i} = \#U_{.i}, i = 1, \ldots, I$, are supposed known and sufficiently large to not create an empty sample in categories $U_{j.}$ and $U_{.i}$. Therefore, these marginal totals constitute the auxiliary information that may be used to estimate a function of interest.

In this population, we select a random sample S. For each cell in the table, it is possible to construct an expansion estimator of the N_{ji}:

$$\hat{N}_{ji} = \sum_{k \in (S \cap U_{ji})} \frac{1}{\pi_k}, \quad \hat{N}_{j.} = \sum_{k \in (S \cap U_{j.})} \frac{1}{\pi_k},$$

$$\hat{N}_{.i} = \sum_{k \in (S \cap U_{.i})} \frac{1}{\pi_k}, \quad \text{and} \quad \hat{N} = \sum_{k \in S} \frac{1}{\pi_k}.$$

The estimated totals with respect to both variables are presented in Table 10.2.

Table 10.1 Population partition.

U_{11}	$\cdots$	U_{1i}	$\cdots$	U_{1I}	$U_{1.}$
$\vdots$		$\vdots$		$\vdots$	$\vdots$
U_{j1}	$\cdots$	U_{ji}	$\cdots$	U_{jI}	$U_{j.}$
$\vdots$		$\vdots$		$\vdots$	$\vdots$
U_{J1}	$\cdots$	U_{Ji}	$\cdots$	U_{JI}	$U_{J.}$
$U_{.1}$	$\cdots$	$U_{.i}$	$\cdots$	$U_{.I}$	U

Table 10.2 Totals with respect to two variables.

$\hat{N}_{11}$	$\cdots$	$\hat{N}_{1i}$	$\cdots$	$\hat{N}_{1I}$	$\hat{N}_{1.}$
$\vdots$		$\vdots$		$\vdots$	$\vdots$
$\hat{N}_{h1}$	$\cdots$	$\hat{N}_{ji}$	$\cdots$	$\hat{N}_{jI}$	$\hat{N}_{j.}$
$\vdots$		$\vdots$		$\vdots$	$\vdots$
$\hat{N}_{J1}$	$\cdots$	$\hat{N}_{Ji}$	$\cdots$	$\hat{N}_{JI}$	$\hat{N}_{J.}$
$\hat{N}_{.1}$	$\cdots$	$\hat{N}_{.i}$	$\cdots$	$\hat{N}_{.I}$	$\hat{N}$

The objective is to estimate the total

$$Y = \sum_{k \in U} y_k. \tag{10.10}$$

To achieve this, we use a homogeneous linear estimator (L) as defined in Expression (9.1):

$$\hat{Y}_L = \sum_{k \in S} w_k(S) y_k,$$

where the weights $w_k(S)$ will depend on $\hat{N}_{ji}$ and the population marginal totals $N_{j.}$ and $N_{.j}$.

Additionally, we want the estimator to be "calibrated" on the margins. This idea of calibration is important and may even be considered as an estimation principle in survey theory. In the context of this problem, this calibration principle can be defined as follows: the marginal totals of population $N_{j.}$ are totals like (10.10) that can be written as

$$N_{j.} = \sum_{k \in U} z_k,$$

where z_k equals 1 if $k \in U_{j.}$ and 0 otherwise. We will say estimator $\hat{N}_{j.}$ is calibrated on $N_{j.}$ if

$$\hat{N}_{j.} = \sum_{k \in S} w_k(S) z_k = N_{j.},$$

i.e. if it is unbiasedly estimated with zero variance for this function of interest. For example, the simple post-stratified estimator is calibrated on the size of the post-strata (except when a post-stratum is empty). Here, we want to construct an estimator that is jointly calibrated on both margins, and we do this by using the calibration algorithm.

10.4.2 Calibration on Marginal Totals

The calibration algorithm on marginal totals was introduced by Deming & Stephan (1940) and Stephan (1942). It was then studied by Friedlander (1961), Ireland & Kullback (1968), Fienberg (1970), Thionet (1959, 1976), Froment & Lenclud (1976), and Durieux & Payen (1976). Many practical applications of this algorithm come from the framework of survey theory.

This algorithm is also called the *Iterative Proportional Fitting Procedure (IPFP)* or raking ratio. We will refer to it as "calibration", which has the merit of being clearer. Consider in a general way Table 10.3 of $a_{ji} \geq 0, i = 1, \ldots, I, j = 1 \ldots, J$, and its margins.

Table 10.3 Calibration, starting table.

a_{11}	$\cdots$	a_{1i}	$\cdots$	a_{1I}	$a_{1.}$
$\vdots$		$\vdots$		$\vdots$	$\vdots$
a_{j1}	$\cdots$	a_{ji}	$\cdots$	a_{jI}	$a_{j.}$
$\vdots$		$\vdots$		$\vdots$	$\vdots$
a_{J1}	$\cdots$	a_{Ji}	$\cdots$	a_{JI}	$a_{J.}$
$a_{.1}$	$\cdots$	$a_{.i}$	$\cdots$	$a_{.I}$	$a_{..}$

We seek a "close" table to the a_{ji} table that contains the margins $b_{j.}, j = 1, \ldots, J$, and $b_{.i}, i = 1, \ldots, I$.

The first step is to initialize the algorithm:

$$b_{ji}^{(0)} = a_{ji}, j = 1, \ldots, J, i = 1, \ldots, I.$$

Then, we repeat the following allocations for $r = 1, 2, 3, \ldots$:

$$b_{ji}^{(2r-1)} = b_{ji}^{(2r-2)} \frac{b_{j.}}{b_{j.}^{(2r-2)}}, j = 1, \ldots, J, i = 1, \ldots, I,$$

$$b_{ji}^{(2r)} = b_{ji}^{(2r-1)} \frac{b_{.i}}{b_{.i}^{(2r-1)}}, j = 1, \ldots, J, i = 1, \ldots, I,$$

where

$$b_{j.}^{(2r-2)} = \sum_{i=1}^{I} b_{ji}^{(2r-2)}, j = 1, \ldots, J,$$

and

$$b_{.i}^{(2r-1)} = \sum_{j=1}^{J} b_{ji}^{(2r-1)}, j = 1, \ldots, J.$$

The algorithm stops when the margins of the table $b_{ji}^{(r)}$ are sufficiently close to the fixed margins. In practice, the algorithm converges very quickly and it is rare to apply more than ten or so steps to achieve a reasonable precision (4 decimals). The algorithm ends at the same solution regardless if we start on the rows or on the columns. It always converges when $a_{ji} > 0, j = 1, \ldots, J, i = 1, \ldots, I$. However, if some a_{ji} are zero, it may have no solution. We will not go into the complex descriptions of cases where the solution does not exist, as the simplest procedure to detect the absence of a solution is to apply the algorithm and see if it converges. If it does not converge, there is no solution.

There also exists a symmetric version of the calibration algorithm, which has the perk of maintaining table symmetry at each step, when the starting table is symmetric and the calibration margins are equal. The first step consists of initializing the algorithm:

$$b_{ji}^{(0)} = a_{ji}, j = 1, \ldots, J, i = 1, \ldots, I.$$

Then, we repeat the following two allocations for $r = 1, 2, 3, \cdots$:

$$b_{ji}^{(2r-1)} = b_{ji}^{(2r-2)} \frac{b_{j.} b_{.i}}{b_{j.}^{(2r-2)} b_{.i}^{(2r-2)}}, j = 1, \ldots, J, i = 1, \ldots, I,$$

$$b_{ji}^{(2r)} = b_{ji}^{(2r-1)} \frac{b_{..}}{b_{..}^{(2r-1)}}, j = 1, \dots, J, i = 1, \dots, I,$$

where

$$b_{..} = \sum_{j=1}^{J} \sum_{i=1}^{I} b_{ji} \quad \text{and} \quad b_{..}^{(2r-1)} = \sum_{j=1}^{J} \sum_{i=1}^{I} b_{ji}^{(2r-1)}.$$

10.4.3 Calibration and Kullback–Leibler Divergence

The calibration algorithm offers a solution close to the starting table and has fixed margins. We can define this "proximity" more precisely. Indeed, the algorithm corresponds to an optimization problem with constraints, where the criteria used is the minimization of the Kullback–Leibler divergence. We seek the table of b_{ji} that minimizes

$$\sum_{j=1}^{J} \sum_{i=1}^{I} b_{ji} \log \frac{b_{ji}}{a_{ji}},$$

subject to

$$\sum_{i=1}^{I} b_{ji} = b_{j.}, j = 1, \dots, J, \tag{10.11}$$

and

$$\sum_{j=1}^{J} b_{ji} = b_{.i}, i = 1, \dots, I. \tag{10.12}$$

We obtain the Lagrange equation:

$$\mathcal{L}(b_{ji}, \lambda_j, \mu_i)$$
$$= \sum_{j=1}^{J} \sum_{i=1}^{I} b_{ji} \log \frac{b_{ji}}{a_{ji}} + \sum_{j=1}^{J} \lambda_j \left(\sum_{i=1}^{I} b_{ji} - b_{j.} \right) + \sum_{i=1}^{I} \mu_i \left(\sum_{j=1}^{J} b_{ji} - b_{.i} \right),$$

where the λ_j and the μ_i are Lagrange multipliers. By canceling the partial derivatives $\mathcal{L}$ with respect to b_{ji}, we obtain

$$\log \frac{b_{ji}}{a_{ji}} + 1 + \lambda_j + \mu_i = 0, \tag{10.13}$$

and by setting to zero the partial derivatives of $\mathcal{L}$ with respect to λ_j and μ_i, we obtain the constraints (10.12) and (10.11). By setting $\alpha_j = \exp(-1/2 - \lambda_j)$ and $\beta_i = \exp(-1/2 - \mu_i)$, in (10.13), we can write

$$b_{ji} = a_{ji}\alpha_j\beta_i, j = 1, \dots, J, i = 1, \dots, I. \tag{10.14}$$

Expression (10.14) shows that the solution can be written as a cell in the starting table a_{ji} multiplied by two coefficients α_j and β_i that only depend on an index. We check by recursion that the proposed algorithm in Section 10.4.2 offers a solution having this form.

10.4.4 Raking Ratio Estimation

Calibration allows us to construct an estimator of Y. By using the same notation introduced in Section 10.4.1, we can start adjusting the $\hat{N}_{ji}$ on the margins $N_{j.}$ and $N_{.i}$. Let $\hat{N}_{RRji}$ denote the estimator of the total obtained by calibration. The $\hat{N}_{RRji}$ are estimators of N_{ji}. We then use the $\hat{N}_{RRji}$ as if we were constructing a post-stratification and we obtain

$$\hat{Y}_{RR} = \sum_{j=1}^{J}\sum_{i=1}^{I}\hat{N}_{RRji}\frac{\hat{Y}_{ji}}{\hat{N}_{ji}},$$

where

$$\hat{Y}_{ji} = \sum_{k\in S\cap U_{ji}}\frac{y_k}{\pi_k},$$

$$\hat{N}_{ji} = \sum_{k\in S\cap U_{ji}}\frac{1}{\pi_k},$$

and $S_{ji} = U_{ji}\cap S$.

Therefore, the estimator is linear as defined in Expression (9.1) and can be written as

$$\hat{Y} = \sum_{k\in S}w_k(S)y_k,$$

where

$$w_k(S) = \frac{\hat{N}_{RRji}}{\pi_k\hat{N}_{ji}}, k\in S\cap U_{ji}.$$

The weights $w_k(S)$ are nevertheless nonlinear functions of $\hat{N}_{ji}$ and the known margins. That is why it is difficult to study the properties of this estimator. These properties have been studied by Arora & Brackstone (1977), Causey (1972), Brackstone & Rao (1979), Konijn (1981), Binder & Théberge (1988), and Skinner (1991). Deville & Särndal (1992) have generalized this estimator in the framework of the general calibration theory shown in Chapter 12 for which it is possible to calculate variance approximations using linearization techniques (see Section 15.6.4, page 313, and Section 15.10.3, page 327).

10.5 Example

In a town of 8000 inhabitants, we select a simple random sample of fixed size. The results are broken down by gender (M, male; F, female) and by marital status (S, single; M, married; WSD, widowed, separated, divorced). Table 10.4 contains estimated average incomes with respect to these categories and Table 10.5 contains the estimated totals using the expansion estimator. From these two tables, we can derive the total income and the average income (variable y).

$$\hat{Y} = 1\ 200\times 4\ 200 + 2\ 000\times 5\ 000 + 600\times 6\ 000$$
$$+1\ 400\times 3\ 500 + 2\ 000\times 4\ 000 + 800\times 3\ 200$$
$$= 34\ 100\ 000,$$

Table 10.4 Salaries in Euros.

	S	M	WSD
M	4200	5000	6000
F	3500	4000	3200

Table 10.5 Estimated totals using simple random sampling without replacement.

	S	M	WSD	Total
M	1200	2000	600	3800
F	1400	2000	800	4200
T	2600	4000	1400	8000

Table 10.6 Known margins using a census.

	S	M	WSD	Total
M				4400
F				3600
T	2200	4500	1300	8000

and

$$\widehat{\overline{Y}} = \frac{34\ 100\ 000}{8\ 000} = 4\ 262.5.$$

After some research at the regional statistical office, we see that the number of males and females are known for the entire population (Table 10.6). We then decide to adjust Table 10.5 using the marginal sample on the known margins at the population level given in Table 10.6. To achieve this, we apply the calibration algorithm.

In the first iteration, we adjust Table 10.5 on the row population totals presented in Table 10.6 to obtain Table 10.7, where the row totals are properly adjusted, but the column totals are not. In the second iteration, we adjust Table 10.7 on the column population totals presented in Table 10.6 to obtain Table 10.8, where the columns are

Table 10.7 Iteration 1: row total adjustment.

	S	M	WSD	Total
M	1389.5	2315.8	694.7	4400
F	1200.0	1714.3	685.7	3600
T	2589.5	4030.1	1380.5	8000

Table 10.8 Iteration 2: column total adjustment.

	S	M	WSD	Total
M	1180.5	2585.8	654.2	4420.6
F	1019.5	1914.2	645.8	3579.4
T	2200.0	4500.0	1300.0	8000.0

Table 10.9 Iteration 3: row total adjustment.

	S	M	WSD	Total
M	1175.0	2573.8	651.2	4400.0
F	1025.4	1925.2	649.5	3600.0
T	2200.4	4499.0	1300.7	8000.0

properly adjusted, but the row totals are not. However, the row totals are much closer to the population totals than in the initial table.

In the third iteration, we again adjust Table 10.8 on the row population totals. In the fourth iteration, we adjust Table 10.9 on the column population totals to obtain Table 10.9. We see that four iterations are sufficient to obtain a good adjustment one decimal away from the target.

We can then estimate the total and mean income using the calibrated estimator. To do this, we use the totals estimated by calibration in Table 10.10, which give

$$\hat{Y}_{RR} = 1174.8 \times 4200 + 2574.4 \times 5000 + 650.9 \times 6000$$
$$+ 1025.2 \times 3500 + 1925.6 \times 4000 + 649.1 \times 3200$$
$$= 35\ 079\ 280,$$

and

$$\hat{\bar{Y}}_{RR} = \frac{35\ 079\ 280}{8000} = 4384.9.$$

The mean income of the calibrated estimator is larger than the expansion estimator, which is explained by the fact that men and married people were under-represented in the sample. Since these two categories tend to have higher incomes, we increase the mean income estimator by calibration.

Table 10.10 Iteration 4: column total adjustment.

	S	M	WSD	Total
M	1174.8	2574.4	650.9	4400.1
F	1025.2	1925.6	649.1	3599.9
T	2200.0	4500.0	1300.0	8000.0

Exercises

10.1 Give the optimal estimator for a stratified design using the variable x and the vector of constants as regressors. Compare it to the regression estimator.

10.2 Adjust the two tables to the indicated margins in bold using the method to adjust a table to its margins. Explain the issue with Table 2.

Table 1

235	78	15	6	**334**
427	43	17	12	**499**
256	32	14	5	**307**
432	27	32	2	**493**
25	**78**	**180**	**1350**	**1633**

Table 2

0	78	0	6	**84**
427	0	17	0	**444**
0	32	0	5	**37**
432	0	32	0	**464**
11	**49**	**110**	**859**	**1029**

10.3 We are interested in the population of first-year students registered in a university. We know the total number of students whose parents have either an elementary school, high school, or university education. We select a survey using a simple random sampling design without replacement of 150 students and obtain the following result:

	Fail	Pass
Elementary	45	15
High	25	25
College	10	30

If the numbers of students whose parents have elementary school, high school or university education are 5000, 3000, and 2000, respectively:
1. Estimate the pass rate of the students using the expansion estimator and give an estimate of the variance for this estimator.
2. Estimate the pass rate of the students using the post-stratified estimator and give an estimate of the variance for this estimator.
3. Estimate the pass rate by parental education using the raking ratio knowing that, in the overall population, the fail rate is 60%.

11

Multiple Regression Estimation

11.1 Introduction

In some design $p(\cdot)$ where the first- and second-order inclusion probabilities are strictly positive, the objective is to estimate a total

$$Y = \sum_{k \in U} y_k$$

using a sample and multivariate auxiliary information. Generally, this auxiliary information appears as J auxiliary variables denoted by $x_1, \ldots, x_j, \ldots, x_J$. The value x_j of unit k is denoted by x_{kj}. Also, let $\mathbf{x}_k = (x_{k1}, \ldots, x_{kj}, \ldots, x_{kJ})^\mathsf{T}$ denote the column vector of values of the J auxiliary variables in unit k. Now,

$$\overline{X}_j = \frac{1}{N} \sum_{k \in U} x_{kj} = \text{and } X_j = \sum_{k \in U} x_{kj}, j = 1, \ldots, J,$$

or with a vector notation

$$\overline{\mathbf{X}} = \frac{1}{N} \sum_{k \in U} \mathbf{x}_k = \text{and } \mathbf{X} = \sum_{k \in U} \mathbf{x}_k.$$

To be able to use the regression estimator, it is sufficient to know the vector of totals for the auxiliary variables $\mathbf{X}$ at the population level. It is not necessary to know all the values x_{kj} for all units in the population. We can then use totals from a census where the individual data are not available.

In the data collection step, the variable of interest y and the J auxiliary variables are measured on all selected units in the sample. We can then calculate the expansion estimator of the variable of interest:

$$\widehat{Y} = \sum_{k \in S} \frac{y_k}{\pi_k}$$

and the vector of J expansion estimators of the auxiliary variables:

$$\widehat{\mathbf{X}} = \sum_{k \in S} \frac{\mathbf{x}_k}{\pi_k}.$$

Obviously the vector of estimators $\widehat{\mathbf{X}}$ is in general not equal to the vector of known totals $\mathbf{X}$. The multiple regression estimator is then used to improve the regression estimator by correcting it using the differences between $\mathbf{X}$ and $\widehat{\mathbf{X}}$.

Sampling and Estimation from Finite Populations, First Edition. Yves Tillé.
© 2020 John Wiley & Sons Ltd. Published 2020 by John Wiley & Sons Ltd.

The multiple regression estimator is part of the class of homogeneous linear estimators, that is to say, estimators that can be written in the form

$$\hat{Y}_L = \sum_{k \in S} w_k(S) y_k,$$

where the weights $w_k(S)$ can depend on both the sampling unit k and the sample. Therefore, the $w_k(S)$ are random variables that depend on the results of the survey. The multiple regression estimator can be seen as a weighting method (see Bethlehem & Keller, 1987).

Since the vector of **X** is known, we would like the estimator to be "calibrated" on this auxiliary information. We refer to the following definition.

Definition 11.1 *A homogeneous linear estimator* $\hat{\mathbf{X}}$ *is said to be calibrated on total* **X** *if and only if*

$$\hat{\mathbf{X}} = \mathbf{X},$$

regardless of the selected sample.

The calibration property has the benefit of giving very coherent results. For example, in official statistics, we usually know very precisely the age pyramids of surveyed populations. An official statistical agency has a vested interest in calibrating on the age pyramid for all surveys on individuals. In addition, we will see in Section 15.6.4, page 313, and in Section 15.10.3, page 327, that calibration techniques allow the variance of estimators to be considerably reduced.

In this chapter, we will examine regression-based methods. We also show an optimal regression estimator under the sampling design. Under certain conditions, this estimator is conditionally unbiased.

11.2 Multiple Regression Estimator

The multiple regression estimator is the result of a long journey (see, among others, Lemel, 1976; Särndal, 1982; Wright, 1983; Särndal, 1980, 1982, 1984; Särndal & Wright, 1984; Deville, 1988; Särndal et al., 1989,1992; Deville & Särndal, 1992; Fuller, 2002). It consists of using the supposed linear relationship between the auxiliary variables and the variable of interest to obtain a more precise estimator. The multiple regression estimator is often called the generalized regression estimator. However, this designation can be confusing as in statistics, generalized regression usually means a set of nonlinear extensions to regression methods. In survey theory, the multiple regression estimator (or generalized) always refers to a linear regression.

We begin by examining the issue as if the entire population was known. If there exists a linear relationship between $\mathbf{x}_k$ and y_k, we seek the vector of regression coefficients $\mathbf{B} \in \mathbb{R}^J$ which minimize the following criteria:

$$\sum_{k \in U} q_k (y_k - \mathbf{x}_k^\mathsf{T} \mathbf{B})^2, \tag{11.1}$$

where q_k is a positive or zero weighting coefficient which grants a particular importance to each unit. The role of the q_k is detailed in Section 11.6. By canceling the derivative of (11.1) with respect to **B**, we find the estimating equation

$$\sum_{k \in U} q_k \mathbf{x}_k (y_k - \mathbf{x}_k^{\mathsf{T}} \mathbf{B}) = \mathbf{0},$$

which gives

$$\sum_{k \in U} q_k \mathbf{x}_k y_k = \sum_{k \in U} q_k \mathbf{x}_k \mathbf{x}_k^{\mathsf{T}} \mathbf{B}.$$

If we define

$$\mathbf{T} = \sum_{k \in U} q_k \mathbf{x}_k \mathbf{x}_k^{\mathsf{T}} \quad \text{and} \quad \mathbf{t} = \sum_{k \in U} q_k \mathbf{x}_k y_k,$$

and we suppose that **T** is invertible, we obtain regression coefficient $\mathbf{B} = \mathbf{T}^{-1}\mathbf{t}$.

Since **T** and **t** are totals, and $d_k = 1/\pi_k$, we simply estimate them by their expansion estimators:

$$\widehat{\mathbf{T}} = \sum_{k \in S} d_k q_k \mathbf{x}_k \mathbf{x}_k^{\mathsf{T}} \quad \text{and} \quad \widehat{\mathbf{t}} = \sum_{k \in S} d_k q_k \mathbf{x}_k y_k,$$

which are unbiased estimators of **T** and **t**. We then estimate **B** by

$$\widehat{\mathbf{B}} = \widehat{\mathbf{T}}^{-1}\widehat{\mathbf{t}},$$

subject to $\widehat{\mathbf{T}}$ being invertible. Estimator $\widehat{\mathbf{B}}$ is not unbiased for **B**. Indeed, the ratio of two unbiased estimators is not generally an unbiased estimator. However, we can show that the bias is asymptotically negligible. Finally, the multiple regression estimator is defined by

$$\widehat{Y}_{\text{REGM}} = \widehat{Y} + (\mathbf{X} - \widehat{\mathbf{X}})^{\mathsf{T}}\widehat{\mathbf{B}}. \tag{11.2}$$

The multiple regression estimator is an expansion estimator corrected using the differences between $\widehat{\mathbf{X}}$ and **X**.

11.3   Alternative Forms of the Estimator

11.3.1   Homogeneous Linear Estimator

The multiple regression estimator can take many alternative forms. First, we can show that $\widehat{Y}_{\text{REGM}}$ belongs to the class of homogeneous linear estimator. In other words it can be written as

$$\widehat{Y}_{\text{REGM}} = \sum_{k \in S} w_k(S) y_k,$$

where

$$w_k(S) = d_k g_k.$$

The quantities g_k are called the g-weights by Särndal et al. (1989). These g-weights are the ratio of the multiple regression estimator weights over the expansion estimator weights:

$$g_k = \frac{w_k(S)}{d_k} = 1 + (\mathbf{X} - \widehat{\mathbf{X}})^{\mathsf{T}}\widehat{\mathbf{T}}^{-1} q_k \mathbf{x}_k. \tag{11.3}$$

The g-weights are close to 1 and represent the distortion of the initial weights caused by the multiple regression estimator. The more the sample needs to be corrected, the more the g-weights will be different to 1. The multiple regression estimator can then be written as

$$\hat{Y}_{\text{REGM}} = \sum_{k \in S} d_k g_k y_k.$$

11.3.2 Projective Form

It is also interesting to create an alternative form of this estimator. From (11.2), we have

$$\hat{Y}_{\text{REGM}} = \mathbf{X}^\top \hat{\mathbf{B}} + \hat{Y} - \hat{\mathbf{X}}^\top \hat{\mathbf{B}} = \mathbf{X}^\top \hat{\mathbf{B}} + \sum_{k \in S} d_k e_k, \tag{11.4}$$

where $e_k = y_k - \mathbf{x}_k^\top \hat{\mathbf{B}}$. The e_k are residuals, i.e. the differences between the observed and predicted values. In certain cases, we have

$$\sum_{k \in S} d_k e_k = 0. \tag{11.5}$$

The multiple regression estimator then has a simpler form.

Result 11.1 *A sufficient condition for*

$$\sum_{k \in S} d_k e_k = 0$$

is that there exists a vector of constants λ such that $\lambda^\top \mathbf{x}_k = 1/q_k$, for all $k \in S$.

Proof:

$$\sum_{k \in S} d_k e_k = \sum_{k \in S} d_k (y_k - \mathbf{x}_k^\top \hat{\mathbf{B}}) = \sum_{k \in S} d_k (y_k - q_k \lambda^\top \mathbf{x}_k \mathbf{x}_k^\top \hat{\mathbf{B}})$$

$$= \hat{Y} - \sum_{k \in S} d_k q_k \lambda^\top \mathbf{x}_k \mathbf{x}_k^\top \hat{\mathbf{T}}^{-1} \hat{\mathbf{t}} = \hat{Y} - \lambda^\top \sum_{k \in S} d_k q_k \mathbf{x}_k y_k = \hat{Y} - \hat{Y} = 0. \quad\blacksquare$$

For example, if $\mathbf{x}_k$ contains an intercept variable, e.g. $x_{k1} = 1$, for all k and $q_k = 1$. Then, the condition (11.5) is satisfied if we take $\lambda = (1, \ldots, 0, \ldots, 0)^\top$.

In the case where the sufficient condition of Result 11.1 is satisfied, we say the multiple regression estimator can be written under a projective form:

$$\hat{Y}_{\text{REGM}} = \mathbf{X}^\top \hat{\mathbf{B}} = \sum_{k \in U} \hat{y}_k = \sum_{k \in U} \mathbf{x}_k^\top \hat{\mathbf{B}}.$$

In this case, the regression estimator is simply the sum of predicted values on the whole population.

11.3.3 Cosmetic Form

Finally, another condition allows us to write the multiple regression estimator as a sum of observed values and unobserved predicted values.

Result 11.2 *If there exists a vector of constants λ such that*

$$q_k \lambda^\top \mathbf{x}_k = (1 - \pi_k), \ \textit{for all } k \in U, \tag{11.6}$$

then the multiple regression estimator can be written as

$$\hat{Y}_{REGM} = \sum_{k \in S} y_k + \sum_{k \in U \setminus S} \mathbf{x}_k^\top \hat{\mathbf{B}}. \tag{11.7}$$

Proof: For Expression (11.7) to be equal to Expression (11.4), it must be that

$$\mathbf{X}^\top \hat{\mathbf{B}} + \sum_{k \in S} d_k e_k - \left(\sum_{k \in S} y_k + \sum_{k \in U \setminus S} \mathbf{x}_k^\top \hat{\mathbf{B}} \right) = 0,$$

and so

$$-\sum_{k \in S} y_k + \sum_{k \in S} \mathbf{x}_k^\top \hat{\mathbf{B}} + \sum_{k \in S} d_k e_k = \sum_{k \in S} d_k e_k (1 - \pi_k) = 0.$$

If $q_k \lambda^\top \mathbf{x}_k / (1 - \pi_k) = 1$, then

$$\sum_{k \in S} d_k e_k (1 - \pi_k) = \sum_{k \in S} d_k (1 - \pi_k) \left(y_k - \mathbf{x}_k^\top \hat{\mathbf{B}} \right)$$

$$= \sum_{k \in S} d_k (1 - \pi_k) \left(y_k - \frac{q_k \lambda^\top \mathbf{x}_k}{1 - \pi_k} \mathbf{x}_k^\top \hat{\mathbf{B}} \right)$$

$$= \sum_{k \in S} d_k (1 - \pi_k) y_k - \sum_{k \in S} d_k q_k \lambda^\top \mathbf{x}_k \mathbf{x}_k^\top \hat{\mathbf{T}}^{-1} \hat{\mathbf{t}}$$

$$= \sum_{k \in S} d_k (1 - \pi_k) y_k - \lambda^\top \sum_{k \in S} d_k q_k \mathbf{x}_k y_k$$

$$= \sum_{k \in S} d_k (1 - \pi_k) y_k - \sum_{k \in S} d_k (1 - \pi_k) y_k = 0.$$

∎

A multiple regression estimator that can be written in the form (11.7) is said to be cosmetic (see Särndal & Wright, 1984; Brewer, 1999). The Expression (11.7) allows the multiple regression estimator to be written in a similar form to the best linear unbiased (BLU) estimator defined in Chapter 13 in the context of model-based estimation. In this framework, regression is used to predict the values of y_k that are not observed. The estimator of the total is simply the sum of known values and predicted values using the regression.

11.4 Calibration of the Multiple Regression Estimator

The multiple regression estimator is calibrated on known auxiliary variables. Indeed, suppose that we calculate the multiple regression estimator for the auxiliary variables x. The regression coefficient matrix is

$$\hat{\mathbf{B}}_{xx} = \left(\sum_{k \in S} d_k q_k \mathbf{x}_k \mathbf{x}_k^\top \right)^{-1} \sum_{k \in S} d_k q_k \mathbf{x}_k \mathbf{x}_k^\top = \mathbf{I}_p,$$

where $\mathbf{I}_p$ is a diagonal matrix of dimension $p \times p$.

The estimator is

$$\hat{\mathbf{X}}_{REGM}^T = \hat{\mathbf{X}}^T + (\mathbf{X} - \hat{\mathbf{X}})^T \hat{\mathbf{B}}_{xx} = \hat{\mathbf{X}}^T + (\mathbf{X} - \hat{\mathbf{X}})^T = \mathbf{X}^T.$$

Therefore, the multiple regression estimator is calibrated on the auxiliary variable totals, in other words,

$$\text{var}_p(\hat{\mathbf{X}}_{REGM}) = \mathbf{0}.$$

If we can write the y_k as a linear combination of auxiliary variables, i.e. that $y_k = \mathbf{x}_k^T \mathbf{B}$, then the regression estimator variance is zero. Therefore, the multiple regression estimator is a precise estimator when the variable of interest is strongly correlated with the auxiliary variables.

11.5 Variance of the Multiple Regression Estimator

It is not possible to calculate the exact variance of the regression estimator, but it is possible to approximate this variance using the linearization methods developed in Section 15.5.4. However, we can quickly end up with a variance approximation using the following heuristic reasoning. The multiple regression estimator can be written as

$$\hat{Y}_{REGM} = \hat{Y} + (\mathbf{X} - \hat{\mathbf{X}})^T \hat{\mathbf{B}}$$
$$= \hat{Y} + (\mathbf{X} - \hat{\mathbf{X}})^T \mathbf{B} + (\mathbf{X} - \hat{\mathbf{X}})^T (\hat{\mathbf{B}} - \mathbf{B}) \approx \hat{Y} + (\mathbf{X} - \hat{\mathbf{X}})^T \mathbf{B}.$$

The term $(\mathbf{X} - \hat{\mathbf{X}})^T (\hat{\mathbf{B}} - \mathbf{B})$ can be ignored, as it is of order N/n in probability. The estimator $\hat{Y}_{REGM}$ is asymptotically unbiased under the design. Indeed,

$$E_p(\hat{Y}_{REGM}) = E_p(\hat{Y}) + [\mathbf{X} - E_p(\hat{\mathbf{X}})]^T \mathbf{B} + E_p[(\mathbf{X} - \hat{\mathbf{X}})^T (\hat{\mathbf{B}} - \mathbf{B})]$$
$$= Y + O(N/n).$$

However, we can simply approach the variance by seeing that

$$\text{var}_p(\hat{Y}_{REGM}) \approx \text{var}_p[\hat{Y} + (\mathbf{X} - \hat{\mathbf{X}})^T \mathbf{B}] = \text{var}_p\left(\sum_{k \in S} d_k E_k\right),$$

where $E_k = y_k - \mathbf{x}_k^T \mathbf{B}$. The E_k are residuals of the regression of y by the variables x realized in the population. We can derive an approximate variance with

$$\text{var}_p(\hat{Y}_{REGM}) \approx \text{var}_p\left(\sum_{k \in S} \frac{E_k}{\pi_k}\right) = \sum_{k \in U} \sum_{\ell \in U} \frac{E_k E_\ell}{\pi_k \pi_\ell} \Delta_{k\ell}.$$

This variance can be estimated by

$$v(\hat{Y}_{REGM}) = \sum_{k \in S} \sum_{\ell \in S} \frac{e_k e_\ell}{\pi_k \pi_\ell} \frac{\Delta_{k\ell}}{\pi_{k\ell}},$$

where $e_k = y_k - \mathbf{x}_k^T \hat{\mathbf{B}}$ is calculated over the sample. An alternate estimator was proposed by Särndal et al. (1989). They proposed to weigh the residuals e_k by the g-weights:

$$v_{alt}(\hat{Y}_{REGM}) = \sum_{k \in S} \sum_{\ell \in S} \frac{g_k e_k g_\ell e_\ell}{\pi_k \pi_\ell} \frac{\Delta_{k\ell}}{\pi_{k\ell}}.$$

In certain special cases, like post-stratification, this alternative estimator is unbiased for the variance of $\hat{Y}_{\text{REGM}}$ conditionally on $\hat{\mathbf{X}}$. In Section 15.5.4, page 309, and Section 15.10.4, page 328, we show that these two estimators correspond to two different ways of linearizing the multiple regression estimator.

11.6 Choice of Weights

The choice of weighting coefficients q_k allows more importance to be given to certain units over others in the calculation of regression coefficients. It takes into account the eventual heteroscedasticity of the relationship between $\mathbf{x}_k$ and y_k. The model-based approach is detailed in Chapter 13, page 263. This approach allows the most appropriate choice of weights q_k depending on the heteroscedasticity.

Suppose that the variables follow a linear model of following type:

$$M : y_k = \mathbf{x}_k^{\mathsf{T}} \boldsymbol{\beta} + \varepsilon_k,$$

where the ε_k are uncorrelated random variables with zero expected value and such that $\text{var}_M(\varepsilon_k) = \sigma^2 v_k^2$. If the v_k^2 are known, the best linear estimator of $\boldsymbol{\beta}$ under this model is

$$\widehat{\boldsymbol{\beta}} = \left(\sum_{k \in S} \frac{\mathbf{x}_k \mathbf{x}_k^{\mathsf{T}}}{v_k^2} \right)^{-1} \sum_{k \in S} \frac{\mathbf{x}_k y_k}{v_k^2}.$$

By setting $q_k = \pi_k / v_k^2$, it is then possible to take into account this heteroscedasticity.

11.7 Special Cases

11.7.1 Ratio Estimator

The multiple regression estimator generalizes many of the estimators we have discussed the use of in the framework of simple designs. The ratio estimator can be obtained by using a single auxiliary variable and by setting $\mathbf{x}_k = x_k, q_k = 1/x_k$. We then have

$$\widehat{B} = \frac{\sum_{k \in S} d_k y_k}{\sum_{k \in S} d_k x_k} = \frac{\hat{Y}}{\hat{X}},$$

and, by (11.2), that

$$\hat{Y}_{\text{REGM}} = \hat{Y} + (X - \hat{X})\frac{\hat{Y}}{\hat{X}} = X \frac{\hat{Y}}{\hat{X}},$$

which is the ratio estimator.

11.7.2 Post-stratified Estimator

The post-stratified estimator is a special case of the multiple regression estimator. The auxiliary variables are the indicators

$$\delta_{kj} = \begin{cases} 1 & \text{if } k \text{ is in post-stratum } U_j \\ 0 & \text{otherwise.} \end{cases}$$

Suppose the totals

$$N_j = \sum_{k \in U} \delta_{kj}$$

of post-strata N_j are known and can be estimated by

$$\widehat{N}_j = \sum_{k \in S} d_k \delta_{kj}.$$

If we define

$$\boldsymbol{\delta}_k = (\delta_{k1}, \dots, \delta_{kj}, \dots, \delta_{kJ})^{\top},$$

$$\mathbf{N} = (N_1, \dots, N_j, \dots, N_J)^{\top}$$

and

$$\widehat{\mathbf{N}} = (\widehat{N}_1, \dots, \widehat{N}_j, \dots, \widehat{N}_J)^{\top},$$

then the generalized regression estimator is

$$\widehat{Y}_{\text{REGM}} = \widehat{Y} + (\mathbf{N} - \widehat{\mathbf{N}})^{\top}\widehat{\mathbf{B}},$$

where $\widehat{\mathbf{B}} = \widehat{\mathbf{T}}^{-1}\widehat{\mathbf{t}}$,

$$\widehat{\mathbf{T}} = \sum_{k \in S} d_k \boldsymbol{\delta}_k \boldsymbol{\delta}_k^{\top} = \begin{pmatrix} \widehat{N}_1 & \cdots & 0 & \cdots & 0 \\ \vdots & \ddots & \vdots & & \vdots \\ 0 & \cdots & \widehat{N}_j & \cdots & 0 \\ \vdots & & \vdots & \ddots & \vdots \\ 0 & \cdots & 0 & \cdots & \widehat{N}_J \end{pmatrix},$$

and

$$\widehat{\mathbf{t}} = \sum_{k \in S} d_k \boldsymbol{\delta}_k y_k = \left(\sum_{k \in S \cap U_1} d_k y_k, \dots, \sum_{k \in S \cap U_j} d_k y_k, \dots, \sum_{k \in S \cap U_J} d_k y_k \right).$$

Therefore,

$$\widehat{\mathbf{B}} = \widehat{\mathbf{T}}^{-1}\widehat{\mathbf{t}} = (\widehat{\overline{Y}}_1, \dots, \widehat{\overline{Y}}_j, \dots, \widehat{\overline{Y}}_J)^{\top},$$

where

$$\widehat{\overline{Y}}_j = \frac{1}{\widehat{N}_j} \sum_{k \in S \cap U_j} d_k y_k.$$

Since $\widehat{\mathbf{N}}^{\top}\widehat{\mathbf{B}} = \widehat{Y}$, we obtain

$$\widehat{Y}_{\text{REGM}} = \widehat{Y} + \mathbf{N}\widehat{\mathbf{B}} - \widehat{Y} = \mathbf{N}\widehat{\mathbf{B}} = \sum_{j=1}^{J} N_j \widehat{\overline{Y}}_j,$$

which is the post-stratified estimator.

11.7.3 Regression Estimation with a Single Explanatory Variable

Let us examine the case where the model consists of an explanatory variable and an intercept variable, i.e. $\mathbf{x}_k = (1 \; x_k)^\mathsf{T}$ with $q_k = 1$. We then obtain

$$\widehat{\mathbf{T}} = \begin{pmatrix} \widehat{N} & \widehat{X} \\ \widehat{X} & \sum_{k \in S} d_k x_k^2 \end{pmatrix} \quad \text{and} \quad \widehat{\mathbf{t}} = \begin{pmatrix} \widehat{Y} \\ \sum_{k \in S} d_k x_k y_k \end{pmatrix},$$

where

$$\widehat{N} = \sum_{k \in S} d_k, \quad \widehat{X} = \sum_{k \in S} d_k x_k \quad \text{and} \quad \widehat{Y} = \sum_{k \in S} d_k y_k.$$

We then obtain

$$\widehat{\mathbf{T}}^{-1} = \frac{1}{\widehat{N}(\widehat{N}-1)\widehat{S}_x^2} \begin{pmatrix} \sum_{k \in S} d_k x_k^2 & -\widehat{X} \\ -\widehat{X} & \widehat{N} \end{pmatrix},$$

$$\widehat{\mathbf{T}}^{-1}\widehat{\mathbf{t}} = \frac{1}{\widehat{N}(\widehat{N}-1)\widehat{S}_x^2} \begin{pmatrix} \widehat{Y}\sum_{k \in S} d_k x_k^2 - \widehat{X}\sum_{k \in S} d_k x_k y_k \\ \widehat{N}(\widehat{N}-1)\widehat{S}_{xy} \end{pmatrix} = \begin{pmatrix} \widehat{\overline{Y}}_{\mathrm{HAJ}} - \widehat{\overline{X}}_{\mathrm{HAJ}}\widehat{B} \\ \widehat{B} \end{pmatrix},$$

where

$$\widehat{\overline{X}}_{\mathrm{HAJ}} = \frac{\widehat{X}}{\widehat{N}}, \quad \widehat{\overline{Y}}_{\mathrm{HAJ}} = \frac{\widehat{Y}}{\widehat{N}}, \quad \widehat{B} = \frac{\widehat{S}_{xy}}{\widehat{S}_x^2},$$

$$\widehat{S}_x^2 = \frac{1}{\widehat{N}-1} \sum_{k \in S} d_k \left(x_k - \frac{\widehat{X}}{\widehat{N}} \right)^2$$

and

$$\widehat{S}_{xy} = \frac{1}{\widehat{N}-1} \sum_{k \in S} d_k \left(x_k - \frac{\widehat{X}}{\widehat{N}} \right) \left(y_k - \frac{\widehat{Y}}{\widehat{N}} \right).$$

Since $\mathbf{X} - \widehat{\mathbf{X}} = (N - \widehat{N} \;\; X - \widehat{X})^\mathsf{T}$, we easily obtain

$$\begin{aligned} \widehat{Y}_{\mathrm{REGM}} &= \widehat{Y} + (\mathbf{X} - \widehat{\mathbf{X}})^\mathsf{T}\widehat{\mathbf{T}}^{-1}\widehat{\mathbf{t}} \\ &= \widehat{Y} + (X - \widehat{X})\widehat{B} + (N - \widehat{N})(\widehat{\overline{Y}}_{\mathrm{HAJ}} - \widehat{\overline{X}}_{\mathrm{HAJ}}\widehat{B}) \\ &= N\widehat{\overline{Y}}_{\mathrm{HAJ}} + (X - N\widehat{\overline{X}}_{\mathrm{HAJ}})\widehat{B}. \end{aligned}$$

In the case where $\widehat{N} = N$, we obtain the classical regression estimator developed in Section 9.4.

11.7.4 Optimal Regression Estimator

The optimal regression estimator was proposed by Montanari (1987) and Rao (1994). Consider the derived general expression of the regression estimator:

$$\widetilde{Y} = \widehat{Y} + (\mathbf{X} - \widehat{\mathbf{X}})^\mathsf{T}\mathbf{B}, \tag{11.8}$$

where $\mathbf{B}$ is not a random vector. The random variable $\tilde{Y}$ is not really an estimator since the vector $\mathbf{B}$ is unknown and must be estimated.

We have $E_p(\tilde{Y}) = Y$ and

$$\text{var}_p(\tilde{Y}) = \text{var}_p(\hat{Y} - \hat{\mathbf{X}}^{\mathsf{T}}\mathbf{B})$$
$$= \text{var}_p(\hat{Y}) + \mathbf{B}^{\mathsf{T}}\text{var}_p(\hat{\mathbf{X}})\mathbf{B} - 2\,\text{cov}_p(\hat{Y}, \hat{\mathbf{X}}^{\mathsf{T}})\mathbf{B}. \tag{11.9}$$

We seek the parameter vector $\mathbf{B}$ which minimizes the variance given in Expression (11.9). To achieve this, the vector of partial derivatives with respect to $\mathbf{B}$ is set to zero:

$$\frac{\partial \text{var}_p(\tilde{Y}_{\mathbf{B}})}{\partial \mathbf{B}} = 2\,\text{var}_p(\hat{\mathbf{X}})\mathbf{B} - 2\,\text{cov}_p(\hat{\mathbf{X}}, \hat{Y}) = \mathbf{0}.$$

We then obtain

$$\mathbf{B}_{\text{OPT}} = \text{var}_p(\hat{\mathbf{X}})^{-1}\text{cov}_p(\hat{\mathbf{X}}, \hat{Y}).$$

By replacing $\mathbf{B}$ by $\mathbf{B}_{\text{OPT}}$ in (11.8), we obtain

$$\tilde{Y}_{\text{REG-OPT}} = \hat{Y} + (\mathbf{X} - \hat{\mathbf{X}})^{\mathsf{T}}\text{var}_p(\hat{\mathbf{X}})^{-1}\text{cov}_p(\hat{\mathbf{X}}, \hat{Y}).$$

Unfortunately, $\tilde{Y}_{\text{REG-OPT}}$ is not an estimator, as it depends on $\text{var}_p(\hat{\mathbf{X}})^{-1}$ and $\text{cov}_p(\hat{Y}, \hat{\mathbf{X}})$, which are unknown quantities.

We can then estimate $\mathbf{B}_{\text{OPT}}$ by $\hat{\mathbf{B}}_{\text{OPT}}$ and so we obtain the estimator of Montanari (1987):

$$\hat{\mathbf{B}}_{\text{OPT}} = v(\hat{\mathbf{X}})^{-1}\,cov(\hat{\mathbf{X}}, \hat{Y}),$$

where $v(\hat{\mathbf{X}})$ and $cov(\hat{Y}, \hat{\mathbf{X}})$ are expansion estimators of $\text{var}_p(\hat{\mathbf{X}})$ and $\text{cov}_p(\hat{Y}, \hat{\mathbf{X}})$ given by

$$v(\hat{\mathbf{X}}) = \sum_{k \in S}\sum_{\ell \in S} \frac{\pi_{k\ell} - \pi_k \pi_\ell}{\pi_{k\ell}} \frac{\mathbf{x}_k \mathbf{x}_\ell^{\mathsf{T}}}{\pi_k \pi_\ell}$$

and

$$cov(\hat{\mathbf{X}}, \hat{Y}) = \sum_{k \in S}\sum_{\ell \in S} \frac{\pi_{k\ell} - \pi_k \pi_\ell}{\pi_{k\ell}} \frac{\mathbf{x}_k y_\ell}{\pi_k \pi_\ell}.$$

We then use the following estimator:

$$\hat{Y}_{\text{REG-OPT}} = \hat{Y} + (\mathbf{X} - \hat{\mathbf{X}})^{\mathsf{T}} v(\hat{\mathbf{X}})^{-1}\,cov(\hat{\mathbf{X}}, \hat{Y}). \tag{11.10}$$

Since we must estimate $\mathbf{B}_{\text{OPT}}$, the obtained estimator is not optimal. Variance estimators are often unstable and the benefit of optimality is penalized by the variability due to the estimation of $\mathbf{B}_{\text{OPT}}$. Therefore, we must apply it with caution and the usefulness of this method must be evaluated using simulations.

The optimal estimator is always a linear estimator in the sense that it can be written as a weighted sum of y_k:

$$\hat{Y}_{\text{REG-OPT}} = \sum_{k \in S} w_k(S)y_k,$$

where the weights are

$$w_k(S) = d_k \left[1 + (\mathbf{X} - \hat{\mathbf{X}})^{\mathsf{T}} v(\hat{\mathbf{X}})^{-1} \left(\sum_{\ell \in S} \frac{\Delta_{k\ell}}{\pi_{k\ell}\pi_\ell} \mathbf{x}_\ell \right) \right].$$

11.7.5   Conditional Estimation

Another relatively simple idea, studied among others by Rao (1994) and Tillé (1998, 1999), looks to reduce the conditional bias of an estimator. Knowing some estimator $\widehat{\theta}$ of the function of interest θ, we can calculate its expectation and conditional bias with respect to some statistic η:

$$B_p(\widehat{\theta}|\eta) = E_p(\widehat{\theta}|\eta) - \theta.$$

We can then apply a correction to remove the conditional bias. Ideally, we would obtain estimator

$$\widehat{\theta}^*_{\mathrm{COND}} = \widehat{\theta} - B_p(\widehat{\theta}|\eta).$$

Result 11.3   *Let $\widehat{\theta}$ be some estimator and $\widehat{\theta}^*_{\mathrm{COND}} = \widehat{\theta} - B[\widehat{\theta}|\eta]$, then*

$$\mathrm{var}_p(\widehat{\theta}^*_{\mathrm{COND}}) = \mathrm{var}_p(\widehat{\theta}) - \mathrm{var}_p[E_p(\widehat{\theta}|\eta)].$$

Proof:

$$\mathrm{var}_p(\widehat{\theta}^*_{\mathrm{COND}}) = E_p\{\mathrm{var}_p[\widehat{\theta} - B(\widehat{\theta}|\eta)|\eta]\} + \mathrm{var}_p\{E_p[\widehat{\theta} - B(\widehat{\theta}|\eta)|\eta]\}$$

$$= E_p[\mathrm{var}_p(\widehat{\theta}|\eta)] = \mathrm{var}_p(\widehat{\theta}) - \mathrm{var}_p[E_p(\widehat{\theta}|\eta)].$$

∎

If we could exactly determine the amount of conditional bias, it would always be better to apply a correction to remove this conditional bias. The corrected estimator has additionally reduced variance the more the estimator $\widehat{\theta}$ is linked to the auxiliary statistic η. Unfortunately, it is not always possible to exactly determine the conditional bias, but we can estimate it. We then have

$$\widehat{\theta}_{\mathrm{COND}} = \widehat{\theta} - \widehat{B}(\widehat{\theta}|\eta).$$

The variance of this estimator is

$$\mathrm{var}_p(\widehat{\theta}_{\mathrm{COND}}) = \mathrm{var}_p\{\widehat{\theta}^*_{\mathrm{COND}} - [\widehat{B}(\widehat{\theta}|\eta) - B(\widehat{\theta}|\eta)]\}$$

$$= \mathrm{var}_p(\widehat{\theta}^*_{\mathrm{COND}}) + \mathrm{var}_p[\widehat{B}(\widehat{\theta}|\eta) - B(\widehat{\theta}|\eta)] - 2\,\mathrm{cov}_p[\widehat{\theta}^*_{\mathrm{COND}}, \widehat{B}(\widehat{\theta}|\eta) - B(\widehat{\theta}|\eta)].$$

If we want to keep the benefits of this approach, we must find a good estimator of the conditional bias which would be a conditionally unbiased estimator with a small conditional variance.

Hájek (1964) and Rosén (1972a,b) have shown that, for conditional Poisson sampling, the expansion estimator is asymptotically normally distributed. Rosen demonstrates it for the univariate case. Suppose that a vector of expansion estimators has an exactly multivariate normal distribution. In this case,

$$\begin{pmatrix} \widehat{X} \\ \widehat{Y} \end{pmatrix} \approx N\left(\begin{pmatrix} X \\ Y \end{pmatrix}, \begin{pmatrix} \mathrm{var}_p(\widehat{X}) & \mathrm{cov}_p(\widehat{X}, \widehat{Y}) \\ \mathrm{cov}_p(\widehat{Y}, \widehat{X}) & \mathrm{var}_p(\widehat{Y}) \end{pmatrix} \right).$$

We then obtain

$$E_p(\widehat{Y}|\widehat{X}) = Y + (\widehat{X} - X)^\top \mathrm{var}_p(\widehat{X})^{-1} \mathrm{cov}_p(\widehat{X}, \widehat{Y}).$$

Again supposing normality, we can construct a conditionally unbiased estimator:

$$\hat{Y}^*_{\text{COND}} = \hat{Y} + (\mathbf{X} - \hat{\mathbf{X}})^\top \text{var}_p(\hat{\mathbf{X}})^{-1} \text{cov}_p(\hat{\mathbf{X}}, \hat{Y}).$$

We find the same estimator as the optimal estimator. Again, the problem is that $\text{var}_p(\hat{\mathbf{X}})$ and $\text{cov}_p(\hat{\mathbf{X}}, \hat{Y})$ must be estimated, which makes the estimator lose its optimal characteristic.

Both estimation principles, optimal and conditional, end up with the same estimator. However, the optimal estimator is difficult in practice except for special cases like sample designs where it can be reduced to a classical regression estimator. The conditional and optimal estimators raise a fundamental problem. Three groups of variables come into estimation during surveys: the variable of interest, the auxiliary variables used *a priori*, and the auxiliary variables used *a posteriori*. The search for an optimal estimator clashes with the interaction of these three groups. Indeed, the auxiliary variables used *a priori* appear in the optimal estimator through the second-order inclusion probabilities. It is possible to show that, for stratified designs (see Tillé, 1999) and for balanced designs, the optimal estimator can be reduced to the classical regression estimator where the auxiliary variables used *a priori* and *a posteriori* are used together as regressors.

11.8 Extension to Regression Estimation

Many generalizations of the multiple regression estimator have been proposed. Breidt & Opsomer (2000), Goga (2005), and Kim et al. (2005) proposed using local polynomials. Breidt et al. (2005) used spline regressions.

There is an abundance of literature on calibration on a distribution function (see, among others, Kovacevic, 1997; Tillé, 2002; Ren, 2000). Robust regression methods have been used by Chambers (1996). These methods are seldom used, as they are often hard to put into practice with many auxiliary variables and there is no simple software to implement them.

Exercise

11.1 Calculate the multiple regression estimator in a stratified design. Suppose that the totals of an auxiliary variable x are known for each stratum and let $q_k = 1/x_k$. The regression estimator is calculated using the following regressors:
1. a single regressor is used by the values x_k of a variable x (no intercept)
2. H regressors are used by $x_k \delta_{kh}$, where H is the number of strata and δ_k equals 1 if k is in the stratum h and 0 otherwise.

How do those estimators differentiate themselves?

12

Calibration Estimation

12.1 Calibrated Methods

Calibration methods have been formalized by Deville & Särndal (1992) and Deville et al. (1993). They are the outcome of older work about survey adjustments using data from a census (see Lemel, 1976; Huang & Fuller, 1978; Bethlehem & Keller, 1987; Deville, 1988). Papers from Särndal (2007), Kott (2009), and Devaud & Tillé (2019) make an inventory of recent developments. The main idea in calibration methods consists of constructing weighting system where the coefficients are close to the weights $d_k = 1/\pi_k$ of the expansion estimator. This system allows us to define calibrated estimators, in the sense that they recover exactly known population totals.

Deville & Särndal (1992) and Deville et al. (1993) give a general framework of calibration estimation and study the properties of these estimators. As in the multiple regression estimator, suppose that auxiliary information is available. Let $\mathbf{x}_k = (x_{k1}, \ldots, x_{kJ})^\top$ be the vector of values taken by J auxiliary variables. Suppose that the vector

$$\sum_{k \in U} \mathbf{x}_k = \mathbf{X} = (X_1, \ldots, X_j, \ldots, X_J)^\top$$

is known. However, it is not necessary to know the $\mathbf{x}_k$ for each unit in the population.

Calibration methods consist of looking for a weighting system to affect individuals. The calibration estimator of the total is a homogeneous linear estimator that is written as

$$\hat{Y}_{\mathrm{CAL}} = \sum_{k \in S} w_k(S) y_k,$$

where the $w_k(S), k \in S$ are weights that depend on the sample. For simplicity, the weights are simply denoted by w_k while not forgetting they depend on a random sample. The weighting system is determined by the way that the estimator $\hat{Y}_{\mathrm{CAL}}$ is calibrated on the auxiliary variable totals. In other words, that the estimators of the auxiliary totals are equal to these totals regardless of the selected sample, which can be written as

$$\sum_{k \in S} w_k \mathbf{x}_k = \sum_{k \in U} \mathbf{x}_k. \tag{12.1}$$

In general, there exists an infinity of weights w_k that satisfy the equation (12.1), so we are looking for weights close to the expansion estimator weights $d_k = 1/\pi_k$, which guarantees small bias. The objective is to look for weights w_k close to d_k that respect the constraints (12.1). Since the resulting weights are close to d_k, the estimator is unbiased.

Sampling and Estimation from Finite Populations, First Edition. Yves Tillé.
© 2020 John Wiley & Sons Ltd. Published 2020 by John Wiley & Sons Ltd.

We still need to define a proximity criterion. To do this, a distance is used, or more precisely a pseudo-distance $G_k(w_k, d_k)$, as symmetry is not required. Suppose that function $G_k(w_k, d_k)$ is positive, differentiable with respect to w_k, strictly convex, and such that $G_k(d_k, d_k) = 0$. The weights $w_k, k \in S$, are obtained by minimizing

$$\sum_{k \in S} \frac{G_k(w_k, d_k)}{q_k}$$

subject to constraint (12.1). The weights q_k, or rather their reciprocals, are weighting coefficients that allow us to determine the importance of each unit in the distance calculation. These weights play the same role as in the regression estimator described in Section 11.2. When there is no reason to grant more importance to certain units, we use $q_k = 1$.

Therefore, this is a minimization problem with constraints. We can then write the Lagrange function:

$$\mathcal{L}(w_k, k \in S, \lambda_j, j = 1, \dots, J) = \sum_{k \in S} \frac{G_k(w_k, d_k)}{q_k} - \lambda^\top \left(\sum_{k \in S} w_k \mathbf{x}_k - \mathbf{X} \right), \qquad (12.2)$$

where $\lambda = (\lambda_1, \dots, \lambda_J)^\top$ is the vector of Lagrange multipliers. By canceling the partial derivatives of (12.2) with respect to w_k, we obtain

$$\frac{\partial \mathcal{L}(w_k, k \in S, \lambda_j, j = 1, \dots, J)}{\partial w_k} = \frac{g_k(w_k, d_k)}{q_k} - \lambda^\top \mathbf{x}_k = 0,$$

where

$$g_k(w_k, d_k) = \frac{\partial G_k(w_k, d_k)}{\partial w_k}.$$

since $G_k(., d_k)$ is strictly convex and positive and

$$G_k(d_k, d_k) = 0,$$

$g_k(., d_k)$ is strictly increasing and $g_k(d_k, d_k) = 0$. We can finally determine the weights by

$$w_k = d_k F_k(\lambda^\top \mathbf{x}_k), \qquad (12.3)$$

where $d_k F_k(\cdot)$ is the inverse function of $g_k(., d_k)/q_k$. Function $F_k(\cdot)$ is called the calibration function. It is strictly increasing and $F_k(0) = 1$. Function $F'_k(\cdot)$, derivative of F_k, is strictly positive. Furthermore, it is supposed that the functions are defined in such a way that $F'_k(0) = q_k$.

By considering Expressions (12.3) and (12.1), we obtain a system of calibration equations:

$$\sum_{k \in S} d_k \mathbf{x}_k F_k (\mathbf{x}_k^\top \lambda) = \mathbf{X}, \qquad (12.4)$$

where $\lambda = (\lambda_1, \dots, \lambda_j, \dots, \lambda_J)^\top$. Calibrations equations are generally nonlinear. We will discuss in Section 12.3.1 how to resolve these equations.

Finally, once λ is determined, we can calculate the weights w_k using Expression (12.3). The calibrated estimator of the total Y is the estimator of the total that uses calibrated weights.

$$\hat{Y}_{CAL} = \sum_{k \in S} w_k y_k. \qquad (12.5)$$

In practice, calibration consists of calculating a weighting system w_k that can be used to estimate any total. To achieve this, one must solve the system of calibration equations in λ, given by Expression (12.4). Then, the weights are calculated using Expression (12.3).

Once the weights are calculated, they can be applied to any variable of interest. Calibration is an extremely operational tool that maintains the coherence of a survey with external sources of auxiliary variables, reduces the variance with respect to the expansion estimator, and reduces nonresponse or coverage errors. There exist numerous variants of this method, depending on the distance used.

12.2 Distances and Calibration Functions

12.2.1 The Linear Method

The simplest calibration is obtained by using a chi-square type function as pseudo-distance:

$$G(w_k, d_k) = \frac{(w_k - d_k)^2}{2d_k}.$$

The derivative of this pseudo-distance is then linear:

$$g(w_k, d_k) = \frac{w_k - d_k}{d_k},$$

and the calibration function is also linear:

$$F_k(u) = 1 + q_k u.$$

Functions $G(w_k, d_k)$, $g(w_k, d_k)$, and $F_k(u)$ are illustrated in Figures 12.1, 12.2, and 12.3.

By using a linear calibration function, the calibration equations given in Expression (12.4) become

$$\sum_{k \in S} d_k \mathbf{x}_k (1 + q_k \boldsymbol{\lambda}^\top \mathbf{x}_k) = \widehat{\mathbf{X}} + \sum_{k \in S} q_k d_k \mathbf{x}_k \mathbf{x}_k^\top \boldsymbol{\lambda} = \mathbf{X}.$$

If matrix $\sum_{k \in S} q_k d_k \mathbf{x}_k \mathbf{x}_k^\top$ is invertible, we can determine λ:

$$\boldsymbol{\lambda} = \left(\sum_{k \in S} q_k d_k \mathbf{x}_k \mathbf{x}_k^\top \right)^{-1} (\mathbf{X} - \widehat{\mathbf{X}}). \tag{12.6}$$

Figure 12.1 Linear method: pseudo-distance $G(w_k, d_k)$ with $q_k = 1$ and $d_k = 10$.

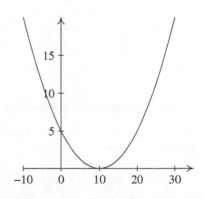

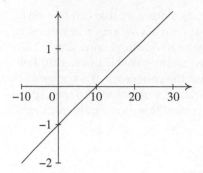

Figure 12.2 Linear method: function $g(w_k, d_k)$ with $q_k = 1$ and $d_k = 10$.

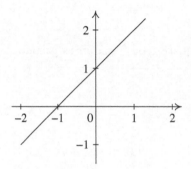

Figure 12.3 Linear method: function $F_k(u)$ with $q_k = 1$.

Then, we obtain the w_k by

$$w_k = d_k(1 + q_k \boldsymbol{\lambda}^\top \mathbf{x}_k) = d_k \left[1 + q_k(\mathbf{X} - \hat{\mathbf{X}})^\top \left(\sum_{\ell \in S} q_\ell d_\ell \mathbf{x}_\ell \mathbf{x}_\ell^\top \right)^{-1} \mathbf{x}_k \right].$$

Finally, the calibrated estimator is written as

$$\hat{Y}_{\text{CAL}} = \sum_{k \in S} w_k y_k = \hat{Y} + (\mathbf{X} - \hat{\mathbf{X}})^\top \left(\sum_{\ell \in S} q_\ell d_\ell \mathbf{x}_\ell \mathbf{x}_\ell^\top \right)^{-1} \sum_{k \in S} q_k d_k \mathbf{x}_k y_k.$$

We recover exactly the regression estimator defined in Expression 11.2, page 227. Two very different approaches, one based on regression and the other on calibration of weights, end up with the same result.

With a linear calibration function, negative weights can be obtained, which can be very problematic in practice, especially when doing small area estimation of totals. However, it is easy to remedy this problem by using a truncated linear calibration function (see Section 12.2.5).

12.2.2 The Raking Ratio Method

The raking ratio method is obtained by using the Kullback–Leibler divergence as a pseudo-distance:

$$G(w_k, d_k) = w_k \log \frac{w_k}{d_k} + d_k - w_k.$$

We then obtain

$$g(w_k, d_k) = \log \frac{w_k}{d_k}$$

and

$$F_k(u) = \exp q_k u.$$

Functions $G(w_k, d_k)$, $g(w_k, d_k)$, and $F_k(u)$ are illustrated in Figures 12.4, 12.5, and 12.6. For the raking ratio method, the weights are

$$w_k = d_k \exp(q_k \lambda^\mathsf{T} x_k),$$

where λ is determined by solving the calibration equations

$$\sum_{k \in S} d_k x_k \exp(q_k x_k^\mathsf{T} \lambda) = X. \qquad (12.7)$$

Figure 12.4 Raking ratio: pseudo-distance $G(w_k, d_k)$ with $q_k = 1$ and $d_k = 10$.

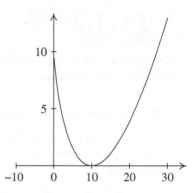

Figure 12.5 Raking ratio: function $g(w_k, d_k)$ with $q_k = 1$ and $d_k = 10$.

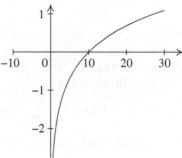

Figure 12.6 Raking ratio: function $F_k(u)$ with $q_k = 1$.

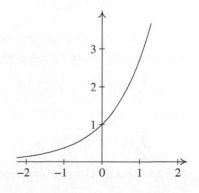

Obviously the system of equations (12.7) is not linear. The Newton method can then be used to solve this system (see Section 12.3.1). Since the calibration function is exponential, this method has the benefit of always giving positive weights. However, it also has a drawback: certain weights can take very large values, which can give enormous importance to certain units in the sample.

A special case of the raking ratio method is the calibration of a contingency table on the margins given by a census (see Section 10.4.2). In this case, the calibration variables x_j are indicator variables equal to 1 or 0 if unit k belongs or not to the subpopulation $U_j \subset U$. If also $q_k = 1, k \in U$, the calibration equations can be simplified and the following weights are obtained:

$$w_k = d_k \prod_{i|U_i \ni k} \beta_i,$$

where $\beta_j = \exp \lambda_j$. The β_j are determined by the equation

$$\sum_{k \in S} d_k x_{kj} \prod_{i|U_i \ni k} \beta_i = X_j, j = 1, \dots, J.$$

The weights w_k are equal to the weights d_k multiplied by coefficients β_i that correspond to the different subpopulations containing unit k.

12.2.3 Pseudo Empirical Likelihood

The pseudo empirical likelihood method, introduced by Chen & Qin (1993), Chen & Sitter (1999), and Chen & Wu (1999), consists of positing the existence of an empirical likelihood function:

$$L(p) = \prod_{k \in U} p_k,$$

where p_k is the probability mass associated with the statistical unit k. This likelihood function is an artifact, as it does not really correspond to the probability of observing the sample. The parameters p_k are unknown and must be estimated.

The log-likelihood is

$$\ell(p) = \log L(p) = \sum_{k \in U} \log p_k.$$

The log-likelihood can be estimated on the sample by:

$$\hat{\ell}(p) = \sum_{k \in S} d_k \log p_k,$$

where $d_k = 1/\pi_k$. Estimating p_k by the pseudo empirical likelihood amounts to finding the probability vector $p_k, k \in S$ which maximizes $\hat{\ell}(p)$ subject to

$$\sum_{k \in S} p_k = 1$$

and

$$\sum_{k \in S} p_k \left(\mathbf{x}_k - \frac{\mathbf{X}}{N} \right) = \mathbf{0},$$

which means that the weighted means of sample S by the p_k are equal to the means in U.

The Lagrange function is written as

$$\mathcal{L}(p,\gamma,\lambda) = \sum_{k \in S} d_k \log p_k - \gamma^\mathsf{T} \sum_{k \in S} p_k \left(\mathbf{x}_k - \frac{\mathbf{X}}{N} \right) - \alpha \left(\sum_{k \in S} p_k - 1 \right).$$

By canceling the partial derivatives with respect to p_k, we obtain

$$\frac{\partial \mathcal{L}(p,\gamma,\lambda)}{\partial p_k} = \frac{d_k}{p_k} - \gamma^\mathsf{T} \left(\mathbf{x}_k - \frac{\mathbf{X}}{N} \right) - \alpha = 0, \qquad (12.8)$$

where α and γ are Lagrange multipliers.

By multiplying (12.8) by p_k and summing over the sample, we obtain

$$\sum_{k \in S} d_k - \gamma^\mathsf{T} \sum_{k \in S} p_k \left(\mathbf{x}_k - \frac{\mathbf{X}}{N} \right) - \alpha \sum_{k \in S} p_k = \sum_{k \in S} d_k - \alpha = 0,$$

which gives

$$\alpha = \sum_{k \in S} d_k.$$

By solving (12.8) with respect to p_k, we have

$$p_k = \frac{d_k}{\sum_{k \in S} d_k + \gamma^\mathsf{T} \left(\mathbf{x}_k - \frac{\mathbf{X}}{N} \right)}.$$

Finally, by setting $\lambda = \gamma / \sum_{k \in S} d_k$, we obtain

$$p_k = \frac{d_k}{\sum_{k \in S} d_k} \frac{1}{1 + \lambda^\mathsf{T} \left(\mathbf{x}_k - \frac{\mathbf{X}}{N} \right)}.$$

The vector λ is obtained by solving the following nonlinear system:

$$\sum_{k \in S} p_k \left(\mathbf{x}_k - \frac{\mathbf{X}}{N} \right) = \mathbf{0} = \sum_{k \in S} \frac{d_k}{\sum_{k \in S} d_k} \frac{\mathbf{x}_k - \frac{\mathbf{X}}{N}}{1 + \lambda^\mathsf{T} \left(\mathbf{x}_k - \frac{\mathbf{X}}{N} \right)}.$$

Any mean can then be estimated by weighting the observed values by the weights p_k:

$$\widehat{\bar{Y}} = \sum_{k \in S} p_k y_k.$$

In particular, we can estimate the distribution function by

$$\widehat{F}(y) = \sum_{k \in S} p_k \mathbb{1}(y \le y_k),$$

where $\mathbb{1}(A)$ equals 1 if A is true and 0 if A is false. The pseudo empirical likelihood method gives weights that sum to 1 and therefore cannot be directly applied to estimate a total. The pseudo empirical likelihood method is a calibration method like any other. Its main use is to directly allow the construction of confidence intervals (see Wu & Rao, 2006).

12.2.4 Reverse Information

Without referring to a likelihood function, Deville & Särndal (1992) proposed using the reverse Kullback–Leibler divergence as a pseudo-distance. This pseudo-distance practically calculates the same weights as the pseudo empirical likelihood method. However, this method is more general, as it allows calculation of weights whose sum is not necessarily equal to 1. The pseudo-distance is defined by

$$G(w_k, d_k) = d_k \log \frac{d_k}{w_k} + w_k - d_k.$$

We then obtain

$$g(w_k, d_k) = 1 - \frac{d_k}{w_k}$$

and

$$F_k(u) = \frac{1}{1 - q_k u}.$$

Functions $G(w_k, d_k)$, $g(w_k, d_k)$, and $F_k(u)$ are illustrated in Figures 12.7, 12.8, and 12.9. The obtained weights are then

$$w_k = \frac{d_k}{1 - q_k \lambda^\top \mathbf{x}_k}$$

and λ is determined by solving the system

$$\sum_{k \in S} \frac{d_k \mathbf{x}_k}{1 - q_k \lambda^\top \mathbf{x}_k} = \mathbf{X}.$$

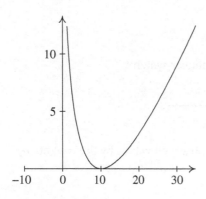

Figure 12.7 Reverse information: pseudo-distance $G(w_k, d_k)$ with $q_k = 1$ and $d_k = 10$.

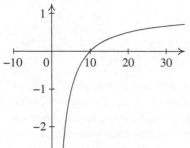

Figure 12.8 Reverse information: function $g(w_k, d_k)$ with $q_k = 1$ and $d_k = 10$.

Figure 12.9 Reverse information: function $F_k(u)$ with $q_k = 1$.

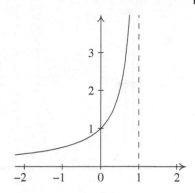

The pseudo empirical likelihood is a special case of reverse information under the condition that vector $\mathbf{x}_k$ implicitly contains at least one constant and that $q_k = 1$. In this case,

$$\sum_{k \in S} w_k = N \text{ and } p_k = \frac{w_k}{N}.$$

The calibration function $F_k(u)$ is hyperbolic and has a discontinuity at $u = 1/q_k$. The line $u = 1/q_k$ is a vertical asymptote of $F_k(u)$, which can cause numerical problems when solving calibration equations (see Section 12.3.3). This calibration function always gives positive weights, but it can produce very large weights w_k when $\mathbf{x}_k^T \lambda$ is close to $1/q_k$.

12.2.5 The Truncated Linear Method

In practice, we often want the w_k not to be too uneven, that is to say that they do not take values that are negative, too large, or zero. It is possible to force the weights w_k to be bounded between Ld_k and Hd_k by using a linear truncated calibration function.

The pseudo-distance is of the truncated chi-square type:

$$G(w_k, d_k) = \begin{cases} \dfrac{(w_k - d_k)^2}{2d_k} & Ld_k < w_k < Hd_k \\ \infty & \text{otherwise.} \end{cases}$$

We then obtain

$$g(w_k, d_k) = \begin{cases} \dfrac{w_k - d_k}{d_k} & Ld_k < w_k < Hd_k \\ -\infty & w_k \leq Ld_k \\ \infty & w_k \geq Hd_k. \end{cases}$$

The calibration function is then a truncated linear function:

$$F_k(u) = \begin{cases} 1 + q_k u & \text{if } (L-1)/q_k \leq u \leq (H-1)/q_k \\ H & \text{if } u > (H-1)/q_k \\ L & \text{if } u < (L-1)/q_k, \end{cases}$$

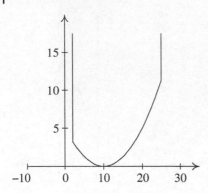

Figure 12.10 Truncated linear method: pseudo-distance $G(w_k, d_k)$ with $q_k = 1, d_k = 10, L = 0.2$, and $H = 2.5$.

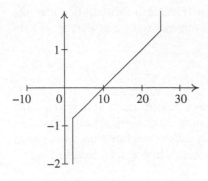

Figure 12.11 Truncated linear method: function $g(w_k, d_k)$ with $q_k = 1, d_k = 10, L = 0.2$, and $H = 2.5$.

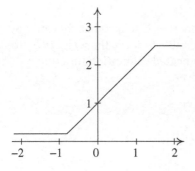

Figure 12.12 Truncated linear method: calibration function $F_k(u)$ with $q_k = 1, d_k = 10, L = 0.2$, and $H = 2.5$.

which guarantees the weights to be between Ld_k and Hd_k. Functions $G(w_k, d_k)$, $g(w_k, d_k)$, and $F_k(u)$ are illustrated in Figures 12.10, 12.11, and 12.12.

Obviously, it is possible to impose bounds on weights w_k by using any calibration function. For example, we can use the raking ratio method and set an upper bound to the weight change to avoid obtaining weights that are too large.

12.2.6 General Pseudo-Distance

Deville & Särndal (1992) reference a general pseudo-distance of which the three previous pseudo-distances are special cases. This pseudo-distance depends on a parameter

$\alpha \in \mathbb{R}$ and does not depend on unit k:

$$
G^\alpha(w_k, d_k) = \begin{cases}
\dfrac{\dfrac{|w_k|^\alpha}{d_k^{\alpha-1}} + (\alpha - 1)d_k - \alpha w_k}{\alpha(\alpha - 1)} & \alpha \in \mathbb{R}\backslash\{0, 1\} \\[4ex]
|w_k| \log \dfrac{|w_k|}{d_k} + d_k - w_k & \alpha = 1 \\[3ex]
d_k \log \dfrac{d_k}{|w_k|} + w_k - d_k & \alpha = 0.
\end{cases}
$$

If we differentiate $G^\alpha(w_k, d_k)$ with respect to w_k, we obtain

$$
g^\alpha(w_k, d_k) = \begin{cases}
\dfrac{1}{(\alpha - 1)}\left(\dfrac{\text{sign}(w_k)|w_k|^{\alpha-1}}{d_k^{\alpha-1}} - 1\right) & \alpha \in \mathbb{R}\backslash\{1\} \\[3ex]
\text{sign}(w_k) \log \dfrac{w_k}{d_k} & \alpha = 1.
\end{cases}
$$

The second derivative of $G^\alpha(w_k, d_k)$ with respect to w_k is

$$
G^{\alpha\prime\prime}(w_k, d_k) = g^{\alpha\prime}(w_k, d_k) = \frac{|w_k|^{\alpha-2}}{d_k^{\alpha-1}}.
$$

We can distinguish two cases:

- If $\alpha > 1$, function $g^\alpha(w_k, d_k)$ is strictly increasing in w_k on $\mathbb{R}$ and then $G^\alpha(w_k, d_k)$ is strictly convex on $\mathbb{R}$.
- If $\alpha \leq 1$, function $g^\alpha(w_k, d_k)$ is strictly increasing in w_k on $\mathbb{R}_+^*$ and then $G^\alpha(w_k, d_k)$ is strictly convex on $\mathbb{R}_+^*$. In this case, function $g^\alpha(., d_k)$ has a point discontinuity at $w_k = 0$. Therefore, we consider that functions $G^\alpha(., d_k)$ and $g^\alpha(., d_k)$ are only defined on $\mathbb{R}_+^*$.

The pseudo-distances $G^\alpha(w_k, d_k)$ for $\alpha = -1, 0, 1/2, 1, 2, 3$ and for $d_k = 2$ are illustrated in Figure 12.13.

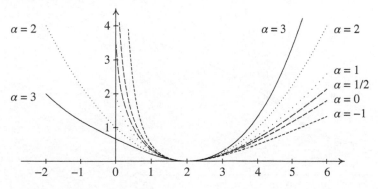

Figure 12.13 Pseudo-distances $G^\alpha(w_k, d_k)$ with $\alpha = -1, 0, 1/2, 1, 2, 3$ and $d_k = 2$.

Table 12.1 Pseudo-distances for calibration.

α	$G^\alpha(w_k, d_k)$	$g^\alpha(w_k, d_k)$	$F_k^\alpha(u)$	Type
2	$\frac{(w_k - d_k)^2}{2d_k}$	$\frac{w_k}{d_k} - 1$	$1 + q_k u$	Chi-square
1	$w_k \log \frac{w_k}{d_k} + d_k - w_k$	$\log \frac{w_k}{d_k}$	$\exp(q_k u)$	Kullback–Leibler
1/2	$2(\sqrt{w_k} - \sqrt{d_k})^2$	$2\left(1 - \sqrt{\frac{d_k}{w_k}}\right)$	$(1 - q_k u/2)^{-2}$	Hellinger distance
0	$d_k \log \frac{d_k}{w_k} + w_k - d_k$	$1 - \frac{d_k}{w_k}$	$(1 - q_k u)^{-1}$	Reverse information
−1	$\frac{(w_k - d_k)^2}{2w_k}$	$\left(1 - \frac{d_k^2}{w_k^2}\right)/2$	$(1 - 2q_k u)^{-1/2}$	Reverse chi-square

Under different values of α, Table 12.1 shows that we obtain many well-known pseudo-distances such as the Kullback–Leibler divergence, the chi-square distance, or the Hellinger distance. The most used pseudo-distances are the cases when $\alpha = 2$ (chi-square), $\alpha = 1$ (Kullback–Leibler divergence), and $\alpha = 0$ (reverse divergence).

We see that if α is positive, extreme positive weights are strongly penalized. If α is small, negative weights become strongly penalized. When $\alpha \leq 1$, negative weights cannot appear.

The inverse $g^\alpha(w_k, d_k)/q_k$ with respect to w_k equals $d_k F_k^\alpha(u)$, where

$$F_k^\alpha(u) = \begin{cases} \text{sign}[1 + q_k u(\alpha - 1)] \; |1 + q_k u(\alpha - 1)|^{1/(\alpha - 1)} & \alpha \in \mathbb{R}\backslash\{1\} \\ \exp q_k u & \alpha = 1. \end{cases}$$

Function $F_k^\alpha(u)$ is illustrated in Figure 12.14 for different values of α.

We must distinguish two cases:

- If $\alpha \geq 1$, function $F_k^\alpha(u)$ is strictly increasing on $\mathbb{R}$.
- If $\alpha < 1$, function $F_k^\alpha(u)$ is strictly increasing on $]-\infty, \{q_k(1 - \alpha)\}^{-1}[$. We will see in Section 12.3.1 that in this last case, solving the calibration equations is more delicate.

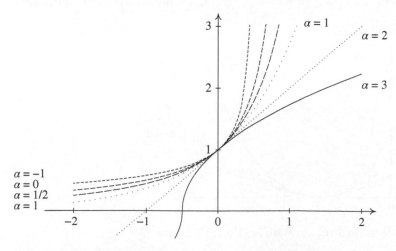

Figure 12.14 Calibration functions $F_k^\alpha(u)$ with $\alpha = -1, 0, 1/2, 1, 2, 3$ and $q_k = 1$.

12.2.7 The Logistic Method

Another way to force weights to be bounded by Ld_k and Hd_k consists of using a logistic calibration function. The pseudo-distance is defined by

$$G(w_k, d_k)$$

$$= \begin{cases} \left(a_k \log \dfrac{a_k}{1-L} + b_k \log \dfrac{b_k}{H-1} \right) \dfrac{d_k}{A} & Ld_k < w_k < Hd_k \\ \infty & \text{otherwise,} \end{cases}$$

where

$$a_k = \frac{w_k}{d_k} - L, b_k = H - \frac{w_k}{d_k}, A = \frac{H-L}{(1-L)(H-1)}.$$

We then obtain

$$g(w_k, d_k) = \begin{cases} \left(\log \dfrac{a_k}{1-L} - \log \dfrac{b_k}{H-1} \right) \dfrac{1}{A} & Ld_k < w_k < Hd_k \\ -\infty & w_k \le Ld_k \\ \infty & w_k \ge Hd_k. \end{cases}$$

The calibration function is then a logistic function:

$$F_k(u) = \frac{L(H-1) + H(1-L)\exp(Aq_k u)}{H - 1 + (1-L)\exp(Aq_k u)}.$$

We have $F_k(-\infty) = L, F_k(\infty) = H$. The weights obtained are then always in the interval $]Ld_k, Hd_k[$. Functions $G(w_k, d_k)$, $g(w_k, d_k)$, and $F_k(u)$ are illustrated in Figures 12.15, 12.16, and 12.17.

12.2.8 Deville Calibration Function

The properties of the weights w_k depend in essence on the calibration function. A good method should give positive weights, but should also avoid giving certain units extremely high importance through weighting. In oral communication, Deville has proposed using a particularly interesting calibration function:

$$F_k(u) = q_k u + \sqrt{1 + q_k^2 u^2}. \tag{12.9}$$

Figure 12.15 Logistic method: pseudo-distance $G(w_k, d_k)$ with $q_k = 1, d_k = 10, L = 0.2,$ and $H = 2.5$.

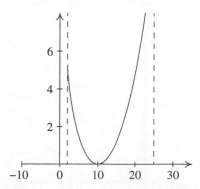

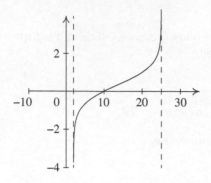

Figure 12.16 Logistic method: function $g(w_k, d_k)$ with $q_k = 1, d_k = 10, L = 0.2$, and $H = 2.5$.

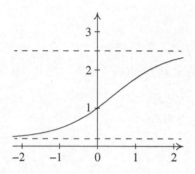

Figure 12.17 Logistic method: calibration function $F_k(u)$ with $q_k = 1, L = 0.2$, and $H = 2.5$.

This function does not have the explosive trait of the exponential function or of the pseudo empirical likelihood calibration function. In fact,

$$\lim_{x \to +\infty} (q_k u + \sqrt{1 + q_k^2 u^2} - 2q_k u) = 0.$$

The line $y = 2q_k u$ is an oblique asymptote of $F_k(u)$.

The calibration function $F_k(u)$ increases much more slowly than an exponential function, which allows us to avoid obtaining extreme weights. The pseudo-distance corresponding to this calibration function is rather curious and equals

$$G(w_k, d_k) = \frac{d_k}{4} \left[\left(\frac{w_k}{d_k} \right)^2 - 1 - 2 \log \frac{w_k}{d_k} \right].$$

Functions $G(w_k, d_k)$ and $F_k(u)$ are represented in Figures 12.18 and 12.19.

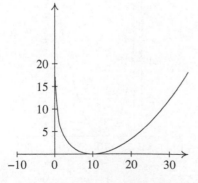

Figure 12.18 Deville calibration: pseudo-distance $G(w_k, d_k)$ with $q_k = 1, d_k = 10$.

Figure 12.19 Deville calibration: calibration function $F_k(u)$ with $q_k = 1$.

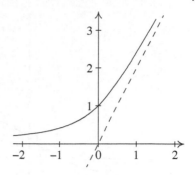

12.2.9 Roy and Vanheuverzwyn Method

Roy & Vanheuverzwyn (2001) proposed using another general distance which can be particularly interesting to find positive weights which do not become too large. The pseudo-distance is an integral which depends on function sinh($\cdot$) (hyperbolic sine):

$$\tilde{G}^a(w_k, d_k) = d_k \int_1^{w_k/d_k} \frac{1}{2\alpha} \sinh\left[\alpha\left(t - \frac{1}{t}\right)\right] dt.$$

This pseudo-distance depends on a parameter $\alpha \in \mathbb{R}^*$.

The derivative of $\tilde{G}^a(w_k, d_k)$ with respect to w_k equals

$$\tilde{g}^a(w_k, d_k) = \frac{1}{2\alpha} \sinh\left[\alpha\left(\frac{w_k}{d_k} - \frac{d_k}{w_k}\right)\right],$$

and the calibration function is

$$\tilde{F}_k^a(u) = \frac{\text{arsinh}(2\alpha q_k u)}{2\alpha} + \sqrt{\left[\frac{\text{arsinh}(2\alpha q_k u)}{2\alpha}\right]^2 + 1},$$

where arsinh($\cdot$) is the inverse function of the hyperbolic sine (area hyperbolic sine):

$$\text{arsinh}(x) = \log(x + \sqrt{x^2 + 1}).$$

The calibration function is always positive. Therefore, the weights are positive given any α. Figures 12.20 and 12.21 shows the effect of the coefficient α. The larger the α, the more weights that are far from the d_k are penalized.

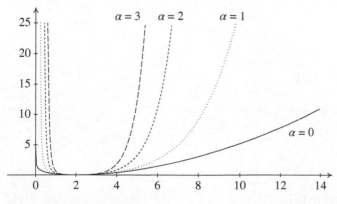

Figure 12.20 Pseudo-distances $\tilde{G}^a(w_k, d_k)$ of Roy and Vanheuverzwyn with $\alpha = 0, 1, 2, 3$, $d_k = 2$, and $q_k = 1$.

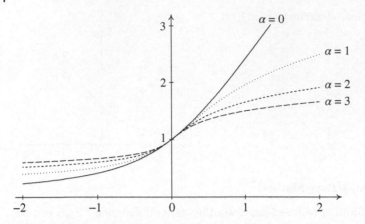

Figure 12.21 Calibration function $\tilde{F}_k^\alpha(u)$ of Roy and Vanheuverzwyn with $\alpha = 0, 1, 2, 3$ and $q_k = 1$.

The three functions $\tilde{G}^\alpha(w_k, d_k)$, $\tilde{g}^\alpha(w_k, d_k)$, and $\tilde{F}_k^\alpha(u)$ are even in α, that is to say $\tilde{F}_k^\alpha(u) = \tilde{F}_k^{-\alpha}(u)$. The method is the same if we change the sign of α. Furthermore, if α tends toward 0, we recover Expression (12.9):

$$\lim_{\alpha \to 0} \tilde{F}_k^\alpha(u) = q_k u + \sqrt{1 + q_k^2 u^2}.$$

Therefore, the Deville method described in Section 12.2.8 is a limiting case of the Roy and Vanheuverzwyn method. The calibration function of Roy and Vanheuverzwyn is a clever solution that allows us to apply more or less strong constraints to the weights without imposing strict ones.

12.3 Solving Calibration Equations

12.3.1 Solving by Newton's Method

One of the more delicate aspects of implementing calibration techniques is to solve the calibration equations given by Expression (12.4). This system of equations is generally nonlinear except when the pseudo-distance used is the chi-square. The problem consists of determining the Lagrange multipliers by solving the following system of equations:

$$\mathbf{X} = \sum_{k \in S} d_k \mathbf{x}_k F_k(\mathbf{x}_k^T \boldsymbol{\lambda}).$$

By setting

$$\boldsymbol{\Phi}(\boldsymbol{\lambda}) = \sum_{k \in S} d_k \mathbf{x}_k [F_k(\mathbf{x}_k^T \boldsymbol{\lambda}) - 1],$$

the calibration equations can be written as

$$\mathbf{X} = \boldsymbol{\Phi}(\boldsymbol{\lambda}) + \hat{\mathbf{X}},$$

or by

$$\boldsymbol{\Phi}(\boldsymbol{\lambda}) = \mathbf{X} - \hat{\mathbf{X}}, \tag{12.10}$$

function $\boldsymbol{\Phi}(\cdot)$ defined from $\mathbb{R}^J$ to $\mathbb{R}^J$.

The system (12.10) can be solved by Newton's method, and the recurrence equation is

$$\lambda^{(m+1)} = \lambda^{(m)} - [\mathbf{\Phi}'(\lambda^{(m)})]^{-1}[\mathbf{\Phi}(\lambda^{(m)}) - \mathbf{X} + \hat{\mathbf{X}}],$$

where

$$\mathbf{\Phi}'(\lambda) = \frac{\partial \mathbf{\Phi}(\lambda)}{\partial \lambda} = \sum_{k \in S} d_k F_k'(\mathbf{x}_k^\mathsf{T} \lambda) \mathbf{x}_k \mathbf{x}_k^\mathsf{T}$$

is the matrix of partial derivatives. If matrix $\mathbf{\Phi}'(\lambda^{(m)})$ is not invertible, we can use the Moore–Penrose inverse (generalized inverse).

We initialize the algorithm with $\lambda^{(0)} = 0$, and the algorithm stops when the obtained weights become stable, that is to say when

$$\max_{k \in S} |w_k^{(m+1)} - w_k^{(m)}| < \varepsilon,$$

where ε is a small positive real number.

To perform a calibration, it is not required to know the pseudo-distance. The calibration function and its derivative are sufficient to apply Newton's method. Table 12.2 contains the distance functions of the various methods that have been examined.

It is interesting to examine the first stage of Newton's method. Since $\lambda^{(0)} = 0$ and $F_k(0) = 1$, we have $\mathbf{\Phi}(\lambda^{(0)}) = \mathbf{0}$. Furthermore, since $F_k'(0) = q_k$

$$\mathbf{\Phi}'(\lambda^{(0)}) = \sum_{k \in S} q_k d_k \mathbf{x}_k \mathbf{x}_k^\mathsf{T},$$

whichever function $F_k(\cdot)$ is used. The first step in the algorithm is always the same:

$$\lambda^{(1)} = \left(\sum_{k \in S} q_k d_k \mathbf{x}_k \mathbf{x}_k^\mathsf{T} \right)^{-1} (\mathbf{X} - \hat{\mathbf{X}}),$$

and gives the value of λ in the linear case given by Expression (12.6). Therefore, the linear case is a first approximation of λ for any calibration function $F_k(\cdot)$.

12.3.2 Bound Management

It is possible to force that weight ratios be bounded, in other words $L \le w_k/d_k \le H$ using any calibration function $F_k(u)$. Managing bounds does not raise any particular problems

Table 12.2 Calibration functions and their derivatives.

Distance	$F_k(u)$	$F_k'(u)$
Chi-square	$1 + q_k u$	q_k
Kullback–Leibler divergence	$\exp(q_k u)$	$q_k \exp(q_k u)$
Hellinger distance	$(1 - q_k u/2)^{-2}$	$q_k(1 - q_k u/2)^{-3}$
Reverse information	$(1 - q_k u)^{-1}$	$q_k(1 - q_k u)^{-2}$
Reverse chi-square	$(1 - 2q_k u)^{-1/2}$	$q_k(1 - 2q_k u)^{-3/2}$
Deville distance	$q_k u + \sqrt{1 + q_k^2 u^2}$	$q_k + \dfrac{q_k^2 u}{\sqrt{1 + q_k u^2}}$
Logistic method	$\dfrac{L(H-1) + H(1-L)\exp(Aq_k u)}{H - 1 + (1-L)\exp(Aq_k u)}$	$\dfrac{q_k(H-L)^2 \exp(Aq_k u)}{[H - 1 + (1-L)\exp(Aq_k u)]^2}$

and is compatible with Newton's method. To do this, we define the bounded calibration function:

$$F_k^B(u) = \begin{cases} F_k(u) & \text{if } L \leq F_k(u) \leq H \\ L & \text{if } F_k(u) < L \\ H & \text{if } F_k(u) > H. \end{cases}$$

The derivative of $F_k^B(u)$ is zero outside the interval $[L, H]$:

$$F^{B'}{}_k(u) = \begin{cases} F_k'(u) & \text{if } L \leq F_k(u) \leq H \\ 0 & \text{if } F_k(u) < L \\ 0 & \text{if } F_k(u) > H. \end{cases}$$

Using a bounded calibration function does not cause any convergence issues in Newton's algorithm. However, setting highly restrictive bounds may cause a situation where there are no solutions to the calibration equations. In this case, Newton's method does not converge toward weights that will satisfy the calibration equations.

Monique Graf (2014) and Rebecq (2019) write the problem as a linear program, which allows us to determine the smallest interval $[L, H]$ in which at least one solution exists. It is then possible to find a solution in this interval for each specific distance by using in R language, the package `icarus` (Rebecq, 2017).

12.3.3 Improper Calibration Functions

A calibration function is improper if it is not increasing over $\mathbb{R}$, but only on a subset of $\mathbb{R}$. In the pseudo empirical likelihood method, presented in Section 12.2.3, the calibration function is a hyperbole $F_k(u) = 1/(1 - q_k u)$ of which the line $u = 1/q_k$ is a vertical asymptote. At each step in Newton's method, the argument of the calibration function $\lambda^\top x_k$ must stay lower than $1/q_k$.

This is also an issue when we use the general pseudo-distance described in Section 12.2.6 of which the pseudo empirical likelihood method is a special case. Indeed, when $\alpha < 1$, $1 + q_k u(\alpha - 1)$ must stay positive. The argument of the calibration function $x_k^\top \lambda$ must stay lower than $[(1 - \alpha)q_k]^{-1}$.

A small modification to Newton's method allows us to ensure convergence for improper calibration functions. If, at the end of iteration m, we observe one of the $x_k^\top \lambda^{(m+1)}$, $k \in S$ is larger than $[(1 - \alpha)q_k]^{-1}$, we change $\lambda^{(m+1)}$ by the mean $(\lambda^{(m+1)} + \lambda^{(m)})/2$. If this new value still does not satisfy the condition, we recalculate the mean between this new value and $\lambda^{(m)}$ until the solution satisfies the condition. Therefore, we reduce the size of the jumps in Newton's method to satisfy the condition at each step.

12.3.4 Existence of a Solution

A calibration problem does not always have a solution. A necessary condition for the existence of a solution is that there exists at least one vector $\mathbf{w} = (w_k) \in \mathbb{R}^n$ such that

$$\sum_{k \in S} w_k x_k = X. \tag{12.11}$$

Therefore, the existence of a solution depends on the selected sample. For a given calibration problem, a solution may exist for certain samples and not for others. Hence, it is a question of chance. The absence of a solution can appear in the most simple calibration problems. For example, in post-stratification which is a special case of calibration, if in a stratum no unit is selected, then the post-stratified sample cannot be calculated.

If there exists at least one solution to Equation (12.11), one of these solutions can be found using the regression estimator if matrix

$$\widehat{\mathbf{T}} = \sum_{k \in S} d_k q_k \mathbf{x}_k \mathbf{x}_k^\top$$

is invertible. If $\widehat{\mathbf{T}}$ is not invertible, we can delete the calibration variables that cause multicolinearity, or use the Moore–Penrose generalized inverse.

Increasing the number of calibration variables decreases the chances that a solution exists. Therefore, we should be parsimonious and only keep useful variables, that is to say those that are correlated with the variables of interest. Setting bounds on the weights w_k also restricts the solution space and reduces the possibility that a solution exists. The narrower the bounds, the more the possibility that a solution exists decreases. Using a calibration function that imposes positive weights also restricts the solution space and decreases the possibility that a solution exists.

12.4 Calibrating on Households and Individuals

In many household surveys, two units of interest are considered: households and individuals. At the population level, totals may be known for household units and individual units. This question, studied among others by Alexander (1987), Lemaître & Dufour (1987), Nieuwenbroek (1993), Isaki et al. (2004), and Steel & Clark (2007), is delicate, as we generally want to avoid individuals belonging to the same household to have different weights.

We are interested in a population U of N individuals grouped in M households $U_i, i = 1, \dots, M$. We first select a sample S_1 of households. Let π_{Ii} denote the probability of selecting household i. If household i is selected, a sample S_{2i} of individuals is selected inside household i. Individual k of household i is selected with inclusion probability $\pi_{k|i}$. Suppose auxiliary information is available both on the households and on individuals. The variables x are defined at the household level and vector $\mathbf{x}_i$ contains I auxiliary variables for household i. The variables z are defined at the individual level and vector $\mathbf{z}_{ki}$ contains the J auxiliary variables for individual k of household i. Suppose that the totals for these variables are known at the population level.

The objective is to obtain a single weighting system w_i which calibrates simultaneously on the household and individual variables, in other words that verifies that

$$\sum_{i \in S_1} w_i \mathbf{x}_i = \sum_{i=1}^{M} \mathbf{x}_i,$$

$$\sum_{i \in S_1} w_i \sum_{k \in S_{2i}} \frac{\mathbf{z}_{ki}}{\pi_{k|i}} = \sum_{i=1}^{M} \sum_{k \in U_i} \mathbf{z}_{ki}.$$

(12.12)

There are many ways to approach this question. A reasonable solution consists of rolling up the individual auxiliary information to the household level. We define a household pseudo-variable that contains an estimator of the total of the individual variables in the household:

$$\hat{\mathbf{z}}_i = \sum_{k \in S_{2i}} \frac{\mathbf{z}_{ki}}{\pi_{k|i}}.$$

Then, the calibration equations given in Expression (12.12) can be written as a calibration only on the household:

$$\sum_{i \in S_1} w_i \mathbf{x}_i = \sum_{i=1}^{M} \mathbf{x}_i,$$

$$\sum_{i \in S_1} w_i \hat{\mathbf{z}}_i = \sum_{i=1}^{M} \sum_{k \in U_i} \mathbf{z}_{ki}.$$

Calibration can then be done in only a single step.

Two special cases merit mention.

- If all the individuals in the selected household are automatically selected, we have a cluster sample. In this case, $\pi_{k|i} = 1$ and the weights of all individuals in the household are equal to the household weight. Therefore, individuals in the same household have the same weights.
- The Kish individual method (Kish, 1965, 1995) consists of selecting households with inclusion probabilities proportional to the number of people in the household. Then, a single individual is selected in each household. If N_i is the number of people in household i, we can use $\pi_{Ii} \propto N_i$ and $\pi_{k|i} = 1/N_i$. All individuals are then selected with equal probabilities.

For these two cases, we can perform the calibration in a single step. More general results for multi-phase surveys with complex auxiliary information on the units and clusters are presented in Le Guennec & Sautory (2002), Sautory & Le Guennec (2003), and Estevao & Särndal (2006).

12.5 Generalized Calibration

12.5.1 Calibration Equations

In more recent work (Deville, 1998a; Estevao & Särndal, 2000; Deville, 2000a, 2002,2004; Kott, 2006; Chang & Kott, 2008; Kott, 2009; Kott & Chang, 2010), calibration methods were generalized to use two sets of variables $\mathbf{x}_k \in \mathbb{R}^J$ and $\mathbf{z}_k \in \mathbb{R}^I$. In principle, I must be equal to J, but it is possible to solve the problem when $I \geq J$. The classical calibration equations can be written as

$$\sum_{k \in S} \mathbf{x}_k d_k F_k(\boldsymbol{\lambda}^\top \mathbf{x}_k) = \sum_{k \in U} \mathbf{x}_k.$$

Once solved, we can determine the weights $w_k = d_k F_k(\boldsymbol{\lambda}^\top \mathbf{x}_k)$.

The idea of generalizing consists of using two groups of variables in the calibration equations:

- The calibration variables $\mathbf{x}_k \in \mathbb{R}^J$ of which the population total must be known to do the calibration.
- The instrumental variables $\mathbf{z}_k \in \mathbb{R}^I$ that must only be known in the sample. Instruments are not auxiliary information.

It is then possible to make the calibration equations asymmetrical:

$$\sum_{k\in S} \mathbf{x}_k d_k F_k(\lambda^\top \mathbf{z}_k) = \sum_{k\in U} \mathbf{x}_k. \tag{12.13}$$

Once the calibration equations are solved, it is possible to determine the weights $d_k F_k(\lambda^\top \mathbf{z}_k)$.

The purpose of generalized calibration is essentially in applications where calibration is used to correct nonresponse. In this case, the variables z are the variables in the nonresponse model, while the variables x are the variables for which the total is known. Lesage et al. (2018) detail exactly the conditions under which generalized calibration corrects nonresponse bias and show that the bias can be large when these conditions are violated.

12.5.2 Linear Calibration Functions

An interesting special case consists of using as calibration function $F_k(u) = 1 + u$. In this case, the calibration equations given in Expression (12.13) become

$$\sum_{k\in S} \mathbf{x}_k d_k(1 + \mathbf{z}_k^\top \lambda) = \sum_{k\in U} \mathbf{x}_k,$$

which can be written as

$$\widehat{\mathbf{X}} + \widehat{\mathbf{T}}_{XZ}\lambda = \mathbf{X},$$

where

$$\widehat{\mathbf{X}} = \sum_{k\in S} d_k \mathbf{x}_k, \quad \mathbf{X} = \sum_{k\in U} \mathbf{x}_k \text{and } \widehat{\mathbf{T}}_{XZ} \sum_{k\in S} d_k \mathbf{x}_k \mathbf{z}_k^\top.$$

If matrix $\widehat{\mathbf{T}}_{XZ}$ is invertible, we obtain

$$\lambda = \widehat{\mathbf{T}}_{XZ}^{-1}(\mathbf{X} - \widehat{\mathbf{X}}).$$

Therefore, the weights are

$$w_k = d_k F_k(\lambda^\top \mathbf{z}_k) = d_k(1 + \lambda^\top \mathbf{z}_k) = d_k[1 + (\mathbf{X} - \widehat{\mathbf{X}})^\top \widehat{\mathbf{T}}_{ZX}^{-1} \mathbf{z}_k].$$

Finally, the calibration estimator with a linear calibration function equals

$$\sum_{k\in S} w_k y_k = \widehat{Y} + (\mathbf{X} - \widehat{\mathbf{X}})^\top \widehat{\mathbf{B}}_{VI}, \tag{12.14}$$

where

$$\widehat{\mathbf{B}}_{VI} = \left(\sum_{k\in S} d_k \mathbf{z}_k \mathbf{x}_k^\top \right)^{-1} \sum_{k\in S} d_k \mathbf{z}_k y_k.$$

We then obtain in Expression (12.14) a regression estimator nearly identical to the one presented in Expression 11.2, page 227, except that the coefficient of regression is calculated differently. In fact, $\widehat{\mathbf{B}}_{VI}$ is a regression coefficient of y in $\mathbf{x}_k$ using the instrumental variables $\mathbf{z}_k$.

Remark 12.1 *The optimal estimator presented in Expression (11.10), page 234, can be presented as a special case of the linear case of the generalized calibration estimator. Indeed, if we take*

$$z_k = \sum_{\ell \in S} \frac{d_\ell \mathbf{x}_\ell \Delta_{k\ell}}{\pi_{k\ell}}.$$

Then,

$$\sum_{k \in S} d_k \mathbf{z}_k \mathbf{x}_k^\top = \sum_{k \in U} \sum_{\ell \in S} \frac{d_k d_\ell \mathbf{x}_k \mathbf{x}_\ell \Delta_{k\ell}}{\pi_{k\ell}} = v(\widehat{\mathbf{X}})$$

and

$$\sum_{k \in S} d_k \mathbf{z}_k y_k^\top = \sum_{k \in U} \sum_{\ell \in S} \frac{d_k d_\ell y_k \mathbf{x}_\ell \Delta_{k\ell}}{\pi_{k\ell}} = cov(\widehat{\mathbf{X}}, \widehat{Y}).$$

We obtain

$$\widehat{\mathbf{B}}_{VI} = [cov(\widehat{\mathbf{X}}, \widehat{Y})]^{-1} v(\widehat{\mathbf{X}}).$$

Therefore, the regression coefficient by the instrumental variables is the same as for the optimal regression estimator.

12.6 Calibration in Practice

In Section 15.6.4, page 313, and Section 15.10.3, page 327, two ways of linearizing the calibrated estimator are developed, which allow us to estimate its variance. The precision gain of calibration essentially depends on the correlation between the calibration variables and the variables of interest.

Too often calibration is used to harmonize survey data and census results, which often leads to increasing the number of calibration variables. Calibration is an estimation method. As in any regression problem, a principle of parsimony should be applied. The number of calibration variables should not uselessly be increased if these are not correlated to the variables of interest. Introducing a new variable leads to a loss of degrees of freedom and can lead to an increase in the variance if the calibration variable is not correlated to the variable of interest.

Setting bounds L and H to the change in the weights allows us to avoid negative, too small, or too large weights. Bounds are often defined by intervals of the type $[1/2, 2]$, $[1/3, 3]$, and $[1/4, 4]$. The more the sample is large, the more it is possible to apply a narrow interval to the weight. If the estimated totals of the sample are very far from the population's, a greater variation of the weights must be considered to satisfy the calibration equations. If the interval defined by the bounds is small, the weights will tend to aggregate on the bounds, which is not recommended. If we restrict the interval further, the calibration problem will not have a solution. It is always interesting to examine the distribution of the weight modifications w_k/d_k, to identify an appropriate weight interval. This question of defining bounds can be avoided by using the Roy and

Vanheuverzwyn distance. It is then possible to modify the distribution of weights by varying the α coefficient of this method.

If the selected sample is bad in the sense that the expansion estimators of the auxiliary variables are very far from the population totals, the weight ratios w_k/d_k are highly dispersed. Indeed, the more the weight ratios are dispersed, the more the sample has to be adjusted (see Chambers, 1996). Therefore, the dispersion of the weights is a quality indicator of the sample. Hence, calibration is not a method that allows a badly selected sample that bears important undercoverage or measurement errors to be recovered.

When many surveys are performed, we can ask whether calibration of certain surveys on a larger survey or on the pooling of many surveys is useful (see, among others, Zieschang, 1986, 1990; Merkouris, 2004, 2010; Guandalini & Tillé, 2017). Calibration can also be used to correct nonresponse (see Section 12.5).

Various software allows calibration to be done by simply choosing the distance and bounds. The software CALMAR (Deville et al., 1993) written in SAS® was developed by the National Institute of Statistics and Economic Studies. Version 2(Le Guennec & Sautory, 2002; Sautory & Le Guennec, 2003) allows the use of generalized calibration. Statbel developed the package gCALIB in SPSS® (Vanderhoeft, 2001; Vanderhoeft et al., 2001). Other institutes also have developed software, such as *G-Est* at Statistics Canada, CLAN at Statistics Sweden, and Bascula at Statistics Netherlands (Nieuwenbroek & Boonstra, 2002). In the R language, the package sampling (Matei & Tillé, 2007) also allows calculation of the weights of a calibrated estimator. Most of this software is free to access on the Internet.

12.7 An Example

The R package sampling developed by Tillé & Matei (2016) contains the dataset belgianmunicipalities concerning the $N = 589$ Belgian municipalities in 2004. A sample of size $n = 50$ is selected with unequal selection probabilities proportional to the number of inhabitants of the municipality in 2004. The weights are calibrated using the linear method, the raking ratio method, and the truncated linear method. The bounds are defined to construct the smallest possible interval. The calibration variables are:

- Men03, the number of men 2003
- Women03, the number of women 2003
- Diffmen, the increase in the number of men between 2003 and 2004
- Diffwom, the increase in the number of women between 2003 and 2004
- TaxableIncome, the total taxable income
- Totaltaxation, the total taxation
- averageincome, the mean income
- medianincome, the median income.

Figure 12.22 contains the values of the g-weights as a function of their rank for the four methods. The linear method produces negative weights. The raking ratio method produces very large weights. The two bounded methods eliminate extreme weights.

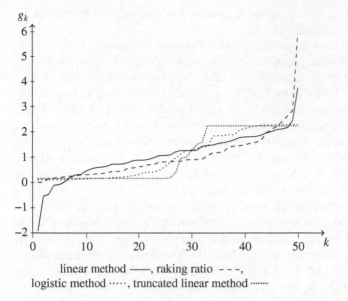

Figure 12.22 Variation of the g-weights for different calibration methods as a function of their rank.

Table 12.3 Minima, maxima, means, and standard deviations of the weights for each calibration method.

Method	Minimum	Maximum	Mean	Standard deviation
Linear method	−1.926	3.773	1.077	0.909
Raking ratio	0.000	5.945	1.028	0.998
Truncated linear	0.158	2.242	1.023	0.981
Logistic	0.108	2.292	1.025	0.889

Table 12.3 contains the minima, maxima, means, and standard deviations of the weights for each calibration method. We see that the means are very close to 1. The standard deviations of the weights are not necessarily smaller when the weights are bounded.

Exercises

12.1 Consider H subpopulations $U_h, h = 1, \ldots, H$, such that $U_h \cap U_\ell = \emptyset, h \neq \ell$. Show that, in a simple design, if we apply a calibration method on the totals $N_h = \#U_h$, that $q_k = 1, k \in U$, and that the pseudo-distance used does not depend on the units k, and the calibration does not depend on the pseudo-distance used.

12.2 Consider a Poisson design with inclusion probabilities $\pi_k, k \in U$. We are interested in the values taken by a variable of interest y of which we want to estimate the total Y. The objective is to construct a calibrated estimator of Y using a single auxiliary calibration variable $x_k = 1, k \in U$. We use the pseudo-distance

$$G^{\alpha}(w_k, d_k) = \begin{cases} \dfrac{\dfrac{w_k^{\alpha}}{d_k^{\alpha-1}} + (\alpha - 1)d_k - \alpha w_k}{\alpha(\alpha - 1)} & \alpha \in \mathbb{R} \backslash \{0, 1\} \\[2em] w_k \log \dfrac{w_k}{d_k} + d_k - w_k & \alpha = 1 \\[1.5em] d_k \log \dfrac{d_k}{w_k} + w_k - d_k & \alpha = 0, \end{cases}$$

by setting $q_k = 1$, for all $k \in U$.

1. Which total will be used to calibrate the estimator?
2. Write the calibration equations.
3. Determine the value of λ for all $\alpha \in \mathbb{R}$.
4. Determine the weights w_k of the estimator.
5. Give the calibrated estimator. Which estimator is this?

12.3 Give the calibration equations for an adjustment problem with two margins in a simple design with $q_k = 1$ using the linear method.

13

Model-Based approach

13.1 Model Approach

A completely different approach to estimation in survey theory was proposed by Ken Brewer (1963b) and mostly developed by Richard Royall (see Royall, 1970a,b, 1976a,b, 1988, 1992b,a; Royall & Herson, 1973a,b; Royall & Cumberland, 1978, 1981a,b; Cumberland & Royall, 1981, 1988). A summary of this approach is detailed in books by Valliant et al. (2000, 2013) and Chambers & Clark (2012).

The general idea of the model-based approach consists of not worrying about the selection process once the sample is selected. Inferences are then made conditionally on the sample, which is equivalent to supposing the sample is not random. Inference is then based on a superpopulation model. The model parameters are estimated using the sample. One then predicts the values of the units that were not observed in the sample.

13.2 The Model

Suppose that population U was generated using a superpopulation model. In what follows, we consider the simplest model: a linear model with uncorrelated error terms. The value of the variable of interest y_k is supposed random in the population. The vector $\mathbf{x}_k = (x_{k1}, \ldots, x_{kj}, \ldots, x_{kJ})^\top$ contains the J values of the auxiliary variables, which are not random. Consider a simple linear model:

$$y_k = \mathbf{x}_k^\top \boldsymbol{\beta} + \varepsilon_k, \tag{13.1}$$

where $\boldsymbol{\beta}$ is the vector of J regression coefficients and ε_k is a random variable with zero expected value under the model.

Let $E_M(\cdot)$ denote the expected value under the model and $\text{var}_M(\cdot)$ the variance under the model. We then have $E_M(\varepsilon_k) = 0$. Moreover, it is supposed that the variance can be written as $\text{var}_M(\varepsilon_k) = v_k^2 \sigma^2$. The model can then be heteroscedastic. Suppose the values v_k are known. The parameter σ^2 is unknown and must be estimated. It follows that $E_M(y_k) = \mathbf{x}_k^\top \boldsymbol{\beta}$ and $\text{var}_M(y_k) = v_k^2 \sigma^2$. Furthermore, suppose that the model error terms are uncorrelated, in other words that $E_M(\varepsilon_k \varepsilon_\ell) = 0$ for all $k \neq \ell \in U$.

In the model-based approach, the population total

$$Y = \sum_{k \in U} y_k$$

Sampling and Estimation from Finite Populations, First Edition. Yves Tillé.
© 2020 John Wiley & Sons Ltd. Published 2020 by John Wiley & Sons Ltd.

is a random variable that we are looking to predict using an estimator. Consider a general linear estimator in $(y_k, k \in S)$ that can be written as

$$\hat{Y}_w = \sum_{k \in S} w_{kS} y_k.$$

We will attempt to determine the "optimal" weights under the model.

For a given set of weights, we can calculate the bias under the model

$$E_M(\hat{Y}_w - Y) = \sum_{k \in S} w_{kS} E_M(y_k) - \sum_{k \in U} E_M(y_k)$$

$$= \sum_{k \in S} w_{kS} \mathbf{x}_k^\top \beta - \sum_{k \in U} \mathbf{x}_k^\top \beta = \left(\sum_{k \in S} w_{kS} \mathbf{x}_k^\top - \sum_{k \in U} \mathbf{x}_k^\top \right) \beta.$$

Therefore, a linear estimator is unbiased under the model if and only if

$$\sum_{k \in S} w_{kS} \mathbf{x}_k = \sum_{k \in U} \mathbf{x}_k. \tag{13.2}$$

Hence, a calibrated estimator is always unbiased under the model.

Since we are looking to predict Y, we can evaluate an estimator as a function of its variance under the model of the difference between the estimator $\hat{Y}_w$ and Y. We can calculate the variance under the model. Therefore,

$$\text{var}_M(\hat{Y}_w - Y) = \text{var}_M \left[\sum_{k \in S} (w_{kS} - 1) y_k - \sum_{k \in U \setminus S} y_k \right]$$

$$= \sum_{k \in S} (w_{kS} - 1)^2 v_k^2 \sigma^2 + \sum_{k \in U \setminus S} v_k^2 \sigma^2. \tag{13.3}$$

The optimal estimator under the model is given by the following result:

Result 13.1 *The best linear unbiased estimator (BLU) under the model, in the sense where it minimizes (13.3) in w_{kS}, is*

$$\hat{Y}_{\text{BLU}} = \sum_{k \in S} y_k + \sum_{k \in U \setminus S} \hat{y}_k,$$

where $\hat{y}_k = \mathbf{x}_k^\top \hat{\beta}$ and

$$\hat{\beta} = \left(\sum_{i \in S} \frac{\mathbf{x}_i \mathbf{x}_i^\top}{v_i^2} \right)^{-1} \sum_{k \in S} \frac{\mathbf{x}_k y_k}{v_k^2}.$$

Proof: We look for the estimator that minimizes the variance $\text{var}_M(\hat{Y}_w - Y)$ given in Expression (13.3) in the w_{kS} under the constraint that is unbiased under the model given in Expression (13.2). To do this, we define the Lagrangian function:

$$\mathcal{L}(w_{kS}, k \in S, \lambda)$$

$$= \sum_{k \in S} (w_{kS} - 1)^2 v_k^2 \sigma^2 + \sum_{k \in U \setminus S} v_k^2 \sigma^2 + 2\lambda^\top \left(\sum_{k \in S} w_{kS} \mathbf{x}_k - \sum_{k \in U} \mathbf{x}_k \right),$$

where 2λ is a vector of the J Lagrange multipliers. By canceling the derivative with respect to w_{kS}, we obtain

$$\frac{\partial \mathcal{L}(w_{kS}, k \in S, \lambda)}{\partial w_{kS}} = 2(w_{kS} - 1)v_k^2\sigma^2 + 2\lambda^\mathsf{T}\mathbf{x}_k = 0.$$

Therefore,

$$w_{kS} = 1 - \frac{\lambda^\mathsf{T}\mathbf{x}_k}{v_k^2\sigma^2}. \tag{13.4}$$

By considering the unbiasedness constraint given in Expression (13.2), we obtain

$$\sum_{k \in S} w_{kS}\mathbf{x}_k^\mathsf{T} = \sum_{k \in S}\mathbf{x}_k^\mathsf{T} - \lambda^\mathsf{T}\sum_{k \in S}\frac{\mathbf{x}_k\mathbf{x}_k^\mathsf{T}}{v_k^2\sigma^2} = \sum_{k \in U}\mathbf{x}_k^\mathsf{T}.$$

Hence,

$$\lambda^\mathsf{T} = \left(\sum_{k \in S}\mathbf{x}_k^\mathsf{T} - \sum_{k \in U}\mathbf{x}_k^\mathsf{T}\right)\left(\sum_{i \in S}\frac{\mathbf{x}_i\mathbf{x}_i^\mathsf{T}}{v_i^2\sigma^2}\right)^{-1} = \sum_{\ell \in U\backslash S}\mathbf{x}_\ell^\mathsf{T}\left(\sum_{i \in S}\frac{\mathbf{x}_i\mathbf{x}_i^\mathsf{T}}{v_i^2\sigma^2}\right)^{-1}.$$

By inserting the value of λ in (13.4), we obtain

$$w_{kS} = 1 - \sum_{\ell \in U\backslash S}\mathbf{x}_\ell^\mathsf{T}\left(\sum_{i \in S}\frac{\mathbf{x}_i\mathbf{x}_i^\mathsf{T}}{v_i^2}\right)^{-1}\frac{\mathbf{x}_k}{v_k^2}. \tag{13.5}$$

Finally,

$$\sum_{k \in S} w_{kS}y_k = \sum_{k \in S}\left[1 - \sum_{\ell \in U\backslash S}\mathbf{x}_\ell^\mathsf{T}\left(\sum_{i \in S}\frac{\mathbf{x}_i\mathbf{x}_i^\mathsf{T}}{v_i^2}\right)^{-1}\frac{\mathbf{x}_k}{v_k^2}\right]y_k = \sum_{k \in S}y_k + \sum_{\ell \in U\backslash S}\mathbf{x}_\ell^\mathsf{T}\hat{\beta}. \quad\blacksquare$$

Result 13.1 has an immediate interpretation. The total is estimated by summing the observed values from the sample and the predicted values of the units that were not observed. The vectors $\mathbf{x}_k$ must be known for all individuals in the population.

Result 13.2 *The variance under the model of the BLU estimator is*

$$\text{var}_M(\hat{Y}_{\text{BLU}} - Y) = \sigma^2\sum_{\ell \in U\backslash S}\mathbf{x}_\ell^\mathsf{T}\mathbf{A}_S^{-1}\sum_{j \in U\backslash S}\mathbf{x}_j + \sum_{k \in U\backslash S}v_k^2\sigma^2, \tag{13.6}$$

where

$$\mathbf{A}_S = \sum_{i \in S}\frac{\mathbf{x}_i\mathbf{x}_i^\mathsf{T}}{v_i^2}.$$

Proof: By replacing w_{kS} given in Expression (13.5) in Expression (13.3), we have

$$\text{var}_M(\hat{Y}_{\text{BLU}} - Y) = \sum_{k \in S}\left(\sum_{\ell \in U\backslash S}\mathbf{x}_\ell^\mathsf{T}\mathbf{A}_S^{-1}\frac{\mathbf{x}_k}{v_k^2}\right)^2 v_k^2\sigma^2 + \sum_{k \in U\backslash S}v_k^2\sigma^2$$

$$= \sum_{k \in S} \sum_{\ell \in U \setminus S} \mathbf{x}_\ell^{\mathsf{T}} \mathbf{A}_S^{-1} \frac{\mathbf{x}_k}{v_k^2} \sum_{j \in U \setminus S} \mathbf{x}_k^{\mathsf{T}} \mathbf{A}_S^{-1} \frac{\mathbf{x}_j}{v_k^2} v_k^2 \sigma^2 + \sum_{k \in U \setminus S} v_k^2 \sigma^2$$

$$= \sum_{\ell \in U \setminus S} \mathbf{x}_\ell^{\mathsf{T}} \mathbf{A}_S^{-1} \left(\sum_{k \in S} \frac{\mathbf{x}_k \mathbf{x}_k^{\mathsf{T}}}{v_k^2} \right) \mathbf{A}_S^{-1} \sigma^2 \sum_{j \in U \setminus S} \mathbf{x}_j + \sum_{k \in U \setminus S} v_k^2 \sigma^2$$

$$= \sigma^2 \sum_{\ell \in U \setminus S} \mathbf{x}_\ell^{\mathsf{T}} \mathbf{A}_S^{-1} \sum_{j \in U \setminus S} \mathbf{x}_j + \sum_{k \in U \setminus S} v_k^2 \sigma^2. \qquad \blacksquare$$

Result 13.3 *An unbiased estimator of σ^2 is*

$$\widehat{\sigma}^2 = \frac{1}{n - J} \sum_{k \in S} \frac{1}{v_k^2} (y_k - \mathbf{x}_k^{\mathsf{T}} \widehat{\boldsymbol{\beta}})^2.$$

Proof: The $n \times n$ matrix of values

$$\frac{\mathbf{x}_k^{\mathsf{T}}}{v_k} \mathbf{A}_S^{-1} \frac{\mathbf{x}_\ell}{v_\ell}, k, \ell, \in S,$$

is idempotent, as

$$\sum_{\ell \in S} \frac{\mathbf{x}_k^{\mathsf{T}}}{v_k} \mathbf{A}_S^{-1} \frac{\mathbf{x}_\ell}{v_\ell} \frac{\mathbf{x}_\ell^{\mathsf{T}}}{v_\ell} \mathbf{A}_S^{-1} \frac{\mathbf{x}_j}{v_j} = \frac{\mathbf{x}_k^{\mathsf{T}}}{v_k} \mathbf{A}_S^{-1} \mathbf{A}_S \mathbf{A}_S^{-1} \frac{\mathbf{x}_j}{v_j} = \frac{\mathbf{x}_k^{\mathsf{T}}}{v_k} \mathbf{A}_S^{-1} \frac{\mathbf{x}_j}{v_j}.$$

This matrix has rank J. For idempotent matrices, the trace equals the rank. Therefore,

$$\sum_{k \in S} \frac{\mathbf{x}_k^{\mathsf{T}}}{v_k} \mathbf{A}_S^{-1} \frac{\mathbf{x}_k}{v_k} = J.$$

We can now calculate the estimated value of the estimator under the model:

$$E_M[(n - J)\widehat{\sigma}^2] = E_M \left[\sum_{k \in S} \frac{1}{v_k^2} (y_k - \mathbf{x}_k^{\mathsf{T}} \widehat{\boldsymbol{\beta}})^2 \right]$$

$$= E_M \left[\sum_{k \in S} \frac{1}{v_k^2} (y_k - \mathbf{x}_k^{\mathsf{T}} \boldsymbol{\beta} - \mathbf{x}_k^{\mathsf{T}} \widehat{\boldsymbol{\beta}} + \mathbf{x}_k^{\mathsf{T}} \boldsymbol{\beta})^2 \right]$$

$$= \sum_{k \in S} \frac{1}{v_k^2} [\text{var}_M(y_k) + \text{var}_M(\mathbf{x}_k^{\mathsf{T}} \widehat{\boldsymbol{\beta}}) - 2\text{cov}_M(\mathbf{x}_k^{\mathsf{T}} \widehat{\boldsymbol{\beta}}, y_k)]$$

$$= \sum_{k \in S} \frac{1}{v_k^2} (v_k^2 \sigma^2 + \sigma^2 \mathbf{x}_k^{\mathsf{T}} \mathbf{A}_S^{-1} \mathbf{x}_k - 2\sigma^2 \mathbf{x}_k^{\mathsf{T}} \mathbf{A}_S^{-1} \mathbf{x}_k)$$

$$= \sum_{k \in S} \sigma^2 - \sigma^2 \sum_{k \in S} \frac{\mathbf{x}_k^{\mathsf{T}} \mathbf{A}_S^{-1} \mathbf{x}_k}{v_k^2} = (n - J)\sigma^2. \qquad \blacksquare$$

The variance under the model can be simply estimated by replacing σ^2 in (13.6) by $\widehat{\sigma}^2$.

The preceding results can obviously be generalized in the case where the error terms are correlated (see Valliant et al., 2000). Many special cases of the general results of this section are described in the next sections. At the end of this chapter, design-based and model-based approaches are put into perspective.

13.3 Homoscedastic Constant Model

Suppose the model is homoscedastic and that a constant is used as an independent variable. We have $v_k = 1$, $\mathbf{x}_k = 1$, and $\boldsymbol{\beta} = \beta$. The model (13.1) becomes

$$y_k = \beta + \varepsilon_k. \tag{13.7}$$

In this case, $\mathbf{A}_S = n$, $\mathbf{A}_S^{-1} = 1/n$ and $\hat{\beta} = \bar{y}$, where

$$\bar{y} = \frac{1}{n}\sum_{k \in S} y_k.$$

The prediction becomes $\hat{y}_k = \bar{y}$ and the estimator is

$$\hat{Y}_{\text{BLU}} = \sum_{k \in S} y_k + \sum_{k \in U \setminus S} \bar{y} = N\bar{y}.$$

The variance can also be simplified.

Result 13.4 *Under model (13.7), the variance of the BLU estimator is*

$$\text{var}_M(\hat{Y}_{\text{BLU}} - Y) = N^2 \sigma^2 \frac{N - n}{Nn}.$$

Proof: According to Expression (13.6):

$$\text{var}_M(\hat{Y}_{\text{BLU}} - Y) = \sigma^2 \sum_{\ell \in U \setminus S} 1 \frac{1}{n} \sum_{j \in U \setminus S} 1 + \sum_{k \in U \setminus S} \sigma^2$$

$$= (N - n)^2 \frac{\sigma^2}{n} + (N - n)\sigma^2 = N^2 \sigma^2 \frac{N - n}{Nn}. \qquad \blacksquare$$

The variance estimator, given in Result 13.3, becomes

$$s_y^2 = \frac{1}{n - 1}\sum_{k \in S}(y_k - \bar{y})^2.$$

It is then possible to estimate the variance under the model:

$$v(\hat{Y}_{\text{BLU}} - Y) = N^2 s_y^2 \frac{N - n}{Nn}. \tag{13.8}$$

The estimator (13.8) is identical to that obtained under simple random sampling without replacement given in Expression 3.3 of Result 3.2, page 30. Even if it is based on a completely different principle, the homoscedastic model with only a constant leads to the same inference to that obtained with an approach based on a simple random sampling without replacement.

13.4 Heteroscedastic Model 1 Without Intercept

Suppose that the model is heteroscedastic and without an intercept. The line goes through the origin. We have $\mathbf{x}_k = x_k > 0$, $\boldsymbol{\beta} = \beta$, $\text{var}_M(\varepsilon_k) = x_k\sigma^2$, and $v_k^2 = x_k$. The model (13.1) becomes

$$y_k = x_k\beta + \varepsilon_k. \tag{13.9}$$

With model (13.9), the observations are aligned along a line going through the origin and the dispersion increases with x, as we can see in the example of Figure 13.1.

Since

$$\bar{x} = \frac{1}{n}\sum_{k\in S}x_k \quad \text{and} \quad \bar{y} = \frac{1}{n}\sum_{k\in S}y_k,$$

we have

$$\mathbf{A}_S = \sum_{k\in S}\frac{x_k^2}{v_k^2} = \sum_{k\in S}x_k = n\bar{x}, \qquad \sum_{k\in S}\frac{x_k y_k}{v_k^2} = \sum_{k\in S}y_k = n\bar{y},$$

and $\hat{\beta} = \bar{y}/\bar{x}$. The prediction becomes $\hat{y}_k = \bar{y}\,x_k/\bar{x}$, and the estimator

$$\hat{Y}_{\text{BLU}} = \sum_{k\in S}y_k + \sum_{k\in U\backslash S}\frac{\bar{y}}{\bar{x}}x_k = n\bar{y} + \frac{\bar{y}}{\bar{x}}(X - n\bar{x}) = \frac{\bar{y}\,X}{\bar{x}},$$

where

$$X = \sum_{k\in U}x_k.$$

This is a ratio estimator.

The variance can also be simplified.

Result 13.5   *Under model (13.9), the variance of the BLU estimator is*

$$\text{var}_M(\hat{Y}_{\text{BLU}} - Y) = (X - n\bar{x})\frac{X\sigma^2}{n\bar{x}} = \left(\frac{X}{n\bar{x}} - 1\right)X\sigma^2.$$

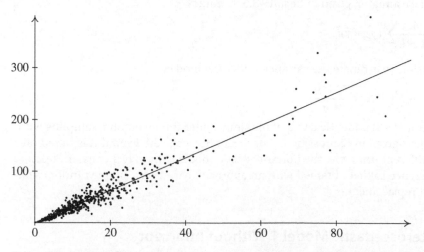

Figure 13.1 Total taxable income in millions of euros with respect to the number of inhabitants in Belgian municipalities of 100 000 people or less in 2004 (Source: Statbel). The cloud of points is aligned along a line going through the origin.

Proof: According to Expression (13.6):

$$\text{var}_M(\hat{Y}_{\text{BLU}} - Y) = \sum_{\ell \in U \backslash S} x_\ell \frac{\sigma^2}{n\bar{x}} \sum_{j \in U \backslash S} x_j + \sum_{k \in U \backslash S} x_k \sigma^2$$

$$= (X - n\bar{x})^2 \frac{\sigma^2}{n\bar{x}} + (X - n\bar{x})\sigma^2 = (X - n\bar{x})\frac{X\sigma^2}{n\bar{x}}.$$

∎

Result 13.5 is much stranger this time. The variance is a decreasing function of $\bar{x}$. The variance may be larger or smaller depending on the selected sample. To obtain a small variance, we must select the sample that has the largest mean $\bar{x}$. In other words, we must select units that have the largest values of x_k. Therefore, the "best sample" is the one containing the n largest values for x. This result leads us not to randomly choose the sample, which can be risky when the model is poorly specified. Indeed, if we only select the largest values, it is impossible to detect that the model does not fit the data. In practice, for robustness, we prefer to select a random sample with perhaps unequal inclusion probabilities.

13.5 Heteroscedastic Model 2 Without Intercept

Suppose again that the model is heteroscedastic and without an intercept. The line goes through the origin. We have $\mathbf{x}_k = x_k > 0$ and $\boldsymbol{\beta} = \beta$. This time, we suppose that the heteroscedasticity is linked to x_k^2, in other words that $\text{var}_M(\varepsilon_k) = x_k^2 \sigma^2$ and $v_k^2 = x_k^2$. The model (13.1) becomes

$$y_k = x_k \beta + \varepsilon_k. \tag{13.10}$$

In this case,

$$A_S = \sum_{k \in S} \frac{x_k^2}{v_k^2} = \sum_{k \in S} 1 = n, \qquad \sum_{k \in S} \frac{\mathbf{x}_k y_k}{v_k^2} = \sum_{k \in S} \frac{y_k}{x_k}, \qquad \text{and} \qquad \hat{\beta} = \frac{1}{n} \sum_{k \in S} \frac{y_k}{x_k}.$$

The prediction becomes

$$\hat{y}_k = \frac{x_k}{n} \sum_{k \in S} \frac{y_k}{x_k},$$

and the estimator is

$$\hat{Y}_{\text{BLU}} = \sum_{k \in S} y_k + \sum_{k \in U \backslash S} \frac{x_k}{n} \sum_{\ell \in S} \frac{y_\ell}{x_\ell}.$$

The variance can also be simplified.

Result 13.6 *Under model (13.9), the variance of the BLU estimator is*

$$\text{var}_M(\hat{Y}_{\text{BLU}} - Y) = \frac{\sigma^2(X - n\bar{x})^2}{n} + \sigma^2 \sum_{k \in U \backslash S} x_k^2.$$

Proof: According to Expression (13.6),

$$\text{var}_M(\hat{Y}_{\text{BLU}} - Y) = \sum_{\ell \in U \backslash S} x_\ell \frac{\sigma^2}{n} \sum_{j \in U \backslash S} x_j + \sum_{k \in U \backslash S} x_k^2 \sigma^2$$

$$= \frac{\sigma^2 (X - n\bar{x})^2}{n} + \sigma^2 \sum_{k \in U \backslash S} x_k^2.$$

∎

The variance under the model is a decreasing function of $\bar{x}$ and $\sigma^2 \sum_{k \in S} x_k^2$. As in the previous section, the sample which produces the smallest variance is the one containing the largest values of x. Again, the optimal sample cannot be selected randomly, which represents a risky strategy in the case of a misspecified model.

13.6 Univariate Homoscedastic Linear Model

Suppose that the model is homoscedastic and a single explanatory variable is used with an intercept, in other words that $v_k^2 \sigma^2 = \sigma^2$, $\mathbf{x}_k = (1, x_k)^\top$ and $\boldsymbol{\beta} = (\beta_0, \beta_1)^\top$. The model (13.1) becomes

$$y_k = \beta_0 + \beta_1 x_k + \varepsilon_k. \tag{13.11}$$

In this case,

$$\mathbf{A}_S = \begin{pmatrix} n & n\bar{x} \\ n\bar{x} & (n-1)s_x^2 + n\bar{x}^2 \end{pmatrix}, \quad \mathbf{A}_S^{-1} = \frac{1}{(n-1)s_x^2} \begin{pmatrix} \frac{n-1}{n}s_x^2 + \bar{x}^2 & -\bar{x} \\ -\bar{x} & 1 \end{pmatrix},$$

$$\sum_{k \in S} \mathbf{x}_k y_k = \begin{pmatrix} n\bar{y} \\ (n-1)s_{xy} + n\bar{x}\bar{y} \end{pmatrix}, \quad \hat{\boldsymbol{\beta}} = \begin{pmatrix} \hat{\beta}_0 \\ \hat{\beta}_1 \end{pmatrix} = \begin{pmatrix} \bar{y} - \frac{s_{xy}}{s_x^2}\bar{x} \\ \frac{s_{xy}}{s_x^2} \end{pmatrix},$$

where

$$\bar{x} = \frac{1}{n}\sum_{k \in S} x_k, \quad \bar{y} = \frac{1}{n}\sum_{k \in S} y_k, \quad s_x^2 = \frac{1}{n-1}\sum_{k \in S}(x_k - \bar{x})^2$$

and

$$s_{xy} = \frac{1}{n-1}\sum_{k \in S}(x_k - \bar{x})(y_k - \bar{y}).$$

The prediction becomes

$$\hat{y}_k = \bar{y} + (x_k - \bar{x})\hat{\beta}_1$$

and the estimator is

$$\hat{Y}_{\text{BLU}} = \sum_{k \in S} y_k + \sum_{k \in U \backslash S} [\bar{y} + (x_k - \bar{x})\hat{\beta}_1] = N\bar{y} + \sum_{k \in U \backslash S}(x_k - \bar{x})\hat{\beta}_1$$

$$= N\bar{y} + \sum_{k \in U}(x_k - \bar{x})\hat{\beta}_1 = N[\bar{y} + (\bar{X} - \bar{x})\hat{\beta}_1],$$

where

$$\overline{X} = \frac{1}{N}\sum_{k\in U} x_k.$$

The variance can also be simplified.

Result 13.7 *Under model (13.11), the variance of the BLU estimator is*

$$\text{var}_M(\hat{Y}_{\text{BLU}} - Y) = \frac{N^2\sigma^2}{n}\left[\frac{N-n}{N} + \frac{(\bar{x}-\overline{X})^2}{s_x^2}\right].$$

Proof: According to Expression (13.6),

$$\text{var}_M(\hat{Y}_{\text{BLU}} - Y) = \sigma^2 \sum_{\ell\in U\backslash S} \mathbf{x}_\ell^{\mathsf{T}} \mathbf{A}_S^{-1} \sum_{j\in U\backslash S} \mathbf{x}_j + \sum_{k\in U\backslash S} \sigma^2$$

$$= \sum_{\ell\in U\backslash S} (1,x_\ell)\frac{\sigma^2}{(n-1)s_x^2}\begin{pmatrix}\frac{n-1}{n}s_x^2+\bar{x}^2 & -\bar{x} \\ -\bar{x} & 1\end{pmatrix}\sum_{j\in U\backslash S}\begin{pmatrix}1\\x_j\end{pmatrix} + \sum_{k\in U\backslash S}\sigma^2$$

$$= \frac{\sigma^2}{(n-1)s_x^2}\left\{\frac{n-1}{n}(N-n)^2s_x^2 + \left[(N-n)\bar{x} - \sum_{k\in U\backslash S}x_k\right]^2\right\} + \sum_{k\in U\backslash S}\sigma^2$$

$$= \frac{(N-n)^2\sigma^2}{n} + \frac{N^2\sigma^2}{ns_x^2}(\bar{x}-\overline{X})^2 + (N-n)\sigma^2$$

$$= \frac{N^2\sigma^2}{n}\left[\frac{N-n}{N} + \frac{(\bar{x}-\overline{X})^2}{s_x^2}\right].$$

∎

Again, the variance depends on the selected sample through $\bar{x}$. Here, we see that it is preferable to select a sample such that $\bar{x}$ is equal or close to $\overline{X}$. In other words, a balanced sample on variable x_k, not necessarily random, allows us to obtain the smallest variance. The sample can obviously still be selected randomly, using, for instance, the cube method (see Section 6.6, page 125).

13.7 Stratified Population

A population is said to be stratified if it is divided in H parts or strata U_h that do not overlap and is generated using the following model:

$$y_k = \mathbf{x}_k^{\mathsf{T}}\beta + \varepsilon_k$$

with

$$\mathbf{x}_k = (x_{k1},\dots,x_{kh},\dots,x_{kH})^{\mathsf{T}},\ x_{kh} = \mathbb{1}(k\in U_h),\ \text{var}_M(\varepsilon_k) = v_h^2\sigma,\ \text{if}\ k\in U_h.$$

The error term variance may vary from one stratum to another.

We can then calculate

$$
\mathbf{A}_S = \begin{pmatrix} \dfrac{n_1}{v_1^2} & \cdots & 0 & \cdots & 0 \\ \vdots & \ddots & \vdots & & \vdots \\ 0 & \cdots & \dfrac{n_h}{v_h^2} & \cdots & 0 \\ \vdots & & \vdots & \ddots & \vdots \\ 0 & \cdots & 0 & \cdots & \dfrac{n_H}{v_H^2} \end{pmatrix}, \quad \sum_{k \in S} \frac{\mathbf{x}_k y_k}{v_{h(k)}^2} = \begin{pmatrix} \dfrac{1}{v_1^2} \sum_{k \in S_1} y_k \\ \vdots \\ \dfrac{1}{v_h^2} \sum_{k \in S_h} y_k \\ \vdots \\ \dfrac{1}{v_H^2} \sum_{k \in S_H} y_k \end{pmatrix}
$$

and $\widehat{\boldsymbol{\beta}} = (\bar{y}_1 \cdots \bar{y}_h \cdots \bar{y}_H)^{\mathsf{T}}$, where $h(k)$ denotes the stratum of unit k, $S_h = U_h \cap S$, $n_h = \#S_h$ and

$$
\bar{y}_h = \frac{1}{n_h} \sum_{k \in S_h} y_k.
$$

The prediction is then $y_k = \bar{y}_{h(k)}$. The estimator becomes

$$
\widehat{Y}_{\mathrm{BLU}} = \sum_{k \in S} y_k + \sum_{k \in U \backslash S} \bar{y}_{h(k)} = \sum_{h=1}^{H} n_h \bar{y}_h + \sum_{h=1}^{H} (N_h - n_h) \bar{y}_h = \sum_{h=1}^{H} N_h \bar{y}_h.
$$

This is exactly equal to the post-stratified estimator in a simple design as described in Expression 10.2, page 212.

The variance given in Expression (13.6) becomes

$$
\mathrm{var}_M(\widehat{Y}_{\mathrm{BLU}} - Y) = \sigma^2 \sum_{\ell \in U \backslash S} \mathbf{x}_\ell^{\mathsf{T}} \mathbf{A}_S^{-1} \sum_{j \in U \backslash S} \mathbf{x}_j + \sum_{k \in U \backslash S} v_k^2 \sigma^2
$$

$$
= \sigma^2 \sum_{h=1}^{H} (N_h - n_h)^2 \frac{v_h^2}{n_h} + \sigma^2 \sum_{h=1}^{H} (N_h - n_h) v_h^2
$$

$$
= \sigma^2 \sum_{h=1}^{H} \frac{N_h^2}{n_h} \frac{N_h - n_h}{N_h} v_h^2. \tag{13.12}
$$

The variance is exactly equal to that in a stratified design as given in Expression 4.2, page 69.

As in the design-based approach, we can look for the allocation which minimizes the variance. Therefore, we look to minimize (13.12) under the constraint of a fixed global sample size. We then find

$$
n_h = \frac{n N_h v_h}{\sum_{\ell \in U} N_\ell v_\ell}.
$$

However, under the model-based approach, the optimal sample does not necessarily need to be random. The model-based approach is nevertheless completely compatible with the design-based approach.

13.8 Simplified Versions of the Optimal Estimator

The previous results may seem disconcerting. Indeed, sometimes it seems clear that the model-based approach is compatible with a design-based approach. In these cases, it is possible to select the sample in a way to obtain exactly the same inferences under the model and under the design. In other cases, as in Section 13.4, the variance under the model leads us to choose a very particular sample, like the units with the largest values of the variable x. This choice is then completely contradictory to a design-based approach. It is a risky choice that does not allow the model to be tested. We prefer to use a more "robust" approach that gives preference to random choice and compatibility with the design-based approach.

We can clarify the situation by isolating the cases where the model-based and design-based approaches are compatible. We refer to the following definition.

Definition 13.1 *The heteroscedasticity of a model is said to be fully explainable if there exist two vectors α_1 and α_2 in $\mathbb{R}^J$ such that*

(i) $\alpha_1^\mathsf{T} \mathbf{x}_k = v_k^2$, *for all* $k \in U$,

(ii) $\alpha_2^\mathsf{T} \mathbf{x}_k = v_k$, *for all* $k \in U$.

A first result refers to point (i) of Definition (13.1).

Result 13.8 *Under hypothesis (i) of Definition 13.1, we have*

$$\widehat{Y}_{\mathrm{BLU}} = \mathbf{X}^\mathsf{T} \widehat{\beta}, \tag{13.13}$$

and

$$\mathrm{var}_M(\widehat{Y}_{\mathrm{BLU}} - Y) = \sigma^2 \mathbf{X}^\mathsf{T} \mathbf{A}_S^{-1} \mathbf{X} - \sum_{k \in U} v_k^2 \sigma^2, \tag{13.14}$$

where

$$\mathbf{X} = \sum_{k \in U} \mathbf{x}_k.$$

Proof: Under hypothesis (i) of Definition 13.1, $\alpha_1^\mathsf{T} \mathbf{x}_k / v_k^2 = 1$, for all $k \in U$. Therefore,

$$\sum_{k \in S} \mathbf{x}_k^\mathsf{T} \widehat{\beta} = \sum_{k \in S} \alpha_1^\mathsf{T} \frac{\mathbf{x}_k \mathbf{x}_k^\mathsf{T}}{v_k^2} \widehat{\beta} = \alpha_1^\mathsf{T} \mathbf{A}_S \widehat{\beta} = \alpha_1^\mathsf{T} \mathbf{A}_S \mathbf{A}_S^{-1} \sum_{k \in S} \frac{\mathbf{x}_k y_k}{v_k^2} = \sum_{k \in S} y_k.$$

Hence,

$$\widehat{Y}_{\mathrm{BLU}} = \sum_{k \in S} y_k + \sum_{k \in U \setminus S} \mathbf{x}_k^\mathsf{T} \widehat{\beta} = \mathbf{X}^\mathsf{T} \widehat{\beta}.$$

Moreover, under hypothesis (i) of Definition 13.1, we have

$$\mathbf{x}_\ell^\mathsf{T} \mathbf{A}_S^{-1} \sum_{j \in S} \mathbf{x}_j = \mathbf{x}_\ell^\mathsf{T} \mathbf{A}_S^{-1} \sum_{j \in S} \frac{\mathbf{x}_j \mathbf{x}_j^\mathsf{T}}{v_k^2} \alpha_1 = \mathbf{x}_\ell^\mathsf{T} \alpha_1 = v_\ell^2.$$

Therefore, Expression (13.14) becomes

$$\sigma^2 \mathbf{X}^\top \mathbf{A}_S^{-1} \mathbf{X} - \sum_{k \in U} v_k^2 \sigma^2$$

$$= \left(\sum_{\ell \in U \setminus S} \mathbf{x}_\ell^\top \mathbf{A}_S^{-1} \sum_{j \in U \setminus S} \mathbf{x}_j + 2 \sum_{\ell \in U \setminus S} \mathbf{x}_\ell^\top \mathbf{A}_S^{-1} \sum_{j \in S} \mathbf{x}_j \right.$$

$$\left. + \sum_{\ell \in S} \mathbf{x}_\ell^\top \mathbf{A}_S^{-1} \sum_{j \in S} \mathbf{x}_j + \sum_{k \in U} v_k^2 \right) \sigma^2.$$

$$= \left(\sum_{\ell \in U \setminus S} \mathbf{x}_\ell^\top \mathbf{A}_S^{-1} \sum_{j \in U \setminus S} \mathbf{x}_j + 2 \sum_{\ell \in U \setminus S} v_k^2 + \sum_{\ell \in S} v_k^2 - \sum_{k \in U} v_k^2 \right) \sigma^2$$

$$= \left(\sum_{\ell \in U \setminus S} \mathbf{x}_\ell^\top \mathbf{A}_S^{-1} \sum_{j \in U \setminus S} \mathbf{x}_j + \sum_{\ell \in U \setminus S} v_k^2 \right) \sigma^2.$$

We recover the variance given in Expression (13.6). ∎

A second result is even more interesting.

Result 13.9 *If the heteroscedasticity of the model is fully explainable, then*

$$\widehat{Y}_{\mathrm{BLU}} = \widehat{Y}_v + (\mathbf{X} - \widehat{\mathbf{X}}_v)^\top \widehat{\boldsymbol{\beta}} \tag{13.15}$$

and

$$\mathrm{var}_M(\widehat{Y}_{\mathrm{BLU}}) = \frac{\sigma^2}{n} \left(\sum_{k \in U} v_k \right)^2 - \sigma^2 \sum_{k \in U} v_k^2 + \sigma^2 (\mathbf{X} - \widehat{\mathbf{X}}_v)^\top \mathbf{A}_S^{-1} (\mathbf{X} - \widehat{\mathbf{X}}_v), \tag{13.16}$$

where

$$\widehat{Y}_v = \sum_{k \in S} \frac{y_k}{n v_k} \sum_{j \in U} v_j, \quad \mathbf{X} = \sum_{k \in U} \mathbf{x}_k \quad and \quad \widehat{\mathbf{X}}_v = \sum_{\ell \in S} \frac{\mathbf{x}_\ell}{n v_\ell} \sum_{j \in U} v_j.$$

Proof: Since $\alpha_2 \mathbf{x}_k = v_k$, we have

$$\widehat{\mathbf{X}}_v^\top \widehat{\boldsymbol{\beta}} = \sum_{k \in S} \frac{\mathbf{x}_k^\top}{n v_k} \sum_{j \in U} v_j \widehat{\boldsymbol{\beta}} = \alpha_2^\top \sum_{k \in S} \frac{\mathbf{x}_k \mathbf{x}_k^\top}{n v_k^2} \sum_{j \in U} v_j \mathbf{A}_S^{-1} \sum_{i \in S} \frac{\mathbf{x}_i y_i}{v_i^2}$$

$$= \alpha_2^\top \sum_{j \in U} v_j \sum_{i \in S} \frac{\mathbf{x}_i y_i}{n v_i^2} = \sum_{j \in U} v_j \sum_{k \in S} \frac{y_k}{n v_k} = \widehat{Y}_v,$$

which shows (13.13) and (13.15) are equal. In addition, since $\alpha_2 \mathbf{x}_k = v_k$, we also have

$$\mathbf{x}_\ell^\top \mathbf{A}_S^{-1} \widehat{\mathbf{X}}_v = \mathbf{x}_\ell^\top \mathbf{A}_S^{-1} \sum_{k \in S} \frac{\mathbf{x}_k}{n v_k^2} \sum_{k \in U} v_k \mathbf{x}_k^\top \alpha_2 = \mathbf{x}_\ell^\top \frac{1}{n} \sum_{k \in U} v_k \alpha_2 = \frac{v_\ell}{n} \sum_{k \in U} v_k.$$

Therefore,

$$\mathbf{X} \mathbf{A}_S^{-1} \widehat{\mathbf{X}}_v = \frac{1}{n} \left(\sum_{k \in U} v_k \right)^2 \quad and \quad \widehat{\mathbf{X}}_v \mathbf{A}_S^{-1} \widehat{\mathbf{X}}_v = \frac{1}{n} \left(\sum_{k \in U} v_k \right)^2.$$

Expression (13.16) can now be written as

$$\frac{\sigma^2}{n}\left(\sum_{k\in U}v_k\right)^2 - \sigma^2\sum_{k\in U}v_k^2 + \sigma^2(\mathbf{X}-\hat{\mathbf{X}}_v)^{\mathsf{T}}\mathbf{A}_S^{-1}(\mathbf{X}-\hat{\mathbf{X}}_v)$$

$$= \frac{\sigma^2}{n}\left(\sum_{k\in U}v_k\right)^2 - \sigma^2\sum_{k\in U}v_k^2 + \sigma^2\hat{\mathbf{X}}_v^{\mathsf{T}}\mathbf{A}_S^{-1}\hat{\mathbf{X}}_v$$
$$- 2\sigma^2\mathbf{X}^{\mathsf{T}}\mathbf{A}_S^{-1}\hat{\mathbf{X}}_v + \sigma^2\mathbf{X}^{\mathsf{T}}\mathbf{A}_S^{-1}\mathbf{X}$$

$$= \frac{\sigma^2}{n}\left(\sum_{k\in U}v_k\right)^2 - \sigma^2\sum_{k\in U}v_k^2 + \frac{\sigma^2}{n}\left(\sum_{k\in U}v_k\right)^2$$
$$- 2\frac{\sigma^2}{n}\left(\sum_{k\in U}v_k\right)^2 + \sigma^2\mathbf{X}^{\mathsf{T}}\mathbf{A}_S^{-1}\mathbf{X}$$

$$= \sigma^2\mathbf{X}^{\mathsf{T}}\mathbf{A}_S^{-1}\mathbf{X} - \sigma^2\sum_{k\in U}v_k^2.$$

We then find Expression (13.14). ∎

Result 13.9 shows that under the hypothesis of Definition 13.1, the model-based estimator can be exactly equal to the regression estimator given in Expression (11.2), page 227. However, it is necessary to select units with inclusion probabilities proportional to the standard deviations of the model error terms, i.e.

$$\pi_k = \min(Cv_k, 1),$$

with C such that

$$\sum_{k\in U}\min(Cv_k, 1) = n.$$

Then, we must set $q_k = v_k^2/\pi_k$. Even though the model-based and design-based approaches are completely different, it is possible to make them coincide and to obtain estimators that are equal under both approaches. However, the design must be conceived on the base of a fully explainable heteroscedasticity. Furthermore, the inclusion probabilities must be proportional to the standard deviations of the error terms. If the design is balanced on the auxiliary variables $\mathbf{x}_k$, then the second term of the variance disappears completely in Expression (13.16). If the heteroscedasticity is fully explainable, it is never necessary to select an extreme sample as in Section 13.4. We must simply select a balanced sample as described in Chapter 6. In this case, the model-based approach is perfectly compatible with a design-based approach, as a balanced sample can be selected randomly with unequal selection probabilities.

If the design is balanced on the explanatory variables of the model, then the estimator variance under the model reduces to the first two terms of Expression (13.16) and becomes

$$\mathrm{var}_M(\hat{Y}_{\mathrm{BLU}}) = \frac{\sigma^2}{n}\left(\sum_{k\in U}v_k\right)^2 - \sigma^2\sum_{k\in U}v_k^2 = \frac{\bar{v}^2\sigma^2N^2(N-n)}{Nn} - \sigma^2(N-1)S_v^2,$$

where

$$\bar{v} = \frac{1}{N}\sum_{k\in U} v_k \text{ and } S_v^2 = \frac{1}{N-1}\sum_{k\in U}(v_k - \bar{v})^2.$$

13.9 Completed Heteroscedasticity Model

Referring to the results of the previous section, we can complete model (13.10) to have fully explainable heteroscedasticity. We then obtain

$$y_k = \beta_1 x_k + \beta_2 x_k^2 + \varepsilon_k,$$

$$\text{var}_M(\varepsilon_k) = x_k^2\sigma^2 \text{ and } \text{cov}_M(\varepsilon_k, \varepsilon_\ell) = 0, k \neq \ell \in U.$$

We can then calculate

$$\mathbf{A}_S = \frac{1}{\sigma^2}\begin{pmatrix} n & \sum_{k\in S} x_k \\ \sum_{k\in S} x_k & \sum_{k\in S} x_k^2 \end{pmatrix}, \quad \mathbf{A}_S^{-1} = \frac{1}{(n-1)s_x^2}\begin{pmatrix} \frac{n-1}{n}s_x^2 + \bar{x}^2 & -\bar{x} \\ -\bar{x} & 1 \end{pmatrix},$$

$$\sum_{k\in S}\frac{\mathbf{x}_k y_k}{v_k^2} = \begin{pmatrix} \sum_{k\in S}\frac{y_k}{x_k} \\ n\bar{y} \end{pmatrix},$$

$$\hat{\beta} = \frac{1}{(n-1)s_x^2}\begin{pmatrix} \left(\frac{n-1}{n}s_x^2 + \bar{x}^2\right)\sum_{k\in S}\frac{y_k}{x_k} - n\bar{y}\bar{x} \\ n\bar{y} - \bar{x}\sum_{k\in S}\frac{y_k}{x_k} \end{pmatrix},$$

$$\hat{\mathbf{X}}_v = \frac{1}{n}\sum_{\ell\in U} v_\ell \begin{pmatrix} \sum_{k\in S}\frac{x_k}{v_k} \\ \sum_{k\in S}\frac{x_k^2}{v_k} \end{pmatrix} = \frac{1}{n}\sum_{\ell\in U} x_\ell \begin{pmatrix} \sum_{k\in S}1 \\ \sum_{k\in S}\frac{x_k^2}{x_k} \end{pmatrix} = \begin{pmatrix} X \\ X\bar{x} \end{pmatrix}$$

and

$$\hat{\mathbf{X}}_v - \mathbf{X} = \begin{pmatrix} 0 \\ \sum_{k\in U} x_k^2 - X\bar{x} \end{pmatrix}.$$

These results allow us to calculate the optimal estimator:

$$\hat{Y}_{BLU} = \hat{Y}_v + (\mathbf{X} - \hat{\mathbf{X}}_v)\hat{\beta} = \frac{X}{n}\sum_{k\in S}\frac{y_k}{x_k} + \left(\sum_{k\in U} x_k^2 - X\bar{x}\right)\frac{n\bar{y} - \bar{x}\sum_{k\in S}\frac{y_k}{x_k}}{(n-1)s_x^2}.$$

The variance under the model becomes

$$\text{var}_M(\hat{Y}_{BLU})$$

$$= \frac{\sigma^2}{n}\left(\sum_{k\in U} v_k\right)^2 - \sigma^2\sum_{k\in U} v_k^2 + \sigma^2(\hat{\mathbf{X}}_v - \mathbf{X})^\top \mathbf{A}_S^{-1}(\hat{\mathbf{X}}_v - \mathbf{X})$$

$$= \frac{\sigma^2}{n}X^2 - \sigma^2\left[(N-1)S_x^2 - \frac{X^2}{N}\right] + \frac{\sigma^2\left(\sum_{k\in U} x_k^2 - X\bar{x}\right)^2}{(n-1)s_x^2}$$

$$= \frac{\sigma^2 X^2(N-n)}{nN} - \sigma^2(N-1)S_x^2 + \frac{\sigma^2\left(\sum_{k\in U}x_k^2 - X\overline{x}\right)^2}{(n-1)s_x^2}. \tag{13.17}$$

This result is compatible with a random sample selection. We must define inclusion probabilities proportional to the x_k and select a sample of fixed size. If in addition the design is balanced on x_k^2, the last term of (13.17) disappears. We can then define an optimal sampling strategy for the completed model.

13.10 Discussion

Model-based estimation is founded on a completely different principle to design-based estimation. If the sample is not selected randomly, the model-based approach is the only way to justify the extrapolation of the results to the complete population. The model-based approach can still be compatible with a design-based approach. Indeed, when the heteroscedasticity of the model is fully explainable, Equation (13.15) shows that model-based estimation consists of calculating a regression estimator. The two approaches end up with the same inference.

In addition, when the heteroscedasticity of the model is fully explainable, Expression (13.16) shows that there exists a sampling design that allows us to obtain the minimum variance under the model. The optimal design is balanced on the explanatory variables of the model and has inclusion probabilities proportional to the standard deviations of the error terms of the model.

Therefore, the model-based approach can be used to devise the sampling design. In the case of surveys that are repeated in time, it is possible to model the population during each wave. At the following wave, the estimated model can be used to optimize the sampling design. Marazzi & Tillé (2017) use this idea to optimize a survey system to verify accounting records and show the practical uses of such a method.

The model-based approach is fundamental for small area estimation. In this case, we generally use mixed models that allow us to take into account the correlation between units that belong to the same domain. The comprehensive book from Rao & Molina (2015) presents state-of-the-art methods for small area estimation.

13.11 An Approach that is Both Model- and Design-based

A mixed or robust approach consists of looking at the inference from both the design and the model. If the approaches match, then the inference is unbiased under the design, even when the model is misspecified. Expression (13.2) shows that an estimator is unbiased under the model if

$$\sum_{k\in S} w_{kS}\mathbf{x}_k = \sum_{k\in U}\mathbf{x}_k,$$

which corresponds to a calibrated estimator in a design-based approach. Therefore, a calibrated estimator is automatically unbiased under the model. An inference is said to be doubly robust if it is unbiased under both the design and the model. Hence, calibration methods, described in Chapter 12, are doubly robust in the sense that the weights stay as close as possible to the weights of the expansion estimator. They produce estimators that are unbiased under the model and approximatively unbiased under the design.

To evaluate the precision of estimators under the model and the design, we can use the anticipated variance (Isaki & Fuller, 1982) defined by

$$\text{Avar}(\widehat{Y}) = E_p E_M (\widehat{Y} - Y)^2$$

to evaluate the estimator. Nedyalkova & Tillé (2008) give a general expression for the anticipated variance for a linear estimator of the form

$$\widehat{Y} = \sum_{k \in S} w_{kS} y_k.$$

Result 13.10 *If $\widehat{Y}_w$ is a linear estimator and the random sample S is independent of the error terms ε_k, then*

$$\text{Avar}(\widehat{Y}) = E_p E_M (\widehat{Y} - Y)^2 = E_p E_M (\widehat{Y}_w - Y)^2$$

$$= E_p \left(\sum_{k \in S} w_{kS} \mathbf{x}_k^\top \boldsymbol{\beta} - \sum_{k \in U} \mathbf{x}_k^\top \boldsymbol{\beta} \right)^2 + \sigma^2 E_p \left[\sum_{k \in S} (w_{kS} - 1)^2 v_k^2 + \sum_{k \in \bar{S}} v_k^2 \right]$$

$$= \text{var}_p \left(\sum_{k \in S} w_{kS} \mathbf{x}_k^\top \boldsymbol{\beta} \right) + \left(\sum_{k \in U} C_k \mathbf{x}_k^\top \boldsymbol{\beta} - \sum_{k \in U} \mathbf{x}_k^\top \boldsymbol{\beta} \right)^2$$

$$+ \sigma^2 \sum_{k \in U} v_k^2 \left[\pi_k \text{var}_p(w_{kS} | a_k = 1) + C_k^2 \frac{1 - \pi_k}{\pi_k} + (C_k - 1)^2 \right],$$

where $C_k = E_p(w_{kS} a_k)$ and a_k is the indicator random variable of the presence of k in the sample.

Proof: First,

$$E_p \left(\sum_{k \in S} w_{kS} \mathbf{x}_k^\top \boldsymbol{\beta} - \sum_{k \in U} \mathbf{x}_k^\top \boldsymbol{\beta} \right)^2$$

$$= \text{var}_p \left(\sum_{k \in S} w_{kS} \mathbf{x}_k^\top \boldsymbol{\beta} \right) + \left(\sum_{k \in U} C_k \mathbf{x}_k^\top \boldsymbol{\beta} - \sum_{k \in U} \mathbf{x}_k^\top \boldsymbol{\beta} \right)^2.$$

Then,

$$\sigma^2 E_p \left[\sum_{k \in S} (w_{kS} - 1)^2 v_k^2 + \sum_{k \in \bar{S}} v_k^2 \right] = \sigma^2 \sum_{k \in U} v_k^2 \{ E_p[(w_{kS} - 1)^2 a_k] + 1 - \pi_k \}.$$

However,

$$E_p[(w_{kS} - 1)^2 a_k] + 1 - \pi_k = \text{var}_p(w_{kS} a_k) + (C_k - 1)^2.$$

In fact,

$$E_p[(w_{kS} - 1)^2 a_k] + 1 - \pi_k = E_p(w_{kS}^2 a_k) - 2E_p(w_{kS} a_k) + \pi_k + 1 - \pi_k$$

$$= E_p(w_{kS}^2 a_k) - E_p^2(w_{kS} a_k) + E_p^2(w_{kS} a_k) - 2E_p(w_{kS} a_k) + 1$$

$$= \text{var}_p(w_{kS} a_k) + (C_k - 1)^2.$$

By the law of total variance, we have

$$
\begin{aligned}
\operatorname{var}_p(w_{kS}a_k) &= E_p\operatorname{var}_p(w_{kS}a_k|a_k) + \operatorname{var}_pE_p(w_{kS}a_k|a_k)\\
&= \pi_k\operatorname{var}_p(w_{kS}|a_k = 1) + \pi_kE_p^2(w_{kS}|a_k = 1) - E_p^2(w_{kS}a_k)\\
&= \pi_k\operatorname{var}_p(w_{kS}|a_k = 1) + \frac{1 - \pi_k}{\pi_k}C_k^2.
\end{aligned}
$$

Finally,

$$
\begin{aligned}
E_pE_M(&\hat{Y}_w - Y)^2\\
&= \operatorname{var}_p\left(\sum_{k\in S}w_{kS}\mathbf{x}_k^\mathsf{T}\beta\right) + \left(\sum_{k\in U}C_k\mathbf{x}_k^\mathsf{T}\beta - \sum_{k\in U}\mathbf{x}_k^\mathsf{T}\beta\right)^2\\
&\quad + \sigma^2\sum_{k\in U}v_k^2\left[\pi_k\operatorname{var}_p(w_{kS}|a_k = 1) + C_k^2\frac{1 - \pi_k}{\pi_k} + (C_k - 1)^2\right].
\end{aligned}
$$

∎

The anticipated variance $E_pE_M(\hat{Y}_w - Y)^2$ contains five terms that can be interpreted separately:

1. The term

$$
\operatorname{var}_p\left(\sum_{k\in S}w_{kS}\mathbf{x}_k^\mathsf{T}\beta\right)
$$

   is the variance under the design of the expectation under the model of the estimator. This term is zero when the estimator is calibrated (or unbiased under the model).

2. The term

$$
\left(\sum_{k\in U}C_k\mathbf{x}_k^\mathsf{T}\beta - \sum_{k\in U}\mathbf{x}_k^\mathsf{T}\beta\right)^2
$$

   is the square of the bias under the design of the expectation under the model. This term is also zero when the estimator is calibrated (or unbiased under the model) or when the estimator is unbiased under the design.

3. The term

$$
\sigma^2\sum_{k\in U}v_k^2\pi_k\operatorname{var}_p(w_{kS}|a_k = 1)
$$

   is the part of the variance due to the fact that the weights w_{kS} are random. This term is zero when the weights are not random, for example in the expansion estimator.

4. The term

$$
\sigma^2\sum_{k\in U}v_k^2C_k^2\frac{1 - \pi_k}{\pi_k}
$$

   depends on the variance of the error terms and the inclusion probabilities. It is not possible to make it go to zero.

5. The term

$$
\sigma^2\sum_{k\in U}v_k^2(C_k - 1)^2
$$

depends on the bias under the design and the variance of the error terms. It is zero if the estimator is unbiased under the design.

The special case of the expansion estimator is given in the following result.

Result 13.11 *Under model (13.1) and under the hypothesis that the random sample S is independent of the ε_k, the anticipated variance of the expansion estimator equals*

$$E_p E_M(\hat{Y} - Y)^2 = \beta^\mathsf{T} \mathrm{var}_p \left(\sum_{k \in S} \frac{\mathbf{x}_k}{\pi_k} \right) \beta + \sigma^2 \sum_{k \in U} \frac{v_k^2}{\pi_k^2} \pi_k (1 - \pi_k). \tag{13.18}$$

Proof: If S is independent of the ε_k, then $E_p E_M(\hat{Y} - Y)^2 = E_M E_p(\hat{Y} - Y)^2 = E_M \mathrm{var}_p(\hat{Y} - Y)^2$. We have

$$\mathrm{var}_p(\hat{Y}) = \sum_{k \in U} \sum_{\ell \in U} \frac{(\mathbf{x}_k^\mathsf{T} \beta + \varepsilon_k)(\mathbf{x}_\ell^\mathsf{T} \beta + \varepsilon_\ell)}{\pi_k \pi_\ell} \Delta_{k\ell}.$$

Hence,

$$E_M \mathrm{var}_p(\hat{Y}) = \beta^\mathsf{T} \mathrm{var}_p \left(\sum_{k \in S} \frac{\mathbf{x}_k}{\pi_k} \right) \beta + \sigma^2 \sum_{k \in U} \frac{v_k^2}{\pi_k^2} \pi_k (1 - \pi_k).$$

∎

The first term of Expression (13.18) is zero when the design is balanced on the explanatory variables $\mathbf{x}_k$. The second term, called the Godambe & Joshi (1965) bound, can then be minimized by looking for optimal inclusion probabilities under the constraints of a global size n. We see again that the inclusion probabilities need to be proportional to the standard deviations of the error terms, in other words that $\pi_k \propto v_k$. Therefore, for each linear model there is a corresponding optimal design. Tillé & Wilhelm (2017) match the usual sampling designs with different models (13.1).

14

Estimation of Complex Parameters

14.1   Estimation of a Function of Totals

Suppose that we have a system of weights that can come from the expansion estimator ($d_k = 1/\pi_k$) or from a calibrated estimator (weight w_k). Such a system allows any total Y to be estimated using the estimator

$$\hat{Y} = \sum_{k \in S} w_k y_k.$$

If we want to estimate a smooth function of totals

$$\theta = f(Y_1, \ldots, Y_j, \ldots, Y_p),$$

we can use the plug-in estimator:

$$\hat{\theta} = f(\hat{Y}_1, \ldots, \hat{Y}_j, \ldots, \hat{Y}_p).$$

Obviously, even if the $\hat{Y}_j$ are unbiased estimators of Y_j, estimator $\hat{\theta}$ is generally biased. The variance calculation of the plug-in estimator is discussed in Chapter 15.

Example 14.1   In order to estimate a ratio

$$R = \frac{Y}{X}$$

one can use

$$\hat{R} = \frac{\hat{Y}}{\hat{X}},$$

where

$$\hat{Y} = \sum_{k \in S} w_k y_k \quad \text{and} \quad \hat{X} = \sum_{k \in S} w_k x_k.$$

Example 14.2   Despite its multiplicative character, the geometric mean can also be written as a function of totals. Indeed,

$$G = \left(\prod_{k \in U} y_k \right)^{1/N} = \exp \log \left(\prod_{k \in U} y_k \right)^{1/N}$$

$$= \exp \frac{1}{N} \sum_{k \in U} \log y_k = \exp \frac{1}{N} \sum_{k \in U} z_k = \exp \frac{Z}{N},$$

Sampling and Estimation from Finite Populations, First Edition. Yves Tillé.
© 2020 John Wiley & Sons Ltd. Published 2020 by John Wiley & Sons Ltd.

where $z_k = \log y_k$ and

$$Z = \sum_{k \in U} z_k.$$

First Z and N can be estimated by

$$\hat{Z} = \sum_{k \in S} w_k z_k \quad \text{and} \quad \hat{N} = \sum_{k \in S} w_k.$$

Therefore, the plug-in estimator is

$$\hat{G} = \exp \frac{\hat{Z}}{\hat{N}} = \exp \frac{1}{\hat{N}} \sum_{k \in S} w_k \log y_k = \prod_{k \in S} y_k^{w_k/\hat{N}}.$$

14.2 Variance Estimation

The population variance can be written in many ways:

$$S_y^2 = \frac{1}{N-1} \sum_{k \in U} (y_k - \overline{Y})^2 = \frac{1}{N-1} \sum_{k \in U} y_k^2 - \frac{Y^2}{N(N-1)}.$$

Therefore, the variance is a function of three totals Y, N, and

$$Y_2 = \sum_{k \in U} y_k^2,$$

which can be estimated by, respectively,

$$\hat{Y} = \sum_{k \in S} w_k y_k, \quad \hat{N} = \sum_{k \in S} w_k, \quad \text{and} \quad \hat{Y}_2 = \sum_{k \in S} w_k y_k^2.$$

Hence, the plug-in estimator is

$$\hat{S}_y^2 = \frac{1}{\hat{N}-1} \sum_{k \in S} w_k y_k^2 - \frac{\hat{Y}^2}{\hat{N}(\hat{N}-1)} = \frac{1}{\hat{N}-1} \sum_{k \in S} w_k (y_k - \hat{\overline{Y}})^2,$$

where $\hat{\overline{Y}} = \hat{Y}/\hat{N}$.

Obviously, there exist other solutions to estimate the variance. For example, by considering that the variance can be written as

$$S_y^2 = \frac{1}{2N(N-1)} \sum_{k \in U} \sum_{\ell \in U} (y_k - y_\ell)^2,$$

and supposing that N is known, we can construct an unbiased estimator by using the second-order inclusion probabilities:

$$\hat{S}_{yalt}^2 = \frac{1}{2N(N-1)} \sum_{k \in S} \sum_{\ell \in S} \frac{(y_k - y_\ell)^2}{\pi_{k\ell}}.$$

This second solution is not a plug-in estimator and it is harder to put into practice, as it necessitates the calculation of a double sum. Estimator $\hat{S}_{yalt}^2$ is nevertheless unbiased, which is not the case for $\hat{S}_y^2$, which is slightly biased.

14.3 Covariance Estimation

The population covariance can also be written as a function of totals:

$$S_{xy} = \frac{1}{N-1} \sum_{k \in U} (x_k - \bar{X})(y_k - \bar{Y}) = \frac{1}{N-1} \sum_{k \in U} x_k y_k - \frac{X\,Y}{N(N-1)}.$$

Therefore, the covariance is a function of four totals X, Y, N, and

$$(XY) = \sum_{k \in U} x_k y_k$$

that can be estimated, respectively, by

$$\hat{X} = \sum_{k \in S} w_k x_k, \quad \hat{Y} = \sum_{k \in S} w_k y_k, \quad \hat{N} = \sum_{k \in S} w_k, \text{ and } \widehat{(XY)} = \sum_{k \in S} w_k x_k y_k.$$

The plug-in estimator is then

$$\hat{S}_{xy} = \frac{1}{\hat{N}-1} \sum_{k \in S} w_k x_k y_k - \frac{\hat{X}\hat{Y}}{\hat{N}(\hat{N}-1)} = \frac{1}{\hat{N}-1} \sum_{k \in S} w_k (x_k - \hat{\bar{X}})(y_k - \hat{\bar{Y}}),$$

where $\hat{\bar{X}} = \hat{X}/\hat{N}$ and $\hat{\bar{Y}} = \hat{Y}/\hat{N}$.

14.4 Implicit Function Estimation

Suppose that a parameter is defined on the population by an implicit function of the type

$$\sum_{k \in U} \phi_k(\theta) = T.$$

We can estimate θ by solving the estimated implicit function using weights:

$$\sum_{k \in S} w_k \phi_k(\hat{\theta}) = T.$$

Example 14.3 The regression coefficient

$$\mathbf{B} = \left(\sum_{k \in U} \mathbf{x}_k \mathbf{x}_k^T \right)^{-1} \sum_{k \in U} \mathbf{x}_k y_k$$

is the solution of the system of equations

$$\sum_{k \in U} \mathbf{x}_k (y_k - \mathbf{x}_k^T \mathbf{B}) = 0.$$

An estimator of $\mathbf{B}$ is the solution to the system given by the implicit function estimator:

$$\sum_{k \in S} w_k \mathbf{x}_k (y_k - \mathbf{x}_k^T \hat{\mathbf{B}}) = 0,$$

which gives

$$\hat{\mathbf{B}} = \left(\sum_{k \in S} w_k \mathbf{x}_k \mathbf{x}_k^T \right)^{-1} \sum_{k \in S} w_k \mathbf{x}_k y_k.$$

14.5   Cumulative Distribution Function and Quantiles

14.5.1   Cumulative Distribution Function Estimation

A cumulative distribution function can be written as a mean. Indeed,

$$F(y) = \frac{1}{N} \sum_{k \in U} \mathbb{1}(y_k \leq y),$$

where $\mathbb{1}(A)$ equals 1 if A is true and 0 if A is false. Therefore, we have a direct plug-in estimator:

$$\widehat{F}_1(y) = \frac{1}{\widehat{N}} \sum_{k \in S} w_k \mathbb{1}(y_k \leq y),$$

where

$$\widehat{N} = \sum_{k \in S} w_k.$$

14.5.2   Quantile Estimation: Method 1

The quantile Q_α of order α is in principle the solution to the implicit equation

$$F(Q_\alpha) = \alpha,$$

and could be estimated by solving equation

$$\widehat{F}_1(\widehat{Q}_\alpha) = \alpha.$$

However, the inverses of functions $F(\cdot)$ and $\widehat{F}(\cdot)$ do not exist, as these functions are step functions and are not strictly increasing. The calculation of a quantile from a step distribution has multiple definitions of which the most common is the following:

$$\widehat{Q}_1(\alpha) = \begin{cases} \dfrac{y_j + y_{j+1}}{2} & \text{if } \widehat{N}\alpha = W_j \\[2mm] y_j & \text{if } \widehat{N}\alpha \neq W_j, \end{cases} \tag{14.1}$$

where the W_k are the cumulative weights w_k from the sorted y_k in nondecreasing order on the sample, $W_0 = 0$, and j is such that $W_{j-1} < \widehat{N}\alpha \leq W_j$.

Example 14.4   Consider Table 14.1, which contains a variable of interest y_k, the weights w_k, the cumulative weights W_k, and the relative cumulative weights p_k.

By using the quantile definition (14.1), we obtain for the first quartile $j = 1$. Since $100 \times 0.25 = W_1$, we have

$$\widehat{Q}_1(1/4) = \frac{y_1 + y_2}{2} = 1.5.$$

For the median, $j = 3$ and

$$\widehat{Q}_1(1/2) = y_3 = 3.$$

Finally, for the third quartile, $j = 5$ and

$$\widehat{Q}_1(3/4) = y_5 = 5.$$

These results are presented in Figure 14.1.

Table 14.1 Sample, variable of interest y_k, weights w_k, cumulative weights W_k, and relative cumulative weights p_k

y_k	w_k	W_k	p_k
1	25	25	0.25
2	10	35	0.35
3	20	55	0.55
4	15	70	0.70
5	30	100	1.00
	100		

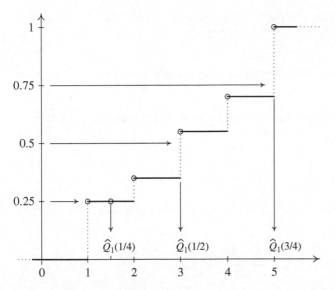

Figure 14.1 Step cumulative distribution function $\hat{F}_1(x)$ with corresponding quartiles.

14.5.3 Quantile Estimation: Method 2

Hyndman & Fan (1996) have presented eight other definitions for the quantile of which six are based on a linear interpolation of the cumulative distribution function. The strictly increasing cumulative distribution function is now inversible. Obviously, depending on the interpolations used, the quantile estimators are only slightly different if the sample size is small.

The simplest linear interpolation of the cumulative distribution function consists of joining by a line the pairs $(y_k, F_1(y_k))$ as on Figure 14.2, which gives

$$\hat{F}_2(y) = \begin{cases} \dfrac{1}{\hat{N}} \left[w_1 + \displaystyle\sum_{k \in S\backslash\{1\}} w_k F_U \left(\dfrac{y - y_{k-1}}{y_k - y_{k-1}} \right) \right] & \text{if } y \geq y_1 \\ 0 & \text{if } y < y_1, \end{cases}$$

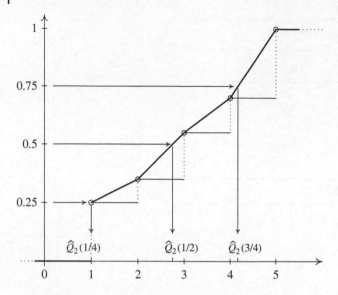

Figure 14.2 Cumulative distribution function $\hat{F}_2(y)$ obtained by interpolation of points $(y_k, F_1(y_k))$ with corresponding quartiles.

where $F_U(\cdot)$ is the cumulative distribution function of a uniform variable on $[0, 1[$:

$$F_U(x) = \begin{cases} 0 & \text{if } x < 0 \\ x & \text{if } 0 \le x < 1 \\ 1 & \text{if } x \ge 1. \end{cases} \tag{14.2}$$

Using this interpolation, it is possible to define a quantile unambiguously. By inverting function $\hat{F}_2(y)$, we obtain

$$\hat{Q}_2(\alpha) = \hat{F}_2^{-1}(\alpha) = y_{j-1} + (y_j - y_{j-1}) \frac{\alpha \hat{N} - W_{j-1}}{w_j},$$

where j is the unit such that $W_{j-1} < \alpha \hat{N} \le W_j$, $y_0 = y_1$, and where the W_k are the cumulative weights w_k given the sorted y_k in nondecreasing order on the sample.

Example 14.5 With the data of Table 14.1 and this second definition, for the first quartile $Q_2(1/4)$, we have $j = 1$ and

$$\hat{Q}_2(1/4) = y_0 + (y_1 - y_0) \frac{0.25 \times 100 - 0}{25} = 1.$$

For the median $Q_2(1/2)$, we have $j = 3$ and

$$\hat{Q}_2(1/2) = y_2 + (y_3 - y_2) \frac{0.5 \times 100 - 35}{20} = 2 + \frac{15}{20} = 2.75.$$

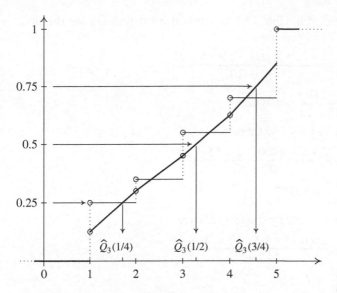

Figure 14.3 Cumulative distribution function $\hat{F}_3(x)$ obtained by interpolating the center of the risers with corresponding quartiles.

For the third quartile $Q_2(3/4)$, we have $j = 5$ and

$$\hat{Q}_2(3/4) = y_4 + (y_5 - y_4)\frac{0.75 \times 100 - 70}{30} = 4 + \frac{5}{30} \approx 4.16666.$$

This result is presented in Figure 14.2.

14.5.4 Quantile Estimation: Method 3

Another solution consists of interpolating the center of the risers, as presented in Figure 14.3:

$$\hat{F}_3(y) = \begin{cases} 0, & y < y_1 \\ \dfrac{1}{\hat{N}}\left[\dfrac{w_1}{2} + \displaystyle\sum_{k\in S\setminus\{1\}}\dfrac{w_{k-1}+w_k}{2}F_u\left(\dfrac{y-y_{k-1}}{y_k-y_{k-1}}\right)\right] & y_1 \le y \le y_n \\ 1 & y > y_n. \end{cases}$$

By inverting function $\hat{F}_3(y)$, we obtain

$$\hat{Q}_3(\alpha) = \hat{F}_3^{-1}(\alpha) = y_{j-1} + (y_j - y_{j-1})\frac{2\alpha\hat{N} - 2W_{j-1} + w_{j-1}}{w_{j-1} + w_j},$$

where j is the unit such that $W_{j-1} - w_{j-1}/2 < \alpha\hat{N} \le W_j - w_j/2$ and $y_0 = y_1$.

Example 14.6 With the data from Table 14.1 and this third definition, for the first quartile $Q_3(1/4)$, we have $j = 2$ and therefore

$$\hat{Q}_3(1/4) = y_1 + (y_2 - y_1)\frac{2 \times 0.25 \times 100 - 2 \times 25 + 25}{25 + 10}$$

$$= 1 + (2 - 1)\frac{25}{25 + 10} \approx 0.7142.$$

For the median $Q_3(1/2)$, we have $j = 4$ and therefore

$$\hat{Q}_3(1/2) = y_3 + (y_4 - y_3)\frac{2 \times 0.5 \times 100 - 2 \times 55 + 20}{20 + 15}$$

$$= 3 + (4 - 3)\frac{10}{35} \approx 3.2857.$$

For the third quartile $Q_3(3/4)$, we have $j = 5$ and therefore

$$\hat{Q}_3(3/4) = y_4 + (y_5 - y_4)\frac{2 \times 0.75 \times 100 - 2 \times 70 + 15}{15 + 30}$$

$$= 4 + (5 - 4)\frac{25}{45} \approx 4.5556.$$

These results are presented in Figure 14.3.

14.5.5 Quantile Estimation: Method 4

A fourth possibility consists of constructing a cumulative distribution function based on a kernel estimator. For example, we can define a cumulative distribution function estimator:

$$\hat{F}_4(y) = \frac{1}{\hat{N}}\sum_{k \in U}\frac{a_k}{\pi_k}\Phi\left(\frac{y - y_k}{h}\right), \tag{14.3}$$

where h is the size of the smoothing window and $\Phi(\cdot)$ is a differentiable cumulative distribution function of a random variable with zero mean and unit variance. Let $\Phi'(\cdot) = \phi(\cdot)$ denote the derivative of $\Phi(\cdot)$. Usually, $\Phi(\cdot)$ can be the cumulative distribution function of a standard normal variable. Then, the quantile is

$$\hat{Q}_4(\alpha) = \hat{F}_4^{-1}(\alpha).$$

14.6 Cumulative Income, Lorenz Curve, and Quintile Share Ratio

14.6.1 Cumulative Income Estimation

Many inequality indices of income are based on the cumulative incomes of a portion of the poorest people. Let $y_1, \dots, y_k, \dots, y_N$ be the incomes of individuals or households in the population and $y_{(1)}, \dots, y_{(k)}, \dots, y_{(N)}$ these same incomes sorted in increasing order.

We can define the total income of the αN poorest individuals by

$$\tilde{Y}(\alpha) = \sum_{k \in U} y_k \mathbb{1}(y_k \le Q_\alpha),$$

where Q_α is the quantile of order α. This definition is unfortunately not very precise, as quantiles have many different definitions when the cumulative distribution function is a step function (see Hyndman & Fan, 1996). We prefer to use the less ambiguous definition of the total income of the αN poorest individuals:

$$Y(\alpha) = \sum_{k \in U} y_{(k)} F_U[\alpha N - (k-1)],$$

where $F_U(\cdot)$ is the cumulative distribution function of a uniform random variable defined in Expression (14.2). The function of interest $Y(\alpha)$ is strictly increasing in α over $]0, 1[$, which is not the case for $\tilde{Y}(\alpha)$.

In order to estimate $Y(\alpha)$, we can use the plug-in estimator

$$\hat{Y}(\alpha) = \sum_{k \in S} w_k y_{(k)} F_U \left(\frac{\alpha \hat{N} - W_{k-1}}{w_k} \right),$$

where the W_k are the cumulative weights w_k according to the sorted y_k by nondecreasing order on the sample and $W_0 = 0$. Estimator $Y(\alpha)$ is also strictly increasing in α over $]0, 1[$.

14.6.2 Lorenz Curve Estimation

The Lorenz (1905) curve is

$$L(\alpha) = \frac{Y(\alpha)}{Y}. \tag{14.4}$$

If y is the income variable, the Lorenz curve $L(\alpha)$ is the proportion of income earned by the proportion α of poorest individuals.

Function $L(\alpha)$ is estimated by

$$\hat{L}(\alpha) = \frac{\hat{Y}(\alpha)}{\hat{Y}}. \tag{14.5}$$

The functions $L(\alpha)$ and $\hat{L}(\alpha)$ are strictly increasing and convex on $]0, 1[$.

14.6.3 Quintile Share Ratio Estimation

The Quintile Share Ratio (QSR) (see, among others, Langel & Tillé, 2011b) is the income share of the 20% richest over the income share of the 20% poorest. We can define this indicator using the Lorenz curve:

$$QSR = \frac{1 - L(4/5)}{L(1/5)}.$$

This quantity can be easily estimated using the Lorenz curve estimator:

$$\widehat{QSR} = \frac{1 - \hat{L}(4/5)}{\hat{L}(1/5)}.$$

14.7 Gini Index

The Gini index (see Gini, 1912, 1914, 1921; Gastwirth, 1972) is equal to twice the surface area between the Lorenz curve and the bisector (see Figure 14.4, page 292):

$$G = 2\int_0^1 [\alpha - L(\alpha)]d\alpha = 1 - 2\int_0^1 L(\alpha)d\alpha.$$

By using the definition of the Lorenz curve given in Expression (14.4), the Gini index can be reformulated for a finite population.

Result 14.1

$$G = \frac{2}{YN}\sum_{k\in U}ky_{(k)} - \frac{N+1}{N} = \frac{\sum_{k\in U}\sum_{\ell\in U}|y_k - y_\ell|}{2NY}.$$

Proof: To calculate the Gini index, we must calculate

$$\int_0^1 L(\alpha)d\alpha = \frac{1}{Y}\int_0^1 Y(\alpha)d\alpha.$$

Yet,

$$\int_0^1 Y(\alpha)d\alpha = \int_0^1 \sum_{k\in U}y_{(k)}F_U[\alpha N - (k-1)]d\alpha$$

$$= \sum_{k\in U}y_{(k)}\int_0^1 F_U[\alpha N - (k-1)]d\alpha$$

$$= \sum_{k\in U}y_{(k)}\int_{-(k-1)}^{N-(k-1)} F_U(\beta)\frac{d\beta}{N}$$

$$= \frac{1}{N}\sum_{k\in U}y_{(k)}\left(\int_0^1 F_U(\beta)d\beta + \int_1^{N-(k-1)} F_U(\beta)d\beta\right)$$

$$= \frac{1}{N}\sum_{k\in U}y_{(k)}\left(\frac{1}{2}+N-k\right)$$

$$= \left(1+\frac{1}{2N}\right)Y - \frac{1}{N}\sum_{k\in U}ky_{(k)}.$$

Therefore, we obtain

$$G = 1 - 2\left(1+\frac{1}{2N}\right) + \frac{2}{YN}\sum_{k\in U}ky_{(k)} = \frac{2}{YN}\sum_{k\in U}ky_{(k)} - \frac{N+1}{N}.$$

∎

Finally, using the Lorenz curve estimator (14.5), we can estimate the Gini index.

Result 14.2

$$\widehat{G} = \frac{2}{\widehat{N}\widehat{Y}}\sum_{k\in S}w_k W_k y_k - \left(1 + \frac{1}{\widehat{N}\widehat{Y}}\sum_{k\in S}w_k^2 y_k\right) = \frac{\sum_{k\in S}\sum_{\ell\in S}w_k w_\ell |y_k - y_\ell|}{2\widehat{N}\widehat{Y}},$$

where the W_k are the cumulative weights w_k according to the sorted y_k by nondecreasing order in the sample.

Proof: To obtain the Gini index estimator, we must calculate

$$\int_0^1 \widehat{L}(\alpha)d\alpha = \frac{1}{\widehat{Y}}\int_0^1 \widehat{Y}(\alpha)d\alpha.$$

Yet,

$$\int_0^1 \widehat{Y}(\alpha)d\alpha = \int_0^1 \sum_{k\in S}w_k y_k F_U\left(\frac{\alpha\widehat{N}-W_{k-1}}{w_k}\right)d\alpha$$

$$= \sum_{k\in S}w_k y_k \int_0^1 F_U\left(\frac{\alpha\widehat{N}-W_{k-1}}{w_k}\right)d\alpha$$

$$= \sum_{k\in S}w_k y_k \int_{-W_{k-1}/w_k}^{(\widehat{N}-W_{k-1})/w_k} F_U(\beta)\frac{w_k d\beta}{\widehat{N}}$$

$$= \sum_{k\in S}w_k^2 y_k \frac{1}{\widehat{N}}\left[\int_0^1 F_U(\beta)d\beta + \int_1^{(\widehat{N}-W_{k-1})/w_k} F_U(\beta)d\beta\right]$$

$$= \frac{1}{\widehat{N}}\sum_{k\in S}w_k^2 y_k\left(\frac{1}{2}+\frac{\widehat{N}-W_k}{w_k}\right)$$

$$= \frac{1}{\widehat{N}}\sum_{k\in S}\frac{w_k^2 y_k}{2} + \widehat{Y} - \frac{1}{\widehat{N}}\sum_{k\in S}w_k W_k y_k.$$

Therefore, we obtain

$$\widehat{G} = 1 - \frac{2}{\widehat{Y}}\left(\frac{1}{\widehat{N}}\sum_{k\in S}\frac{w_k^2 y_k}{2} + \widehat{Y} - \frac{1}{\widehat{N}}\sum_{k\in S}w_k W_k y_k\right)$$

$$= \frac{2}{\widehat{N}\widehat{Y}}\sum_{k\in S}w_k W_k y_k - \left(1 + \frac{1}{\widehat{N}\widehat{Y}}\sum_{k\in S}w_k^2 y_k\right). \tag{14.6}$$

∎

The variance calculation of a Gini index calculated under a sampling design is a relatively complex problem which is dealt with in Sections 15.7.8 and 15.10.6.

14.8 An Example

Table 14.2 contains data on a sample of five individuals sorted in increasing order of income. Each individual has a weight w_k and an income y_k. The other columns in

Table 14.2 Table of fictitious incomes y_k, weights w_k, cumulative weights W_k, relative cumulative weights p_k, cumulative incomes $\widehat{Y}(p_k)$, and the Lorenz curve $\widehat{L}(p_k)$.

k	w_k	y_k	W_k	p_k	$w_k y_k$	$\widehat{Y}(p_k)$	$\widehat{L}(p_k)$
1	10	10	10	0.20	100	100	0.05
2	19	20	29	0.58	380	480	0.24
3	10	40	39	0.78	400	880	0.44
4	9	80	48	0.96	720	1600	0.80
5	2	200	50	1.00	400	2000	1.00
	50				2000		

Table 14.2 contain the cumulative weights W_k, the relative cumulative weights

$$p_k = \frac{W_k}{\sum_{\ell \in S} w_\ell},$$

the product of incomes and weights $w_k y_k$, and these same cumulated quantities

$$\widehat{Y}(p_k) = \sum_{\ell=1}^{k} w_\ell y_\ell.$$

Finally, the Lorenz curve is calculated for values p_k:

$$\widehat{L}(p_k) = \frac{\sum_{\ell=1}^{k} w_\ell y_\ell}{\sum_{\ell \in U} w_\ell y_\ell} = \frac{\widehat{Y}(p_k)}{\widehat{Y}}.$$

The Lorenz curve that connects the points

$$(p_k, \widehat{L}(p_k)), k \in S \text{ with}(p_0, \widehat{L}(p_0)) = (0, 0)$$

is presented in Figure 14.4. The shaded area between the diagonal and the Lorenz curve is equal to half the estimated Gini index.

To estimate the Gini index, we must again calculate the totals contained in Table 14.3.

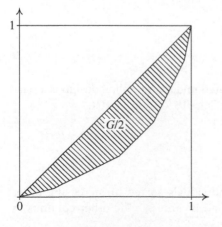

Figure 14.4 Lorenz curve and the surface associated with the Gini index.

Table 14.3 Totals necessary to estimate the Gini index.

w_k	$w_k y_k$	$w_k y_k W_k$	$w_k^2 y_k$
10	100	1000	1000
19	380	11 020	7220
10	400	15 600	4000
9	720	34 560	6480
2	400	20 000	800
50	2000	82 180	19 500

We obtain

$$\hat{G} = \frac{2}{\hat{N}\hat{Y}} \sum_{k\in S} w_k W_k y_k - \left(1 + \frac{1}{\hat{N}\hat{Y}} \sum_{k\in S} w_k^2 y_k\right)$$

$$= \frac{2}{50 \times 2\,000} \times 82\,180 - \left(1 + \frac{1}{50 \times 2\,000} 19\,500\right) = \hat{G} \approx 44.86\%.$$

Finally, we can calculate the QSR estimator. From Table 14.2 we obtain

$$\hat{L}(0.20) = 0.05.$$

In addition,

$$\hat{Y}_{0.80} = \sum_{k\in S} w_k y_k F_U \left(\frac{0.80\hat{N} - W_{k-1}}{w_k}\right)$$

$$= \sum_{k\in S, k\leq 3} w_k y_k + w_4 y_4 F_U \left(\frac{0.80\hat{N} - W_3}{w_4}\right)$$

$$= 880 + 720 \times F_U \left(\frac{0.80 \times 50 - 39}{9}\right) = 880 + \frac{720}{9} = 960.$$

Therefore,

$$\hat{L}(0.80) = \frac{\hat{Y}_{0.80}}{\hat{Y}} = \frac{960}{2\,000} = 0.48.$$

The QSR estimator is equal to

$$\widehat{QSR} = \frac{1 - \hat{L}(0.80)}{\hat{L}(0.20)} = \frac{1 - 0.48}{0.05} = 10.4.$$

15

Variance Estimation by Linearization

15.1 Introduction

Estimated functions of interest during a sample survey are rarely limited to simple linear functions of totals. The simplest functions are often proportions on subsets of the population (or domains). These are in fact ratios and are nonlinear in the values taken by the variables. Often, significantly more complex problems need to be dealt with, such as variance estimation for a regression, a correlation coefficient, a variance, or inequality indices. In addition, the use of auxiliary information allows us to calibrate estimators on known variables, which gives even more complex forms to estimators.

This chapter seeks to give a preview of the linearization method that allows the estimation of the precision of complex statistics. The idea is to approach a complex statistic by the total of a new so-called linearized variable. We can then approach the variance of this complex statistic by calculating the variance of the estimator of a simple total. Linearization techniques have been introduced by Woodruff (1971). Their application to survey theory is the object of many publications: Krewski & Rao (1981b), Binder (1983), Wolter (1985), Binder & Patak (1994), Binder (1996), Stukel et al. (1996), Kovacevic & Binder (1997), Demnati & Rao (2004), Kott (2004), Deville (1999), Osier (2009), Goga et al. (2009), Graf (2011), and Vallée & Tillé (2019).

Linearization techniques consist of approaching a complex statistic by a linear approximation. We then estimate the variance of this approximation. Obviously, the validity of these approximations is based on asymptotic considerations that are only valid for large samples. We first review some tools for asymptotic statistics. The reader can consult Fuller (1976, p. 179) or Serfling (1980) on this topic. We then summarize linearization techniques while giving many practical applications.

15.2 Orders of Magnitude in Probability

Linearization techniques are based on limited Taylor series expansion methods. However, a special difficulty appears, as the expansion is calculated with respect to a random variable. Therefore, the determination of the remainder will be a bit more delicate. To deal with these problems, we define the notion of order of magnitude and orders of magnitude in probability.

Sampling and Estimation from Finite Populations, First Edition. Yves Tillé.
© 2020 John Wiley & Sons Ltd. Published 2020 by John Wiley & Sons Ltd.

Definition 15.1 *A sequence of real numbers f_n, $n = 1, 2, \ldots$ is said to be of lower order than $h_n > 0$, $n = 1, 2, \ldots$, if*

$$\lim_{n \to \infty} \frac{f_n}{h_n} = 0.$$

We then write

$$f_n = o(h_n).$$

Definition 15.2 *A sequence of real numbers f_n, $n = 1, 2, \ldots$ is said to be at most of order $h_n > 0$, $n = 1, 2, \ldots$, or bounded with respect to $h_n > 0$, if there exists $M > 0$ such that*

$$|f_n| \leq M h_n,$$

for all $n = 1, 2, \ldots$. We then write

$$f_n = O(h_n).$$

The notion of orders of magnitude can also be applied to random variables. However, the definition is somewhat different as it is linked to the notion of convergence in probability.

Definition 15.3 *A sequence of random variables X_n converges in probability towards a random variable X if, for all $\varepsilon > 0$,*

$$\lim_{n \to \infty} \Pr(|X_n - X| > \varepsilon) = 0.$$

We then write

$$\lim_{n \to \infty} p X_n = X,$$

or more simply

$$X_n \overset{P}{\to} X.$$

Convergence in probability allows the introduction of the notion of order in probability also called the stochastic order of magnitude.

Definition 15.4 *Let X_n be a sequence of random variables and h_n a deterministic sequence. The sequence X_n is said to be of lower order than $h_n > 0$ in probability if*

$$\lim_{n \to \infty} p \frac{X_n}{h_n} = 0.$$

We then write

$$X_n = o_p(h_n).$$

Definition 15.5 *Let X_n be a sequence of random variables. X_n is said to be of order at most h_n in probability or bounded in probability by $h_n > 0$ if for all $\varepsilon > 0$, there exists a number $M_\varepsilon > 0$ such that*

$$\Pr[|X_n| \geq M_\varepsilon h_n] \leq \varepsilon$$

for all n = 1, 2, 3, ... We then write

$$X_n = O_p(h_n).$$

Result 15.1   *Let X_n and Y_n be two sequences of random variables such that*

$$X_n = o_p(h_n) \quad and \quad Y_n = o_p(g_n).$$

If a is a real number and $\alpha > 0$, then

i) $aX_n = o_p(h_n)$,
ii) $|X_n|^\alpha = o_p(h_n^\alpha)$,
iii) $X_n Y_n = o_p(h_n g_n)$,
iv) $X_n + Y_n = o_p(\max(h_n, g_n))$.

Proof:
  i) If $X_n = o_p(h_n)$ and $Y_n = o_p(g_n)$, then

$$\lim_{n \to \infty} \Pr\left(\left|\frac{X_n}{h_n}\right| > \varepsilon\right) = 0$$

and

$$\lim_{n \to \infty} \Pr\left(\left|\frac{Y_n}{g_n}\right| > \varepsilon\right) = 0$$

for all $\varepsilon > 0$. This directly implies that

$$aX_n = o_p(h_n).$$

 ii) Since

$$\Pr\left[\left|\frac{X_n}{h_n}\right| > \varepsilon\right] = \Pr\left[\left|\frac{X_n}{h_n}\right|^\alpha > \varepsilon^\alpha\right],$$

we obtain $|X_n|^\alpha = o_p(h_n^\alpha)$.
iii) Then, we have for all $\varepsilon > 0$,

$$\Pr\left[\left|\frac{X_n}{h_n}\right| > \varepsilon\right] + \Pr\left[\left|\frac{Y_n}{g_n}\right| > \varepsilon\right] \geq \Pr\left[\left|\frac{X_n}{h_n}\right| > \varepsilon \ ou \ \left|\frac{Y_n}{g_n}\right| > \varepsilon\right]$$

$$\geq \Pr\left[\left|\frac{X_n Y_n}{h_n g_n}\right| > \varepsilon^2\right],$$

which directly implies that

$$\lim_{n \to \infty} \Pr\left[\left|\frac{X_n Y_n}{h_n g_n}\right| > \varepsilon^2\right] = 0,$$

where

$$X_n Y_n = o_p(h_n g_n).$$

iv) Finally, $X_n + Y_n = o_p(\max(h_n, g_n))$ is obvious. ∎

Result 15.2   *Let X_n and Y_n be two sequences of random variables such that*

$$X_n = O_p(h_n) \quad and \quad Y_n = O_p(g_n).$$

If a is a real number and $\alpha > 0$, then

$$aX_n = O_p(h_n),$$

$$|X_n|^\alpha = O_p(h_n^\alpha),$$

$$X_n Y_n = O_p(h_n g_n),$$

$$X_n + Y_n = O_p[\max(h_n, g_n)].$$

The proofs are similar to the previous ones.

Theorem 15.1 *Bienaymé–Chebyshev inequality (direct case). Let $\alpha > 0$ and X be a discrete random variable such that $E(X^\alpha) < \infty$, then for all $\varepsilon > 0$ and for all $A \in \mathbb{R}$,*

$$\Pr[|X - A| \geq \varepsilon] \leq \frac{E(|X - A|^\alpha)}{\varepsilon^\alpha}.$$

Proof: Let $X_1, \ldots, X_i, \ldots, X_I$, denote the possible values of X. We can write

$$E(|X - A|^\alpha) = \sum_{i=1}^{I} |X_i - A|^\alpha \Pr[X = X_i]$$

$$= \sum_{\substack{i=1 \\ |X_i - A| < \varepsilon}}^{I} |X_i - A|^\alpha \Pr[X = X_i] + \sum_{\substack{i=1 \\ |X_i - A| \geq \varepsilon}}^{I} |X_i - A|^\alpha \Pr[X = X_i]$$

$$\geq \varepsilon^\alpha \sum_{\substack{i=1 \\ |X_i - A| \geq \varepsilon}}^{I} \Pr[X = X_i] = \varepsilon^\alpha \Pr[|X - A| \geq \varepsilon].$$

∎

Result 15.3 *Let X_n be a sequence of random variables such that*

$$E(X_n^2) = O(h_n),$$

then $X_n = O_p(\sqrt{h_n})$.

Proof: On one hand, since $E(X_n^2) = O(h_n)$, there exists $M_A > 0$ such that $E(X_n^2) \leq M_A h_n$, for all n. On the other hand, let $\delta > 0$. With $\alpha = 2, A = 0$, and $\varepsilon = \sqrt{M_B h_n}$ (M_B some positive real number), by the Bienaymé–Chebyshev inequality, we obtain

$$\Pr(|X_n| \geq \sqrt{M_B h_n}) \leq \frac{E(X_n^2)}{M_B h_n} \leq \frac{M_A}{M_B}.$$

By taking $M_B \geq M_A/\delta$, we finally obtain

$$\Pr(|X_n| \geq \sqrt{M_B h_n}) \leq \delta$$

and therefore $X_n = O_p(\sqrt{h_n})$.

∎

Result 15.4 *Let X_n be a sequence of random variables such that*

$$E\{[X_n - E(X_n)]^2\} = O(h_n)$$

and that

$$E(X_n) = O(\sqrt{h_n}),$$

then $X_n = O_p(\sqrt{h_n})$.

Proof: Since

$$E(X_n^2) = E\{[X_n - E(X_n)]^2\} + E(X_n)^2 = O(h_n),$$

the result follows from Result 15.3. ∎

Example 15.1 Let $X_1, \ldots, X_n$, n be independent identically distributed random variables of mean μ and standard deviation σ. The variable

$$\overline{X}_n = \frac{1}{n} \sum_{i=1}^{n} X_n$$

is of variance

$$\text{var}(\overline{X}_n) = \frac{\sigma^2}{n},$$

and therefore $\overline{X}_n - \mu = O_p(n^{-1/2})$.

Result 15.5 *Let h_n be a sequence of positive numbers such that $\lim_{n\to\infty} h_n = 0$, X_n a sequence of random variables such that $X_n = x_0 + O_p(h_n)$, and $f(x)$ a function differentiable α times with its derivatives continuous at point $x = x_0$. We have*

$$f(X_n) = f(x_0) + \sum_{i=1}^{\alpha-1} (X_n - x_0)^i \frac{f^{(i)}(x_0)}{i!} + O_p(h_n^\alpha),$$

where $f^{(i)}(x_0)$ is the ith derivative off $(\cdot)$ evaluated at point $x = x_0$.

Proof: By applying a Taylor expansion, we have at once

$$f(X_n) = f(x_0) + \sum_{i=1}^{\alpha-1} (X_n - x_0)^i \frac{f^{(i)}(x_0)}{i!} + (X_n - x_0)^\alpha \frac{f^{(\alpha)}(b)}{\alpha!},$$

where b is between x_0 and X_n. Function $f^{(\alpha)}(\cdot)$ is continuous at $x = x_0$ and $|b - x_0| = o_p(1)$. Indeed, $|b - x_0| \leq |X_n - x_0| = O_p(h_n)$ and $h_n \to 0$ where $n \to \infty$. Therefore, we have that $f^{(\alpha)}(b)$ is bounded in probability, in other words that $f^{(\alpha)}(b) = O_p(1)$. We finally obtain

$$(X_n - x_0)^\alpha \frac{f^{(\alpha)}(b)}{\alpha!} = O_p(h_n^\alpha).$$

∎

Result 15.6 *Let h_n be a sequence of positive numbers such that $\lim_{n \to \infty} h_n = 0$, $X_{1n}, \ldots,$ $X_{jn}, \ldots, X_{pn}$, a sequence of p random variables such that $X_{jn} = x_{j0} + O_p(h_n), j = 1, \ldots, p$, and $f(x_1, \ldots, x_p)$ a function whose second-order partial derivatives exist and are continuous at point $(x_1, \ldots, x_p) = (x_{10}, \ldots, x_{p0})$, then*

$$f(X_{1n}, \ldots, X_{pn}) = f(x_{10}, \ldots, x_{p0})$$
$$+ \sum_{j=1}^{p} (X_{jn} - x_{j0}) \left. \frac{\partial f(x_1, \ldots, x_p)}{\partial x_j} \right|_{(x_1, \ldots, x_p) = (x_{10}, \ldots, x_{p0})}$$
$$+ O_p(h_n^2).$$

Proof: By applying a Taylor expansion, we obtain

$$f(X_{1n}, \ldots, X_{pn})$$
$$= f(x_{10}, \ldots, x_{p0}) + \sum_{j=1}^{p} (X_{jn} - x_{j0}) \left. \frac{\partial f(x_1, \ldots, x_p)}{\partial x_j} \right|_{(x_1, \ldots, x_p) = (x_{10}, \ldots, x_{p0})}$$
$$+ \sum_{j=1}^{p} \sum_{i=1}^{p} \frac{(X_{jn} - x_{j0})(X_{in} - x_{i0})}{2!} \left. \frac{\partial^2 f(x_1, \ldots, x_p)}{\partial x_j \partial x_i} \right|_{(x_1, \ldots, x_p) = (b_1, \ldots, b_p)},$$

where the b_j are between X_{jn} and x_{j0}. As in the previous result, $(X_{jn} - x_{j0})(X_{in} - x_{i0}) = O_p(h_n^2)$, which is a multiple of a quantity bounded in probability. ∎

15.3 Asymptotic Hypotheses

There does not really exist, properly speaking, a general asymptotic survey theory. However, we have a few noteworthy results that show, under certain conditions, the asymptotic normality of mean estimators. A result concerning simple designs comes from Madow (1948). Certain unequal probability designs have been dealt with by Rosén (1972a,b) and Berger (1996, 1998a,b,c). In many recent publications, asymptotic results are proposed by (Rubin-Bleuer, 2011; Bonnéry et al., 2012; Brändén & Jonasson, 2012; Boistard et al., 2012, 2017; Jonasson et al., 2013; Saegusa, 2015; Rebecq, 2019; Bertail et al., 2017).

The main difficulty comes from the fact that, for finite populations, the construction of an asymptotic theory cannot just limit itself to make the sample size tend towards infinity. If we construct an asymptotic theory, we must first define a sequence of populations of increasing size. Then, we define a sequence of sampling designs to apply on each of these populations. We can then construct a sequence of estimators and discuss their asymptotic properties.

We then use the asymptotic hypotheses of Isaki & Fuller (1982) that are also used in Deville & Särndal (1992). Consider p totals $Y_1, \ldots, Y_j, \ldots, Y_p$, where

$$Y_j = \sum_{k \in U} y_{kj}.$$

Suppose we have homogeneous linear estimators of Y_j that can be written as

$$\hat{Y}_j = \sum_{k \in S} w_k y_{kj}, j = 1, \ldots, p.$$

The weights w_k can be the expansion weights $d_k = 1/\pi_k$, but also the weights of a regression estimator or a calibrated estimator. Suppose also that we have a variance estimator of each of the $\widehat{Y}_j$ which are denoted by $v(\widehat{Y}_j)$.

Suppose that the $\widehat{Y}_j$ satisfy the following conditions:

1. The quantities Y_j/N have a limit.
2. The quantities $(\widehat{Y}_j - Y_j)/N$ converge in probability towards 0 and more specifically

$$\frac{\widehat{Y}_j - Y_j}{N} = O_p\left(\frac{1}{\sqrt{n}}\right), j = 1, \ldots, p. \tag{15.1}$$

3. The vector $n^{-1/2}N^{-1}(\widehat{Y}_1, \ldots, \widehat{Y}_j, \ldots, \widehat{Y}_p)$ has a multidimensional normal distribution.

These hypotheses are relatively realistic and are true for simple designs and by extension for stratified designs (if the number of strata does not increase too quickly in n) and for cluster designs (if the number of clusters increases in n).

Consider a function of interest θ that is a function of y_{kj}. This function of interest can be a function of totals $\theta = f(Y_1, \ldots, Y_j, \ldots, Y_p)$ and can be estimated by $\theta = f(\widehat{Y}_1, \ldots, \widehat{Y}_j, \ldots, \widehat{Y}_p)$. We then consider functions of interest that are more general than functions of totals such as quantiles or inequality indices which are presented in Chapter 14.

Definition 15.6 *Let $\alpha \geq 0$. If $N^{-\alpha}\theta$ is bounded for all values of N, then we say that θ is of degree α.*

For example, $R = Y/X$ is of degree 0, Y is of degree 1, $\overline{Y} = Y/N$ is of degree 0 (as it is a ratio of two functions of degree 1), and $\text{var}_p(\widehat{Y})$ is of degree 2. The objective is to calculate the variance of the estimators of θ.

Definition 15.7 *The variable z_k is called the linearized statistic $\widehat{\theta}$ of degree α if*

$$\frac{\widehat{\theta} - \theta}{N^\alpha} = \frac{\sum_{k \in S} w_k z_k - \sum_{k \in U} z_k}{N^\alpha} + O_p\left(\frac{1}{n}\right).$$

The linearized variable can also be used to construct an approximate variance of $\widehat{\theta}$:

$$\text{var}_p\left(\frac{\widehat{\theta}}{N^\alpha}\right) \approx \text{var}_p\left(\frac{\sum_{k \in S} w_k z_k}{N^\alpha}\right).$$

15.3.1 Linearizing a Function of Totals

A first objective is to estimate the variance of a function of p totals whose second-order derivatives exist and are continuous:

$$\theta = f(Y_1, \ldots, Y_j, \ldots, Y_p).$$

To estimate this function, we use the plug-in estimator:

$$\widehat{\theta} = f(\widehat{Y}_1, \ldots, \widehat{Y}_j, \ldots \widehat{Y}_p),$$

where

$$\hat{Y}_j = \sum_{k \in S} w_k y_{kj}$$

are estimators (potentially biased) of the Y_j. Generally the $\hat{Y}_j$ are expansion estimators ($w_k = 1/\pi_k$), but they can also be more complex estimators.

Result 15.7   *Variable*

$$v_k = \sum_{j=1}^{p} y_{kj} \frac{\partial f(c_1, \dots, c_p)}{\partial c_j}\bigg|_{c_1 = Y_1, \dots, c_p = Y_p} \quad , k \in U, \tag{15.2}$$

is a linearized variable of $\hat{\theta}$.

Proof: Let a function of interest θ of degree α be estimated by $\hat{\theta}$. By setting $\overline{\hat{Y}}_j = \hat{Y}_j/N$, we have

$$N^{-\alpha}\hat{\theta} = N^{-\alpha}f(\hat{Y}_1, \dots, \hat{Y}_j, \dots, \hat{Y}_p) = N^{-\alpha}f(N\overline{\hat{Y}}_1, \dots, N\overline{\hat{Y}}_j, \dots, N\overline{\hat{Y}}_p).$$

Condition (15.1) gives $\overline{\hat{Y}}_j = \overline{Y}_j + O_p(n^{-1/2})$ and in accordance with Result 15.6, we directly have

$$N^{-\alpha}\hat{\theta} = N^{-\alpha}f(N\overline{\hat{Y}}_1, \dots, N\overline{\hat{Y}}_j, \dots, N\overline{\hat{Y}}_p)$$

$$= N^{-\alpha}f(N\overline{Y}_1, \dots, N\overline{Y}_j, \dots, N\overline{Y}_p)$$

$$+ N^{-\alpha} \sum_{j=1}^{p} (\overline{\hat{Y}}_j - \overline{Y}_j) \frac{\partial f(Nc_1, \dots, Nc_p)}{\partial c_j}\bigg|_{c_1 = \overline{Y}_1, \dots, c_p = \overline{Y}_p} + O_p\left(\frac{1}{n}\right)$$

$$= N^{-\alpha}\theta$$

$$+ N^{-\alpha} \sum_{j=1}^{p} (\hat{Y}_j - Y_j) \frac{\partial f(c_1, \dots, c_p)}{\partial c_j}\bigg|_{c_1 = Y_1, \dots, c_p = Y_p} + O_p\left(\frac{1}{n}\right)$$

$$= N^{-\alpha}\theta + N^{-\alpha}(v - V) + O_p\left(\frac{1}{n}\right),$$

where

$$V = \sum_{k \in U} v_k \quad \text{and} \quad v = \sum_{k \in S} w_k v_k.$$

■

If $N^{-\alpha}(v - V) = O_p(n^{-1/2})$, the variance of the estimator of the function of interest can be simply approximated. In fact,

$$\text{var}_p(N^{-\alpha}\hat{\theta}) = \text{var}_p\left[N^{-\alpha}\theta + N^{-\alpha}(v - V) + O_p\left(\frac{1}{n}\right)\right]$$

$$= \text{var}_p(N^{-\alpha}v) + 2E_p\left[N^{-\alpha}(v - V) \times O_p\left(\frac{1}{n}\right)\right] + E_p\left[O_p\left(\frac{1}{n}\right)^2\right]$$

$$= \text{var}_p\left(\frac{v}{N^{\alpha}}\right) + E_p O_p\left(\frac{1}{n^{3/2}}\right).$$

By considering that $E_p O_p(n^{-3/2})$ is negligible, we can construct an approximation of the variance:

$$\text{var}_p(\hat{\theta}) \approx \text{var}_p(v).$$

15.3.2 Variance Estimation

To estimate the variance, we cannot directly use the v_k, as they depend on unknown population totals Y_j. We approach the v_k by replacing the unknown totals by their estimators. Let $\hat{v}_k$ denote the approximation of the linearized variable. In order to estimate the variance of $\hat{\theta}$, we use a variance estimator $v(.)$. If the $\hat{Y}_j$ are expansion estimators, we can in general use the expansion variance estimator:

$$v(\hat{\theta}) = \sum_{k \in S} \frac{\hat{v}_k^2}{\pi_k^2}(1 - \pi_k) + \sum_{k \in S}\sum_{\substack{\ell \in S \\ \ell \neq k}} \frac{\hat{v}_k \hat{v}_\ell}{\pi_k \pi_\ell} \frac{\pi_{k\ell} - \pi_k \pi_\ell}{\pi_{k\ell}}.$$

However, for each particular design, like simple designs, stratified or balanced designs, there exists simpler estimators that can be directly used with the linearized variable.

15.4 Linearization of Functions of Interest

15.4.1 Linearization of a Ratio

The most classical problem is to estimate a ratio $R = Y/X$ and its variance using simple random sampling without replacement. We use the classical notation introduced in Section 9.1. We first define $f(c_1, c_2) = c_1/c_2$ and therefore

$$R = f(Y, X).$$

The function of interest R is a function of totals that can be simply estimated by

$$\hat{R} = f(\hat{Y}, \hat{X}) = \frac{\hat{Y}}{\hat{X}}.$$

We then calculate the partial derivatives:

$$\frac{\partial f(c_1, c_2)}{\partial c_1}\bigg|_{c_1=Y, c_2=X} = \frac{1}{X} \quad \text{and} \quad \frac{\partial f(c_1, c_2)}{\partial c_2}\bigg|_{c_1=Y, c_2=X} = -\frac{Y}{X^2},$$

and by (15.2) we obtain

$$v_k = \frac{y_k}{X} - \frac{Y}{X^2}x_k = \frac{1}{X}(y_k - R\,x_k). \tag{15.3}$$

In a given design, the approximate variance of the ratio estimator is

$$\text{var}_p(\hat{R}) \approx \text{var}_p\left[\sum_{k \in S} \frac{1}{X}(y_k - R\,x_k)\right]$$

$$= \frac{1}{X^2}[\text{var}_p(\hat{Y}) + R^2 \ \text{var}_p(\hat{X}) - 2R \ \text{cov}_p(\hat{X}, \hat{Y})],$$

and is estimated by

$$v(\widehat{R}) = \frac{1}{\widehat{X}}[v(\widehat{Y}) + \widehat{R}^2 \, v(\widehat{X}) - 2\widehat{R} \, cov(\widehat{X}, \widehat{Y})],$$

where $\widehat{R} = \widehat{Y}/\widehat{X}$.

In sampling random sampling without replacement we obtain

$$\text{var}_p(\widehat{R}) \approx N^2 \left(1 - \frac{n}{N}\right) \frac{1}{n} \frac{1}{X^2} (S_y^2 - 2R \, S_{xy} + R^2 \, S_x^2),$$

which is estimated by

$$v(\widehat{R}) = N^2 \left(1 - \frac{n}{N}\right) \frac{1}{n} \frac{1}{n-1} \sum_{k \in S} \widehat{v}_k^2$$

$$= N^2 \left(1 - \frac{n}{N}\right) \frac{1}{n} \frac{1}{\widehat{X}^2} (s_y^2 - 2\widehat{R} \, s_{xy} + \widehat{R}^2 \, s_x^2).$$

15.4.2 Linearization of a Ratio Estimator

In Section 9.2 we saw that the variance of the ratio estimator cannot be directly calculated. This estimator is defined by

$$\widehat{Y}_Q = \frac{X \, \widehat{Y}}{\widehat{X}},$$

where $\widehat{X}$ and $\widehat{Y}$ are the expansion estimators of X and Y. This estimator consists of using the function of totals $f(c_1, c_2) = X(c_1/c_2)$. We can then calculate the partial derivatives:

$$\left.\frac{\partial f(c_1, c_2)}{\partial c_1}\right|_{c_1=Y, c_2=X} = X\frac{1}{X} = 1,$$

$$\left.\frac{\partial f(c_1, c_2)}{\partial c_2}\right|_{c_1=Y, c_2=X} = -X\frac{Y}{X^2} = -\frac{Y}{X},$$

and by (15.2) we obtain

$$v_k = y_k - \frac{Y}{X}x_k = y_k - R \, x_k.$$

In some designs, the approximate variance of the ratio estimator is

$$\text{var}_p(\widehat{Y}_Q) \approx \text{var}_p \left[\sum_{k \in S}(y_k - R \, x_k)\right]$$

$$= \text{var}_p(\widehat{Y}) + R^2 \, \text{var}_p(\widehat{X}) - 2 \, R \, cov_p(\widehat{X}, \widehat{Y}), \tag{15.4}$$

and is estimated by

$$v(\widehat{Y}_Q) = v(\widehat{Y}) + \widehat{R}^2 \, v(\widehat{X}) - 2\widehat{R} \, cov(\widehat{X}, \widehat{Y}), \tag{15.5}$$

where $\widehat{R} = \widehat{Y}/\widehat{X}$.

Under simple random sampling without replacement we obtain

$$\text{var}_p(\widehat{Y}_Q) \approx N^2 \left(1 - \frac{n}{N}\right) \frac{1}{n}(S_y^2 - 2R \, S_{xy} + R^2 \, S_x^2),$$

which is estimated by

$$v(\widehat{Y}_Q) = N^2 \left(1 - \frac{n}{N}\right) \frac{1}{n} \frac{1}{n-1} \sum_{k \in S} \widehat{v}_k^2$$

$$= N^2 \left(1 - \frac{n}{N}\right) \frac{1}{n}(s_y^2 - 2\widehat{R}\, s_{xy} + \widehat{R}^2\, s_x^2).$$

15.4.3 Linearization of a Geometric Mean

The geometric mean can be written as a function of a total. Indeed,

$$G = \left(\prod_{k \in U} y_k\right)^{1/N} = \exp \frac{1}{N} \sum_{k \in U} z_k = \exp \frac{Z}{N},$$

with $z_k = \log y_k$ and $Z = \sum_{k \in U} z_k$, and $\overline{Z} = Z/N$. We estimate Z by the expansion estimator:

$$\widehat{Z} = \sum_{k \in S} \frac{z_k}{\pi_k}.$$

Therefore, the plug-in estimator is

$$\widehat{G} = \exp \frac{\widehat{Z}}{N} = \left(\prod_{k \in S} y_k\right)^{1/(N\pi_k)}.$$

Linearizing gives

$$v_k = \frac{z_k}{N} \exp \overline{Z} - \frac{\overline{Z}}{N} \exp \overline{Z} = \frac{1}{N}(z_k - \overline{Z})G,$$

which is estimated by

$$\widehat{v}_k = \frac{1}{\widehat{N}}(z_k - \widehat{\overline{Z}}_{\mathrm{HAJ}}) \exp \widehat{\overline{Z}}_{\mathrm{HAJ}} = \frac{1}{\widehat{N}}(z_k - \widehat{\overline{Z}}_{\mathrm{HAJ}})\widehat{G},$$

where

$$\widehat{N} = \sum_{k \in S} d_k$$

and $\widehat{\overline{Z}}_{\mathrm{HAJ}}$ is the Hájek estimator

$$\widehat{\overline{Z}}_{\mathrm{HAJ}} = \frac{1}{\widehat{N}} \sum_{k \in S} d_k z_k.$$

For a design of fixed size, the variance estimator is

$$v(\widehat{G}) = -\frac{1}{2} \frac{\widehat{G}^2}{\widehat{N}^2} \sum_{k \in S} \sum_{\substack{\ell \in S \\ \ell \neq k}} [d_k(z_k - \widehat{\overline{Z}}_{\mathrm{HAJ}}) - d_\ell(z_\ell - \widehat{\overline{Z}}_{\mathrm{HAJ}})]^2 \frac{\pi_{k\ell} - \pi_k \pi_\ell}{\pi_{k\ell}}.$$

15.4.4 Linearization of a Variance

The variance can be written as a function of totals:

$$S_y^2 = \frac{1}{(N-1)} \sum_{k \in U} \left(y_k - \frac{Y}{N}\right)^2 = \frac{Y_2}{N-1} - \frac{Y^2}{N(N-1)},$$

where

$$N = \sum_{k \in U} 1, \quad Y = \sum_{k \in U} y_k, \quad \text{and} \quad Y_2 = \sum_{k \in U} y_k^2.$$

Therefore, we can construct a plug-in estimator by estimating the totals:

$$\widehat{S}_y^2 = \frac{\widehat{Y}_2}{\widehat{N} - 1} - \frac{\widehat{Y}^2}{\widehat{N}(\widehat{N} - 1)} = \frac{1}{\widehat{N} - 1} \sum_{k \in S} d_k (y_k - \widehat{\overline{Y}}_{\mathrm{HAJ}})^2,$$

where

$$\widehat{N} = \sum_{k \in S} d_k, \quad \widehat{Y} = \sum_{k \in S} d_k y_k, \quad \widehat{Y}_2 = \sum_{k \in S} d_k y_k^2, \quad \text{and} \quad \widehat{\overline{Y}}_{\mathrm{HAJ}} = \frac{\widehat{Y}}{\widehat{N}}.$$

The estimator $\widehat{S}_y^2$ is slightly biased, even in simple random sampling without replacement. The linearized variable is

$$\begin{aligned} v_k &= y_k^2 \frac{1}{N-1} - Y_2 \frac{1}{(N-1)^2} - 2y_k Y \frac{1}{N(N-1)} + Y^2 \frac{2N-1}{N^2(N-1)^2} \\ &= \frac{1}{N-1} \left[(y_k - \overline{Y})^2 - S_y^2 \right], \end{aligned} \tag{15.6}$$

and is estimated by

$$\hat{v}_k = \frac{1}{\widehat{N} - 1} \left[(y_k - \widehat{\overline{Y}}_{\mathrm{HAJ}})^2 - \widehat{S}_y^2 \right].$$

Under simple random sampling without replacement we obtain estimator

$$v\left(\widehat{S}_y^2\right) = N^2 \left(1 - \frac{n}{N}\right) \frac{1}{n} \frac{1}{n-1} \sum_{k \in S} (\hat{v}_k - \overline{\hat{v}})^2,$$

where

$$\overline{\hat{v}} = \frac{1}{n} \sum_{k \in S} \hat{v}_k.$$

After some calculations we obtain

$$\begin{aligned} v\left(\widehat{S}_y^2\right) &= \frac{N(N-n)}{(N-1)^2 n(n-1)} \sum_{k \in S} \left[(y_k - \widehat{\overline{Y}}_{\mathrm{HAJ}})^2 - \frac{N-1}{N} \widehat{S}_y^2 \right]^2 \\ &= \left(1 - \frac{n}{N}\right) \frac{1}{n} \left[\frac{N^2}{(N-1)^2} m_4 - \frac{n}{n-1} \widehat{S}_y^4 \right], \end{aligned}$$

where

$$m_4 = \frac{1}{n-1} \sum_{k \in S} (y_k - \widehat{\overline{Y}}_{\mathrm{HAJ}})^4.$$

15.4.5 Linearization of a Covariance

The covariance can also be written as a function of totals:

$$S_{xy} = \frac{1}{N-1} \sum_{k \in U} \left(x_k - \frac{X}{N}\right) \left(y_k - \frac{Y}{N}\right) = \frac{(XY)}{N-1} - \frac{X}{N(N-1)} \frac{Y}{N},$$

where

$$N = \sum_{k \in U} 1, \quad X = \sum_{k \in U} x_k, \quad Y = \sum_{k \in U} y_k, \quad (XY) = \sum_{k \in U} x_k y_k.$$

We can then estimate the covariance by simply estimating these four totals:

$$\hat{S}_{xy} = \frac{\widehat{(XY)}}{\hat{N} - 1} - \frac{\hat{X}\hat{Y}}{\hat{N}(\hat{N} - 1)} = \frac{1}{\hat{N} - 1} \sum_{k \in S} d_k (x_k - \hat{\bar{X}}_{\mathrm{HAJ}})(y_k - \hat{\bar{Y}}_{\mathrm{HAJ}}),$$

where

$$\hat{N} = \sum_{k \in S} d_k, \quad \hat{X} = \sum_{k \in S} d_k x_k, \quad \hat{Y} = \sum_{k \in S} d_k y_k,$$

$$\widehat{(XY)} = \sum_{k \in S} d_k x_k y_k, \quad \hat{\bar{X}}_{\mathrm{HAJ}} = \frac{\hat{X}}{\hat{N}}, \quad \text{and} \quad \hat{\bar{Y}}_{\mathrm{HAJ}} = \frac{\hat{Y}}{\hat{N}}.$$

The linearized variable equals

$$
\begin{aligned}
v_k &= x_k y_k \frac{1}{N-1} - \frac{(XY)}{(N-1)^2} \\
&\quad - x_k \frac{Y}{N(N-1)} - y_k \frac{X}{N(N-1)} + X\,Y \frac{2N-1}{N^2(N-1)^2} \\
&= \frac{1}{N-1}[(y_k - \bar{Y})(x_k - \bar{X}) - S_{xy}]
\end{aligned}
\tag{15.7}
$$

and is estimated by

$$\hat{v}_k = \frac{1}{\hat{N} - 1}\left[\left(x_k - \hat{\bar{X}}_{\mathrm{HAJ}}\right)\left(y_k - \hat{\bar{Y}}_{\mathrm{HAJ}}\right) - \hat{S}_{xy}\right].$$

15.4.6 Linearization of a Vector of Regression Coefficients

In a complex design in which the first- and second-order inclusion probabilities are positive, we want to estimate the variance of the vector of regression coefficients in the multiple regression estimator:

$$\hat{\mathbf{B}} = \left(\sum_{k \in S} d_k q_k \mathbf{x}_k \mathbf{x}_k^{\mathsf{T}}\right)^{-1} \sum_{k \in S} d_k q_k \mathbf{x}_k y_k.$$

The function of interest to estimate is

$$\mathbf{B} = \left(\sum_{k \in U} q_k \mathbf{x}_k \mathbf{x}_k^{\mathsf{T}}\right)^{-1} \sum_{k \in U} q_k \mathbf{x}_k y_k.$$

If

$$\mathbf{T} = \sum_{k \in U} q_k \mathbf{x}_k \mathbf{x}_k^{\mathsf{T}},$$

the vector of linearized variables is

$$
\begin{aligned}
\mathbf{v}_k &= \mathbf{T}^{-1} q_k \mathbf{x}_k y_k - \mathbf{T}^{-1} q_k \mathbf{x}_k \mathbf{x}_k^{\mathsf{T}} \mathbf{T}^{-1} \sum_{k \in U} q_k \mathbf{x}_k y_k \\
&= \mathbf{T}^{-1} \mathbf{x}_k q_k (y_k - \mathbf{x}_k^{\mathsf{T}} \mathbf{B}).
\end{aligned}
$$

If we set $E_k = y_k - \mathbf{x}_k^{\mathsf{T}}\mathbf{B}$, we have

$$\mathbf{v}_k = \mathbf{T}^{-1}\mathbf{x}_k q_k E_k. \tag{15.8}$$

Finally, we estimate $\mathbf{v}_k$ by $\widehat{\mathbf{v}}_k = \widehat{\mathbf{T}}^{-1}\mathbf{x}_k q_k e_k$, where

$$\widehat{\mathbf{T}} = \sum_{k\in S} d_k q_k \mathbf{x}_k \mathbf{x}_k^{\mathsf{T}} \quad \text{and} \quad e_k = y_k - \mathbf{x}_k^{\mathsf{T}}\widehat{\mathbf{B}}.$$

15.5 Linearization by Steps

15.5.1 Decomposition of Linearization by Steps

Linearization can be applied by steps. Suppose that $\theta = f(Y_1, \dots, Y_j, \dots, Y_p, \lambda)$, where λ is also a function of totals for which we know the linearized variable u_k, it is then easy to show that a linearized variable of θ is

$$v_k = \sum_{j=1}^{p} y_{kj} \left.\frac{\partial f(c_1, \dots, c_p, \lambda)}{\partial c_j}\right|_{c_1=Y_1,\dots,c_p=Y_p} + u_k \left.\frac{\partial f(Y_1, \dots, Y_p, z)}{\partial z}\right|_{z=\lambda}.$$

This result follows directly from the formula of the derivative of a function composition.

Example 15.2 For some design for which we know the first- and second-order inclusion probabilities, we want to calculate the variance of the square of the Hájek estimator:

$$\widehat{\theta} = \widehat{\overline{Y}}_{\mathrm{HAJ}}^2 = \left[\left(\sum_{k\in S}\frac{1}{\pi_k}\right)^{-1}\sum_{k\in S}\frac{y_k}{\pi_k}\right]^2.$$

A linearized variable of the Hájek estimator $\widehat{\overline{Y}}_{\mathrm{HAJ}}$ directly comes from the linearized variable of the ratio (15.3) and is

$$u_k = \frac{1}{N}(y_k - \overline{Y}).$$

By applying the linearization method by steps, the linearized variable of $\overline{Y}^2$ is then

$$v_k = 2\overline{Y}u_k = \frac{2\overline{Y}}{N}(y_k - \overline{Y}).$$

We estimate v_k by

$$\widehat{v}_k = \frac{2\widehat{\overline{Y}}_{\mathrm{HAJ}}}{\widehat{N}}(y_k - \widehat{\overline{Y}}_{\mathrm{HAJ}}).$$

15.5.2 Linearization of a Regression Coefficient

The coefficient of a regression with one explanatory variable and one intercept is the ratio of the covariance between the two variables over the variance of the explanatory variable:

$$B = \frac{S_{xy}}{S_x^2}.$$

By applying the linearization method by steps, we obtain z_k, a linearized variable of B:

$$z_k = \frac{v_k}{S_x^2} - \frac{u_k S_{xy}}{S_x^4} = \frac{1}{S_x^2}(v_k - u_k B),$$

where v_k is a linearized variable of S_{xy} and u_k is a linearized variable of S_x^2. By using the results in Expressions (15.6) and (15.7), we obtain

$$z_k = \frac{1}{S_x^2} \left\{ \frac{1}{N-1}[(y_k - \overline{Y})(x_k - \overline{X}) - S_{xy}] \right.$$

$$- \frac{1}{N-1}\left[(x_k - \overline{X})^2 - S_x^2]B \right\}$$

$$= \frac{x_k - \overline{X}}{(N-1)S_x^2}[(y_k - \overline{Y}) - (x_k - \overline{X})B].$$

15.5.3 Linearization of a Univariate Regression Estimator

The regression estimator is

$$\widehat{Y}_{\mathrm{REG}} = \widehat{Y} + (X - \widehat{X})\widehat{B} = f(\widehat{Y}, \widehat{X}, \widehat{B}),$$

where

$$f(a, b, c) = a + (X - b)c.$$

Since we know the linearized variable of $\widehat{B}$, we can linearize by steps:

$$v_k = y_k - x_k B + z_k(X - X) = y_k - x_k B,$$

where z_k is a linearized variable of $\widehat{B}$ that we finally do not need in order to calculate the linearized variable of $\widehat{Y}_{\mathrm{REG}}$. The approximation of the variance of the regression estimator is then

$$\mathrm{var}_p(\widehat{Y}_{\mathrm{REG}}) \approx \mathrm{var}_p \left[\sum_{k \in S} d_k(y_k - x_k B) \right]$$

$$= \mathrm{var}_p(\widehat{Y} - \widehat{X}B)$$

$$= \mathrm{var}_p(\widehat{Y}) - 2B\,\mathrm{cov}_p(\widehat{X}, \widehat{Y}) + B^2\,\mathrm{var}_p(\widehat{X}). \tag{15.9}$$

We simply estimate this variance by

$$v(\widehat{Y}_{\mathrm{REG}}) = v(\widehat{Y}) - 2\widehat{B}\mathrm{cov}_p(\widehat{X}, \widehat{Y}) + \widehat{B}^2 v(\widehat{X}).$$

15.5.4 Linearization of a Multiple Regression Estimator

The multiple regression estimator is defined in Expression (11.2):

$$\widehat{Y}_{\mathrm{REGM}} = \widehat{Y} + (\mathbf{X} - \widehat{\mathbf{X}})^{\mathsf{T}} \widehat{\mathbf{B}}.$$

It can be written as

$$\widehat{Y}_{\mathrm{REGM}} = f(\widehat{Y}, \widehat{\mathbf{X}}, \widehat{\mathbf{B}}).$$

This estimator depends on two totals $\widehat{Y}$ and $\widehat{\mathbf{X}}$, and on $\widehat{\mathbf{B}}$ for which we know the linearized variable (15.8). If we use linearization by steps for $f(Y, \mathbf{X}, \mathbf{B})$, we obtain

$$u_k = y_k - \mathbf{x}_k^\top \mathbf{B} + (\mathbf{X} - \mathbf{X})^\top v_k = y_k - \mathbf{x}_k^\top \mathbf{B} = E_k,$$

where v_k is a linearized variable of $\mathbf{B}$ that is not involved in the expression of the linearized variable u_k. Therefore, the estimator of the linearized variable is

$$\widehat{u}_k = e_k = y_k - \mathbf{x}_k^\top \widehat{\mathbf{B}}.$$

The variance is simply estimated using the regression residuals of y in x.

We can then construct an approximation of the variance:

$$\mathrm{var}_p(\widehat{Y}_{\mathrm{REGM}}) \approx \mathrm{var}_p\left(\sum_{k \in S} d_k E_k \right) = \mathrm{var}_p(\widehat{Y} - \widehat{\mathbf{X}}^\top \mathbf{B})$$

$$= \mathrm{var}_p(\widehat{Y}) - 2\mathbf{B}^\top \mathrm{cov}_p(\widehat{\mathbf{X}}, \widehat{Y}) + \mathbf{B}^\top \mathrm{var}_p(\widehat{X})\mathbf{B}$$

and an estimator of this approximation is

$$v(\widehat{Y}_{\mathrm{REGM}}) = v(\widehat{Y}) - 2\widehat{\mathbf{B}}^\top \, cov(\widehat{\mathbf{X}}, \widehat{Y}) + \widehat{\mathbf{B}}^\top v(\widehat{X})\widehat{\mathbf{B}}$$

or equivalently

$$v(\widehat{Y}_{\mathrm{REGM}}) = \sum_{k \in S}\sum_{\ell \in S} \frac{e_k e_\ell}{\pi_k \pi_\ell} \frac{\Delta_{k\ell}}{\pi_{k\ell}}.$$

15.6 Linearization of an Implicit Function of Interest

15.6.1 Estimating Equation and Implicit Function of Interest

We saw in Section 14.4, page 283, that the function of interest θ is often defined in an implicit manner by an estimating equation of type

$$T(\theta) = \sum_{k \in U} \phi_k(\theta) = T_0. \tag{15.10}$$

In such cases, Binder (1983) has shown that it is also possible to construct a linearized variable that allows the estimation of the variance of an estimator of θ even if the function $T(\cdot)$ is nonlinear. For simplicity, we give a brief and somewhat empirical derivation.

In addition, suppose that ϕ_k are continuous differentiable functions. If θ is defined by (15.10), we then estimate θ by $\widehat{\theta}$, which is the solution of

$$\widehat{T}(\widehat{\theta}) = \sum_{k \in S} d_k \phi_k(\widehat{\theta}) = T_0. \tag{15.11}$$

If $\widehat{T}(\cdot)$ is continuous and differentiable, we can approximate (15.11) by

$$\widehat{T}(\widehat{\theta}) \approx \widehat{T}(\theta^*) + \left.\frac{d\widehat{T}(\theta)}{d\theta}\right|_{\theta=\theta^*} (\widehat{\theta} - \theta^*) = T_0,$$

where θ^* is the value of θ which solves (15.10). We obtain

$$\widehat{T}(\theta^*) \approx -\left.\frac{d\widehat{T}(\theta)}{d\theta}\right|_{\theta=\theta^*} (\widehat{\theta} - \theta^*) + T_0,$$

and therefore

$$(\hat{\theta} - \theta^*) \approx - \left. \frac{d\hat{T}(\theta)}{d\theta} \right|_{\theta=\theta^*}^{-1} (\hat{T}(\theta^*) - T_0).$$

Finally, supposing that the bias of $\hat{\theta}$ is negligible and the variance of

$$\left. \frac{d\hat{T}(\theta)}{d\theta} \right|_{\theta=\theta^*}$$

is negligible compared to $\hat{T}(\theta^*)$'s variance for large sample sizes, we obtain the variance of $\hat{\theta}$:

$$\mathrm{var}_p(\hat{\theta}) \approx \mathrm{var}_p \left[- \left. \frac{dT(\theta)}{d\theta} \right|_{\theta=\theta^*}^{-1} \hat{T}(\theta^*) \right]$$

$$= \mathrm{var}_p \left[- \left. \frac{dT(\theta)}{d\theta} \right|_{\theta=\theta^*}^{-1} \sum_{k \in S} d_k \phi_k(\theta^*) \right].$$

The linearized variable is obtained by

$$u_k = - \left. \frac{dT(\theta)}{d\theta} \right|_{\theta=\theta^*}^{-1} \phi_k(\theta^*).$$

We then estimate this linearized variable by replacing the totals in u_k by their estimators. This reasoning is also valid in the case where θ is a vector with multiple components. Then, $dT(\theta)/d\theta$ is a square matrix.

Example 15.3 The ratio can be defined by an estimating equation. In fact, set $\phi_k(\theta) = y_k - \theta x_k$, $T_0 = 0$ and therefore

$$T(\theta) = Y - \theta X = 0.$$

We then obtain $\theta^* = R = Y/X$. Since

$$\left. \frac{dT(\theta)}{d\theta} \right|_{\theta=\theta^*} = -X,$$

we obtain

$$u_k = \frac{1}{X}(y_k - Rx_k),$$

which is the same result as in (15.3).

15.6.2 Linearization of a Logistic Regression Coefficient

Suppose that the variable of interest follows the model

$$y_k = \begin{cases} 1 & \text{with probability } f(\mathbf{x}_k^\top \mathbf{B}) \\ 0 & \text{with probability } 1 - f(\mathbf{x}_k^\top \mathbf{B}), \end{cases}$$

where

$$f(u) = \frac{\exp u}{1 + \exp u}.$$

Usually, to determine **B**, we use the likelihood function which is defined at the population level by

$$L(\mathbf{B}|y) = \prod_{k \in U} \{f(\mathbf{x}_k^\top \mathbf{B})^{y_k}[1 - f(\mathbf{x}_k^\top \mathbf{B})]^{1-y_k}\}.$$

By canceling the derivative of the logarithm of $L(\mathbf{B}|y)$, we obtain

$$\frac{\partial \log L(\mathbf{B}|y)}{\partial \mathbf{B}} = T(\mathbf{B}) = \sum_{k \in U} \mathbf{x}_k[y_k - f(\mathbf{x}_k^\top \mathbf{B})] = \mathbf{0}. \tag{15.12}$$

The regression coefficient is then the solution of the estimating Equation (15.12). If we define

$$\boldsymbol{\phi}_k(\mathbf{B}) = \mathbf{x}_k\{y_k - f(\mathbf{x}_k^\top \mathbf{B})\},$$

the System (15.12) is

$$T(\mathbf{B}) = \sum_{k \in U} \boldsymbol{\phi}_k(\mathbf{B}) = \mathbf{0}.$$

Let $\mathbf{B}^*$ denote the vector that verifies this system. To estimate $\mathbf{B}^*$ at the sample level, we must solve the system:

$$\widehat{T}(\mathbf{B}) = \sum_{k \in S} d_k \boldsymbol{\phi}_k(\mathbf{B}) = \mathbf{0}$$

and let $\widehat{\mathbf{B}}$ denote the solution. The linearized variable is

$$\mathbf{u}_k = -\left.\frac{\partial T(\mathbf{B})}{\partial \mathbf{B}}\right|_{\mathbf{B}=\mathbf{B}^*}^{-1} \times \boldsymbol{\phi}_k(\mathbf{B}^*).$$

Yet,

$$\left.\frac{\partial T(\mathbf{B})}{\partial \mathbf{B}}\right|_{\mathbf{B}=\mathbf{B}^*} = -\sum_{k \in U} \mathbf{x}_k f'(\mathbf{x}_k^\top \mathbf{B}^*)\mathbf{x}_k^\top.$$

In addition, we have

$$f'(u) = \frac{\exp u}{1 + \exp u} - \frac{\exp u}{(1 + \exp u)^2} \exp u = f(u)[1 - f(u)].$$

Therefore,

$$\left.\frac{\partial T(\mathbf{B})}{\partial \mathbf{B}}\right|_{\mathbf{B}=\mathbf{B}^*} = -\sum_{k \in U} \mathbf{x}_k f(\mathbf{x}_k^\top \mathbf{B}^*)[1 - f(\mathbf{x}_k^\top \mathbf{B}^*)]\mathbf{x}_k^\top,$$

and

$$\mathbf{u}_k = \left\{\sum_{k \in U} \mathbf{x}_k \mathbf{x}_k^\top f(\mathbf{x}_k^\top \mathbf{B}^*)[1 - f(\mathbf{x}_k^\top \mathbf{B}^*)]\right\}^{-1} \mathbf{x}_k[y_k - f(\mathbf{x}_k^\top \mathbf{B}^*)].$$

Finally, the estimation of the linearized variable is

$$\widehat{\mathbf{u}}_k = \left\{\sum_{k \in S} d_k \mathbf{x}_k \mathbf{x}_k^\top f(\mathbf{x}_k^\top \widehat{\mathbf{B}})[1 - f(\mathbf{x}_k^\top \widehat{\mathbf{B}})]\right\}^{-1} \mathbf{x}_k[y_k - f(\mathbf{x}_k^\top \widehat{\mathbf{B}})].$$

15.6.3 Linearization of a Calibration Equation Parameter

Calibration equations are defined in Expression (12.4), page 238:

$$\sum_{k \in S} d_k \mathbf{x}_k F_k(\mathbf{x}_k^\top \lambda) = \mathbf{X},$$

and we are looking for the variance-covariance matrix of λ. If we set

$$\phi_k(\lambda) = \mathbf{x}_k F_k(\mathbf{x}_k^\top \lambda),$$

then the calibration equation is

$$\hat{T}(\lambda) = \sum_{k \in S} d_k \phi_k(\lambda) = \mathbf{X}.$$

The solution of system

$$T(\lambda) = \sum_{k \in U} \phi_k(\lambda) = \mathbf{X}$$

is $\lambda^* = \mathbf{0}$, as $F_k(0) = 1$. The linearized variable of λ is

$$u_k = -\left[\sum_{k \in U} \phi'(\lambda^*) \right]^{-1} \phi_k(\lambda^*)$$

$$= -\left[\sum_{k \in U} \mathbf{x}_k F_k'(\mathbf{x}_k^\top \lambda^*) \mathbf{x}_k^\top \right]^{-1} \mathbf{x}_k F_k(\mathbf{x}_k^\top \lambda^*)$$

$$= -\left(\sum_{k \in U} q_k \mathbf{x}_k \mathbf{x}_k^\top \right)^{-1} \mathbf{x}_k, \qquad (15.13)$$

as $F_k(\mathbf{x}_k^\top \lambda^*) = F_k(0) = 1$ and $F_k'(\mathbf{x}_k^\top \lambda^*) = F_k'(0) = q_k$. The estimator of this linearized variable is then

$$\hat{u}_k = -\left(\sum_{k \in S} d_k q_k \mathbf{x}_k \mathbf{x}_k^\top \right)^{-1} \mathbf{x}_k.$$

15.6.4 Linearization of a Calibrated Estimator

Calibrated estimators can all be written in the form (see Expression (12.5)):

$$\hat{Y}_{\text{CAL}} = \sum_{k \in S} d_k y_k F_k(\mathbf{x}_k^\top \lambda).$$

This estimator is a special function of totals. In fact, we can write

$$\hat{h}(\lambda) = \hat{Y}_{\text{CAL}},$$

where $\hat{h}(\lambda)$ is an estimator of the total which depends on a parameter λ:

$$\hat{h}(\lambda) = \sum_{k \in S} d_k y_k F_k(\mathbf{x}_k^\top \lambda),$$

$$h(\lambda) = \sum_{k \in U} y_k F_k(\mathbf{x}_k^\top \lambda),$$

and λ is defined by the estimating equations and whose linearized variable is given in Expression (15.13). Also note that $\lambda^* = 0$, $F_k(0) = 1$, $F_k'(0) = q_k$ and that

$$h(\lambda^*) = \sum_{k \in U} y_k F_k(\mathbf{x}_k^\top \lambda^*) = Y.$$

We simply apply linearization by steps to obtain the linearized variable of $\hat{Y}_{\mathrm{CAL}}$

$$
\begin{aligned}
v_k &= y_k \left. \frac{\partial h(\lambda)}{\partial h(\lambda)} \right|_{h(\lambda)=Y, \lambda=\lambda^*} + \mathbf{u}_k^\top \left. \frac{\partial h(\lambda)}{\partial \lambda} \right|_{\lambda=\lambda^*} \\
&= y_k + \mathbf{u}_k^\top \sum_{k \in U} \mathbf{x}_k F_k'(\mathbf{x}_k^\top \lambda^*) y_k \\
&= y_k + \mathbf{u}_k^\top \sum_{k \in U} q_k \mathbf{x}_k y_k,
\end{aligned}
$$

where $\mathbf{u}_k$ is a linearized variable of λ given in Expression (15.13), which gives

$$v_k = y_k - \mathbf{x}_k^\top \left(\sum_{k \in U} q_k \mathbf{x}_k \mathbf{x}_k^\top \right)^{-1} \sum_{k \in U} q_k \mathbf{x}_k y_k = y_k - \mathbf{x}_k^\top \mathbf{B} = E_k.$$

The linearized variable of any calibrated estimator is equal to the regression residuals of y_k in $\mathbf{x}_k$. The residuals do not depend on the calibration function. In Section 15.10.3 another linearization method for the calibration estimator is presented. This other method gives a slightly different result which depends on the calibration function.

15.7 Influence Function Approach

15.7.1 Function of Interest, Functional

In robust statistics, the influence function proposed by Hampel (1974) and Hampel et al. (1985) measures the effect of a small perturbation of the data on an estimator. This tool allows us to study the robustness of a function of interest. The influence function can also be used to calculate an approximation of the variance.

Deville (1999) gave a little different definition from that of Hampel to take into account the fact that the estimator of the population size can be random and therefore can increase the variance. To define the influence function, Deville uses a measure M having unit mass at each point x_k of the population. According to Deville's definition, the measure M is positive, discrete, and has total mass N, while the total mass of the measure is equal to 1 for the influence function proposed by Hampel (1974) and Hampel et al. (1985).

A function of interest can be presented as a functional $T(M)$ which associates to any measure a real number or a vector. For example, a total Y can be written as

$$Y = \int y \, dM = \sum_{k \in U} y_k.$$

In addition, we suppose that all the considered functionals are linear and homogeneous in the sense that there always exists a real number α such that

$$T(tM) = t^\alpha T(M), \quad \text{for all } t \in \mathbb{R}.$$

The coefficient α is called the degree of the functional. For example, a total has degree 1, a mean of degree 0, and a double sum of degree 2.

The measure M is estimated by measure $\widehat{M}$ having a mass equal to w_k in each point y_k of sample S. The weights w_k can be the expansion weights $d_k = 1/\pi_k$, but can also be the weights obtained by calibration. The plug-in estimator of a functional $T(M)$ is simply $T(\widehat{M})$. For example, the estimator of a total is

$$\int y d\widehat{M} = \sum_{k\in S} w_k y_k.$$

An estimating equation can also be written using a functional. In fact, the equation

$$T(M, \lambda) = 0$$

is a functional parameterized by a function of interest λ. The plug-in estimator $\widehat{\lambda}$ of λ is then the solution to the equation

$$T(\widehat{M}, \widehat{\lambda}) = 0.$$

The functional $T(M, \lambda)$ and λ can be vectors but must have the same dimension.

15.7.2 Definition

The influence function in the sense of Deville (1999) is defined by

$$IT(M, x) = \lim_{t\to 0} \frac{T(M + t\delta_x) - T(M)}{t},$$

when this limit exists, where δ_x is the Dirac measure at point x. The influence function is none other then the Gâteaux derivative of the functional in the direction of the Dirac mass at point x.

Deville (1999) has shown that the influence function $z_k = IT(M, x_k)$ is a linearized $T(\widehat{M})$ in the sense that it can approximate the function of interest $T(M)$:

$$\frac{T(\widehat{M}) - T(M)}{N^\alpha} \approx \frac{1}{N^\alpha}\left(\sum_{k\in S} w_k z_k - \sum_{k\in U} z_k\right).$$

The variance approximation of $T(\widehat{M})$ can be obtained by calculating the variance of the weighted sum of the z_k on the sample using the weights w_k:

$$\mathrm{var}_p[T(\widehat{M})] \approx \mathrm{var}_p\left(\sum_{k\in S} w_k z_k\right).$$

The function of interest generally depends on unknown population parameters. We simply estimate the influence function by replacing these unknown parameters by their plug-in estimators. We then obtain $\hat{z}_k$, the estimator of the linearized variable z_k.

The calculation of influence follows the rules of differential calculus. Deville (1999) establishes many results that allow the function of interest to be calculated for the usual statistics. The rules allow all the results obtained for totals, functions of totals, and implicit equations to be reconstructed. The use of the influence function also allows a linearized variable to be calculated for statistics that are not functions of totals like the Gini index.

15.7.3 Linearization of a Total

Result 15.8 *If*

$$T(M) = \sum_{k \in U} y_k = \int y \, dM(y),$$

then the influence function is $IT(M, x_k) = y_k$.

The proof directly follows from the definition of the influence function. The total is a linear function. Therefore, it is logical that the linearized variable of a total is equal to the variable of interest itself.

15.7.4 Linearization of a Function of Totals

Functions of interest like the ratio, means, and cumulative distribution functions are functions of totals that we can linearize thanks to the following result.

Result 15.9 *If* $g(T)$ *is a differentiable vectorial function on the space of values of the vectorial functional* T, *we have*

$$Ig(T)(M, x) = \frac{\partial g(T)}{\partial T} IT(M, x),$$

where $\partial g(T)/\partial T$ *is the vector of partial derivatives.*

Proof:

$$Ig(T)(M, x) = \lim_{t \to 0} \frac{g(T)(M + t\delta_x) - g(T)(M)}{t}$$

$$= \frac{\partial g(T)}{\partial T} \lim_{t \to 0} \frac{T(M + t\delta_x) - T(M)}{t} = \frac{\partial g(T)}{\partial T} IT(M, x). \quad \blacksquare$$

In the special case where T is a vector of totals then $IT(M, x) = x$ and therefore

$$Ig(T)(M, x) = \frac{\partial g(T)}{\partial T} x.$$

We then obtain the results of Section 15.3.1, which allows us to calculate a linearized variable of a function of totals.

Example 15.4 For example, if the function of interest is ratio

$$f(Y, X) = R = \frac{Y}{X},$$

then

$$z_k = IT(M, \mathbf{x}_k) = \left(\frac{1}{X} - \frac{Y}{X^2} \right) \begin{pmatrix} y_k \\ x_k \end{pmatrix} = \frac{1}{X}(y_k - Rx_k).$$

The linearized variable is estimated by

$$\hat{z}_k = \frac{1}{\hat{X}} \left(y_k - \hat{R}x_k \right),$$

where

$$\hat{Y} = \sum_{k \in S} w_k y_k, \quad \hat{X} = \sum_{k \in S} w_k x_k, \quad \text{and} \quad \hat{R} = \frac{\hat{Y}}{\hat{X}}.$$

Example 15.5   A mean $\overline{Y} = Y/N$ is a ratio of two totals Y and N. Therefore, its linearized variable directly follows from (15.3):

$$z_k = \frac{1}{N}(y_k - \overline{Y}).$$

Example 15.6   The cumulative distribution function can be presented as mean

$$F(x) = \frac{1}{N} \sum_{k \in U} \mathbb{1}(y_k \leq x).$$

Therefore, its linearized variable is

$$z_k = \frac{1}{N}[\mathbb{1}(y_k \leq x) - F(x)].$$

15.7.5   Linearization of Sums and Products

Result 15.10   *If S and T are two functionals, then*

$$I(S + T) = IS + IT \quad and \quad I(ST) = S \cdot IT + T \cdot IS.$$

Proof: The property $I(S + T) = IS + IT$ directly follows from the definition of the influence function. By Result 15.9, we have $I \log(ST) = I(ST)/(ST)$, and therefore

$$I(ST) = ST\, I\log(ST) = ST\, I(\log S + \log T)$$
$$= ST \left(\frac{IS}{S} + \frac{IT}{T} \right) = SIT + TIS.$$
∎

Result 15.11   *Let $S_i, i = 1, \dots, q$ be a sequence of scalar functionals and*

$$S = \prod_{i=1}^{q} S_i$$

then

$$IS = S \sum_{i=1}^{q} \frac{IS_i}{S_i}.$$

Proof: By Result 15.9, we have $I \log S = IS/S$, and therefore

$$IS = SI \log S = SI \left(\sum_{i=1}^{q} \log S_i \right) = S \sum_{i=1}^{q} I \log S_i = S \sum_{i=1}^{q} \frac{IS_i}{S_i}.$$
∎

This last result allows us to calculate a linearized variable of a geometric mean, for example (see Section 15.4.3).

15.7.6 Linearization by Steps

The calculation of the influence function can be done in steps.

Result 15.12 *Let* **S** *be a functional in* $\mathbb{R}^q$, T_λ *a family of functionals that depend on* $\lambda \in \mathbb{R}^q$, *then*

$$I(T_S) = I(T_{\lambda=S}) + \left.\frac{\partial T}{\partial \lambda}\right|_{\lambda=S} IS.$$

Proof: Since $T_S = T(S(M), M)$, we have

$$IT_S = IT(S(M), M) = I(T_{\lambda=S}) + \left.\frac{\partial T}{\partial \lambda}\right|_{\lambda=S} IS. \qquad \blacksquare$$

We obtain the result developed in Section 15.5.1. This result allows us to simply calculate the linearized variable of the regression estimator using the linearized variable of the regression coefficient. More generally, this result allows the linearized variable of the calibrated estimator to be calculated from the linearized variable of the Lagrange multiplier vector λ defined by the calibration equations.

15.7.7 Linearization of a Parameter Defined by an Implicit Function

The function of influence can also be calculated when the function of interest is defined by an implicit function. Suppose that $T(M, \lambda)$ is a functional in $\mathbb{R}^q$. Also suppose that the matrix $\partial T/\partial \lambda$ is invertible and the application of $T(M, \lambda)$ has an inverse. The estimating equation

$$T(M, \lambda) = T_0$$

has a unique solution that defines a functional $\lambda(M)$.

Result 15.13 *If* $\lambda(M)$ *is defined by the estimating equation* $T(\lambda, M) = T_0$, *then*

$$I\lambda(M, x) = -\left[\frac{\partial T(M, \lambda)}{\partial \lambda}\right]^{-1} IT(M, \lambda, x). \tag{15.14}$$

Proof: Since

$$IT(M, \lambda(M), x) = IT_0 = 0,$$

and, by Result 15.12,

$$IT(M, \lambda(M), x) = IT(M, \lambda, x) + \frac{\partial T(M, \lambda)}{\partial \lambda} I(\lambda(M), x),$$

we obtain Expression (15.14), which is none other than the empirical result from Section 15.6. $\qquad \blacksquare$

Example 15.7 Quantile Q_α of order α is in principle obtained using the estimating equation:

$$F(Q_\alpha) = \alpha. \tag{15.15}$$

Since Equation (15.15) is an implicit function, by applying Result 15.13, we directly obtain a linearized variable:

$$z_k = -\frac{1}{F'(Q_a)}\frac{1}{N}(\mathbb{1}(y_k \leq Q_a) - \alpha).$$

Obviously, the calculation is not that simple, as the function $F(x)$ is not necessarily differentiable (see Section 14.5). The linear interpolations presented in Section 14.5 can be differentiated but are highly sensitive to sampling fluctuations. Deville (1999) advocates the use of a kernel estimator to obtain a smooth cumulative distribution function like the one proposed in Expression (14.3), page 288. This method has been used by Osier (2009) to calculate linearized variables of quantiles.

15.7.8 Linearization of a Double Sum

The use of an influence function allows the linearized variables of more complex functions of interest than functions of totals or parameters defined by an implicit function to be calculated. Double sums of the form

$$S = \sum_{k \in U}\sum_{\ell \in U}\phi(y_k, y_\ell)$$

can also be linearized.

Result 15.14 *If*

$$S(M) = \int\int \phi(x, y)dM(x)dM(y),$$

where $\phi(.,.)$ is a function from $\mathbb{R}^2$ to $\mathbb{R}$, then

$$IS(M, \xi) = \int \phi(x, \xi)dM(x) + \int \phi(\xi, y)dM(y).$$

Proof: We have

$$C(t) = \frac{1}{t}[S(M + t\delta_\xi) - S(M)]$$

$$= \frac{1}{t}\left[\int\int \phi(x, y)d(M + t\delta_\xi)(x)d(M + t\delta_\xi)(y)\right.$$

$$\left. - \int\int \phi(x, y)dM(x)dM(y)\right]$$

$$= \frac{1}{t}\int\left[\int \phi(x, y)d(M + t\delta_\xi)(x) - \int \phi(x, y)dM(x)\right]dM(y)$$

$$+ \frac{1}{t}\int\int \phi(x, y)d(M + t\delta_\xi)(x)d(t\delta_\xi)(y)$$

$$= \frac{1}{t}\left[\int\int \phi(x, y)dM(y)d(M + t\delta_\xi)(x) - \int\int \phi(x, y)dM(y)dM(x)\right]$$

$$+ \int \phi(x, \xi)d(M + t\delta_\xi)(x).$$

We obtain

$$IS(M, \xi) = \lim_{t \to 0} C(t) = \int \phi(\xi, y)dM(y) + \int \phi(x, \xi)dM(x).$$

∎

If $\phi(x, y) = \phi(y, x)$ for all x, y then the influence is simply

$$IS(M, \xi) = 2 \int \phi(x, \xi)dM(x).$$

Example 15.8 The Gini index is a delicate function of interest to linearize (see Lomnicki, 1952; Sandström et al., 1985; Deville, 1996; Dell et al., 2002; Dell & d'Haultfoeuille, 2006; Leiten, 2005; Langel & Tillé, 2013). Result 15.14 allows a very rapid calculation. We have seen, in Section 14.7, that the Gini index can be written as

$$G = \frac{A}{2NY},$$

where A is a double sum:

$$A = \sum_{k \in U} \sum_{\ell \in U} |y_k - y_\ell|.$$

By applying Result 15.14, we obtain a linearized variable of A:

$$z_k = 2 \sum_{\ell \in U} |y_k - y_\ell| = 2[2N_k(y_k - \overline{Y}_k) + Y - Ny_k],$$

where

$$\overline{Y}_k = \frac{\sum_{\ell \in U} y_\ell \mathbb{1}(y_\ell \le y_k)}{N_k}$$

and

$$N_k = \sum_{\ell \in U} \mathbb{1}(y_\ell \le y_k).$$

By now applying a linearization by steps, we obtain a linearized variable of the Gini index

$$v_k = \frac{z_k}{2NY} - \frac{A}{2N^2Y} - y_k \frac{A}{2NY^2} = \frac{1}{NY}\left[\frac{z_k}{2} - G(Y + y_k N)\right]$$

$$= \frac{1}{NY}[2N_k(y_k - \overline{Y}_k) + Y - Ny_k - G(Y + y_k N)].$$

We estimate this linearized variable by

$$\hat{v}_k = \frac{1}{\hat{N}\hat{Y}}[2\hat{N}_k(y_k - \widehat{\overline{Y}}_k) + \hat{Y} - \hat{N}y_k - \hat{G}(\hat{Y} + y_k \hat{N})], \tag{15.16}$$

where

$$\widehat{\overline{Y}}_k = \frac{\sum_{\ell \in S} w_\ell y_\ell \mathbb{1}(y_\ell \le y_k)}{\hat{N}_k}$$

and

$$\hat{N}_k = \sum_{\ell \in S} w_\ell \mathbb{1}(y_\ell \le y_k).$$

Example 15.9 We have seen in Section 15.4.4 that the variance is a function of totals. However, this variance can also be written as

$$S_y^2 = \frac{A}{2N(N-1)},$$

where A is a double sum:

$$A = \sum_{k \in U} \sum_{\ell \in U} (y_k - y_\ell)^2.$$

By applying Result 15.14, we obtain a linearized variable of A:

$$z_k = 2 \sum_{\ell \in U} (y_k - y_\ell)^2 = 2N(y_k - \overline{Y})^2 + 2(N-1)S_y^2.$$

By now applying linearization by steps, we obtain a linearized variable of the variance:

$$v_k = \frac{z_k}{2N(N-1)} - \frac{A(2N-1)}{2N^2(N-1)^2} = \frac{z_k}{2N(N-1)} - \frac{S_y^2(2N-1)}{N(N-1)}$$

$$= \frac{2N(y_k - \overline{Y})^2 + 2(N-1)S_y^2}{2N(N-1)} - \frac{S_y^2(2N-1)}{N(N-1)}$$

$$= \frac{1}{N-1}[(y_k - \overline{Y})^2 - S_y^2].$$

This result is identical to the one obtained in Section 15.4.4 when considering the variance as a function of totals.

15.8 Binder's Cookbook Approach

In the "cookbook" approach, Binder (1996) proposed directly linearizing the estimator and not the function of interest. In the classical approach obtained in Expression (15.2), the function of interest $\theta = f(Y_1, \ldots, Y_p)$ is linearized. This linearized variable is then estimated. In Binder's direct approach, we directly linearize the estimator $\hat{\theta} = f(\hat{Y}_1, \ldots, \hat{Y}_p)$:

$$\hat{v}_k = \sum_{j=1}^{p} y_{kj} \left. \frac{\partial f(c_1, \ldots, c_p)}{\partial c_j} \right|_{c_1 = \hat{Y}_1, \ldots, c_p = \hat{Y}_p}, \quad k \in S. \tag{15.17}$$

Results obtained by the direct approach are different from those with the classical approach when the function of interest is an estimator using auxiliary information. We would then obtain different results for ratio, regression, and calibration estimators. Binder (1996) also proposed a method of direct linearization for a parameter defined by an implicit function.

Example 15.10 For example, consider the ratio estimator

$$\hat{Y}_Q = X \frac{\hat{Y}}{\hat{X}}.$$

This estimator is a function of two estimators of totals $\hat{Y}$ and $\hat{X}$ and can be written as

$$\hat{Y}_Q = f(\hat{Y}, \hat{X}).$$

By using the classical approach, we obtain the following linearized variable:

$$v_k = y_k \left.\frac{\partial f(a,b)}{\partial a}\right|_{a=Y} + x_k \left.\frac{\partial f(a,b)}{\partial b}\right|_{b=X}$$

$$= y_k X \frac{1}{X} - x_k X \frac{Y}{X^2} = y_k - \frac{Y}{X} x_k.$$

Then, we estimate this linearized variable by

$$\hat{v}_k = y_k - \frac{\hat{Y}}{\hat{X}} x_k, \tag{15.18}$$

which allows a variance estimator to be constructed (see Expression (15.5)):

$$v(\hat{Y}_Q) = v(\hat{Y}) + \hat{R}^2 \, v(\hat{X}) - 2\hat{R} \, cov(\hat{X}, \hat{Y}).$$

With the direct approach, we obtain in one step

$$\hat{v}_k = y_k \left.\frac{\partial f(a,b)}{\partial a}\right|_{a=\hat{Y}} + x_k \left.\frac{\partial f(a,b)}{\partial b}\right|_{b=\hat{X}}$$

$$= y_k X \frac{1}{\hat{X}} - x_k X \frac{\hat{Y}}{\hat{X}^2} = \frac{X}{\hat{X}} \left(y_k - \frac{\hat{Y}}{\hat{X}} x_k \right). \tag{15.19}$$

Estimations of the linearized variable (15.19) differ from (15.18) by the product of the ratio $X/\hat{X}$, which is close to 1. The variance estimator is then slightly different to the one obtained by the classical method:

$$v_{alt}(\hat{Y}_Q) = \frac{X^2}{\hat{X}^2} v(\hat{Y}_Q).$$

15.9 Demnati and Rao Approach

Let $\hat{\theta}$ be an estimator of θ. This estimator depends on weights w_k that can be the expansion weights $1/\pi_k$ or the calibrated weights. We define $\omega_k = w_k a_k$. Demnati & Rao (2004) recommend simply differentiating the estimator with respect to the ω_k to obtain the linearized variable. This method allows a linearized variable to be calculated for any estimator which is a differentiable function of weights $\omega_k = w_k a_k$.

For a function of totals, the Demnati & Rao (2004) method directly gives Expression (15.17). Indeed, if

$$\hat{Y}_j = \sum_{k \in S} w_k y_{kj} = \sum_{k \in U} \omega_k y_{kj}, j = 1, \dots, p,$$

we have

$$\frac{\partial \hat{Y}_j}{\partial \omega_k} = y_{kj}.$$

Therefore,

$$\hat{v}_k = \frac{\partial \hat{\theta}}{\partial \omega_k} = \frac{\partial f(\hat{Y}_1, \dots, \hat{Y}_p)}{\partial \omega_k}$$

$$= \sum_{j=1}^{p} \frac{\partial f(\hat{Y}_1, \dots, \hat{Y}_p)}{\partial \hat{Y}_j} \frac{\partial \hat{Y}_j}{\partial \omega_k} = \sum_{j=1}^{p} \frac{\partial f(\hat{Y}_1, \dots, \hat{Y}_p)}{\partial \hat{Y}_j} y_{kj}.$$

Example 15.11 The ratio estimator can be written as

$$\hat{Y}_Q = X\frac{\hat{Y}}{\hat{X}} = X\frac{\sum_{k\in U}\omega_k y_k}{\sum_{k\in U}\omega_k x_k}.$$

The linearized variable can be directly calculated:

$$z_k = \frac{\partial \hat{Y}_Q}{\partial \omega_k} = X\left(\frac{y_k}{\hat{X}} - \frac{x_k\hat{Y}}{\hat{X}^2}\right) = \frac{X}{\hat{X}}(y_k - x_k\hat{R}),$$

where $\hat{R} = \hat{Y}/\hat{X}$.

Example 15.12 By referring to Expression (14.6), the Gini index estimator can be written as

$$\hat{G} = \frac{2}{\hat{N}\hat{Y}}\sum_{k\in U}\omega_k\hat{N}_k y_k - \left(1 + \frac{1}{\hat{N}\hat{Y}}\sum_{k\in U}\omega_k^2 y_k\right),$$

where

$$\hat{N}_k = \sum_{\ell\in U}\omega_\ell \mathbb{1}(y_\ell \leq y_k).$$

Result 15.15 *The estimator of the linearized variable of the Gini index in the sense of Demnati and Rao is*

$$\hat{v}_j = \frac{\partial \hat{G}}{\partial \omega_j} = \frac{1}{\hat{N}\hat{Y}}[2\hat{N}_j(y_j - \hat{\overline{Y}}_j) + \hat{Y} - y_j\hat{N} - \hat{G}(y_j\hat{N} + \hat{Y})],$$

where $\hat{\overline{Y}}_j$ is the estimated mean of values lower than y_j:

$$\hat{\overline{Y}}_j = \frac{1}{\hat{N}_j}\sum_{\ell\in S}w_\ell y_\ell \mathbb{1}(y_\ell \leq y_j).$$

Proof: We can write the Gini index estimator in the form $\hat{G} = A/(\hat{N}\hat{Y})$, where

$$A = 2\sum_{k\in S}w_k \sum_{\ell\in S}^{k} w_\ell y_k \mathbb{1}(y_\ell \leq y_k) - \hat{N}\hat{Y} - \sum_{k\in S}w_k^2 y_k.$$

Therefore, we have

$$\frac{\partial \hat{G}}{\partial w_j} = \frac{1}{\hat{N}\hat{Y}}\frac{\partial A}{\partial w_j} - \frac{A}{\hat{N}^2\hat{Y}} - y_j\frac{A}{\hat{N}\hat{Y}^2} = \frac{1}{\hat{N}\hat{Y}}\left[\frac{\partial A}{\partial w_j} - \hat{G}(y_j\hat{N} + \hat{Y})\right].$$

In addition,

$$\frac{\partial A}{\partial w_j} = 2\hat{N}_j y_j + 2\sum_{\ell\in S}w_\ell y_\ell \mathbb{1}(y_\ell \geq y_j) - \hat{Y} - y_j\hat{N} - 2w_j y_j$$

$$= 2\hat{N}_j(y_j - \hat{\overline{Y}}_j) + \hat{Y} - y_j\hat{N},$$

where

$$\hat{\overline{Y}}_j = \frac{1}{\hat{N}_j}\sum_{\ell\in S}w_\ell y_\ell \mathbb{1}(y_\ell \leq y_j).$$

This allows us to obtain Result 15.15, which is the same as that presented in Expression (15.16). ∎

15.10 Linearization by the Sample Indicator Variables

15.10.1 The Method

Monique Graf (2011) has proposed a method that is simultaneously simple, intuitive, and coherent. It consists of differentiating the estimator with respect to the variables a_k which indicate sample adhesion. This approach is justified by the fact that, in the design-based approach, the only source of randomness is vector $\mathbf{a}$ of a_k. Therefore, we linearize with respect to the random components of the estimator. If the estimator can be approximated by a linear combination of a_k, then the calculation of the expectation and variance is easy to realize.

Graf's method differs from the Demnati and Rao approach. If the weights w_k are obtained by calibration, then they depend on the a_k and therefore are random. With Graf's approach, the weights must also be differentiated. This method allows variances to be calculated for complex estimators like quantiles which are not necessarily functions of totals. However, the estimator must be a differentiable function in the a_k. This method also allows a calibrated estimator to be linearized.

It is also possible to directly calculate, in one step, a linearized variable for an estimator of a complex, nonlinear function that uses weights obtained through calibration. Vallée & Tillé (2019) have even shown that this method can be extended to calculate the variance for estimators tainted by nonresponse (see Section 16.7, page 347).

We can calculate an approximation of the estimator in two ways by using a Taylor expansion around sample indicators, either around the estimator $\widehat{\theta}$ or around the population parameter θ.

The following is the development around the parameter:

$$\widehat{\theta}_{approx} = \widehat{\theta}(\mathbf{y}, \boldsymbol{\pi}) + \sum_{\ell \in U} \tilde{z}_\ell (a_\ell - \pi_\ell), \tag{15.20}$$

the linearized variable is

$$\tilde{z}_\ell = \left. \frac{\partial \widehat{\theta}}{\partial a_\ell} \right|_{\mathbf{a}=\boldsymbol{\pi}},$$

and remainder

$$R = \frac{1}{2} \sum_{k \in U} \sum_{\ell \in U} \left. \frac{\partial^2 \widehat{\theta}}{\partial a_k \partial a_\ell} \right|_{\mathbf{a}=\tau\mathbf{a}+(1-\tau)\boldsymbol{\pi}} (a_k - \pi_k)(a_\ell - \pi_\ell),$$

with $\tau \in [0, 1]$ (see, for example, Edwards, 1994, p. 133).

The following is the linearized variable around the estimator:

$$\widehat{\theta}(\mathbf{y}, \boldsymbol{\pi}) \approx \widehat{\theta} + \sum_{\ell \in U} z_\ell (\pi_\ell - a_\ell),$$

where the linearized variable is $z_\ell = \partial \widehat{\theta} / \partial a_\ell$. We obtain approximation

$$\widehat{\theta} \approx \widehat{\theta}(\mathbf{y}, \boldsymbol{\pi}) + \sum_{\ell \in U} z_\ell (a_\ell - \pi_\ell). \tag{15.21}$$

The approximations (15.20) and (15.21) can be used to approach and estimate the variance of $\widehat{\theta}$, so we can write

$$\widehat{\theta} - \widehat{\theta}(\mathbf{y}, \boldsymbol{\pi}) \approx \sum_{\ell \in U} \tilde{z}_\ell (a_\ell - \pi_\ell) \tag{15.22}$$

and

$$\hat{\theta} - \hat{\theta}(\mathbf{y}, \boldsymbol{\pi}) \approx \sum_{\ell \in U} z_\ell (a_\ell - \pi_\ell). \tag{15.23}$$

The linearized variable $\tilde{z}_\ell$ is not random. We can use it to construct an approximate variance of $\hat{\theta}$:

$$\text{var}_p(\hat{\theta}) \approx \sum_{k \in U}\sum_{\ell \in U} \Delta_{k\ell} \tilde{z}_k \tilde{z}_\ell.$$

Variable $\tilde{z}_\ell$ depends on population values. $\tilde{z}_k$ must be estimated by an estimator $\hat{\tilde{z}}_\ell$ to construct a variance estimator:

$$v(\hat{\theta}) = \sum_{k \in U}\sum_{\ell \in U} a_k a_\ell \frac{\Delta_{k\ell}}{\pi_{k\ell}} \hat{\tilde{z}}_k \hat{\tilde{z}}_\ell.$$

The linearized variable z_ℓ can depend on a_ℓ and therefore be random. If we consider (15.22) and (15.23), the linearized variable $\tilde{z}_\ell$ is estimated by variable z_ℓ. We can also construct a variance estimator using the z_k:

$$v(\hat{\theta}) = \sum_{k \in U}\sum_{\ell \in U} a_k a_\ell \frac{\Delta_{k\ell}}{\pi_{k\ell}} z_k z_\ell. \tag{15.24}$$

We see that sometimes the z_k can be a little different from the $\hat{\tilde{z}}_k$.

Example 15.13 If the total

$$Y = \sum_{k \in U} y_k$$

is estimated by $\hat{Y} = \sum_{k \in U} a_k d_k y_k$, where $d_k = \pi_k^{-1}$, then $z_\ell = \hat{\tilde{z}}_\ell = \tilde{z}_\ell = d_\ell y_\ell$.

Example 15.14 Let

$$\bar{Y} = N^{-1}\sum_{k \in U} y_k$$

be the population mean estimated by the Hájek estimator, $\hat{\bar{Y}} = \hat{N}^{-1}\hat{Y}$, where $\hat{N} = \sum_{k \in U} a_k d_k$. Then,

$$z_\ell = \frac{y_\ell - \hat{\bar{Y}}}{\hat{N}\pi_\ell}, \quad \tilde{z}_\ell = \frac{y_\ell - \bar{Y}}{N\pi_\ell}, \quad \hat{\tilde{z}}_\ell = z_\ell.$$

In this case, $\hat{\tilde{z}}_\ell = z_\ell$ and the two variance estimators are equal.

Example 15.15 If the total Y is estimated by $\hat{Y} = N\hat{\bar{Y}}$, then

$$z_\ell = \frac{N(y_\ell - \hat{\bar{Y}})}{\hat{N}\pi_\ell}, \quad \tilde{z}_\ell = \frac{y_\ell - \bar{Y}}{\pi_\ell} \quad \text{and} \quad \hat{\tilde{z}}_\ell = \frac{y_\ell - \hat{\bar{Y}}}{\pi_\ell}.$$

In this case, z_ℓ is not equal to $\hat{\tilde{z}}_\ell$.

Example 15.16 Let $R = Y/X$ be a ratio where

$$X = \sum_{k \in U} x_k,$$

and estimated by $\widehat{R} = \widehat{Y}/\widehat{X}$, where

$$\widehat{X} = \sum_{k \in U} a_k d_k x_k.$$

Then,

$$z_\ell = \widehat{z}_\ell = d_\ell \frac{y_\ell - x_\ell \widehat{R}}{\widehat{X}} \quad \text{and} \quad \check{z}_\ell = d_\ell \frac{y_\ell - x_\ell R}{X}.$$

Example 15.17 Let the geometric mean

$$G = \left(\prod_{k \in U} y_k \right)^{1/N},$$

with $y_k > 0$, be estimated by

$$\widehat{G} = \prod_{k \in s} y_k^{1/(\widehat{N}\pi_k)} = \exp\left(\frac{1}{\widehat{N}} \sum_{k \in U} \frac{a_k}{\pi_k} \log y_k \right).$$

We obtain the linearized variables

$$z_\ell = \widehat{z}_\ell = \widehat{N}^{-1} \widehat{G} d_\ell \log(y_\ell/\widehat{G}) \text{and} \; \check{z}_\ell = N^{-1} G d_\ell \log(y_\ell/G).$$

15.10.2 Linearization of a Quantile

Quantiles are complex parameters that cannot be written as functions of totals. However, to calculate a linearized variable, the cumulative distribution function must be differentiable with respect to the a_k. We can use the cumulative distribution function defined in Expression (14.3), page 288:

$$\widehat{F}(y) = \frac{1}{\widehat{N}} \sum_{k \in U} \frac{a_k}{\pi_k} \Phi\left(\frac{y - y_k}{h} \right).$$

The derivative of $\widehat{F}(y)$ is the kernel estimator of the density function:

$$\widehat{f}(y) = \widehat{F}'(y) = \frac{1}{h\widehat{N}} \sum_{k \in U} \frac{a_k}{\pi_k} \phi\left(\frac{y - y_k}{h} \right).$$

Estimator $\widehat{Q}_\alpha$ of the quantile of order α is obtained by solving equation

$$\widehat{F}(\widehat{Q}_\alpha) = \alpha. \tag{15.25}$$

By differentiating Equation (15.25) with respect to a_ℓ, we obtain

$$\frac{\partial \widehat{F}(\widehat{Q}_p)}{\partial a_\ell} = \frac{1}{\pi_\ell \widehat{N}} \Phi\left(\frac{\widehat{Q}_\alpha - y_k}{h} \right)$$

$$+ \frac{1}{\pi_\ell \widehat{N}} \sum_{k \in U} \frac{a_k}{\pi_k} \phi\left(\frac{\widehat{Q}_\alpha - y_k}{h} \right) \frac{\partial \widehat{Q}}{h \partial a_\ell} - \frac{1}{\pi_\ell \widehat{N}^2} \sum_{k \in U} \frac{a_k}{\pi_k} \Phi\left(\frac{\widehat{Q}_\alpha - y_k}{h} \right)$$

$$= \frac{1}{\pi_\ell \widehat{N}} \Phi\left(\frac{\widehat{Q}_\alpha - y_k}{h} \right) + \widehat{f}(\widehat{Q}_\alpha) \frac{\partial \widehat{Q}}{\partial a_\ell} - \frac{1}{\pi_\ell \widehat{N}} \widehat{F}(\widehat{Q}_\alpha) = \frac{\partial p}{\partial a_\ell} = 0.$$

It then suffices to solve this equation to find the linearized variable of the quantile estimator:

$$z_\ell = \frac{\partial \widehat{Q}}{\partial a_\ell} = -\frac{1}{\pi_\ell \widehat{f}(\widehat{Q}_\alpha)\widehat{N}} \left[\Phi\left(\frac{\widehat{Q}_\alpha - y_\ell}{h}\right) - \widehat{F}(\widehat{Q}_\alpha) \right]$$

$$= -\frac{1}{\pi_\ell \widehat{f}(\widehat{Q}_\alpha)\widehat{N}} \left[\Phi\left(\frac{\widehat{Q}_\alpha - y_\ell}{h}\right) - \alpha \right].$$

To estimate the variance, we put this linearized variable in Expression (15.24). We recover an analogous result to the one obtained for the influence function presented in Example 15.7, page 319.

15.10.3 Linearization of a Calibrated Estimator

Suppose we want to estimate a total Y by a calibrated estimator described in Chapter 12:

$$\widehat{Y}_{CAL} = \sum_{k \in S} w_k y_k,$$

where the weights are defined by

$$w_k = d_k F_k(\mathbf{x}_k^\top \lambda). \tag{15.26}$$

The vector λ is determined by solving the calibration equations:

$$\sum_{k \in U} a_k F_k(\mathbf{x}_k^\top \lambda)\mathbf{x}_k = \sum_{k \in U} \mathbf{x}_k. \tag{15.27}$$

Graf (2015) has calculated a linearized variable of the estimator by differentiating with respect to the indicators a_k.

Result 15.16 *The linearized variable of $\widehat{Y}_{CAL}$ by linearizing with respect to a_ℓ is*
$z_\ell = w_\ell e_\ell$, where $e_\ell = y_\ell - \mathbf{x}_\ell^\top \widehat{\mathbf{B}}_{y|x}$, $\widehat{\mathbf{B}}_{y|x} = \widehat{\mathbf{T}}_{xx}^{-1} \widehat{\mathbf{t}}_{xy}$,

$$\widehat{\mathbf{T}}_{xx} = \sum_{k \in U} a_k d_k F_k'(\mathbf{x}_k^\top \lambda)\mathbf{x}_k \mathbf{x}_k^\top, \quad \widehat{\mathbf{t}}_{xy} = \sum_{k \in U} a_k d_k F_k'(\mathbf{x}_k^\top \lambda)\mathbf{x}_k y_k. \tag{15.28}$$

Proof: We differentiate the calibration equations (15.27) with respect to a_ℓ and we obtain

$$\frac{\partial}{\partial a_\ell} \sum_{k \in U} a_k d_k F_k(\mathbf{x}_k^\top \lambda)\mathbf{x}_k = \frac{\partial}{\partial a_\ell} \sum_{k \in U} \mathbf{x}_k$$

$$d_\ell F_\ell(\mathbf{x}_\ell^\top \lambda)\mathbf{x}_\ell + \sum_{k \in U} a_k d_k F_k'(\mathbf{x}_k^\top \lambda)\mathbf{x}_k \mathbf{x}_k^\top \frac{\partial \lambda}{\partial a_\ell} = 0,$$

which gives

$$\frac{\partial \lambda}{\partial a_\ell} = -\widehat{\mathbf{T}}_{xx}^{-1} d_\ell F_\ell(\mathbf{x}_\ell^\top \lambda)\mathbf{x}_\ell = -\widehat{\mathbf{T}}_{xx}^{-1} w_\ell \mathbf{x}_\ell.$$

Therefore, the derivative of the estimator is

$$\frac{\partial \widehat{Y}}{\partial a_\ell} = w_\ell y_\ell - \sum_{k \in U} a_k d_k F_k'(\mathbf{x}_k^\top \lambda)y_k \mathbf{x}_k^\top \widehat{\mathbf{T}}_{xx}^{-1} w_\ell \mathbf{x}_\ell = w_\ell e_\ell.$$

∎

The linearized variable z_ℓ is then inserted in the variance estimator given in Expression (15.24). The linearized variable is a residual of a regression of the variable of interest by the auxiliary variables. However, the vector of regression coefficients is weighted by the derivatives of the calibration functions $F'_\ell(\mathbf{x}_\ell^\top \lambda)$.

The linearized variable obtained in Result 15.16 is the same as the one obtained in Demnati & Rao (2004). However, this result is slightly different from the linearized variable proposed in Section 15.6.4, page 313, which is also the one proposed in the original article by Deville & Särndal (1992). They proposed using

$$\xi_\ell = y_\ell - \mathbf{x}_k^\top \left(\sum_{k \in U} a_k w_k q_k \mathbf{x}_k \mathbf{x}_k^\top \right)^{-1} \sum_{k \in U} a_k w_k q_k \mathbf{x}_k y_k. \tag{15.29}$$

There are two differences between the ξ_ℓ of Expression (15.29) and the e_ℓ of Result 15.16. The residual e_ℓ incorporates the g-weights $g_k = F(\mathbf{x}_k^\top \lambda)$. In addition, the regression coefficient of e_k is weighted by $F'_k(\mathbf{x}_k^\top \lambda)$ while for ξ_ℓ it is weighted by $F'_k(0) = q_k$. The ξ_ℓ do not depend on the calibration function while the e_ℓ do depend on it through $F_k(\cdot)$ and $F'_k(\cdot)$. In practice, these two linearized variables give very similar estimators. There are no definitive arguments to use one over the other.

15.10.4 Linearization of a Multiple Regression Estimator

The multiple regression estimator is a special case of the calibration estimator that corresponds to the case where the calibration function is $F_k(u) = 1 + q_k u$. Therefore, $F'_k(u) = q_k$. Result 15.16 then becomes

$$z_\ell = w_\ell y_\ell - w_\ell \mathbf{x}_\ell^\top \left(\sum_{k \in U} a_k d_k q_k \mathbf{x}_k \mathbf{x}_k^\top \right)^{-1} \sum_{k \in U} a_k d_k q_k \mathbf{x}_k y_k = w_\ell e_\ell = d_\ell g_\ell e_\ell, \tag{15.30}$$

where $g_\ell = \pi_\ell w_\ell = w_\ell / d_\ell$ is called the g-weight. This g-weight represents the distortion of the w_k with respect to the base weights $d_k = 1/\pi_k$. The linearized variable given in Expression (15.30) is again different from the one given by (15.29) because of the presence of this development in the residuals. Särndal (1982) recommends using these g-weights to estimate the variance of the regression estimator. He shows that in certain special cases the use of g-weights secures good conditional properties for the variance estimators.

In the initial article on calibration, Deville & Särndal (1992) have also proposed imposing bounds on the weights (see Section 12.2.5, page 245). To do this, they suggested using a truncated linear calibration function:

$$F_k(u) = \max(L, \min(1 + q_k u, H)),$$

where $L < 1 < H$ are the bounds for the g-weights. In this case, the derivative of $F_k(u)$ is zero when u is outside the interval $[(L - 1)/q_k, (H - 1)/q_k]$. With the linearized variable proposed in Result 15.16, units with bounded weights do not contribute to the calculation of regression coefficients.

15.10.5 Linearization of an Estimator of a Complex Function with Calibrated Weights

Let $\mathbf{d} = (d_1, \cdots, d_k, \cdots, d_N)^\mathsf{T}$ be a vector. Suppose we can estimate parameter θ by estimator $\hat{\theta}_d = \hat{\theta}(\mathbf{y}, \mathbf{a}, \mathbf{d})$. If the a_k are always multiplied by their weights d_k in the estimator, then the linearized variable of $\hat{\theta}_d$ is $z_\ell^d = \partial\hat{\theta}_d / \partial a_\ell = d_\ell h_{\ell 1}(\mathbf{y}, \mathbf{a}, \mathbf{d})$. We also note that $h_{\ell 2}(\mathbf{y}, \mathbf{a}, \mathbf{d}) = \partial\hat{\theta}_d / \partial d_\ell = a_\ell h_{\ell 1}(\mathbf{y}, \mathbf{a}, \mathbf{d})$. The vector of calibrated weights $\mathbf{w} = (w_1 \cdots w_k \cdots w_N)^\mathsf{T}$, w_k, defined in Expression (15.26), is used to estimate θ by $\hat{\theta}_{\text{CAL}} = \hat{\theta}(\mathbf{y}, \mathbf{a}, \mathbf{w})$. The linearized variable of $\hat{\theta}_{\text{CAL}}$ is given by the following result:

Result 15.17 *The linearized variable of $\hat{\theta}_w$ obtained by linearizing with respect to a_ℓ is $z_\ell^w = w_\ell v_\ell^w$, where*

$$v_\ell^w = h_{\ell 1}(\mathbf{y}, \mathbf{a}, \mathbf{w}) - \mathbf{x}_\ell^\mathsf{T} \hat{\mathbf{B}}_{h1|x},$$

$$\hat{\mathbf{B}}_{h1|x} = \hat{\mathbf{T}}_{xx}^{-1} \sum_{k \in U} a_k d_k F_k'(\mathbf{x}_k^\mathsf{T} \lambda) \mathbf{x}_k h_{k1}(\mathbf{y}, \mathbf{a}, \mathbf{w}),$$

and $\hat{\mathbf{T}}_{xx}$ is defined in Expression (15.28).

Proof: Since vector $\mathbf{w}$ depends on $\mathbf{a}$, the linearized variable is $z_\ell^w = w_\ell v_\ell^w$, with

$$v_\ell^w = h_{\ell 1}(\mathbf{y}, \mathbf{a}, \mathbf{w}) + \sum_{k \in U} a_k h_{k2}(\mathbf{y}, \mathbf{a}, \mathbf{w}) \frac{\partial w_k}{\partial a_\ell}.$$

The derivative of w_k is obtained by differentiating Equation (15.27) with respect to a_ℓ:

$$w_\ell \mathbf{x}_\ell + \sum_{k \in U} a_k d_k \mathbf{x}_k \mathbf{x}_k^\mathsf{T} F_k'(\mathbf{x}_k^\mathsf{T} \lambda) \frac{\partial \lambda}{\partial a_\ell} = 0.$$

Therefore,

$$\frac{\partial \lambda}{\partial a_\ell} = -w_\ell \hat{\mathbf{T}}_{xx}^{-1} \mathbf{x}_\ell,$$

$$\frac{\partial w_k}{\partial a_\ell} = d_k \frac{\partial F_k(\mathbf{x}_k^\mathsf{T} \lambda)}{\partial a_\ell} = -d_k F_k'(\mathbf{x}_k^\mathsf{T} \lambda) \mathbf{x}_k^\mathsf{T} \hat{\mathbf{T}}_{xx}^{-1} w_\ell \mathbf{x}_\ell$$

and

$$v_\ell^w = h_{\ell 1}(\mathbf{y}, \mathbf{a}, \mathbf{w}) - \mathbf{x}_\ell^\mathsf{T} \hat{\mathbf{T}}_{xx}^{-1} \sum_{k \in U} a_k d_k F_k'(\mathbf{x}_k^\mathsf{T} \lambda) \mathbf{x}_k h_{k2}(\mathbf{a}, \mathbf{w}, \mathbf{y})$$

$$= h_{\ell 1}(\mathbf{y}, \mathbf{a}, \mathbf{w}) - \mathbf{x}_\ell^\mathsf{T} \hat{\mathbf{B}}_{h1|x}. \qquad \blacksquare$$

Result 15.17 is limited to the cases where w_k and a_k appear in $\hat{\theta}_w$ like a product $w_k a_k$. Therefore, the linearization can always be done in two steps. First, we linearize the complex function by considering that the weights are not random. Then, we simply apply a residual technique to this first linearization.

15.10.6 Linearization of the Gini Index

Suppose the Gini index with nonrandom weights of the expansion estimator $d_k = 1/\pi_k$ is

$$\hat{G} = \frac{1}{2\hat{N}\hat{Y}} \sum_{i \in U} \sum_{k \in U} a_i a_k d_i d_k |y_i - y_k|,$$

where

$$\hat{N} = \sum_{k \in U} a_k d_k \quad \text{and} \quad \hat{N} = \sum_{k \in U} a_k d_k y_k.$$

Langel & Tillé (2013) give an overview of the literature concerning the variance calculation of the Gini index. However, we quickly obtain a linearized variable by differentiating with respect to the indicators. Indeed,

$$\frac{\partial \hat{G}}{\partial a_\ell} = \frac{d_\ell}{\hat{N}\hat{Y}} \left[\sum_{k \in U} a_k d_k |y_\ell - y_k| - \hat{G}(\hat{Y} + \hat{N}y_\ell) \right]$$

$$= \frac{d_\ell}{\hat{N}\hat{Y}} \left[\hat{Y} - \hat{N}y_\ell + 2\hat{N}_\ell(y_\ell - \overline{\hat{Y}}_\ell) - \hat{G}(\hat{Y} - \hat{N}y_\ell) \right],$$

where

$$\hat{N}_\ell = \sum_{i \in U} a_i d_i \mathbb{1}_{y_i \le y_\ell}, \quad \overline{\hat{Y}}_\ell = \frac{1}{\hat{N}_\ell} \sum_{i \in U} a_i w_i y_i \mathbb{1}_{\hat{N}_i \le \hat{N}_\ell}.$$

The result is analogous to the one obtained in Expression (15.16), page 320, and Result 15.15, page 323.

We now consider a Gini estimator calculated with calibrated weights:

$$\hat{G}_{\text{CAL}} = \frac{1}{2\hat{N}_{\text{CAL}}\hat{Y}_{\text{CAL}}} \sum_{i \in U} \sum_{k \in U} a_i a_k w_i w_k |y_i - y_k|,$$

where

$$\hat{N}_{\text{CAL}} = \sum_{k \in U} a_k w_k \quad \text{and} \quad \hat{Y}_{\text{CAL}} = \sum_{k \in U} a_k w_k y_k.$$

Therefore, we have

$$h_{\ell 1}(\mathbf{y}, \mathbf{a}, \mathbf{w}) = \frac{1}{\hat{N}_{\text{CAL}}\hat{Y}_{\text{CAL}}} \left[\hat{Y}_{\text{CAL}} - \hat{N}_{\text{CAL}}y_\ell + 2\hat{N}_\ell(y_\ell - \overline{\hat{Y}}_\ell) - \hat{G}(\hat{Y}_{\text{CAL}} - \hat{N}_{\text{CAL}}y_\ell) \right],$$

where

$$\hat{N}_\ell = \sum_{i \in U} a_i w_i \mathbb{1}_{y_i \le y_\ell}, \quad \overline{\hat{Y}}_\ell = \frac{1}{\hat{N}_\ell} \sum_{i \in U} a_i w_i y_i \mathbb{1}_{y_i \le y_\ell}.$$

The linearized variable of the Gini index calculated with calibrated weights is simply

$$z_\ell^w = w_\ell \left[h_{\ell 1}(\mathbf{y}, \mathbf{a}, \mathbf{w}) - \mathbf{x}_\ell^\top \hat{\mathbf{B}}_{h1|x} \right].$$

15.11 Discussion on Variance Estimation

There exist numerous variations of linearization techniques that give relatively similar results. The simplest method to apply is the Graf method. It allows linearization by directly taking into account the complexity of the estimator and the gain in precision due to calibration.

It is also possible to estimate the variance using resampling techniques like the *bootstrap* and *jackknife*. However, it is not possible to use these methods without adapting them to finite populations. Indeed, the variance estimators depend on the sampling design.

The bootstrap was initially proposed by Bradley Efron (1979) (see also Efron & Tibshirani, 1993). Extensive scientific literature is dedicated to adapting the bootstrap to finite populations (Bickel & Freedman, 1984; Chao & Lo, 1985; Mac Carthy & Snowden, 1985; Kuk, 1989; Kovar et al., 1988; Sitter, 1992b,a; Chen & Sitter, 1993; Barbe & Bertail, 1995; Nigam & Rao, 1996; Bertail & Combris, 1997; Kaufman, 1998, 1999; Holmberg, 1998; Saigo et al., 2001; Shao, 2003; Funaoka et al., 2006; Giorgi et al., 2006). The pseudo-population technique consists of reconstructing a population from the sample and then resampling this pseudo-population (Booth et al., 1994; Booth & Hall, 1994). It is also possible to conceive of specific bootstrap designs for each sampling design (Antal & Tillé, 2011, 2014; Beaumont & Patak, 2012).

Similarly, the jackknife, originally proposed by Maurice Quenouille (1949), has been the subject of numerous adaptations to be applied to finite populations (see, among others, Jones, 1974; Krewski & Rao, 1981; Fay, 1985; Rao & Shao, 1992; Yung & Rao, 1996; Rao & Tausi, 2004; Berger, 2007; Berger & Skinner, 2005; Berger & Rao, 2006).

Exercises

15.1 After a stage in the Tour de France, a simple random sampling of fixed size of n cyclists was carried out. For each selected cyclist, their mean speed for the stage was obtained. Estimate the mean speed of all the cyclists and then estimate the variance of this estimator.

15.2 For a Poisson design with unequal probabilities, give the variance of the expansion estimator of the total. Then give an unbiased estimator of this variance. Again, for this same design, use the multiple regression estimator with a single auxiliary variable $x_k = 1, k \in U$ and by taking $q_k = 1, k \in U$.
   1. Simplify the multiple regression estimator. Which estimator is this? Is it unbiased?
   2. Using the linearization technique, give the linearized variable of this function of interest.
   3. Give an approximation of the variance using the linearized variable. Simplify the expression.
   4. Give an estimator of this variance.

15.3 **1.** Recall the linearized variable of variance and covariance.

$$S_y^2 = \frac{1}{N-1} \sum_{k \in U} (y_k - \overline{Y})^2$$

$$S_{xy} = \frac{1}{N-1} \sum_{k \in U} (x_k - \overline{X})(y_k - \overline{Y})$$

2. By using the linearization in stages technique, give the linearized variable of the determination coefficient (supposing N is unknown) given by

$$r^2 = \frac{S_{xy}^2}{S_x^2 S_y^2},$$

by starting with the linearized variables for variance and covariance.

3. Show that the determination coefficient can also be written as a function of regression coefficients.

4. By using linearization in stages, give the linearized variable of the determination coefficient by starting with the linearized variable of the regression coefficient.

5. Do both methods end up with the same result?

16

Treatment of Nonresponse

This chapter contains a brief introduction to the methodology of the treatment of non-response. Nonresponse is one of the many sources of errors in surveys (Section 16.1). The other sources are sampling errors, coverage errors (Section 16.2), and measurement errors. One can distinguish questionnaire nonresponse, which applies to the entirety of a questionnaire, from item nonresponse, which applies only to a limited number of variables (Section 16.3). To be able to "treat" nonresponse, we must reasonably presume that there exists a true value of the variable for which response is missing, which is obviously not the case for variables that represent opinions. The base hypothesis consists of supposing that nonresponse is the result of a random experiment (Section 16.4). We mainly present methods for which nonresponse is supposed to depend on auxiliary variables and not on the variable of interest.

Treatment of questionnaire nonresponse (Section 16.5) is realized by supposing that nonresponse is a second sampling phase. The estimation of inclusion probabilities for the second phase can be done by either modeling or calibration.

Item nonresponse (Section 16.6) is generally treated by imputation techniques. Classic techniques (*hot deck*, nearest neighbor) are presented, but we show that variance due to imputation can be reduced by selecting the values by a balanced sampling. Finally, an introduction to variance estimation is given in Section 16.7. This chapter is based on work by Rubin (1976), Särndal & Swensson (1987), Rao (1990), Fay (1991), Rao & Sitter (1995), Särndal (1992), Deville & Dupont (1993), Caron (1996), Dupont (1996), Deville (1998b, 2000a), Shao & Steel (1999), Folsom & Singh (2000), Lee et al. (2002), Haziza & Rao (2003), Särndal & Lundström (2005), Kim & Kim (2007), Kim & Rao (2009), Lundström & Särndal (1999), Haziza (2005, 2009), Beaumont (2005), Kott (2006), Beaumont et al. (2007), Skinner (2011), Haziza & Lesage (2016), Beaumont & Haziza (2016), and Haziza & Beaumont (2017).

16.1   Sources of Error

Four main sources of error can be distinguished in survey sampling:

1. *Coverage error.* This comes from the fact that the survey frame rarely corresponds to the target population and it is impossible to exactly apply the chosen sampling design.
2. *Sampling error.* This comes from the random choice of units and is therefore inherent to sampling techniques. It is the main topic of the previous chapters.

Sampling and Estimation from Finite Populations, First Edition. Yves Tillé.
© 2020 John Wiley & Sons Ltd. Published 2020 by John Wiley & Sons Ltd.

3. *Nonresponse error.* Questionnaire nonresponse can be distinguished from item nonresponse. Questionnaire nonresponse happens when all the values are missing for a unit. Item nonresponse happens if one or more (but not all) values are missing for a unit.

4. *Measurement error.* This can have many diverse causes attributable to the interviewer or the interviewee: bad questionnaire design, an inability to respond, a willingness to sabotage the survey, coding problems, or faulty work from the investigators.

Only sampling error is specific to surveys. The other errors also apply to censuses. Sampling error is in some way voluntary and planned. The knowledge of the sampling design can therefore be used to conduct inference. The other errors are incurred. It is not possible to conduct inference without formulating a hypothesis on the occurrence of these errors. Errors can cause a bias in the estimators. The main challenge is to correct this bias. Measurement errors and their corrections have an impact on the variance of estimators.

16.2   Coverage Errors

Many types of coverage errors can be distinguished:

1. *Overcoverage.* The survey frame contains units that are not part of the target population. The target population is therefore a domain of the survey frame and we can do an estimation of parameters.

2. *Undercoverage.* The survey frame is a subpopulation of the target population. This error will mar all estimators with bias, as certain units have a null inclusion probability.

3. *Duplication of units in the frame.* This problem is definitely the most delicate to deal with. In many administrative survey frames, certain units are coded multiple times. A simple typographic error is sometimes enough to create a new record. This problem is particularly difficult to deal with, as it necessitates comparing units with each other, which often implies considerable computer resources.

Detection of coverage errors is generally done by control surveys.

16.3   Different Types of Nonresponse

Nonresponse is an absence of response to one or many questionnaire variables. Questionnaire nonresponse, which is the complete absence of information for a unit, is distinguished from item nonresponse, which is the absence of information for a unit which is limited to only certain variables.

• Questionaire nonresponse can be due to a refusal to answer from the interviewee, an inability to contact the person, the inability of the interviewee to respond, a loss or destruction of the questionnaire, or the abandonment of the questionnaire at the beginning of the survey.

• Item nonresponse can result from the refusal to answer certain questions, of an incomprehension of questions or responses, an abandonment during the survey, or an invalidation of parts of the questionnaire due to some incoherence in the collected data.

It is important to not confuse nonresponse and response "not applicable". Indeed, we cannot talk about nonresponse for variables for which there is no value. For example, an individual may not have income because he does not have a job, while another may refuse to report his income. Nonresponse can also appear as a possible response to opinion questions. The choice "no opinion" is then a valid response to the question. It is therefore not a measurement error if we agree that someone may realistically not have an opinion on a topic. Nonresponse is a measurement error when we can reasonably assume that behind a nonresponse is hidden a true value for the variable of interest. This is the case for factual variables but not for opinion variables. The questionnaire must be well designed and anticipate these problems. For example, if a question has no value, for the clarity of the interview it should not be asked. The value "no opinion" must exist for opinion questions. A bad questionnaire will inevitably lead to confusion between "no value", "nonresponse", and "no opinion".

There exist multiple techniques to "treat" nonresponse. The general idea of treatment is to construct estimators that are not marred by nonresponse bias. All treatments of nonresponse imply some implicit modeling of the nonresponse mechanism. The validity of the treatment will depend on the adequacy of the model. A descriptive analysis of the missing responses is therefore recommended rather than trying to understand the nonresponse mechanism. In general, there will be important links between certain variables of interest and nonresponse. In addition, we most often observe that if a value is missing for some variable of a unit, there is a higher chance of obtaining a missing value for the other variables of this unit. Nonresponse therefore tends to be concentrated in very specific subgroups.

Correcting nonresponse requires knowledge of auxiliary information known for nonrespondents, for either the sample or the population. It is then possible to use this auxiliary information to model nonresponse.

16.4 Nonresponse Modeling

In all possible cases, we suppose that the nonresponse mechanism can be modeled using a random variable. Response or failure to respond is then considered as the result of a random experiment for which the model will determine the treatment of the nonresponse. Rubin (1976) proposed a typology for the different types of missing data (see also Little & Rubin, 1987, 2014):

- Nonresponse is called *uniform* or *missing completely at random (MCAR)* if it does not depend on either the variable of interest or auxiliary variables. The response probability is then constant for all the units in the population.
- Nonresponse is called *ignorable* or *missing at random (MAR)* if the response probability depends on auxiliary variables, which are not affected by nonresponse, but do not depend on the variable of interest.
- Nonresponse is called *nonignorable* or *missing not at random (MNAR)* if it depends on the variable of interest which is affected by nonresponse. For example, if the nonresponse of an income variable depends on the income itself.

The term *nonconfounding* groups MCAR and MAR. The term *nonuniform* groups MAR and MNAR.

16.5   Treating Nonresponse by Reweighting

16.5.1   Nonresponse Coming from a Sample

To treat the issue of questionnaire nonresponse, the classic hypothesis consists of sup-posing that the decision to respond is the result of a random experiment. The nonre-sponse can then be interpreted as a sampling phase. Any survey can be decomposed into a double sampling:

1. The selection of n units of sample S in the population using a sampling design $p(s)$.
2. The selection of the random set of respondents using the response design.

There exists, in general, two ways to model.

1. In the classic approach, called *two phases*, described in Särndal & Swensson (1987) and presented in Figure 16.1, we suppose that nonresponse is a second phase of sampling. The problem is described as a classic two-phase design, as presented in Section 7.5, except that the response probabilities $\phi_k = \Pr(k \in R|S)$ and $\phi_{k\ell} = \Pr(k \in R$ and $\ell \in R|S)$ are unknown and we need to estimate them.
2. In the approach called *reversed*, represented in Figure 16.2, we suppose that the sampling and response design are two independent designs. In this case, accepting to respond or not does not depend on the selection of the sample. In this case, $\phi_k = \Pr(k \in R|S) = \Pr(k \in R)$ and $\Pr(k \in R \cap S) = \Pr(k \in R)\Pr(k \in S) = \phi_k \pi_k$, for all $k \in U$. This hypothesis, proposed by Fay (1991) and Shao & Steel (1999), consider-ably simplifies variance estimation. In fact, having two independent samples allows the variance to be calculated and estimated by conditioning with respect to S or R (see Sections 7.5 and 7.6).

The response design is often associated with a Poisson design. In this case, each unit "decides" to respond or not, independently of any other unit in the population. This hypothesis excludes a cluster effect which could happen if, for example, a superintendent blocks access to an entire apartment building. Cluster effects in nonresponse are studied by Haziza & Rao (2003) and Skinner (2011).

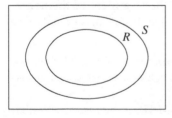

Figure 16.1 Two-phase approach for nonresponse. The set of respondents R is a subset of sample S.

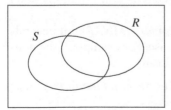

Figure 16.2 The reversed approach for nonresponse. The sample of nonrespondents R is independent of the selected sample S.

In a Poisson design, the knowledge of the first-order inclusion probabilities allows the sampling design to be completely identified and therefore the second-order inclusion probabilities, which are

$$\Pr(k \in R \quad \text{and} \quad \ell \in R | S) = \Pr(k \in R | S)\Pr(\ell \in R | S).$$

Under the hypothesis of a Poisson design, knowledge of response probabilities $\Pr(k \in R | S)$ completely determines the response design.

16.5.2 Modeling the Nonresponse Mechanism

In the case where nonresponse is ignorable (*missing at random*), we can suppose that nonresponse on a vector of auxiliary variables $\mathbf{x}_k$. To model the nonresponse mechanism, we suppose that vectors $\mathbf{x}_k$ are available for all the units of S. Estimating the response probabilities ϕ_k is generally done using a parametric or nonparametric model (see, for example, da Silva & Opsomer, 2006, 2009). Let r_k denote the Bernoulli random variable, which equals 1 if the individual responds and 0 otherwise.

Let also $\mathbf{r} = (r_1, \dots, r_N)^\top$ denote the vector of response indicators for all the units of the population. Even in the two-phase approach, we can reasonably suppose that r_k also exists for units which are nonselected in the sample. A unit is therefore observed if $r_k a_k = 1$.

If we suppose that nonresponse is random and that it does not depend on the variable of interest, we can use the following general model:

$$E(r_k | S) = \Pr(k \in R | S) = h(\mathbf{x}_k^\top \mathbf{b}),$$

where $\mathbf{x}_k = (x_{k1}, \dots, x_{kJ})^\top$ is a vector in $\mathbb{R}^J$ containing the J auxiliary variable values, $h(.)$ is a positive, increasing, differentiable function of $\mathbb{R}$ into $]0, 1]$ such that $h(0) = 1$, and $\mathbf{b}$ is a vector of J parameters that must be estimated.

For example, we can use a logistic function. In this case

$$\phi_k = h(\mathbf{x}_k^\top \mathbf{b}) = \frac{\exp(\mathbf{x}_k^\top \mathbf{b})}{1 + \exp(\mathbf{x}_k^\top \mathbf{b})}. \tag{16.1}$$

We therefore look to estimate the model parameter $\Pr(k \in R | S) = h(\mathbf{x}_k^\top \mathbf{b})$, which can be done in multiple ways.

1. We can estimate $\mathbf{b}$ by simply applying the least squares method. We then seek the vector $\mathbf{b}$ which minimizes

$$\sum_{k \in U} a_k [r_k - h(\mathbf{x}_k^\top \mathbf{b})]^2,$$

which is found by finding the solution of the system:

$$\sum_{k \in U} a_k \mathbf{x}_k h'(\mathbf{x}_k^\top \mathbf{b})[r_k - h(\mathbf{x}_k^\top \mathbf{b})] = \mathbf{0}, \tag{16.2}$$

where $h'(.)$ is the derivative of $h(.)$. The equality (16.2) can also be written as

$$\sum_{k \in U} a_k r_k \mathbf{x}_k h'(\mathbf{x}_k^\top \mathbf{b}) = \sum_{k \in U} a_k \mathbf{x}_k h'(\mathbf{x}_k^\top \mathbf{b}) h(\mathbf{x}_k^\top \mathbf{b}). \tag{16.3}$$

Calibration weights can also be used in the estimating equation:

$$\sum_{k \in U} \frac{a_k r_k \mathbf{x}_k h'(\mathbf{x}_k^\top \mathbf{b})}{\pi_k} = \sum_{k \in U} \frac{a_k \mathbf{x}_k h'(\mathbf{x}_k^\top \mathbf{b}) h(\mathbf{x}_k^\top \mathbf{b})}{\pi_k}. \tag{16.4}$$

Using weights or not in the estimating equation to estimate response probabilities is still debated (see Haziza & Beaumont, 2017, p. 221). If sample S is independent of R, then the use of weights or not should lead to a similar result.

2. Another solution could consist of using the maximum likelihood method. Since the response design is Poisson, the probability of obtaining a particular set of respondents is:

$$L(\mathbf{b}|\mathbf{r}) = \left\{ \prod_{k \in U} h(\mathbf{x}_k^\top \mathbf{b})^{a_k r_k} \right\} \left\{ \prod_{k \in U} [1 - h(\mathbf{x}_k^\top \mathbf{b})]^{a_k(1-r_k)} \right\}.$$

By canceling the derivative of $\log L(\mathbf{b}|\mathbf{r})$ with respect to $\mathbf{b}$, we obtain

$$\sum_{k \in U} \frac{a_k r_k \mathbf{x}_k h'(\mathbf{x}_k^\top \mathbf{b})}{h(\mathbf{x}_k^\top \mathbf{b})} = \sum_{k \in U} \frac{a_k(1 - r_k)\mathbf{x}_k h'(\mathbf{x}_k^\top \mathbf{b})}{1 - h(\mathbf{x}_k^\top \mathbf{b})},$$

which can also be written as

$$\sum_{k \in U} \frac{a_k r_k \mathbf{x}_k h'(\mathbf{x}_k^\top \mathbf{b})}{h(\mathbf{x}_k^\top \mathbf{b})[1 - h(\mathbf{x}_k^\top \mathbf{b})]} = \sum_{k \in U} \frac{a_k \mathbf{x}_k h'(\mathbf{x}_k^\top \mathbf{b})}{1 - h(\mathbf{x}_k^\top \mathbf{b})}. \tag{16.5}$$

Again, we can introduce survey weights in the estimating equation:

$$\sum_{k \in U} \frac{a_k r_k \mathbf{x}_k h'(\mathbf{x}_k^\top \mathbf{b})}{\pi_k h(\mathbf{x}_k^\top \mathbf{b})[1 - h(\mathbf{x}_k^\top \mathbf{b})]} = \sum_{k \in U} \frac{a_k \mathbf{x}_k h'(\mathbf{x}_k^\top \mathbf{b})}{\pi_k[1 - h(\mathbf{x}_k^\top \mathbf{b})]}. \tag{16.6}$$

3. Finally, it is possible to estimate $\mathbf{b}$ based only on a calibration criteria (Iannacchione et al., 1991; Deville, 2000a). Since we know the total of the $\mathbf{x}_k$ on the sample, we want the $\Pr(k \in R|S)$ to be estimated in such a way that the estimator of the total on the respondents be calibrated on an estimator of the total:

$$\sum_{k \in U} \frac{a_k r_k \mathbf{x}_k}{h(\mathbf{x}_k^\top \mathbf{b})\pi_k} = \sum_{k \in U} \frac{a_k \mathbf{x}_k}{\pi_k}. \tag{16.7}$$

We therefore have an estimating function that needs to be solved.

In all cases, $\mathbf{b}$ is estimated by solving an estimating equation of the type

$$\sum_{k \in U} \frac{a_k r_k \mathbf{u}_k}{h(\mathbf{x}_k^\top \mathbf{b})} = \sum_{k \in U} a_k \mathbf{u}_k,$$

where

1. $\mathbf{u}_k = \mathbf{x}_k h(\mathbf{x}_k^\top \mathbf{b}) h'(\mathbf{x}_k^\top \mathbf{b})$ in Expression (16.3)
2. $\mathbf{u}_k = \mathbf{x}_k h(\mathbf{x}_k^\top \mathbf{b}) h'(\mathbf{x}_k^\top \mathbf{b})/\pi_k$ in Expression (16.4)
3. $\mathbf{u}_k = \mathbf{x}_k h'(\mathbf{x}_k^\top \mathbf{b})/[1 - h(\mathbf{x}_k^\top \mathbf{b})]$ in Expression (16.5)
4. $\mathbf{u}_k = \mathbf{x}_k h'(\mathbf{x}_k^\top \mathbf{b})/\{[1 - h(\mathbf{x}_k^\top \mathbf{b})]\pi_k\}$ in Expression (16.6)
5. $\mathbf{u}_k = \mathbf{x}_k/\pi_k$ in Expression (16.7).

Estimating by a calibration technique has the benefit of producing an estimator suitably calibrated on the estimator of the total of sample S.

If the model is logistic, $h(\mathbf{x}_k^\top \mathbf{b})$ is defined in Expression (16.1) and

$$h'(\mathbf{x}_k^\top \mathbf{b}) = h(\mathbf{x}_k^\top \mathbf{b})[1 - h(\mathbf{x}_k^\top \mathbf{b})].$$

In the six cases, the values $\mathbf{u}_k$ become:

1. $\mathbf{u}_k = \mathbf{x}_k h^2(\mathbf{x}_k^\top \mathbf{b})[1 - h(\mathbf{x}_k^\top \mathbf{b})]$ in Expression (16.3)
2. $\mathbf{u}_k = \mathbf{x}_k h^2(\mathbf{x}_k^\top \mathbf{b})[1 - h(\mathbf{x}_k^\top \mathbf{b})]/\pi_k$ in Expression (16.4)
3. $\mathbf{u}_k = \mathbf{x}_k h(\mathbf{x}_k^\top \mathbf{b})[1 - h(\mathbf{x}_k^\top \mathbf{b})]/[1 - h(\mathbf{x}_k^\top \mathbf{b})] = \mathbf{x}_k h(\mathbf{x}_k^\top \mathbf{b})$ in Expression (16.5)
4. $\mathbf{u}_k = \mathbf{x}_k h(\mathbf{x}_k^\top \mathbf{b})[1 - h(\mathbf{x}_k^\top \mathbf{b})]/\{[1 - h(\mathbf{x}_k^\top \mathbf{b})]\pi_k\} = \mathbf{x}_k h(\mathbf{x}_k^\top \mathbf{b})/\pi_k$ in Expression (16.6)
5. $\mathbf{u}_k = \mathbf{x}_k/\pi_k$ in Expression (16.7).

If the probabilities ϕ_k are known, we could use an estimator for a two-phase design:

$$\check{Y} = \sum_{k \in U} \frac{a_k r_k y_k}{\pi_k \phi_k}.$$

This estimator would be unbiased and its variance and variance estimator are described in Sections 7.5 and 7.6 for the so-called reverse approach.

However, since the response probabilities are unknown, we must resign ourselves to estimate them by the previously described methods. We therefore use

$$\hat{Y}_{\mathrm{NR}} = \sum_{k \in U} \frac{a_k r_k y_k}{\pi_k \hat{\phi}_k},$$

where $\hat{\phi}_k = h(\mathbf{x}_k^\top \hat{\mathbf{b}})$ and $\hat{\mathbf{b}}$ is an estimator of $\mathbf{b}$. This estimator is slightly biased when the model is well specified. It is obviously biased if the model is misspecified or incomplete. Beaumont (2005) and Kim & Kim (2007) have shown that estimating the response probability reduces the variance with respect to $\check{Y}$. The variance of $\hat{Y}_{\mathrm{NR}}$ is in general smaller than the variance of $\check{Y}$. We therefore cannot simply apply the theory developed in Sections 7.5 and 7.6 to estimate the variance.

In practice, the calculation of the variance is more complex because the weights $1/(\pi_k \hat{\phi}_k)$ are generally calibrated on known population totals. The topic of variance estimation is treated in Section 16.7.

16.5.3 Direct Calibration of Nonresponse

For simplicity of notation, we pose $H(\mathbf{x}_k^\top \mathbf{b}) = 1/[1 - h(\mathbf{x}_k^\top \mathbf{b})]$. The function $H(.)$ is therefore increasing and defined from $\mathbb{R}$ into $[1, \infty)$. If $h(u) = \exp(u)/[1 + \exp(u)]$ is a logistic function, then $H(u) = 1 + \exp(u)$.

If $\mathbf{b}^*$ represents the "true" value of the unknown parameter, the expansion estimator of the total of the auxiliary variables in a two-phase design is then

$$\hat{X}_{\mathrm{NR}} = \sum_{k \in U} \frac{a_k r_k \mathbf{x}_k H(\mathbf{x}_k^\top \mathbf{b}^*)}{\pi_k}.$$

However, this estimator could be improved in the case where the vector of totals

$$\mathbf{X} = \sum_{k \in U} \mathbf{x}_k$$

is known. Auxiliary information can be used to model nonresponse but also to recalibrate the expansion estimator using the calibration technique described in Section 12. Calibration equations given in 12.4 then become:

$$\sum_{k \in U} \frac{a_k r_k \mathbf{x}_k H(\mathbf{x}_k^{\mathsf{T}} \mathbf{b}^*) F_k(\mathbf{x}_k^{\mathsf{T}} \lambda)}{\pi_k} = \mathbf{X},$$

where $F_k(.)$ is the calibration function.

If we also suppose that we can write

$$J_k(\mathbf{x}_k^{\mathsf{T}} \gamma) = H(\mathbf{x}_k^{\mathsf{T}} \mathbf{b}^*) F_k(\mathbf{x}_k^{\mathsf{T}} \lambda), \tag{16.8}$$

where γ is a vector of J parameters, we come back to a classic calibration problem. Dupont (1996) describes the cases where the simplification given in Expression (16.8) is exactly applied, for example when the functions $H(\cdot)$ and $F_k(\cdot)$ are exponential or when the variables $\mathbf{x}_k$ are of stratum membership indicators . For the other cases, we suppose we have a function $J_k(\mathbf{x}_k^{\mathsf{T}} \gamma)$ which is a good approximation of $H(\mathbf{x}_k^{\mathsf{T}} \mathbf{b}^*) F_k(\mathbf{x}_k^{\mathsf{T}} \lambda)$. In this case, the problem is reduced to simple calibration using the calibration function $J_k(\cdot)$:

$$\sum_{k \in U} \frac{a_k r_k \mathbf{x}_k J_k(\mathbf{x}_k^{\mathsf{T}} \gamma)}{\pi_k} = \mathbf{X}. \tag{16.9}$$

Estimation of $\mathbf{b}$ is not necessary to identify calibration weights.

If $\hat{\gamma}$ denotes the vector which satisfies (16.9), then since the sampling error is negligible with respect to nonresponse error, we can estimate the nonresponse probability by

$$\widehat{\Pr}(k \in R|S) = \frac{1}{J_k(\mathbf{x}_k^{\mathsf{T}} \hat{\gamma})}.$$

Calibration therefore allows both the sampling error and the nonresponse error to be corrected simultaneously, and the response probability to be estimated without needing to know the value of $\mathbf{x}_k$ on the nonrespondents of sample S. Knowing an estimator of the $\Pr(k \in R|S)$, it is possible to estimate the variance of the estimator.

We can pretty freely choose functions for calibration $F_k(.)$ and for nonresponse $H(.)$. However, we prefer functions that are always positive. This choice will have little influence on the estimator variance, as is the case for all calibration techniques, in the case where we have large sample sizes.

For example, we can take an exponential calibration function $F_k(\mathbf{x}_k^{\mathsf{T}} \lambda) = \exp(\mathbf{x}_k^{\mathsf{T}} \lambda)$ and a nonresponse model that is also written exponentially

$$H(\mathbf{x}_k^{\mathsf{T}} \mathbf{b}) = \exp(\mathbf{x}_k^{\mathsf{T}} \mathbf{b}^*).$$

We then have

$$J_k(\mathbf{x}_k^{\mathsf{T}} \mathbf{b}^*) = H(\mathbf{x}_k^{\mathsf{T}} \mathbf{b}^*) F_k(\mathbf{x}_k^{\mathsf{T}} \lambda) = \exp[\mathbf{x}_k^{\mathsf{T}} (\mathbf{b}^* + \lambda)],$$

and therefore $J_k(u) = \exp(u)$ and $\gamma = \lambda + \mathbf{b}^*$. Using an exponential calibration function consists of using a nonresponse model of type

$$\Pr(k \in R|S) = \frac{1}{\exp(\mathbf{x}_k^{\mathsf{T}} \mathbf{b})}.$$

The calibration approach is quite advantageous as calibration is applied directly on population totals. It is not necessary to know the auxiliary variable values on every unit of the sample, as is necessary in the two-step approach, where we first estimate the response probability and then do calibration.

The direct calibration approach is recommended by Lundström & Särndal (1999), Särndal & Lundström (2005), and Folsom & Singh (2000). On the other hand, Haziza & Lesage (2016) and Haziza & Beaumont (2017) suggest using the two-step approach. To be able to use the direct approach, the auxiliary variables need to be correlated with the variable of interest and adequately model nonresponse.

Example 16.1 Poststratification to treat nonresponse. Suppose that $\gamma = (\gamma_1, \ldots, \gamma_j, \ldots, \gamma_J)^T$, $x_{kj} = 1$ if $k \in U_j$, and 0 otherwise. The U_j of size N_j define a partition of population U. It is a homogeneous nonresponse model in the post-strata. The calibration equations are written as

$$\sum_{k \in U_j} \frac{a_k r_k J(\gamma_j)}{\pi_k} = N_j, j = 1, \ldots, J.$$

By noting that

$$\widehat{N}_{Rj} = \sum_{k \in U_j} \frac{a_k r_k}{\pi_k} \quad \text{and} \quad \widehat{Y}_{Rj} = \sum_{k \in U_j} \frac{a_k r_k y_k}{\pi_k},$$

we obtain

$$J(\gamma_j) = \frac{N_j}{\widehat{N}_{Rj}}.$$

Estimation of the nonresponse probability is therefore

$$\widehat{\mathrm{Pr}}(k \in R|S) = \frac{\widehat{N}_{Rj}}{N_j}, k \in U_j.$$

Finally, the estimator is

$$\widehat{Y}_{\mathrm{NR}} = \sum_{j=1}^{J} \frac{N_j}{\widehat{N}_{Rj}} \widehat{Y}_{Rj}.$$

It is a post-stratified estimator which is used to correct nonresponse and sampling errors simultaneously.

16.5.4 Reweighting by Generalized Calibration

Generalized calibration, described in Section 12.5, page 256, can be used to treat questionnaire nonresponse by direct calibration (Estevao & Särndal, 2000; Deville, 1998a, 2000a, 2002, 2004; Kott, 2006; Chang & Kott, 2008; Kott, 2009; Kott & Chang, 2010). In this case, the calibration equation is written as

$$\sum_{k \in U} \frac{a_k r_k \mathbf{x}_k F_k(\lambda^T \mathbf{z}_k)}{\pi_k} = \sum_{k \in U} \mathbf{x}_k.$$

This calibration equation breaks the symmetry of the calibration since two sets of variables are used. The variables $\mathbf{z}_k$ must be known only on the respondents and are explanatory variables of nonresponse. The variables contained in $\mathbf{x}_k$ are calibration variables for which population totals must be known.

It is even possible to use one or more variables of interest y_k in vector $\mathbf{z}_k$. This method allows the treatment of nonignorable nonresponse that depends on the variable of interest. However, a misspecification of the model can considerably amplify the bias and variance. Lesage et al. (2018) describe precisely situation when this method can be dangerous.

16.6 Imputation

16.6.1 General Principles

Imputation is used for item nonresponse and consists of attaching a value to nonresponse (Rao, 1990; Rao & Sitter, 1995; Lee et al., 2002; Beaumont et al., 2007; Haziza, 2009; Kim & Rao, 2009). This value can be generated randomly or not. It can be chosen from among existing respondent values or be predicted using a model. Item nonresponse can be treated by reweighting, but then each variable would have a different weighting system, which makes the analysis of relations between variables difficult. Imputing variables affected by item nonresponse allows a complete file where the same weighting system is used for all variables to be obtained.

Imputation has an impact on the variance of estimators. A specific method must be used to estimate the variance of imputed data or risk badly estimating the variance. Certain imputation methods consist of imputing a central value, like a median or mean or a predicted value. The distribution of the imputed value is then affected by a reduction in its dispersion. Relations between variables may also be affected.

A good imputation technique must at least impute a coherent value to missing data. For example, we would hesitate to impute a status of "working" to someone over 70 years old. Imputation methods obviously introduce bias if the underlying hypothesis is false. Certain imputations can be done completely manually using logical deductions. All the methods above are only useful for ignorable nonresponse.

16.6.2 Imputing From an Existing Value

A first group of imputation methods is based on choosing a value to impute from existing values. This method is called this donor imputation. For donor imputation techniques, even if we do not describe explicitly the nonresponse mechanism, there always exists an implicit model. Here are two methods:

1. *Historical imputation.* This method consists of taking the value from a previous survey for this unit.
2. *Nearest-neighbor imputation.* A distance is defined thanks to auxiliary variables. We simply impute the value from the nearest neighbor.

16.6.3 Imputation by Prediction

A second type of method consists of predicting the missing value using a general model. Again, since nonresponse is supposed to be random and only dependent on auxiliary variables, the following general model can be used:

$$y_k = f(\mathbf{x}_k^{\mathsf{T}} \boldsymbol{\beta}) + \varepsilon_k, k \in U.$$

This model can be linear ($f(u) = u$) or, if y_k only takes the values 0 and 1, logistic ($f(u) = \exp(u)/(1 + \exp u)$), or probit ($f(u)$ is the cumulative distribution function of a standard normal distribution). The ε_k are residuals of mean zero and variance $v_k^2 \sigma^2$.

The imputed value is denoted by y_k^*. Once imputation is realized, any population parameter can be estimated. To estimate the total, one simply uses the following estimator:

$$\hat{Y}_{\mathrm{IMP}} = \sum_{k \in U} w_k a_k r_k y_k + \sum_{k \in U} w_k a_k (1 - r_k) y_k^*, \qquad (16.10)$$

where the w_k are weights that can be those of the expansion estimator $w_k = 1/\pi_k$.

We estimate the vector of regression coefficients with the units for which we know the value of y_k. This estimator is denoted by $\hat{\boldsymbol{\beta}}$. Then, the y_k are predicted using the model:

- *Imputation by prediction* For all missing values, we simply impute the value

$$y_k^* = f(\mathbf{x}_k^{\mathsf{T}} \hat{\boldsymbol{\beta}}), k \in S \backslash R.$$

As special cases, we have regression imputation ($f(u) = u$). For categorical variables, we use a probit or logit model and impute a proportion.
- *Ratio imputation* This method is just a special case of imputation by prediction. We impute the value $x_i \hat{Y}_R / \hat{X}_R$, where $\hat{Y}_R$ and $\hat{X}_R$ are estimators of totals of the variable of interest and of auxiliary variable x calculated on the respondents.
- *Mean imputation* This method is also a special case of imputation by prediction. The mean of the respondents is imputed to the missing values. This method is generally applied to relatively small and homogeneous categories.
- A typical nonparametric imputation technique consists of considering neighbouring responding units to the units that needs imputation. This method is often called *k nearest neighbors* or kNN. A mean or weighted mean of the k nearest neighbors can then be imputed.
- A variant of the previous method consists of randomly choosing a responding unit that is in the neighborhood of the unit to impute.

16.6.4 Link Between Regression Imputation and Reweighting

If we use regression imputation to estimate a total, we have

$$\hat{Y}_{\mathrm{IMP}} = \sum_{k \in U} \frac{a_k r_k y_k}{\pi_k} + \sum_{\ell \in U} \frac{a_\ell (1 - r_\ell) y_\ell^*}{\pi_\ell},$$

where $y_\ell^* = \mathbf{x}_\ell^{\mathsf{T}} \hat{\boldsymbol{\beta}}$. Generally, we can estimate $\boldsymbol{\beta}$ by

$$\hat{\boldsymbol{\beta}} = \hat{\mathbf{T}}^{-1} \sum_{k \in U} \frac{a_k r_k q_k \mathbf{x}_k y_k}{v_k^2 \pi_k} \quad \text{where} \quad \hat{\mathbf{T}} = \sum_{k \in U} \frac{a_k r_k q_k \mathbf{x}_k \mathbf{x}_k^{\mathsf{T}}}{v_k^2 \pi_k},$$

and q_k is a positive weight.

We can also write

$$\hat{Y}_{\mathrm{IMP}} = \sum_{k \in U} a_k r_k \left[1 + \sum_{\ell \in U} \frac{a_\ell (1 - r_\ell) \mathbf{x}_\ell^\top}{\pi_\ell} \hat{\mathbf{T}}^{-1} \frac{q_k \mathbf{x}_k}{v_k^2} \right] \frac{y_k}{\pi_k}.$$

The imputed regression estimator can therefore be written as a weighted estimator where the weights are given by

$$w_k = \frac{1}{\pi_k} \left[1 + \frac{q_k \mathbf{x}_k^\top}{v_k^2} \hat{\mathbf{T}}^{-1} \sum_{\ell \in U} \frac{a_\ell (1 - r_\ell) \mathbf{x}_\ell}{\pi_\ell} \right]. \tag{16.11}$$

In certain cases, the imputed regression estimator even allows a projective form (Section 11.3.2, page 228):

Result 16.1 *If there exists a vector λ such that $q_k \mathbf{x}_k^\top \lambda = v_k^2$ then*

$$\hat{Y}_{\mathrm{IMP}} = \sum_{\ell \in U} a_\ell \mathbf{x}_\ell^\top \hat{\beta}.$$

Proof: The weight given in Expression (16.11) is:

$$\begin{aligned}
w_k &= \frac{1}{\pi_k} \left[1 + \frac{q_k \mathbf{x}_k^\top}{v_k^2} \hat{\mathbf{T}}^{-1} \left(\sum_{\ell \in U} \frac{a_\ell \mathbf{x}_\ell}{\pi_\ell} - \sum_{\ell \in U} \frac{a_\ell r_\ell \mathbf{x}_\ell}{\pi_\ell} \right) \right] \\
&= \frac{1}{\pi_k} \left[1 + \frac{q_k \mathbf{x}_k^\top}{v_k^2} \hat{\mathbf{T}}^{-1} \left(\sum_{\ell \in U} \frac{a_\ell \mathbf{x}_\ell}{\pi_\ell} - \sum_{\ell \in U} \frac{a_\ell r_\ell \mathbf{x}_\ell}{\pi_\ell} \frac{q_\ell \mathbf{x}_\ell^\top \lambda}{v_\ell^2} \right) \right] \\
&= \frac{1}{\pi_k} \left[1 + \frac{q_k \mathbf{x}_k^\top}{v_k^2} \hat{\mathbf{T}}^{-1} \left(\sum_{\ell \in U} \frac{a_\ell \mathbf{x}_\ell}{\pi_\ell} - \hat{\mathbf{T}} \lambda \right) \right] \\
&= \frac{1}{\pi_k} \left(1 + \frac{q_k \mathbf{x}_k^\top}{v_k^2} \hat{\mathbf{T}}^{-1} \sum_{\ell \in U} a_\ell \mathbf{x}_\ell - \frac{q_k \mathbf{x}_k^\top}{v_k^2} \lambda \right) = \frac{q_k \mathbf{x}_k^\top}{\pi_k v_k^2} \hat{\mathbf{T}}^{-1} \sum_{\ell \in U} a_\ell \mathbf{x}_\ell.
\end{aligned}$$

Therefore

$$\hat{Y}_{\mathrm{IMP}} = \sum_{k \in U} w_k a_k r_k y_k = \sum_{k \in U} \frac{q_k \mathbf{x}_k^\top}{\pi_k v_k^2} \hat{\mathbf{T}}^{-1} \sum_{\ell \in U} a_\ell \mathbf{x}_\ell a_k r_k y_k = \sum_{\ell \in U} a_\ell \mathbf{x}_\ell^\top \hat{\beta}. \qquad \blacksquare$$

Example 16.2 Suppose we have a single auxiliary variable $\mathbf{x}_k = 1$. Define also

$$\hat{Y}_R = \sum_{k \in U} \frac{a_k r_k y_k}{\pi_k}, \hat{N} = \sum_{k \in S} \frac{1}{\pi_k} \quad \text{and} \quad \hat{N}_R = \sum_{k \in U} \frac{a_k r_k}{\pi_k}.$$

In this case, if $v_k = 1$,

$$\hat{\mathbf{T}} = \sum_{k \in U} \frac{a_k r_k}{\pi_k}, \hat{\beta} = \frac{\sum_{k \in U} \frac{a_k r_k y_k}{\pi_k}}{\sum_{k \in U} \frac{a_k r_k}{\pi_k}} = \frac{\hat{Y}_R}{\hat{N}_R} \quad \text{and} \quad \mathbf{x}_k \hat{\beta} = \frac{\hat{Y}_R}{\hat{N}_R}.$$

The imputed estimator of the total is therefore

$$\hat{Y}_{IMP} = \sum_{k\in U}\frac{a_k r_k y_k}{\pi_k} + \sum_{\ell\in U}a_\ell(1 - r_\ell)\frac{y_\ell^*}{\pi_\ell}$$

$$= \sum_{k\in U}\frac{a_k r_k y_k}{\pi_k} + \frac{\hat{Y}_R}{\hat{N}_R}\sum_{\ell\in U}a_\ell(1 - r_\ell)\frac{1}{\pi_\ell}$$

$$= \hat{Y}_R + \frac{\hat{Y}_R}{\hat{N}_R}(\hat{N} - \hat{N}_R) = \frac{\hat{N}\hat{Y}_R}{\hat{N}_R}.$$

In this case, using an imputed estimator is the same as using a ratio estimator.

Methods based on regression imputation include mean imputation, mean imputation within strata, or ratio imputation. This methods all have an important drawback: they bring about false inferences on quantiles and on parameters of dispersion, form, or asymmetry. This drawback results from the fact that a predicted central value is imputed. This method is therefore only valid to estimate totals or means.

16.6.5 Random Imputation

In order to not affect the distribution of variable y, it is recommended that a random imputation method is used. The main methods are as follows:

- *Imputation by random prediction*
  - If the variable is quantitative, we impute a prediction to which we add a residual

$$y_k^* = f(\mathbf{x}_k^\top\hat{\beta}) + v_k\dot{\varepsilon}_k, k \in S\backslash R,$$

where the $\dot{\varepsilon}_k$ are also imputed values that we can either randomly generate or randomly choose among the estimated residuals on R:

$$\hat{\varepsilon}_k = \frac{y_k - f(\mathbf{x}_k^\top\hat{\beta})}{v_k}, k \in R.$$

The issue of incorrect estimation of quantiles and of parameters of dispersion, form or asymmetry is solved. However, another drawback appears: the variance of estimators increases. This problem comes from the fact that

$$\sum_{k\in U}a_k(1 - r_k)\dot{\varepsilon}_k \tag{16.12}$$

is usually different to 0. By cleverly selecting the residuals, we can avoid uselessly increasing the variance of estimators of totals. Chauvet et al. (2011) proposed selecting residuals for imputation using a balanced design to not increase the variance of estimators of totals. The solution consists of choosing imputed residuals $\dot{\varepsilon}_k \in S\backslash R$, among the estimated residuals $\varepsilon_k \in R$, in such a way that (16.12) is zero, which can be done using balanced sampling (see Chapter 6, page 119) of $\#(S\backslash R)$ estimated residuals in R. The cancelation of Expression (16.12) then represents the first balancing equation, the second being the fixed size of the sample.

- If the variable y is an indicator variable (for a qualitative variable), then $f(\mathbf{x}_k^\top \widehat{\boldsymbol{\beta}})$ represents an estimation of the probability that y_k equals 1:

$$f(\mathbf{x}_k^\top \widehat{\boldsymbol{\beta}}) = \widehat{\Pr}(y_k = 1) = 1 - \widehat{\Pr}(y_k = 0).$$

We then only impute values 1 or 0. We do the imputation according to

$$y_k^* = \begin{cases} 1 & \text{with probability } f(\mathbf{x}_k^\top \widehat{\boldsymbol{\beta}}) \\ 0 & \text{with probability } 1 - f(\mathbf{x}_k^\top \widehat{\boldsymbol{\beta}}). \end{cases}$$

The classic technique consists of applying a Poisson sampling design. This method has the drawback of increasing the variance of the estimator. This problem can be resolved by sampling such that

$$\sum_{k \in S \setminus R} y_k^* = \sum_{k \in S \setminus R} f(\mathbf{x}_k^\top \widehat{\boldsymbol{\beta}}).$$

This constraint can be simply satisfied by sampling with unequal probabilities $f(\mathbf{x}_k^\top \widehat{\boldsymbol{\beta}})$ with fixed size equal to

$$\sum_{k \in S \setminus R} f(\mathbf{x}_k^\top \widehat{\boldsymbol{\beta}})$$

in $S \setminus R$, the selected units have value 1 and the others value 0.

- *Hot deck.* This technique simply consists of randomly choosing a unit in the sample and assigning the value of this unit to the unit that did not respond. The *hot deck* is based on the implicit independence hypothesis between nonresponse and the variable marred with nonresponse.
- *Hot deck in homogeneous strata.* A simple method that consists of constructing small homogeneous strata and then randomly selecting a unit in a homogeneous stratum to impute a unit belonging to the same stratum.
- *Random nearest-neighbor imputation* The kNN method can also be applied randomly (*random kNN or RkNN*). In this case, we randomly select a unit among the k nearest neighbors. This method has the benefit of imputing an existing value. Hasler & Tillé (2016) have proposed selecting donors with a balanced design to control the increasing of variance.

In practice, it is often difficult to find variables which are never affected by nonresponse. Nonresponse can potentially affect any variable. Certain authors have called *Swiss cheese nonresponse* a configuration where all variables can be affected by nonresponse. These authors probably falsely believed all Swiss cheeses have holes, which is obviously not the case.

Raghunathan et al. (2001, 2002, 2016) proposed the program IVEWARE (Imputation and Variance Estimation Software) to treat Swiss cheese nonresponse. The method first consists of doing a relatively simply imputation starting with the variable with the least nonresponse. Once a first imputation is realized, each variable is imputed again using all other variables as regressors. This operation is repeated until a stable configuration is obtained.

Tillé & Vallée (2017a) proposed a donor imputation method in the cases of Swiss nonresponse by imposing coherence constraints on the donor in such a way that totals are not affected if the donors are also used to impute the nonmissing variables of the imputed units.

16.7 Variance Estimation with Nonresponse

16.7.1 General Principles

The calculation and estimation of the variance of estimators where nonresponse has been mitigated by either reweighting or imputation is certainly one of the more complex questions in the theory of survey sampling. The variance of estimators can depend on four sources of randomness.

1. The random sample S or the vector $\mathbf{a} = (a_1, \dots, a_N)^\top$ of the sample membership indicators is always a source of randomness. In the case where the estimator is not linear in the a_k, we can linearize the estimator as proposed in Section 15.10. Let $E_p(.)$ and $var_p(.)$ denote the expectation and the variance under the sample design.
2. The random set of respondents R that is generated by the nonresponse mechanism is the second source of randomness. Let $\mathbf{r} = (r_1, \dots, r_N)^\top$ denote the vector of response indicators with $E_q(r_k) = \phi_k$. Let $E_q(.)$ and $var_q(.)$ denote the expectation and variance under the nonresponse mechanism.
3. A superpopulation model can also be used to base inference in survey theory. As presented in Section 13, the model is generally linear: $y_k = \mathbf{x}_k^\top \beta + \varepsilon_k$. Let $E_M(.)$ and $var_M(.)$ denote the expectation and the variance under the model.
4. In the case of random imputation, imputation itself can add randomness to the estimator. Let $E_I(.)$ and $var_I(.)$ denote the expectation and the variance under the nonresponse mechanism.

In the following, we will only take into account the first three sources of randomness. To estimate the variance, two approaches can be used: the approach based on the nonresponse model and the approach based on the imputation model.

In the approach based on the nonresponse model, the vector of indicators $\mathbf{a}$ and the vector of response indicators $\mathbf{r}$ are assumed to be the only sources of randomness. The y_k are assumed to not be random. If the response design is Poisson, the r_k are independent. The response probabilities ϕ_k are generally unknown and estimated by $\hat{\phi}_k$. Suppose we have an estimator $\hat{\theta}_{NR}$ of a parameter θ. The variance of $\hat{\theta}_{NR}$ can be calculated by means of the classic approach

$$var(\hat{\theta}_{NR}) = E_p var_q(\hat{\theta}_{NR}) + var_p E_q(\hat{\theta}_{NR}),$$

or the variance by the reverse approach (see Section 7.6)

$$var(\hat{\theta}_{NR}) = E_q var_p(\hat{\theta}_{NR}) + var_q E_p(\hat{\theta}_{NR}).$$

In the approach based on the imputation model, the vectors $\mathbf{a}$, $\mathbf{r}$ as well as the $\mathbf{y} = (y_1, \dots, y_N)$ are random. We simply assume that the nonresponse mechanism is ignorable, in the sense that it does not depend on the variable of interest conditionally on the auxiliary variables. For example, if we have a linear model $M : y_k = \mathbf{x}_k \beta + \varepsilon_k$, the randomness of the y_k comes from the error terms ε_k. Ignorability means that the ε_k are independent of the r_k. The two-phase variance is the following:

$$var(\hat{\theta}_{NR} - \theta) = E_M E_p E_q (\hat{\theta}_{NR} - \theta)^2 = E_M E_p E_q (\hat{\theta}_{NR} - \hat{\theta} + \hat{\theta} - \theta)^2$$
$$= E_M E_p var_q [(\hat{\theta}_{NR} - \hat{\theta}) + (\hat{\theta} - \theta)] + E_M var_p E_q [(\hat{\theta}_{NR} - \hat{\theta}) + (\hat{\theta} - \theta)]$$
$$+ E_p E_q var_M [(\hat{\theta}_{NR} - \hat{\theta}) + (\hat{\theta} - \theta)]$$

$$= E_M E_p E_q(\widehat{\theta} - \theta)^2 + E_M E_p E_q(\widehat{\theta}_{NR} - \widehat{\theta})^2 + 2E_M E_p E_q(\widehat{\theta}_{NR} - \widehat{\theta})(\widehat{\theta} - \theta)$$
$$= E_M \mathrm{var}_p(\widehat{\theta}) + E_p E_q \mathrm{var}_M(\widehat{\theta}_{NR} - \widehat{\theta}) + 2E_p E_q E_M[(\widehat{\theta}_{NR} - \widehat{\theta})(\widehat{\theta} - \theta)],$$

where $\widehat{\theta}$ is the estimator of θ when there is no nonresponse. The order of the expectations can be switched since the nonresponse mechanism is ignorable. If $\tilde{\theta}_{NR} = E_p(\widehat{\theta}_{NR})$, the variance of the reverse approach is:

$$\mathrm{var}(\widehat{\theta}_{NR} - \theta) = E_q E_M E_p(\widehat{\theta}_{NR} - \tilde{\theta}_{NR} + \tilde{\theta}_{NR} - \theta)^2$$
$$= E_q E_M \mathrm{var}_p(\widehat{\theta}_{NR} - \tilde{\theta}_{NR}) + E_q \mathrm{var}_M E_p(\tilde{\theta}_{NR} - \theta)$$
$$+ 2E_q E_M E_p[(\widehat{\theta}_{NR} - \tilde{\theta}_{NR})(\tilde{\theta}_{NR} - \theta)]$$
$$= E_q E_M \mathrm{var}_p(\widehat{\theta}_{NR}) + E_q \mathrm{var}_M E_p(\widehat{\theta}_{NR} - \theta).$$

In nearly all cases, it is necessary to proceed with a linearization even for estimators of totals. Since the estimator can depend on three sources of randomness r_k, a_k and ε_k, it is possible to linearize with respect to its different sources of randomness to obtain an approximation of the variance and an estimator according to the methodology of Section 15.10. In the following, we examine a few special cases from the very abundant literature.

For example, Kott (2006) linearizes with respect to weights. Kim & Kim (2007) and Kim & Rao (2009) use estimation equations with respect to nonresponse and imputation models. The two-phase approach for nonresponse was developed bySärndal & Swensson (1987) in the framework of the approach based on the nonresponse mechanism. It has been applied for imputation bySärndal (1992) in the framework of the approach based on the imputation model.

The reverse approach was proposed by Fay (1991). This approach is possible if the response design is independent of the sampling design, which Beaumont & Haziza (2016) call the strong invariance property (see also Tillé & Vallée, 2017b). Shao & Steel (1999) use the reverse approach for the two inferences.

16.7.2 Estimation by Direct Calibration

The simplest case has been studied by Kott (2006). Direct calibration can be used to simultaneously correct questionnaire nonresponse error and sampling error (see Section 16.5.3). In this case, the weights are calculated using the calibration equation

$$\sum_{k \in U} a_k r_k w_k \mathbf{x}_k = \sum_{k \in U} \mathbf{x}_k,$$

where $w_k = J_k(\mathbf{x}_k^\top \lambda)/\pi_k$. The estimator of the total of Y is

$$\widehat{Y}_R = \sum_{k \in U} a_k r_k w_k y_k.$$

This case is easy to treat, as the sample membership indicators a_k and response indicators r_k always appear as a product $\delta_k = a_k r_k$. In Kott (2006), the nonresponse mechanism follows a Poisson design and the estimated response probabilities are $\widehat{\phi}_k = 1/J_k(\mathbf{x}_k^\top \lambda)$.

The calibration equations can be written as

$$\sum_{k \in U} \delta_k w_k \mathbf{x}_k = \sum_{k \in U} \mathbf{x}_k. \tag{16.13}$$

The estimator is

$$\widehat{Y}_R = \sum_{k \in U} \delta_k w_k y_k.$$

The linearization is given by the following result.

Result 16.2 *The linearization of $\widehat{Y}_R$ with respect to $\delta_\ell = a_\ell r_\ell$ is $w_\ell e_\ell$, where $e_\ell = y_\ell - \mathbf{x}_\ell^\top \widehat{\mathbf{B}}_{y|x}$,*

$$\widehat{\mathbf{B}}_{y|x} = \widehat{\mathbf{T}}^{-1} \sum_{k \in U} \frac{\delta_k J_k'(\mathbf{x}_k^\top \lambda) \mathbf{x}_k y_k}{\pi_k} \quad and \quad \widehat{\mathbf{T}} = \sum_{k \in U} \frac{\delta_k J_k'(\mathbf{x}_k^\top \lambda) \mathbf{x}_k \mathbf{x}_k^\top}{\pi_k}.$$

Proof: By canceling the partial derivatives of (16.13) with respect to δ_ℓ, we have

$$\frac{J_\ell(\mathbf{x}_\ell^\top \lambda) \mathbf{x}_\ell}{\pi_\ell} + \sum_{k \in U} \frac{\delta_k J_k'(\mathbf{x}_k^\top \lambda) \mathbf{x}_k \mathbf{x}_k^\top}{\pi_k} \frac{\partial \lambda}{\partial \delta_\ell} = 0.$$

We then obtain $\partial \lambda / \partial \delta_\ell = -\widehat{\mathbf{T}}_{xx}^{-1} w_\ell \mathbf{x}_\ell$. The linearization of $\widehat{Y}_R$ is then

$$z_\ell = \frac{\partial \widehat{Y}_R}{\partial \delta_\ell} = w_\ell y_\ell - w_\ell \mathbf{x}_\ell^\top \widehat{\mathbf{T}}_{xx}^{-1} \sum_{k \in U} \frac{\delta_k J_k(\mathbf{x}_k^\top \lambda) \mathbf{x}_k y_k}{\pi_k} = w_\ell e_\ell.$$ ∎

To estimate the variance, it is then enough to replace $y_k^2/(\pi_{ak}\pi_{bk})$ by the linearization $z_k = w_k e_k$ in the expression of the two-phase variance estimator given in Expression (7.9), page 163.

16.7.3 General Case

In the general case, the sample indicators a_k and the response indicators r_k only appear as a product $a_k r_k$. Each term in the variance must be linearized separately for each indicator.

Consider an estimator $\widehat{\theta}_{NR}$. In the reverse approach,

$$\mathrm{var}(\widehat{\theta}_{NR}) = +E_q \mathrm{var}_p(\widehat{\theta}_{NR}) + \mathrm{var}_q E_p(\widehat{\theta}_{NR}). \tag{16.14}$$

To estimate the variance, we proceed in the following way:

- linearization of $\widehat{\theta}_{NR}$ with respect to a_k, which allows an estimator of $\mathrm{var}_p(\widehat{\theta}_{NR})$ to be obtained and therefore an estimator of $E\,\mathrm{var}_p(\widehat{\theta}_{NR})$
- computation of $\breve{\theta}_{NR} \approx E_p(\widehat{\theta}_{NR})$ that is linearized with respect to r_k. An estimator of $\mathrm{var}_q E_p(\widehat{\theta}_{NR})$ can then be calculated.

This linearization in two steps allows the variance to be calculated. This method can also be adapted to variance estimation by taking into account the model (Vallée & Tillé, 2019). In the following, we give a few examples by limiting ourselves to the reverse approach under the nonresponse mechanism, which is the simplest way to obtain a variance.

16.7.4 Variance for Maximum Likelihood Estimation

First, consider the case in Section 16.5.2 where the nonresponse probability is estimated using the maximum likelihood by solving the equations

$$A = \sum_{k \in U} a_k r_k \mathbf{x}_k = \sum_{k \in U} a_k \mathbf{x}_k h(\mathbf{x}_k^\top \mathbf{b}), \tag{16.15}$$

where $h(.)$ is a logistic function. The estimator is then

$$\hat{Y}_{NR} = \sum_{k \in U} \frac{a_k r_k y_k}{h(\mathbf{x}_k^\top \mathbf{b}) \pi_k}. \tag{16.16}$$

Start by linearizing $\hat{Y}_{NR}$ with respect to a_ℓ.

Result 16.3 *If* $\mathbf{b}$ *is estimated by the maximum likelihood method and the nonresponse function is logistic, then*

$$\frac{\hat{Y}_{NR}}{\partial a_\ell} = r_\ell \left[\frac{y_\ell}{\pi_\ell h(\mathbf{x}_\ell^\top \mathbf{b})} - \mathbf{x}_\ell^\top \boldsymbol{\gamma}_1 \right] + h(\mathbf{x}_\ell^\top \mathbf{b}) \mathbf{x}_\ell^\top \boldsymbol{\gamma}_1,$$

where $\boldsymbol{\gamma}_1 = \mathbf{T}_1^{-1} \mathbf{t}_1$,

$$\mathbf{T}_1 = \sum_{k \in U} a_k \mathbf{x}_k \mathbf{x}_k^\top h'(\mathbf{x}_k^\top \mathbf{b}) \quad and \quad \mathbf{t}_1 = \sum_{k \in U} \frac{a_k r_k y_k \mathbf{x}_k h'(\mathbf{x}_k^\top \mathbf{b})}{\pi_k h^2(\mathbf{x}_k^\top \mathbf{b})}.$$

Proof: We differentiate Equation (16.15) with respect to a_ℓ and we obtain

$$\frac{\partial A}{\partial a_\ell} = r_\ell \mathbf{x}_\ell = \mathbf{x}_\ell h(\mathbf{x}_\ell^\top \mathbf{b}) + \sum_{k \in U} a_k \mathbf{x}_k \mathbf{x}_k^\top h'(\mathbf{x}_k^\top \mathbf{b}) \frac{\partial \mathbf{b}}{\partial a_\ell}.$$

Therefore,

$$\frac{\partial \mathbf{b}}{\partial a_\ell} = \mathbf{T}_1^{-1} \mathbf{x}_\ell [r_\ell - h(\mathbf{x}_\ell^\top \mathbf{b})].$$

We then differentiate the estimator given in Expression (16.16):

$$\frac{\hat{Y}_{NR}}{\partial a_\ell} = \frac{r_\ell y_\ell}{\pi_\ell h(\mathbf{x}_\ell^\top \mathbf{b})} - \sum_{k \in U} \frac{a_k r_k y_k h'(\mathbf{x}_k^\top \mathbf{b})}{\pi_k h^2(\mathbf{x}_k^\top \mathbf{b})} \mathbf{x}_k^\top \frac{\partial \mathbf{b}}{\partial a_\ell}$$

$$= \frac{r_\ell y_\ell}{\pi_\ell h(\mathbf{x}_\ell^\top \mathbf{b})} - [r_\ell - h(\mathbf{x}_\ell^\top \mathbf{b})] \mathbf{x}_\ell^\top \mathbf{T}_1^{-1} \mathbf{t}. \qquad \blacksquare$$

Result 16.3 allows us to write

$$\hat{Y}_{NR} \approx \sum_{k \in U} \frac{y_k}{h(\mathbf{x}_k^\top \mathbf{b})} + \sum_{\ell \in U} v_{1\ell} (a_\ell - \pi_\ell),$$

where

$$v_{1\ell} = r_\ell \left[\frac{y_\ell}{\pi_\ell h(\mathbf{x}_\ell^\top \mathbf{b})} - \mathbf{x}_\ell^\top \boldsymbol{\gamma}_1 \right] + h(\mathbf{x}_\ell^\top \mathbf{b}) \mathbf{x}_\ell^\top \boldsymbol{\gamma}_1,$$

which allows the first term of the variance given in Expression (16.14) to be estimated:

$$\widehat{\mathrm{var}}_p(\hat{Y}_{\mathrm{NR}}) = \sum_{k \in U} \sum_{\ell \in U} \frac{a_k a_\ell v_{1k} v_{1\ell} \Delta_{k\ell}}{\pi_{k\ell}}. \tag{16.17}$$

To calculate the second term of the variance given in Expression (16.14), we suppose the nonresponse mechanism is well specified. Therefore

$$E_p(\hat{Y}_{\mathrm{NR}}) \approx \check{Y} = \sum_{k \in U} \frac{r_k y_k}{h(\mathbf{x}_k^\mathsf{T} \mathbf{b}^*)},$$

where $\mathbf{b}^*$ is obtained by solving the approximate expected value under the design of Equation (16.15):

$$\sum_{k \in U} r_k \mathbf{x}_k \pi_k = \sum_{k \in U} \mathbf{x}_k \pi_k h(\mathbf{x}_k^\mathsf{T} \mathbf{b}^*). \tag{16.18}$$

Result 16.4 *The linearization of $\check{Y}$ with respect to r_ℓ is*

$$\frac{\partial \check{Y}}{\partial r_\ell} = \frac{y_\ell}{h(\mathbf{x}_\ell^\mathsf{T} \mathbf{b}^*)} - \pi_\ell \mathbf{x}_\ell^\mathsf{T} \gamma_2,$$

where

$$\gamma_2 = \mathbf{T}_2^{-1} \mathbf{t}_2, \mathbf{T}_2 = \sum_{k \in U} \mathbf{x}_k \mathbf{x}_k^\mathsf{T} \pi_k h'(\mathbf{x}_k^\mathsf{T} \mathbf{b}^*) \quad and \quad \mathbf{t}_2 = \sum_{k \in U} \frac{r_k \mathbf{x}_k y_k h'(\mathbf{x}_k^\mathsf{T} \mathbf{b}^*)}{h^2(\mathbf{x}_k^\mathsf{T} \mathbf{b}^*)}.$$

Proof: By differentiating Expression (16.18) with respect to r_ℓ, we have

$$\mathbf{x}_\ell \pi_\ell = \sum_{k \in U} \mathbf{x}_k \mathbf{x}_k^\mathsf{T} \pi_k h'(\mathbf{x}_k^\mathsf{T} \mathbf{b}^*) \frac{\partial \mathbf{b}^*}{\partial r_\ell}.$$

Therefore,

$$\frac{\partial \mathbf{b}^*}{\partial r_\ell} = \mathbf{T}_2^{-1} \mathbf{x}_\ell \pi_\ell.$$

The linearization is therefore

$$\frac{\partial \check{Y}}{\partial r_\ell} = \frac{y_\ell}{h(\mathbf{x}_\ell^\mathsf{T} \mathbf{b}^*)} - \sum_{k \in U} \frac{r_k y_k \mathbf{x}_k^\mathsf{T} h'(\mathbf{x}_k^\mathsf{T} \mathbf{b}^*)}{h^2(\mathbf{x}_k^\mathsf{T} \mathbf{b}^*)} \frac{\partial \mathbf{b}^*}{\partial r_\ell} = \frac{y_\ell}{h(\mathbf{x}_\ell^\mathsf{T} \mathbf{b}^*)} - \pi_\ell \mathbf{x}_\ell^\mathsf{T} \mathbf{T}_2^{-1} \mathbf{t}_2. \qquad \blacksquare$$

We can then write

$$\check{Y} \approx Y + \sum_{\ell \in U} v_{2k}(r_k - \phi_k),$$

where

$$v_{2\ell} = \frac{y_\ell}{h(\mathbf{x}_\ell^\mathsf{T} \mathbf{b}^*)} - \pi_\ell \mathbf{x}_\ell^\mathsf{T} \gamma_2.$$

Since the nonresponse mechanism follows a Poisson design, we can approximate the variance by

$$\mathrm{var}_q(\check{Y}) \approx \sum_{\ell \in U} v_{2\ell}^2 \phi_\ell (1 - \phi_\ell).$$

This variance is then estimated by

$$\widehat{\text{var}}_q(\check{Y}) \approx \sum_{\ell \in U} \frac{r_\ell a_\ell \hat{v}_{2\ell}^2 \hat{\phi}_\ell (1 - \hat{\phi}_\ell)}{\hat{\phi}_\ell \pi_\ell}, \tag{16.19}$$

where $\hat{v}_{2\ell}$ is an estimation of $v_{2\ell}$ by

$$\hat{v}_{2\ell} = \frac{y_\ell}{\hat{\phi}_\ell} - \pi_\ell \mathbf{x}_\ell^\mathsf{T} \boldsymbol{\gamma}_1.$$

In fact, $\boldsymbol{\gamma}_1$ can be used to estimate $\boldsymbol{\gamma}_2$.

By grouping the terms (16.17) and (16.19), we finally obtain a variance estimator

$$v(\hat{Y}_{\text{NR}}) = \sum_{k \in U} \sum_{\ell \in U} \frac{a_k a_\ell \hat{v}_{1k} \hat{v}_{1\ell} \Delta_{k\ell}}{\pi_{k\ell}} + \sum_{\ell \in U} \frac{r_\ell a_\ell (y_\ell - \hat{\phi}_\ell \pi_\ell \mathbf{x}_\ell^\mathsf{T} \boldsymbol{\gamma}_1)^2 (1 - \hat{\phi}_\ell)}{\hat{\phi}_\ell^2 \pi_\ell},$$

where

$$\hat{v}_{1\ell} = \frac{r_\ell}{\pi_\ell \hat{\phi}_\ell} (y_\ell - \pi_\ell \hat{\phi}_\ell \mathbf{x}_\ell^\mathsf{T} \boldsymbol{\gamma}_1) + \hat{\phi}_\ell \mathbf{x}_\ell^\mathsf{T} \boldsymbol{\gamma}_1.$$

Example 16.3 Suppose that a sample is selected by simple random sampling without replacement and that nonresponse is missing completely at random. For the nonresponse model $\mathbf{x}_k = 1, k \in U$. In this case, $\pi_k = n/N$. If $m = \#(R \cap S)$, then $h(\mathbf{x}_k^\mathsf{T} \mathbf{b}) = \hat{\phi}_k = m/n$,

$$h'(\mathbf{x}_k^\mathsf{T} \mathbf{b}) = \frac{m(n-m)}{n^2}, \mathbf{T}_1 = n \frac{m(n-m)}{n^2},$$

$$\mathbf{t}_1 = \frac{Nn^2}{nm^2} \frac{m(n-m)}{n^2} \sum_{k \in U} a_k r_k y_k,$$

$$\boldsymbol{\gamma}_1 = \frac{N}{m^2} \sum_{k \in U} a_k r_k y_k = \frac{N}{m} \hat{\bar{Y}}_R,$$

where

$$\hat{\bar{Y}}_R = \frac{1}{m} \sum_{k \in U} a_k r_k y_k,$$

$$v_{1\ell} = \frac{r_\ell N}{m} (y_\ell - \hat{\bar{Y}}_R) + \frac{N}{n} \hat{\bar{Y}}_R,$$

and

$$\hat{v}_{2\ell} = \frac{n}{m} (y_\ell - \hat{\bar{Y}}_R).$$

We see that

$$\hat{\bar{v}}_1 = \frac{1}{n} \sum_{k \in U} a_k v_{1k} = \frac{N}{n} \hat{\bar{Y}}.$$

Therefore

$$v_{1\ell} - \hat{\bar{v}}_1 = \frac{r_\ell N}{m} (y_\ell - \hat{\bar{Y}}_R).$$

By taking into account that we have simple random sampling, the first term comes directly:

$$v(\hat{Y}_{NR}) = N^2 \frac{N-n}{Nn} \frac{1}{n-1} \sum_{k\in U} a_k \left[\frac{n}{N}(v_{1k} - \hat{\bar{v}}_1) \right]^2$$

$$+ \sum_{\ell\in U} \frac{r_\ell a_\ell \frac{n^2}{m^2}(y_\ell - \hat{\bar{Y}}_R)^2 \frac{m}{n}\left(1 - \frac{m}{n}\right)}{\frac{m}{n}\frac{n}{N}}$$

$$= N^2 \frac{N-n}{N} \frac{1}{n-1} \frac{n}{m^2} \sum_{k\in U} r_k a_k (y_k - \hat{\bar{Y}}_R)^2$$

$$+ \frac{N^2}{m^2} \frac{n-m}{N} \sum_{\ell\in U} r_\ell a_\ell (y_\ell - \hat{\bar{Y}}_R)^2$$

$$= N^2 \frac{N-m}{Nm} \frac{1}{m-1} \sum_{k\in U} r_k a_k (y_k - \hat{\bar{Y}}_R)^2$$

$$- \frac{(n-m)(N-1)N}{(m-1)m^2(n-1)} \sum_{k\in U} r_k a_k (y_k - \hat{\bar{Y}}_R)^2$$

$$= \left[1 - \frac{(n-m)(N-1)}{m(n-1)(N-m)} \right] N^2 \frac{N-m}{Nm} \frac{1}{m-1} \sum_{k\in U} r_k a_k (y_k - \hat{\bar{Y}}_R)^2.$$

With a negligible difference, we obtain the variance of a simple random sampling design of size m in a population of size N.

16.7.5 Variance for Estimation by Calibration

We can also consider the case covered in Section 16.5.2 where the probability of nonresponse is estimated using calibration by solving the equations

$$A = \sum_{k\in U} \frac{a_k r_k \mathbf{x}_k}{h(\mathbf{x}_k^\top \mathbf{b})} = \sum_{k\in U} a_k \mathbf{x}_k, \tag{16.20}$$

where $h(.)$ is a logistic function. The estimator is then

$$\hat{Y}_{NR} = \sum_{k\in U} \frac{a_k r_k y_k}{h(\mathbf{x}_k^\top \mathbf{b})\pi_k}. \tag{16.21}$$

We start by linearizing $\hat{Y}_{NR}$ with respect to a_ℓ.

Result 16.5 *If $\mathbf{b}$ is estimated by calibration and the nonresponse function if logistic, then*

$$\frac{\hat{Y}_{NR}}{\partial a_\ell} = r_\ell \frac{y_\ell - \mathbf{x}_\ell^\top \boldsymbol{\gamma}_3}{\pi_\ell h(\mathbf{x}_\ell \mathbf{b})} + \frac{\mathbf{x}_\ell^\top \boldsymbol{\gamma}_3}{\pi_\ell},$$

where $\boldsymbol{\gamma}_3 = \mathbf{T}_3^{-1}\mathbf{t}_3$,

$$\mathbf{T}_3 = \sum_{k\in U} \frac{a_k r_k \mathbf{x}_k \mathbf{x}_k^\top h'(\mathbf{x}_k^\top \mathbf{b})}{\pi_k h^2(\mathbf{x}_k^\top \mathbf{b})} \quad and \quad \mathbf{t}_3 = \sum_{k\in U} \frac{a_k r_k y_k \mathbf{x}_k h'(\mathbf{x}_k^\top \mathbf{b})}{\pi_k h^2(\mathbf{x}_k^\top \mathbf{b})}.$$

Proof: By differentiating the calibration equation given in Expression (16.20), we obtain

$$\frac{\partial A}{\partial a_\ell} = \frac{r_\ell \mathbf{x}_\ell}{\pi_\ell h(\mathbf{x}_\ell^\top \mathbf{b})} - \sum_{k \in U} \frac{a_k r_k \mathbf{x}_k \mathbf{x}_k^\top h'(\mathbf{x}_k^\top \mathbf{b})}{\pi_k h^2(\mathbf{x}_k^\top \mathbf{b})} \frac{\partial \mathbf{b}}{\partial a_\ell} = \frac{\mathbf{x}_\ell}{\pi_\ell}.$$

Therefore

$$\frac{\partial \mathbf{b}}{\partial a_\ell} = \mathbf{T}_3^{-1} \frac{\mathbf{x}_\ell}{\pi_\ell} \left[\frac{r_\ell}{h(\mathbf{x}_\ell^\top \mathbf{b})} - 1 \right].$$

By differentiating the estimator given in Expression (16.21), we obtain

$$\frac{\partial \hat{Y}_{\mathrm{NR}}}{\partial a_\ell} = \frac{r_\ell y_\ell}{\pi_\ell h(\mathbf{x}_\ell^\top \mathbf{b})} - \sum_{k \in U} \frac{a_k r_k y_k h'(\mathbf{x}_k^\top \mathbf{b})}{\pi_k h^2(\mathbf{x}_k^\top \mathbf{b})} \mathbf{x}_k^\top \frac{\partial \mathbf{b}}{\partial a_\ell}$$

$$= \frac{r_\ell y_\ell}{\pi_\ell h(\mathbf{x}_\ell^\top \mathbf{b})} - \left[\frac{r_\ell}{h(\mathbf{x}_\ell^\top \mathbf{b})} - 1 \right] \frac{\mathbf{x}_\ell}{\pi_\ell}^\top \mathbf{T}_3^{-1} \mathbf{t}.$$ ∎

Let $\mathbf{b}^*$ be the solution of the system

$$\sum_{k \in U} \frac{r_k \mathbf{x}_k}{h(\mathbf{x}_k^\top \mathbf{b})} = \sum_{k \in U} \mathbf{x}_k,$$

and

$$\check{Y}_{\mathrm{NR}} = \sum_{k \in U} \frac{r_k y_k}{h(\mathbf{x}_k^\top \mathbf{b})}.$$

Then, we can write

$$\hat{Y}_{\mathrm{NR}} - \check{Y}_{\mathrm{NR}} \approx \sum_{\ell \in U} \left[r_\ell \frac{y_\ell - \mathbf{x}_\ell \boldsymbol{\gamma}_3}{\pi_\ell h(\mathbf{x}_\ell^\top \mathbf{b})} + \frac{\mathbf{x}_\ell^\top \boldsymbol{\gamma}_3}{\pi_\ell} \right] (a_\ell - \pi_\ell).$$

Therefore,

$$E_p(\hat{Y}_{\mathrm{NR}} | \mathbf{r}) \approx \check{Y}_{\mathrm{NR}}$$

and

$$\mathrm{var}_p(\hat{Y}_{\mathrm{NR}} | \mathbf{r}) = \sum_{k \in U} \sum_{\ell \in U} z_k z_\ell \Delta_{k\ell}, \qquad (16.22)$$

where

$$z_\ell = r_\ell \frac{y_\ell - \mathbf{x}_\ell \boldsymbol{\gamma}_3}{\pi_\ell h(\mathbf{x}_\ell^\top \mathbf{b})} + \frac{\mathbf{x}_\ell^\top}{\pi_\ell} \boldsymbol{\gamma}_3.$$

To calculate the second term of the variance given in Expression (16.14), we suppose that the nonresponse mechanism is well specified. Therefore

$$E_p(\hat{Y}_{\mathrm{NR}}) \approx \check{Y} = \sum_{k \in U} \frac{r_k y_k}{h(\mathbf{x}_k^\top \mathbf{b}^*)},$$

where $\mathbf{b}^*$ is obtained by solving the approximate expected value under the design of Equation (16.20):

$$\sum_{k \in U} \frac{r_k \mathbf{x}_k \pi_k}{h(\mathbf{x}_k^\top \mathbf{b}^*)} = \sum_{k \in U} \mathbf{x}_k \pi_k. \qquad (16.23)$$

Result 16.6 *If b^* is estimated by solving (16.23) and the nonresponse function is logistic, then*

$$\frac{\partial \check{Y}_{\mathrm{NR}}}{\partial r_\ell} = \frac{y_\ell - \mathbf{x}_\ell^\top \gamma_4}{h(\mathbf{x}_\ell^\top \mathbf{b}^*)},$$

where $\gamma_4 = \mathbf{T}_4^{-1} \mathbf{t}_4$,

$$\mathbf{T}_4 = \sum_{k \in U} \frac{r_k \mathbf{x}_k \mathbf{x}_k^\top h'(\mathbf{x}_k^\top \mathbf{b}^*)}{h^2(\mathbf{x}_k^\top \mathbf{b}^*)} \quad and \quad \mathbf{t}_4 = \sum_{k \in U} \frac{r_k y_k \mathbf{x}_k h'(\mathbf{x}_k^\top \mathbf{b}^*)}{h^2(\mathbf{x}_k^\top \mathbf{b}^*)}.$$

Proof: We differentiate Equation (16.23) with respect to r_k:

$$A = \sum_{k \in U} \frac{r_k \mathbf{x}_k}{h(\mathbf{x}_k^\top \mathbf{b}^*)} = \sum_{k \in U} \mathbf{x}_k.$$

We have

$$\frac{\partial A}{\partial r_\ell} = \frac{\mathbf{x}_\ell}{h(\mathbf{x}_\ell^\top \mathbf{b}^*)} - \sum_{k \in U} \frac{r_k \mathbf{x}_k \mathbf{x}_k^\top h'(\mathbf{x}_k^\top \mathbf{b}^*)}{h^2(\mathbf{x}_k^\top \mathbf{b}^*)} \frac{\partial \mathbf{b}^*}{\partial r_\ell} = 0.$$

Therefore

$$\frac{\partial \mathbf{b}^*}{\partial r_\ell} = \mathbf{T}_4^{-1} \frac{\mathbf{x}_\ell}{h(\mathbf{x}_\ell^\top \mathbf{b}^*)}.$$

By differentiating the estimator with respect to r_ℓ, we have:

$$\frac{\partial \check{Y}_{\mathrm{NR}}}{\partial r_\ell} = \frac{y_\ell}{h(\mathbf{x}_\ell^\top \mathbf{b}^*)} - \sum_{k \in U} \frac{r_k y_k h'(\mathbf{x}_k^\top \mathbf{b}^*)}{h^2(\mathbf{x}_k^\top \mathbf{b}^*)} \mathbf{x}_k^\top \frac{\partial \mathbf{b}^*}{\partial a_\ell} = \frac{y_\ell - \mathbf{x}_\ell^\top \gamma_4}{h(\mathbf{x}_\ell^\top \mathbf{b}^*)}.$$

∎

We can then write

$$\frac{\partial \check{Y}_{\mathrm{NR}}}{\partial r_\ell} - Y \approx \sum_{\ell \in U} \frac{y_\ell - \mathbf{x}_\ell^\top \gamma_4}{h(\mathbf{x}_\ell^\top \mathbf{b}^*)} (r_\ell - \phi_\ell).$$

Therefore

$$\mathrm{var}_q E_p(\check{Y}_{\mathrm{NR}}) \approx \sum_{k \in U} \frac{(y_k - \mathbf{x}_k^\top \gamma_4)^2}{h^2(\mathbf{x}_k^\top \mathbf{b}^*)} \phi_k (1 - \phi_k). \tag{16.24}$$

By considering the Expressions (16.22) and (16.24) and that γ_3 is an estimator of γ_4

$$v(\hat{Y}_{\mathrm{NR}})$$

$$= \sum_{k \in U} \sum_{\ell \in U} a_k a_\ell \left(r_k \frac{y_k - \mathbf{x}_k^\top \gamma_3}{\pi_k \hat{\phi}_k} + \frac{\mathbf{x}_k^\top \gamma_3}{\pi_k} \right) \left(r_\ell \frac{y_\ell - \mathbf{x}_\ell^\top \gamma_3}{\pi_\ell \hat{\phi}_\ell} + \frac{\mathbf{x}_\ell^\top \gamma_3}{\pi_\ell} \right) \frac{\Delta_{k\ell}}{\pi_{k\ell}}$$

$$+ \sum_{k \in U} a_k r_k \frac{(y_k - \mathbf{x}_k^\top \gamma_3)^2}{\pi_k \hat{\phi}_k^2} (1 - \hat{\phi}_k). \tag{16.25}$$

16.7.6 Variance of an Estimator Imputed by Regression

Consider the estimator given in Expression (16.10),

$$\widehat{Y}_{\text{IMP}} = \sum_{k \in U} \frac{a_k r_k y_k}{\pi_k} + \sum_{k \in U} \frac{a_k (1 - r_k) y_k^*}{\pi_k},$$

where imputation is realized using a multiple regression

$$y_k^* = \mathbf{x}_k^\top \widehat{\beta}, k \in S \backslash R.$$

$$\widehat{\beta} = \widehat{\mathbf{T}}^{-1} \widehat{\mathbf{t}}, \quad \widehat{\mathbf{t}} = \sum_{k \in U} \frac{a_k r_k q_k \mathbf{x}_k y_k}{v_k^2 \pi_k} \quad \text{and} \quad \widehat{\mathbf{T}} = \sum_{k \in U} \frac{a_k r_k q_k \mathbf{x}_k \mathbf{x}_k^\top}{v_k^2 \pi_k},$$

since

$$\frac{\partial \widehat{\beta}}{\partial a_\ell} = \widehat{\mathbf{T}}^{-1} \frac{r_\ell q_\ell \mathbf{x}_\ell y_\ell}{v_\ell^2 \pi_\ell} - \widehat{\mathbf{T}}^{-1} \frac{r_\ell q_\ell \mathbf{x}_\ell \mathbf{x}_\ell^\top}{v_\ell^2 \pi_\ell} \widehat{\mathbf{T}}^{-1} \widehat{\mathbf{t}} = \widehat{\mathbf{T}}^{-1} \frac{r_\ell q_\ell \mathbf{x}_\ell e_\ell}{v_\ell^2 \pi_\ell},$$

where $e_\ell = y_\ell - \mathbf{x}_\ell^\top \widehat{\beta}$. The derivative of the imputed estimator with respect to a_ℓ is

$$
\begin{aligned}
v_{1\ell} &= \frac{\partial \widehat{Y}_{\text{IMP}}}{\partial a_\ell} = \frac{r_\ell y_\ell}{\pi_\ell} + \frac{(1 - r_\ell) y_\ell^*}{\pi_\ell} + \sum_{k \in U} a_k (1 - r_k) \mathbf{x}_k^\top \frac{\partial \widehat{\beta}}{\partial a_\ell} \\
&= \frac{r_\ell e_\ell}{\pi_\ell} + \frac{y_\ell^*}{\pi_\ell} + \frac{q_\ell r_\ell \mathbf{x}_\ell^\top e_\ell}{v_\ell^2 \pi_\ell} \widehat{\mathbf{T}}^{-1} \sum_{k \in U} a_k (1 - r_k) \mathbf{x}_k \\
&= \frac{y_\ell^*}{\pi_\ell} + \frac{r_\ell e_\ell}{\pi_\ell} \left[1 + \frac{q_\ell \mathbf{x}_\ell^\top}{v_\ell^2} \widehat{\mathbf{T}}^{-1} \sum_{k \in U} a_k (1 - r_k) \mathbf{x}_k \right] = \frac{y_\ell^*}{\pi_\ell} + w_\ell r_\ell e_\ell,
\end{aligned}
$$

where w_ℓ is defined in Expression (16.11).

In addition, we can approximate the expected value under the design by

$$\check{Y}_{\text{IMP}} \approx E_p(\widehat{Y}_{\text{IMP}}) = \sum_{k \in U} r_k y_k + \sum_{k \in U} (1 - r_k) \mathbf{x}_k^\top \beta,$$

where

$$\beta = \mathbf{T}^{-1} \mathbf{t}, \quad \mathbf{t} = \sum_{k \in U} \frac{r_k q_k \mathbf{x}_k y_k}{v_k^2} \quad \text{and} \quad \mathbf{T} = \sum_{k \in U} \frac{r_k q_k \mathbf{x}_k \mathbf{x}_k^\top}{v_k^2}.$$

We therefore have

$$
\begin{aligned}
v_{2\ell} &= \frac{\partial \check{Y}_{\text{IMP}}}{\partial r_\ell} \\
&= y_\ell - \mathbf{x}_\ell^\top \beta + \sum_{k \in U} (1 - r_k) \mathbf{x}_k^\top \frac{\partial \beta}{\partial r_\ell} \\
&= \check{e}_k \left[1 + \frac{q_k \mathbf{x}_k}{v_k^2} \mathbf{T}^{-1} \sum_{k \in U} (1 - r_k) \mathbf{x}_k \right],
\end{aligned}
$$

where $\check{e}_k = y_k - \mathbf{x}_k^\top \beta$. This linearization can be estimated by $\hat{v}_{2\ell} = e_\ell \pi_\ell w_\ell$. We can then construct a variance estimator

$$v(\hat{Y}_{\text{NR}}) = \sum_{k \in U} \sum_{\ell \in U} a_k a_\ell \left(\frac{y_k^*}{\pi_k} + w_k r_k e_k \right) \left(\frac{y_\ell^*}{\pi_\ell} + w_\ell r_\ell e_\ell \right) \frac{\Delta_{k\ell}}{\pi_{k\ell}}$$

$$+ \sum_{k \in U} a_k r_k \frac{(e_k \pi_k w_k)^2}{\pi_k} (1 - \hat{\phi}_k).$$

If we estimate ϕ_k by $\hat{\phi}_k = 1/(w_k \pi_k)$, we then have $w_k = 1/(\hat{\phi}_k \pi_k)$. We obtain the estimator

$$v(\hat{Y}_{\text{NR}}) = \sum_{k \in U} \sum_{\ell \in U} a_k a_\ell \left(\frac{y_k^*}{\pi_k} + \frac{r_k e_k}{\pi_k \hat{\phi}_k} \right) \left(\frac{y_\ell^*}{\pi_\ell} + \frac{r_\ell e_\ell}{\pi_\ell \hat{\phi}_\ell} \right) \frac{\Delta_{k\ell}}{\pi_{k\ell}} + \sum_{k \in U} \frac{a_k r_k e_k^2}{\hat{\phi}_k^2 \pi_k} (1 - \hat{\phi}_k).$$

This variance estimator is very close to the variance estimator of nonresponse treated by calibration given in Expression (16.25). This similarity comes from the link between imputation by regression and reweighting developed in Section 16.6.4.

16.7.7 Other Variance Estimation Techniques

Linearization by the sample and response indicators allows to tackle even more complex cases such as estimators calibrated both at the sample and population levels. Many examples are given in Vallée & Tillé (2019). It is also possible to linearize with respect to the imputation model to estimate the variance, as in Deville & Särndal (1994).

The variance can also be estimated by means of resampling techniques but these methods must be adapted to nonresponse (Rao, 1990; Rao & Shao, 1992; Rao & Sitter, 1995; Yung & Rao, 2000; Skinner & Rao, 2001) as well as to the design.

17

Summary Solutions to the Exercises

Exercise 2.1

$$\frac{1}{2N(N-1)} \sum_{k \in U} \sum_{\substack{\ell \in U \\ \ell \neq k}} (y_k - y_\ell)^2 = \frac{1}{2N(N-1)} \sum_{k \in U} \sum_{\ell \in U} (y_k - y_\ell)^2$$

$$= \frac{1}{N-1} \sum_{k \in U} y_k^2 - \frac{1}{N(N-1)} \sum_{k \in U} \sum_{\ell \in U} y_k y_\ell$$

$$= \frac{1}{N-1} \sum_{k \in U} y_k^2 - \frac{N}{N-1} \overline{Y}^2 = S_y^2.$$

Exercise 2.2 $\pi_1 = 3/4, \pi_2 = 3/4, \pi_3 = 1/2, \Delta_{11} = 3/16, \Delta_{12} = -1/16, \Delta_{13} = -1/8,$ $\Delta_{22} = 3/16, \Delta_{23} = -1/8, \Delta_{33} = 1/4.$

Exercise 2.3 The inclusion probabilities are

$$\pi_1 = 3/4, \pi_2 = 3/4, \pi_3 = 1/2, \pi_{12} = 1/2, \pi_{13} = 1/4, \pi_{23} = 1/4.$$

The probability distributions of the expansion estimator and the Hájek estimator are

$$\widehat{\overline{Y}} = \begin{cases} 4(y_1 + y_2)/9 & \text{if } S = \{1,2\} \\ (4y_1 + 6y_3)/9 & \text{if } S = \{1,3\} \\ (4y_2 + 6y_3)/9 & \text{if } S = \{2,3\} \end{cases}, \widehat{\overline{Y}}_{HAJ} = \begin{cases} (y_1 + y_2)/2 & \text{if } S = \{1,2\} \\ (2y_1 + 3y_3)/5 & \text{if } S = \{1,3\} \\ (2y_2 + 3y_3)/5 & \text{if } S = \{2,3\}. \end{cases}$$

The sum of the weights assigned to the observations is equal to one for the Hájek estimator. However, the estimator is biased. Indeed,

$$E_p(\widehat{\overline{Y}}_{HAJ}) = \frac{1}{2} \times \frac{1}{2}(y_1 + y_2) + \frac{1}{4} \times \frac{1}{5}(2y_1 + 3y_3) + \frac{1}{4} \times \frac{1}{5}(2y_2 + 3y_3)$$

$$= \frac{1}{20}(7y_1 + 7y_2 + 6y_3) = \overline{Y} + \frac{1}{60}(y_1 + y_2 - 2y_3).$$

If $y_k = \pi_k, k \in U$, then $y_k/\pi_k = 1, k \in U$, and

$$\widehat{\overline{Y}} = \frac{1}{N} \sum_{k \in S} \frac{y_k}{\pi_k} = \frac{1}{N} \sum_{k \in S} 1 = \frac{n}{N},$$

Sampling and Estimation from Finite Populations, First Edition. Yves Tillé.
© 2020 John Wiley & Sons Ltd. Published 2020 by John Wiley & Sons Ltd.

whatever the selected sample. Therefore, we have $\text{var}_p(\widehat{\overline{Y}}) = 0$. Now, we calculate the two variance estimators. The Sen–Yates–Grundy estimator is

$$v_{\mathrm{SYG}}(\widehat{\overline{Y}}) = \frac{1}{2N^2} \sum_{k \in S} \sum_{\substack{\ell \in S \\ \ell \neq k}} \left(\frac{y_k}{\pi_k} - \frac{y_\ell}{\pi_\ell} \right)^2 \frac{\pi_k \pi_\ell - \pi_{k\ell}}{\pi_{k\ell}}.$$

Since $y_k / \pi_k = 1, k \in U$,

$$\left(\frac{y_k}{\pi_k} - \frac{y_\ell}{\pi_\ell} \right) = 0,$$

for all k, ℓ and therefore $v_{\mathrm{SYG}}(\widehat{\overline{Y}}) = 0$.

The Horvitz–Thompson estimator of the variance is

$$v_{\mathrm{HT}}(\widehat{\overline{Y}}) = \frac{1}{N^2} \sum_{k \in S} \frac{y_k^2}{\pi_k^2} (1 - \pi_k) + \frac{1}{N^2} \sum_{k \in S} \sum_{\substack{\ell \in S \\ \ell \neq k}} \frac{y_k y_\ell}{\pi_k \pi_\ell} \frac{\pi_{k\ell} - \pi_k \pi_\ell}{\pi_{k\ell}}.$$

If $S = \{1, 2\}$ and knowing that $y_k = \pi_k, k \in U$, we obtain

$$v_{\mathrm{HT}}(\widehat{\overline{Y}}) = \frac{1}{N^2} \left[(1 - \pi_1) + (1 - \pi_2) + 2 \times \left(1 - \frac{\pi_1 \pi_2}{\pi_{12}} \right) \right]$$

$$= \frac{1}{N^2} \left[4 - \pi_1 - \pi_2 - 2 \times \frac{\pi_1 \pi_2}{\pi_{12}} \right]$$

$$= \frac{1}{9} \left[4 - \frac{3}{4} - \frac{3}{4} - 2 \times \frac{3/4 \times 3/4}{1/2} \right] = \frac{1}{36}.$$

If $S = \{1, 3\}$, inspired by the previous result, we have

$$v_{\mathrm{HT}}(\widehat{\overline{Y}}) = \frac{1}{N^2} \left[4 - \pi_1 - \pi_3 - 2 \times \frac{\pi_1 \pi_3}{\pi_{13}} \right]$$

$$= \frac{1}{9} \left[4 - \frac{3}{4} - \frac{1}{2} - 2 \times \frac{3/4 \times 1/2}{1/4} \right] = -\frac{1}{36}.$$

Finally, if $S = \{2, 3\}$, we obtain

$$v_{\mathrm{HT}}(\widehat{\overline{Y}}) = \frac{1}{N^2} \left[4 - \pi_2 - \pi_3 - 2 \times \frac{\pi_2 \pi_3}{\pi_{23}} \right]$$

$$= \frac{1}{9} \left[4 - \frac{3}{4} - \frac{1}{2} - 2 \times \frac{3/4 \times 1/2}{1/4} \right] = -\frac{1}{36}.$$

Therefore, the probability distribution of $v_{\mathrm{HT}}(\widehat{\overline{Y}})$ is

$$v_{\mathrm{HT}}(\widehat{\overline{Y}}) = \begin{cases} 1/36 \text{ with probability } 1/2 \text{ (if } S = \{1,2\}) \\ -1/36 \text{ with probability } 1/2 \text{ (if } S = \{1,3\} \text{ or if } S = \{2,3\}). \end{cases}$$

It is obviously preferable to use $v_{\mathrm{SYG}}(\widehat{\overline{Y}})$, which accurately estimates the variance of the mean estimator. Indeed, $v_{\mathrm{HT}}(\widehat{Y})$ is unbiased but can take a negative value.

Exercise 2.4

$$E_p(\hat{\theta}) = \frac{1}{N^2} \sum_{k \in U} \frac{y_k E_p(a_k)}{\pi_k} + \frac{1}{N^2} \sum_{k \in U} \sum_{\substack{\ell \in U \\ \ell \neq k}} \frac{y_\ell E_p(a_k a_\ell)}{\pi_{k\ell}} = \frac{1}{N^2} \sum_{k \in U} \sum_{\ell \in U} y_\ell = \overline{Y}.$$

Exercise 2.5 Since we can write

$$S_y^2 = \frac{1}{2N(N-1)} \sum_{k \in U} \sum_{\substack{\ell \in U \\ \ell \neq k}} (y_k - y_\ell)^2,$$

we directly obtain the estimator of S_y^2 by

$$\widehat{S}_{y1}^2 = \frac{1}{2N(N-1)} \sum_{k \in S} \sum_{\substack{\ell \in S \\ \ell \neq k}} \frac{(y_k - y_\ell)^2}{\pi_{k\ell}}.$$

Exercise 2.6
1. In order to obtain Result (2.11), we multiply Expression (2.10) by a_ℓ and we obtain

$$\sum_{k \in U} \frac{z_k}{\pi_k} a_k a_\ell = \sum_{k \in U} z_k a_\ell.$$

   Then, we calculate the expectation of the two members of this equality.
2. If $z_k = \pi_k$, then the balanced design is reduced to a design with fixed sample size. Expression (2.11) becomes

$$\sum_{\substack{k \in U \\ k \neq \ell}} \pi_{k\ell} = \pi_\ell(n-1).$$

3. The proof is obtained by developing the square of Expression (2.12).
4. When $z_k = \pi_k$, the variance of Yates–Grundy is obtained for fixed-size designs.
5. The estimator is

$$v(\widehat{Z}) = \frac{1}{2} \sum_{k \in S} \sum_{\substack{\ell \in S \\ \ell \neq k}} \left(\frac{y_k}{z_k} - \frac{y_\ell}{z_\ell} \right)^2 z_k z_\ell \frac{\pi_k \pi_\ell - \pi_{k\ell}}{\pi_k \pi_\ell \pi_{k\ell}}.$$

Exercise 2.7 The design is not of fixed sample size because the sums of the rows and columns of the $\Delta_{k\ell}$ array are not zero. As $\pi_k(1 - \pi_k) = 6/25$, for all $k \in U$, $\pi_1 = \pi_2 = \pi_3 = 3/5 > \pi_4 = \pi_5 = 2/5$. The matrix of the joint inclusion probabilities is

$$\Pi = \begin{pmatrix} 1 & 1 & 1 & -1 & -1 \\ 1 & 1 & 1 & -1 & -1 \\ 1 & 1 & 1 & -1 & -1 \\ -1 & -1 & -1 & 1 & 1 \\ -1 & -1 & -1 & 1 & 1 \end{pmatrix} \times \frac{6}{25} + \begin{pmatrix} 9 & 9 & 9 & 6 & 6 \\ 9 & 9 & 9 & 6 & 6 \\ 9 & 9 & 9 & 6 & 6 \\ 6 & 6 & 6 & 4 & 4 \\ 6 & 6 & 6 & 4 & 4 \end{pmatrix} \times \frac{1}{25}$$

$$= \begin{pmatrix} 3 & 3 & 3 & 0 & 0 \\ 3 & 3 & 3 & 0 & 0 \\ 3 & 3 & 3 & 0 & 0 \\ 0 & 0 & 0 & 2 & 2 \\ 0 & 0 & 0 & 2 & 2 \end{pmatrix} \times \frac{1}{5}.$$

The design is reduced to randomly selecting one of the two samples:

$$p(\{1,2,3\}) = 3/5, p(\{4,5\}) = 2/5.$$

Exercise 3.1 The expectation of Expression (3.25) is

$$E_p(s_y^2) = \frac{1}{2n(n-1)} \sum_{k \in U} \sum_{\substack{\ell \in U \\ \ell \neq k}} (y_k - y_\ell)^2 E_p(a_k a_\ell)$$

$$= \frac{1}{2N(N-1)} \sum_{k \in U} \sum_{\substack{\ell \in U \\ \ell \neq k}} (y_k - y_\ell)^2 = S_y^2.$$

The expectation of Expression (3.26) is

$$E_p(s_y^2) = E_p \left(\frac{1}{n-1} \sum_{k \in U} y_k^2 a_k - \frac{n}{n-1} \widehat{\overline{Y}}^2 \right)$$

$$= \frac{1}{n-1} \sum_{k \in U} y_k^2 E_p(a_k) - \frac{n}{n-1} E_p(\widehat{\overline{Y}}^2)$$

$$= \frac{n}{n-1} \frac{1}{N} \sum_{k \in U} y_k^2 - \frac{n}{n-1} \left[\mathrm{var}_p(\widehat{\overline{Y}}) + E_p^2(\widehat{\overline{Y}}) \right]$$

$$= \frac{n}{n-1} \frac{1}{N} \sum_{k \in U} y_k^2 - \frac{n}{n-1} \left(\frac{N-n}{Nn} S_y^2 + \overline{Y}^2 \right)$$

$$= \frac{n}{n-1} \frac{N-1}{N} S_y^2 - \frac{n}{n-1} \frac{N-n}{Nn} S_y^2 = S_y^2.$$

Exercise 3.2

$$\Pr(k \in S) = 1 - \Pr(k \notin S) = 1 - \left(1 - \frac{1}{N}\right)^n = 1 - \sum_{j=0}^{n} \binom{n}{j} \left(-\frac{1}{N}\right)^{n-j}$$

$$= 1 - 1 + \frac{n}{N} - \sum_{z=2}^{n} \binom{n}{2} \left(-\frac{1}{N}\right)^z = \frac{n}{N} + O\left(\frac{1}{N^2}\right).$$

Exercise 3.3

1. Simple random sampling with replacement.
   The length of the confidence interval for a mean is

$$IC(0.95) = \left[\widehat{\overline{Y}} - z_{1-\alpha/2} \sqrt{\frac{s_y^2}{m}}, \widehat{\overline{Y}} + z_{1-\alpha/2} \sqrt{\frac{s_y^2}{m}} \right].$$

If $\widehat{P}_{AR}$ denotes the proportion estimator under simple random sampling with replacement, we can write

$$IC(0.95) = \left[\widehat{P}_{AR} - z_{1-\alpha/2} \sqrt{\frac{\widehat{P}_{AR}(1-\widehat{P}_{AR})}{m-1}}, \widehat{P}_{AR} + z_{1-\alpha/2} \sqrt{\frac{\widehat{P}_{AR}(1-\widehat{P}_{AR})}{m-1}} \right].$$

In this case,

$$\mathrm{var}_p(\widehat{P}_{AR}) = \frac{\widehat{P}_{AR}(1 - \widehat{P}_{AR})}{(m-1)}.$$

In order that the total length of the confidence interval does not exceed 0.02, we must have

$$2z_{1-\alpha/2}\sqrt{\frac{\hat{P}_{AR}(1-\hat{P}_{AR})}{m-1}} \leq 0.02.$$

By dividing by two and taking the square, we obtain

$$z_{1-\alpha/2}^2 \frac{\hat{P}_{AR}(1-\hat{P}_{AR})}{m-1} \leq 0.0001,$$

which gives

$$m-1 \geq z_{1-\alpha/2}^2 \times \frac{\hat{P}_{AR}(1-\hat{P}_{AR})}{0.0001},$$

$$m = 1 + 1.96^2 \times \frac{0.3 \times 0.7}{0.0001} \approx 8\ 068.36.$$

The sample size ($n = 8\ 069$) is larger than the size of the population.

2. Simple random sampling without replacement.

The length of the confidence interval for a mean is

$$IC(0.95) = \left[\hat{\bar{Y}} - z_{1-\alpha/2}\sqrt{\frac{N-n}{N}\frac{s_y^2}{n}},\ \hat{\bar{Y}} + z_{1-\alpha/2}\sqrt{\frac{N-n}{N}\frac{s_y^2}{n}}\right].$$

For a proportion P, we have

$$IC(0.95) = \left[\hat{P}_{SR} - z_{1-\alpha/2}\sqrt{\frac{N-n}{N}\frac{\hat{P}_{SR}(1-\hat{P}_{SR})}{n-1}},\ \hat{P}_{SR} + z_{1-\alpha/2}\sqrt{\frac{N-n}{N}\frac{\hat{P}_{SR}(1-\hat{P}_{SR})}{n-1}}\right].$$

In order that the total length of the confidence interval does not exceed 0.02, we must have

$$2z_{1-\alpha/2}\sqrt{\frac{N-n}{N}\frac{\hat{P}_{SR}(1-\hat{P}_{SR})}{n-1}} \leq 0.02.$$

By dividing by two and taking the square, we obtain

$$z_{1-\alpha/2}^2 \frac{N-n}{N}\frac{\hat{P}_{SR}(1-\hat{P}_{SR})}{n-1} \leq 0.0001,$$

which gives

$$(n-1) \times 0.0001 - z_{1-\alpha/2}^2 \times \frac{N-n}{N}\hat{P}_{SR}(1-\hat{P}_{SR}) \geq 0.$$

Therefore

$$n\left\{0.0001 + z_{1-\alpha/2}^2 \times \frac{1}{N}\hat{P}_{SR}(1-\hat{P}_{SR})\right\}$$

$$\geq 0.0001 + z_{1-\alpha/2}^2 \times \hat{P}_{SR}(1-\hat{P}_{SR}),$$

$$n \geq \frac{0.0001 + z_{1-\alpha/2}^2 \times \hat{P}_{SR}(1-\hat{P}_{SR}).}{\left\{0.0001 + z_{1-\alpha/2}^2 \times \frac{1}{N}\hat{P}_{SR}(1-\hat{P}_{SR})\right\}},$$

and

$$n \geq \frac{0.0001 + 1.96^2 \times 0.30 \times 0.70.}{\left\{ 0.0001 + 1.96^2 \times \frac{1}{1\,500} \times 0.30 \times 0.70 \right\}} \approx 1\,264.98.$$

Here, a sample size of 1265 is enough. Therefore, the impact of the correction of finite population can be decisive when the population size is small.

Exercise 3.4   The distribution is a coupon collector problem. If X is the number of draws needed to get r distinct individuals, we have

$$E(X) = \sum_{j=0}^{r-1} \frac{N}{N-j} = \sum_{j=0}^{349} \frac{350}{350-j} \approx 2\,251.8.$$

Exercise 3.5   The probability of obtaining r distinct rats by selecting m rats at random with replacement on the island is

$$f_N(r) = \frac{N!}{(N-r)!N^m} \left\{ \begin{matrix} r \\ m \end{matrix} \right\}, r = 1, \dots, \min(m, N). \tag{17.1}$$

Therefore, we minimize function $f_N(r)$ in N. Now, maximizing $f_N(r)$ is equivalent to maximizing

$$\frac{N!}{(N-r)!N^m} = \frac{\prod_{i=0}^{r-1}(N-i)}{N^m}$$

because $\left\{ \begin{matrix} r \\ m \end{matrix} \right\}$ does not depend on N. With $r = 42$, the size of the population is estimated by $\hat{N} = 136$ by maximizing the likelihood function. We must proceed by trial and error to find the maximum.

Exercise 3.6
1. The success rate of men is $\hat{R}_H = 35/55$ and for women is $\hat{R}_F = 25/45$. These two estimators are ratios. Indeed, the denominators of these estimators are random.

	Men	Women	Total
Success	P_{11}	P_{12}	$P_{1.}$
Failure	P_{21}	P_{22}	$P_{2.}$
Total	$P_{.1}$	$P_{.2}$	1

2. The bias of a ratio is

$$B_p(\hat{R}) = E_p(\hat{R}) - R \approx R \left(\frac{S_x^2}{\overline{X}^2} - \frac{S_{xy}}{\overline{X}\,\overline{Y}} \right) \frac{1-f}{n},$$

where x_k equals 1 if individual k is a man (resp. a woman) and 0 otherwise and y_k equals 1 if unit k is a man who succeeded and 0 otherwise. Therefore, we have, for the men,

$$S_x^2 = \frac{1}{N-1} \left(\sum_{k \in U} x_k^2 - N \overline{X}^2 \right) = \frac{N}{N-1} P_{.1}(1 - P_{.1})$$

and

$$S_{xy} = \frac{1}{N-1}\left(\sum_{k \in U} x_k y_k - N\, \overline{X}\, \overline{Y}\right) = \frac{N}{N-1} P_{11}(1 - P_{.1}),$$

which gives

$$\frac{S_x^2}{\overline{X}^2} - \frac{S_{xy}}{\overline{X}\,\overline{Y}} = \frac{1}{P_{.1}^2}\frac{N}{N-1}P_{.1}(1 - P_{.1}) - \frac{1}{P_{.1}P_{11}}\frac{N}{N-1}P_{11}(1 - P_{.1}) = 0.$$

The bias is approximately zero: $B_p(\widehat{R}) \approx 0$.

3. The mean square error is

$$\mathrm{MSE}(\widehat{R}) = \frac{1-f}{n\overline{X}^2}(S_y^2 - 2R\,S_{xy} + R^2\,S_x^2),$$

where $R = \overline{Y}/\overline{X}$. For the men, we obtain

$$\mathrm{MSE}(\widehat{R}_H) = \frac{1-f}{nP_{.1}^2}\frac{N}{N-1}\left[P_{11}(1 - P_{11}) - 2\frac{P_{11}}{P_{.1}}P_{11}(1 - P_{.1}) + \frac{P_{11}^2}{P_{.1}^2}P_{.1}(1 - P_{.1})\right]$$

$$= \frac{1-f}{nP_{.1}}\frac{N}{N-1}\frac{P_{11}}{P_{.1}}\left(1 - \frac{P_{11}}{P_{.1}}\right).$$

The estimator comes directly:

$$\widehat{\mathrm{MSE}}(\widehat{R}_H) = \frac{1-f}{n\widehat{P}_{.1}}\frac{N}{N-1}\frac{\widehat{P}_{11}}{\widehat{P}_{.1}}\left(1 - \frac{\widehat{P}_{11}}{\widehat{P}_{.1}}\right).$$

For the men, we obtain

$$\widehat{\mathrm{MSE}}(\widehat{R}_H) \approx 0.0046327,$$

and for the women, we obtain

$$\widehat{\mathrm{MSE}}(\widehat{R}_F) \approx 0.0060417.$$

Exercise 3.7 If k denotes the order number and j the number of selected units already selected at the beginning of step k, we have

k	u_k	j	$\dfrac{n-j}{N-(k-1)}$
1	0.375489	0	$4/10 = 0.4000$
2	0.624004	1	$3/9 = 0.3333$
3	0.517951	1	$3/8 = 0.3750$
4	0.045450	1	$3/7 = 0.4286$
5	0.632912	2	$2/6 = 0.3333$
6	0.246090	2	$2/5 = 0.4000$
7	0.927398	3	$1/4 = 0.2500$
8	0.325950	3	$1/3 = 0.3333$
9	0.645951	4	$0/2 = 0.0000$
10	0.178048	4	$0/1 = 0.0000$

Therefore, the sample is $\{1, 4, 6, 8\}$.

Exercise 4.1

1. The parameters are $\overline{Y} = 0$, $S_y^2 = 2/3$. Under simple random sampling without replacement, $\mathrm{var}_p(\widehat{\overline{Y}}_{SIMPL}) = 1/6$.
2. We also have $\overline{Y}_1 = 0$, $\overline{Y}_2 = 0$, $S_{y1}^2 = 0$, and $S_{y2}^2 = 2$. In the stratified design, $\mathrm{var}_p(\widehat{\overline{Y}}_{STRAT}) = 1/4$. The variance of the stratified design is larger.

This particular result is due to the fact that, in this example, the inter-strata variance is zero and the population size is small.

Exercise 4.2

1. We have $\overline{Y} = 5.2$, $S_{y1}^2 = 4$, $S_{y2}^2 = 2.25$, $S_y^2 \approx 4.2399$.
2. Under simple random sampling without replacement, $\mathrm{var}_p(\widehat{Y}) \approx 3\,815.91$.
3. If a stratified design with proportional allocation is used, then $n_1 = 6$, $n_2 = 4$, and $\mathrm{var}_p(\widehat{Y}_{PROP}) = 2\,970$.
4. If a stratified design with optimal allocation is used, then $N_1 S_{y1} = 120$, $N_2 S_{y2} = 60$, $n_1 \approx 6.66667$, and $n_2 \approx 3.33333$. If we round to the nearest integer, $n_1 = 7$ and $n_2 = 3$. The variance of the expansion estimator of the total is then $\mathrm{var}_p(\widehat{Y}_{OPT}) \approx 2\,927.4$. The gain in precision is not very important.

Exercise 4.3 If X is a uniform random variable in $[a, b]$, then $E(X) = (a + b)/2$, $\mathrm{var}(X) = (b - a)^2/12$, and $\sqrt{\mathrm{var}(X)} = (b - a)/(2\sqrt{3})$. Thus, $S_{y1}^2 \approx 0.0834168$, $S_{y2}^2 \approx 6.81818$, $S_{y3}^2 = 750$.

For the stratification with proportional allocation, $\mathrm{var}_p(\widehat{\overline{Y}}) \approx 0.06037$. For optimal stratification, $N_1 S_{y1} \approx 288.82$, $N_2 S_{y2} \approx 261.116$, $N_3 S_{y3} \approx 273.861$, $n_1 \approx 38.9161$, $n_2 \approx 35.1833$, and $n_3 \approx 36.9006$. The size of the third stratum $n_3 \approx 36.906$ is larger than $N_3 = 10$. In this case, we select all the units of the third stratum $n_3 = N_3 = 10$ and then select (optimally) 101 units among the 1100 units of strata one and two. So we have $n_1 \approx 53$, $n_2 \approx 48$, and $\mathrm{var}_p(\widehat{\overline{Y}}) \approx 0.0018092$. The gain in precision is very large.

Exercise 4.4 In a stratified design with proportional allocation, we have

$$\pi_{k\ell} = \begin{cases} \dfrac{n_h(n_h - 1)}{N_h(N_h - 1)} & \text{if } k, \ell \in U_k \\[2ex] \dfrac{n_h^2}{N_h^2} & \text{if } k \in U_h, \ell \in U_i, h \neq i. \end{cases}$$

The estimator is

$$\widehat{S}_y^2 = \frac{1}{2N(N-1)} \sum_{k \in S} \sum_{\substack{\ell \in S \\ \ell \neq k}} \frac{(y_k - y_\ell)^2}{\pi_{k\ell}}$$

$$= \frac{1}{2N(N-1)} \sum_{h=1}^{H} \sum_{k \in S_h} \sum_{\substack{\ell \in S_h \\ \ell \neq k}} (y_k - y_\ell)^2 \frac{N_h(N_h - 1)}{n_h(n_h - 1)}$$

$$+ \frac{1}{2N(N-1)} \sum_{h=1}^{H} \sum_{\substack{i=1 \\ i \neq h}}^{H} \sum_{k \in S_h} \sum_{\ell \in S_h} (y_k - y_\ell)^2 \frac{N_h^2}{n_h^2}$$

$$= \frac{1}{2N(N-1)} \sum_{h=1}^{H} \sum_{i=1}^{H} \sum_{k \in S_h} \sum_{\substack{\ell \in S_h \\ \ell \neq k}} (y_k - y_\ell)^2 \frac{N_h^2}{n_h^2}$$

$$+ \frac{1}{2N(N-1)} \sum_{h=1}^{H} \sum_{k \in S_h} \sum_{\substack{\ell \in S_h \\ \ell \neq k}} (y_k - y_\ell)^2 \left[\frac{N_h(N_h - 1)}{n_h(n_h - 1)} - \frac{N_h^2}{n_h^2} \right].$$

Since

$$\frac{N_h^2}{n_h^2} = \frac{N^2}{n^2}$$

and

$$\frac{N_h(N_h - 1)}{n_h(n_h - 1)} - \frac{N_h^2}{n_h^2} = \frac{N_h(N_h - n_h)}{n_h^2(n_h - 1)} = \frac{N - n}{n} \frac{N_h}{n_h(n_h - 1)},$$

we obtain

$$\widehat{S}_y^2 = \frac{N}{2n^2(N-1)} \sum_{h=1}^{H} \sum_{i=1}^{H} \sum_{k \in S_h} \sum_{\substack{\ell \in S_h \\ \ell \neq k}} (y_k - y_\ell)^2$$

$$+ \frac{1}{2N(N-1)} \sum_{h=1}^{H} \sum_{k \in S_h} \sum_{\substack{\ell \in S_h \\ \ell \neq k}} (y_k - y_\ell)^2 \frac{N - n}{n} \frac{N_h}{n_h(n_h - 1)}$$

$$= s_y^2 \frac{nN}{(n-1)(N-1)} + \frac{N - n}{nN(N-1)} \sum_{h=1}^{H} N_h s_{yh}^2$$

$$= s_y^2 \frac{nN}{(n-1)(N-1)} + \frac{N}{N-1} v(\widehat{\overline{Y}}_{\mathrm{PROP}}).$$

We have

$$E_p(\widehat{S}_y^2) = S_y^2 = E_p(s_y^2) \frac{Nn}{(n-1)(N-1)} + \frac{N}{N-1} \mathrm{var}_p(\widehat{\overline{Y}}_{\mathrm{PROP}}).$$

Thus,

$$E_p(s_y^2) = \frac{n-1}{n} \left[\frac{N-1}{N} S_y^2 - \mathrm{var}_p(\widehat{\overline{Y}}_{\mathrm{PROP}}) \right] = S_y^2 + O\left(\frac{1}{n}\right).$$

Exercise 4.5 The first-order inclusion probabilities are

$$\pi_1 = 1/2, \pi_2 = 2/3, \pi_3 = 1/2, \pi_4 = 2/3, \pi_5 = 2/3,$$

and the matrix of second-order inclusion probabilities is

$$\Pi = \begin{pmatrix} 1/2 & 1/3 & 0 & 1/3 & 1/3 \\ 1/3 & 2/3 & 1/3 & 1/3 & 1/3 \\ 0 & 1/3 & 1/2 & 1/3 & 1/3 \\ 1/3 & 1/3 & 1/3 & 2/3 & 1/3 \\ 1/3 & 1/3 & 1/3 & 1/3 & 2/3 \end{pmatrix}$$

Finally, the matrix of $\Delta_{k\ell}$ is

$$\Delta = \begin{pmatrix} 1/4 & 0 & -1/4 & 0 & 0 \\ 0 & 2/9 & 0 & -1/9 & -1/9 \\ -1/4 & 0 & 1/4 & 0 & 0 \\ 0 & -1/9 & 0 & 2/9 & -1/9 \\ 0 & -1/9 & 0 & -1/9 & 2/9 \end{pmatrix}.$$

A large number of $\Delta_{k\ell}$ are null, which is characteristic of a stratified design. Considering that two units that belong to the same stratum cannot have a $\Delta_{k\ell}$ null, both strata are necessarily

$$\{1,3\}, \{2,4,5\}.$$

In stratum $\{1,3\}$, one unit is selected (which explains $\pi_{13} = 0$) and, in stratum $\{2,4,5\}$, two units are selected.

Exercise 4.6

1. Since

$$D_i = \overline{Y}_i \left(1 - \frac{N_i}{N}\right) - \frac{1}{N}\sum_{h\neq i} N_h \overline{Y}_h,$$

   we have

$$\widehat{D}_i = \widehat{\overline{Y}}_i \left(1 - \frac{N_i}{N}\right) - \frac{1}{N}\sum_{h\neq i} N_h \widehat{\overline{Y}}_h.$$

2. The variance is

$$\mathrm{var}_p(\widehat{D}_i) = \left(1 - \frac{N_i}{N}\right)^2 \frac{N_i - n_i}{n_i N_i} S_{yi}^2 + \frac{1}{N^2}\sum_{h\neq i} N_h^2 \frac{N_h - n_h}{n_h N_h} S_{yh}^2.$$

3. If we set

$$z_k = \begin{cases} y_k(N/N_i - 1) & \text{if } k \in U_i \\ -y_k & \text{sinon,} \end{cases}$$

   the optimal allocation is

$$n_h = \frac{N_h S_{zh}}{\sum_{j=1}^{H} N_i S_{zi}},$$

   where

$$S_{zh} = \begin{cases} S_{yi}(N/N_i - 1), & \text{if } h = i \\ S_{yh}, & \text{otherwise.} \end{cases}$$

4. With respect to the usual optimal allocation, we "over-represent" stratum U_i by a factor of $(N/N_i - 1)$.

Exercise 4.7 The cost of a survey in the metropolitan population is C_1 and in the overseas population it is $C_2 = 2C_1$. Let N_h denote the size of stratum h, n_h the sample size in stratum h, $\widehat{\overline{Y}}_h$ the sample mean in stratum h, C_h the cost of a survey in stratum h, and C the total cost of the survey. Since both surveys are independent, we must minimize

$$\text{var}_p(\widehat{\overline{Y}}_1) + \text{var}_p(\widehat{\overline{Y}}_2) = \frac{N_1 - n_1}{N_1 n_1} S_{y1}^2 + \frac{N_2 - n_2}{N_2 n_2} S_{y2}^2$$

subject to

$$n_1 C_1 + n_2 C_2 = C.$$

After some calculations, we have $n_h = S_h / \sqrt{\lambda C_h}, h = 1, 2$, where λ is the Lagrange multiplier, and therefore

$$n_h = \frac{S_{hy} C}{\sqrt{C_h} [\sqrt{C_1}(S_{y1} + S_{y2}\sqrt{2})]}.$$

Exercise 4.8

1. Criterion W can be simplified in

$$W = 2(H - 1) \sum_{h=1}^{H} \frac{N_h - n_h}{N_h n_h} S_h^2.$$

Therefore, we have $C = 2(H - 1)$.

2. The optimal allocation is

$$n_h = \frac{n S_h}{\sum_{\ell=1}^{H} S_\ell}.$$

The optimal allocation does not depend on the size of the strata in the population.

Exercise 5.1 Look at the permutations $\{1, 1, 6, 6, 6\}$ and $\{1, 6, 1, 6, 6\}$. For these two cases, it is impossible to select jointly the two small units with systematic sampling. Then, we can show that the second-order inclusion probabilities of all the other permutations are given by one of these two particular cases.

Exercise 5.2

1. The vector $(a_1, \dots, a_k, \dots, a_N)^\top$ has a multinomial distribution.
2. $E_p(a_k) = 3p_k$, $\text{var}_p(a_k) = 3p_k(1 - p_k)$, and $\text{cov}_p(a_k, a_\ell) = -3p_k p_\ell$.
3. $\Pr(k \notin S) = (1 - p_k)^3$, $\Pr(S = \{k\}) = p_k^3$, $\Pr(S = \{k, \ell\}) = 3p_k^2 p_\ell + 3p_k p_\ell^2 = 3p_k p_\ell(p_k + p_\ell), k \neq \ell$, $\Pr(S = \{k, \ell, m\}) = 6p_k p_\ell p_m, k \neq \ell \neq m$.
4. The inclusion probability is $\pi_k = 1 - (1 - p_k)^3$.
5. $E_p(a_k \mid S = \{k\}) = 3, k \in U$, $E_p(a_k \mid S = \{k, \ell\}) = \frac{2p_k + p_\ell}{p_k + p_\ell}, k \neq \ell \in U$, $E_p(a_k \mid S = \{k, \ell, m\}) = 1, k \neq \ell \neq m \in U$.
6. The Hansen–Hurwitz estimator is $\widehat{Y}_{HH} = \frac{1}{3} \sum_{k \in U} \frac{y_k a_k}{p_k}$.
7. The Rao–Blackwellized estimator is

$$\text{if } S = \{k\}, \quad \widehat{Y}_{RB} = \frac{y_k}{p_k},$$

$$\text{if } S = \{k, \ell\}, \quad \widehat{Y}_{RB} = \frac{1}{3}\left(\frac{y_k}{p_k} + \frac{y_\ell}{p_\ell} + \frac{y_k + y_\ell}{p_k + p_\ell}\right),$$

$$\text{if } S = \{k, \ell, m\}, \quad \widehat{Y}_{RB} = \frac{1}{3}\left(\frac{y_k}{p_k} + \frac{y_\ell}{p_\ell} + \frac{y_m}{p_m}\right).$$

Exercise 5.3 The inclusion probabilities are $\pi_1 = 1, \pi_2 = 1, \pi_3 = 0.8, \pi_4 = 0.1,$ $\pi_5 = 0.05, \pi_6 = 0.05$. The selected sample is $\{1, 2, 4\}$.

Exercise 5.4

1. Since $X = \sum_{k \in U} x_k = 400$, we compute $n x_1 / X = 4 \times 200/400 = 2 > 1$. We eliminate unit 1 from the population and we still have to select 3 units from the remaining 5 units. Then we calculate $\sum_{k \in U \setminus \{1\}} = 200$. Since $3 \times 80/200 = 1.2 > 1$, we eliminate unit 2 from the population and we still have to select 2 out of the 4 remaining units. Finally, we have $\sum_{k \in U \setminus \{1,2\}} = 120$. $\pi_3 = \pi_4 = 2 \times 50/120 = 5/6$ and $\pi_5 = \pi_6 = 2 \times 10/120 = 1/6$. The cumulated probabilities are $V_0 = 0, V_1 = 1,$ $V_2 = 2, V_3 = 17/6, V_4 = 22/6, V_5 = 23/6, V_6 = 4$. We thus select sample $\{1, 2, 3, 4\}$.
2. The matrix of the second-order inclusion probabilities is

$$\begin{pmatrix} - & 1 & 5/6 & 5/6 & 1/6 & 1/6 \\ 1 & - & 5/6 & 5/6 & 1/6 & 1/6 \\ 5/6 & 5/6 & - & 4/6 & 1/6 & 0 \\ 5/6 & 5/6 & 4/6 & - & 0 & 1/6 \\ 1/6 & 1/6 & 1/6 & 0 & - & 0 \\ 1/6 & 1/6 & 0 & 1/6 & 0 & - \end{pmatrix}$$

Exercise 5.5 The inclusion probabilities are $1, 1, 0.75, 0.5, 0.25, 0.25, 0.125, 0.125$. The cumulated inclusion probabilities V_i are $0, 1, 2, 2.75, 3.25, 3.50, 3.75, 3.875, 4$. We select units $1, 2, 3, 5$ and $\pi_{35} = 0.25$.

Exercise 5.6

1. $\pi_k = (1 - \alpha_k) n / (N - 1), \pi_{k\ell} = (1 - \alpha_k - \alpha_\ell) n(n-1)/\{(N-1)(N-2)\}$.
2. Therefore $\pi_k \pi_\ell - \pi_{k\ell} \geq 0$.
3. $\alpha_k = 1 - \pi_k (N - 1)/n$.
4. The method is without replacement and of fixed sample size but it is only applicable if $\pi_k \leq n/(N - 1)$.

Exercise 5.7 We have $\pi_k = \alpha_k (N - n)/(N - 1) + (n - 1)/(N - 1)$ and

$$\pi_{k\ell} = (\alpha_k + \alpha_\ell) \frac{(n-1)(N-n)}{(N-1)(N-2)} + \frac{(n-1)(n-2)}{(N-1)(N-2)}.$$

We calculate the α_k in function of the π_k by $\alpha_k = \pi_k (N-1)(N-n) - (n-1)(N-n)$. The second-order inclusion probabilities can be written as

$$\pi_{k\ell} = \frac{(n-1)}{N-2} \left(\pi_k + \pi_\ell - \frac{n}{N-1} \right).$$

The second-order inclusion probabilities satisfy the Yates–Grundy conditions. Indeed, we have

$$\pi_k \pi_\ell - \pi_{k\ell} = (1 - \alpha_k - \alpha_\ell + \alpha_k \alpha_\ell) \frac{(n-1)(N-n)}{(N-1)^2(N-2)}$$

$$+ \alpha_k \alpha_\ell \left[\frac{(N-n)^2}{(N-1)^2} - \frac{(n-1)(N-n)}{(N-1)^2(N-2)} \right] \geq 0.$$

The method is of fixed sample size and without replacement, but for it to be applicable, it is necessary that $\alpha_k = \pi_k(N-1)/(N-n) - (n-1)/(N-n) \geq 0$ and therefore that $\pi_k \geq (n-1)/(N-1)$, which is rarely the case.

Exercise 5.8

1. We have $\pi_k^* = 1 - \pi_k$, $\pi_{k\ell}^* = 1 - \pi_k - \pi_\ell + \pi_{k\ell}$.
2.

$$\pi_k^* \pi_\ell^* - \pi_{k\ell}^* = (1 - \pi_k)(1 - \pi_k) - (1 - \pi_k - \pi_\ell + \pi_{k\ell}) = \pi_k \pi_k - \pi_{k\ell} \geq 0.$$

3. Brewer's method gives the split vectors

$$\pi_k^{(j)} = \begin{cases} 1 & k = j \\ \pi_k \frac{n-1}{n-\pi_j} & k \neq j \end{cases}$$

with λ_j such that

$$\sum_{j=1}^N \lambda_j \pi_k^{(j)} = \pi_k.$$

The complementary design is

$$\pi_k^{(j)*} = \begin{cases} 0 & k = j \\ 1 - (1 - \pi_k^*)\frac{N-n^*-1}{N-n^*-1+\pi_j^*} & k \neq j \end{cases}$$

with the same λ_j.
4. The method now consists of eliminating units at each step (eliminatory method).
5. The decomposition method in simple random sampling gives split vectors $\pi_k^{(1)} = n/N$, $\pi_k^{(2)} = (\pi_k - \lambda n/N)/(1 - \lambda)$. Therefore, the complementary design consists of splitting $\pi_k^{(1)*} = 1 - \pi_k^{(1)} = 1 - n/N = n^*/N$, $\pi_k^{(2)*} = (\pi_k^* - \lambda n/N)/(1 - \lambda)$.

Exercise 5.9 The samples can be selected using the following code in the R language.

```
library(sampling)
n=8
X=c(28,32,45,48,52,54,61,78,125,147, 165,258,356,438,462,
685,729,739,749,749)
pik=inclusionprobabilities(X,n)
UPsystematic(pik)
UPtille(pik)
UPbrewer(pik)
UPmaxentropy(pik)
UPminimalsupport(pik)
```

Exercise 6.1 A sample can be selected using the following code:

```
library(sampling)
N=20;n=8
Const=rep(1,20)
```

```
Gender=c(rep(1,10),rep(2,10))
Age=c(18,18,19,19,20,20,20,21,22,25,18,
19,19,19,20,22,24,25,24,28)
Grade=c(5,13,10,12,14,6,12,16,15,3,8,19,
14,9,7,19,18,15,11,14)
X=cbind(Const,Gender,Age,Grade)
pik=rep(n/N,N)
samplecube(X,pik)
```

Exercise 7.1

$$\text{var}_p(\widehat{Y}) = \frac{N-1}{N} \frac{M-m}{M-1} \frac{S^2_{inter}}{m}.$$

Exercise 7.2 It is a survey with a cluster design where the clusters are selected with equal probabilities with $M = 60, m = 3, N = 5\,000$. The inclusion probabilities are $\pi_k = m/M = 3/60 = 1/20$.

1. The mean income is

$$\widehat{\overline{Y}} = \frac{1}{5\,000} \left(\frac{1\,500}{1/20} + \frac{2\,000}{1/20} + \frac{2\,100}{1/20} \right) = \frac{5\,600 \times 20}{5\,000} = 22.4.$$

Therefore, the estimator of the total income is

$$\widehat{Y} = N\widehat{\overline{Y}} = 5\,000 \times 22.4 = 112\,000.$$

2. The variance of this estimator of the mean is

$$v(\widehat{\overline{Y}}) = \frac{M-m}{m-1} \frac{M}{m} \sum_{i \in S_I} \left(\frac{\overline{Y}_i N_i}{N} - \frac{\widehat{Y}}{M} \right)^2$$

$$= \frac{60-3}{3-2} \frac{60}{3} \left[\left(\frac{1\,500}{5\,000} - \frac{22.4}{60} \right)^2 + \left(\frac{2\,000}{5\,000} - \frac{22.4}{60} \right)^2 \right.$$

$$\left. + \left(\frac{2\,100}{5\,000} - \frac{22.4}{60} \right)^2 \right]$$

$$\approx 4.712.$$

3. The Hàjek estimator is

$$\widehat{\overline{Y}}_{HAJ} = \left(\frac{120}{1/20} + \frac{100}{1/20} + \frac{80}{1/20} \right)^{-1} \left(\frac{1\,500}{1/20} + \frac{2\,000}{1/20} + \frac{2\,100}{1/20} \right)$$

$$= \frac{5\,600 \times 20}{300 \times 20} \approx 18.666667.$$

Exercise 7.3

1. The first-order inclusion probabilities are

$$\pi_1 = \pi_2 = \pi_3 = 1/3, \quad \pi_4 = \pi_5 = \pi_6 = \pi_7 = \pi_8 = \pi_9 = 1/6.$$

2. The design is created as a function of the following population partition:

$$\{1,2,3\}, \{4,5,6\}, \{7,8,9\}.$$

The design consists of choosing one of three parts with probabilities $1/2$, $1/4$, and $1/4$, respectively. Then, in the selected part, we do a simple design without replacement. Hence, it is a two-stage design, where the primary units (the parts) are selected with unequal probabilities and the secondary units are selected following a simple design without replacement of size $n = 2$.

Exercise 9.1 $\widehat{\overline{Y}} = 10$, $v(\widehat{\overline{Y}}) \approx 0.01998$, $\widehat{\overline{Y}}_D = 11$, $v(\widehat{\overline{Y}}_D) \approx 0.014985$, $\widehat{\overline{Y}}_{\text{RATIO}} \approx 10.7142$, $v(\widehat{\overline{Y}}_{\text{RATIO}}) \approx 0.0113152$, $\widehat{\overline{Y}}_{\text{REG}} \approx 10.6$. $v(\widehat{\overline{Y}}_{\text{REG}}) \approx 0.010989$.

Exercise 9.2

1. $\widehat{\overline{Y}} = 24$, $\widehat{\overline{Y}}_{\text{RATIO}} \approx 23.86$, $\widehat{\overline{Y}}_D = 23$.
2. $v(\widehat{\overline{Y}}) \approx 0.0399$, $v(\widehat{\overline{Y}}_{\text{RATIO}}) \approx 0.0176$, $v(\widehat{\overline{Y}}_D) \approx 0.7398$.
3. We advise, of course, the ratio estimator, which has a much smaller variance than the other two estimators.

Exercise 9.3

1. $\widehat{\overline{Y}} = 150$, $\widehat{\overline{Y}}_D = 150 + 38 - 40 = 148$, $\widehat{\overline{Y}}_{\text{RATIO}} = \frac{38}{40} \times 150 = 142.5$.
2. The variances are

$$\text{var}_p(\widehat{\overline{Y}}) = \frac{N-n}{Nn}s_y^2 = \frac{1\,000\,000 - 400}{1\,000\,000 \times 400} \times 900 \approx 2.2491,$$

$$\text{var}_p(\widehat{\overline{Y}}_D) = \frac{N-n}{Nn}(s_y^2 - 2s_{xy} + s_x^2) \approx 1.7493,$$

$$\text{var}_p(\widehat{\overline{Y}}_{\text{RATIO}}) = \frac{N-n}{Nn}\left(s_y^2 - 2\frac{\widehat{\overline{Y}}}{\widehat{\overline{X}}}s_{xy} + \frac{\widehat{\overline{Y}}^2}{\widehat{\overline{X}}^2}s_x^2\right) \approx 2.9519.$$

Exercise 10.1 The optimal estimator is

$$\hat{Y}_{\text{REG-OPT}} = \hat{Y} + (X - \hat{X})\frac{\sum_{h=1}^{H}\frac{N_h - n_h}{N_h}\frac{1}{n_h - 1}\sum_{k \in S_h}(x_k - \widehat{\overline{X}}_h)(y_k - \widehat{\overline{Y}}_h)}{\sum_{h=1}^{H}\frac{N_h - n_h}{N_h}\frac{1}{n_h - 1}\sum_{k \in S_h}(x_k - \widehat{\overline{X}}_h)^2}.$$

The regression estimator is

$$\hat{Y}_{\text{REG}} = \hat{Y} + (X - \hat{X})\frac{s_{xy}^2}{s_x^2}.$$

The variances s_x^2 and s_{xy}^2 do not unbiasedly estimate S_x^2 and s_{xy}^2 in a stratified design. The regression coefficient of the optimal estimator is more unstable than the one belonging to the regression estimator.

Exercise 10.2

After ten algorithm iterations, we obtain Table 1.

Table 1

2.54	25.40	17.81	288.25	334
3.74	11.36	16.37	467.54	499
3.14	11.85	18.89	273.11	307
15.58	29.39	126.93	321.09	493
25	78	180	1 350	1 633

By reordering the rows and columns we get Table 2.

Table 2

78	6	0	0	**84**
32	5	0	0	**37**
0	0	427	17	**444**
0	0	432	32	**464**
49	**859**	**11**	**110**	**1 029**

The adjustment method conserves zeros in the table. The method then adjusts separately the two tables 2×2 of the diagonal. Since $84 + 37 \neq 49 + 859$ and $11 + 110 \neq 444 + 464$, it is impossible to adjust these two tables.

Exercise 10.3

1. The table margins are

	Fail	Pass	Total
Elementary	45	15	60
High	25	25	50
University	10	30	40
Total	80	70	150

The π-estimator of the pass rate is $\widehat{P} = 70/150 \approx 0.466666$. When a variable takes on values 0 or 1, we have

$$s_y^2 = \frac{1}{n-1} \sum_{k \in S} (y_k - \overline{Y})^2 = \frac{n\widehat{P}(1 - \widehat{P})}{n - 1}.$$

So,

$$v(\widehat{P}) = \frac{N - n}{N} \frac{\widehat{P}(1 - \widehat{P})}{n - 1} = \frac{10\,000 - 150}{10\,000} \times \frac{7}{15} \times \frac{8}{15} \times \frac{1}{149} \approx 0.00164.$$

2. The post-stratified estimator is

$$\hat{P}_{post} = \frac{1}{10\,000}\left(5\,000 \times \frac{15}{60} + 3\,000 \times \frac{25}{50} + 2\,000 \times \frac{30}{40}\right) \approx 0.4250.$$

The variance estimators inside the strata are the following:

$$s_{y1}^2 = \frac{\hat{P}_1(1-\hat{P}_1)}{n_1-1}n_1 = \frac{15}{60} \times \frac{45}{60} \times \frac{60}{59} \approx 0.1906779,$$

$$s_{y2}^2 = \frac{\hat{P}_2(1-\hat{P}_2)}{n_2-1}n_2 = \frac{25}{50} \times \frac{25}{50} \times \frac{50}{49} \approx 0.255102,$$

$$s_{y3}^2 = \frac{\hat{P}_3(1-\hat{P}_3)}{n_3-1}n_3 = \frac{30}{40} \times \frac{10}{40} \times \frac{40}{39} \approx 0.192376.$$

We can then estimate the variance

$$v(\hat{P}_{post}) = \frac{N-n}{nN^2}\sum_{h=1}^{H}N_h s_{yh}^2 + \frac{N-n}{n^2(N-1)}\sum_{h=1}^{H}\frac{N-N_h}{N}s_{yh}^2$$
$$\approx 0.0013999.$$

3. After ten iterations, we obtain

Stage 10	Fail	Pass	Total	Margins
Elementary	3 861.2	1 138.9	5 000.1	5 000
High	1 591.6	1 408.4	3 000.0	3 000
University	547.2	1 452.7	2 000.0	2 000
Total	6 000	4 000	10 000	
Margins	6 000	4 000		10 000

The pass rates given the parental education level are for elementary school $1\,138.9/5\,000.1 = 0.22778$, for high school $1\,408.4/3\,000 \approx 0.4694666$, and for university $1\,452.7/2\,000 \approx 0.72635$.

Exercise 11.1
1. If we use a single regressor given values x_k of variable x, we obtain $\hat{Y}_{REGM} = \hat{Y} + (X - \hat{X})\hat{Y}/\hat{X} = X\hat{Y}/\hat{X}$. This is the ratio estimator.
2. If we use H regressors given by $x_k\delta_{kh}$, where H is the number of strata and δ_k equals 1 if k is in the stratum h and 0 otherwise we obtain

$$\hat{Y}_{REGM} = \sum_{h=1}^{H}\frac{\sum_{k\in S_h}\frac{y_k}{\pi_k}}{\sum_{k\in S_h}\frac{x_k}{\pi_k}}\sum_{k\in U_h}x_k.$$

Therefore, it is a sum of ratio estimators calculated inside each stratum.

Exercise 12.1 The H auxiliary variables take the values $x_{k1}, \ldots, x_{kh}, \ldots, x_{kH}$, $k \in U$, where $x_{kh} = 1$ if $k \in U_h$ and 0 otherwise. In addition, we have $x_{kh}x_{k\ell} = 0$, for all h

with $k \neq \ell$. If $q_k = 1, k \in U$, and the pseudo-distance does not depend on k, then the calibration equations are written as

$$\sum_{k \in S} d_k x_{kh} F(\lambda^\top \mathbf{x}_k) = X_h.$$

However, $\lambda^\top \mathbf{x}_k = \lambda_{h(k)}$, where $h(k)$ gives the stratum of unit k, which gives us $\sum_{k \in S} x_{kh} F(\lambda_{h(k)}) N/n = N_h$, and hence $F(\lambda_h) = (N_h n)/(n_h N)$. The weights are $w_k = N_h/n_h, k \in U_h$, and the estimator that is obtained is the post-stratified estimation whatever the pseudo-distance used.

Exercise 12.2   We calibrate on the population size. Indeed, $\sum_{k \in U} 1 = N$. The calibration equation is $N = \sum_{k \in S} d_k F_k^\alpha(\lambda \times 1)$. Since

$$F_k^\alpha(u) = F^\alpha(u) = \begin{cases} \sqrt[\alpha-1]{1 + u(\alpha - 1)} & \alpha \in \mathbb{R}\backslash\{1\} \\ \exp u & \alpha = 1, \end{cases}$$

and $\sum_{k \in S} d_k = \widehat{N}$, the calibration equation becomes $N = F^\alpha(\lambda)\widehat{N}$. Hence,

$$\lambda = \begin{cases} \dfrac{\left(\frac{N}{\widehat{N}}\right)^{\alpha-1} - 1}{\alpha - 1} & \alpha \in \mathbb{R}\backslash\{1\} \\ \log \frac{N}{\widehat{N}} & \alpha = 1. \end{cases}$$

Since $F^\alpha(\lambda) = N/\widehat{N}$, whatever the value of α, the weights are written as $w_k = d_k N/\widehat{N}$. Then, the calibrated estimator is $\sum_{k \in S} d_k y_k N/\widehat{N} = \widehat{Y}N/\widehat{N}$. This is a ratio estimator calibrated on the sample size, which is the Hájek estimator.

Exercise 12.3   The auxiliary variables can be decomposed in two parts:

$$x_{k1}, \ldots, x_{kh}, \ldots, x_{kH} \text{ and } z_{k1}, \ldots, z_{ki}, \ldots, z_{kI},$$

where $x_{kh} = 1$ if $k \in U_{h.}$ and $x_{kh} = 0$ otherwise, and $z_{ki} = 1$ if $k \in U_{.i}$ and $z_{ki} = 0$ otherwise. The calibration equations become $N_{h.} = n_{h.} \frac{N}{n}(1 + \lambda_{h.}) + \sum_{i=1}^{I} n_{hi} \frac{N}{n} \lambda_{.i}$, $N_{.i} = n_{.i} \frac{N}{n}(1 + \lambda_{.i}) + \sum_{h=1}^{H} n_{hi} \frac{N}{n} \lambda_{h.}$, therefore we need to solve a system of $H + I$ equations with $H + I$ unknowns.

Exercise 15.1   We calculate the harmonic mean

$$M = \frac{N \times L}{\sum \text{times by cyclist}} = \frac{N}{\sum_{k \in U} 1/z_k}.$$

by setting $y_k = 1/z_k$. $M = N/Y$, therefore the linearized variable of M is

$$v_k = (1/Y - y_k N/Y^2) = (1 - y_k N/Y)/Y,$$

and is estimated by $\hat{v}_k = (1 - y_k \widehat{N}/\widehat{Y})/\widehat{Y}$, therefore $\widehat{M} = \widehat{N}/\widehat{Y}$ and

$$v(\widehat{M}) = \frac{N-n}{Nn} \frac{1}{n-1} \sum_{k \in S} (\hat{v}_k - \widehat{\overline{V}})^2,$$

where $\widehat{\overline{V}} = \sum_{k \in S} \hat{v}_k$.

Exercise 15.2 We use the Horvitz–Thompson variance, knowing that for a Poisson design we have $\pi_{kl} = \pi_k \pi_\ell$ for all $k \neq \ell$, which gives:

$$\text{var}_p(\widehat{Y}) = \sum_{k \in U} \frac{y_k^2}{\pi_k}(1 - \pi_k).$$

This variance can be estimated by

$$v(\widehat{Y}) = \sum_{k \in S} \frac{y_k^2}{\pi_k^2}(1 - \pi_k).$$

The multiple regression estimator is $\widehat{Y}_{\text{REGM}} = \widehat{Y} + (N - \widehat{N})\widehat{B}$, where $\widehat{B} = \widehat{Y}/\widehat{N}$, and therefore $\widehat{Y}_{\text{REGM}} = \widehat{Y}N/\widehat{N}$. This is the Hájek estimator, which is a special case of the ratio estimator where the auxiliary variable is $x_k = 1$.

The linearized variable of $\widehat{Y}_{\text{REGM}}$ is $v_k = y_k - \widehat{Y}/N = y_k - \overline{Y}$ and its estimator is $\hat{v}_k = y_k - \widehat{\overline{Y}}_{\text{HAJ}}$, where $\overline{Y}_{\text{HAJ}}$ is the Hájek estimator. The variance approximation is then

$$\text{var}_p(\widehat{Y}) = \sum_{k \in U} \frac{v_k^2}{\pi_k}(1 - \pi_k) = \sum_{k \in U} \frac{1 - \pi_k}{\pi_k}(y_k - \overline{Y})^2$$

and its estimator is

$$v(\widehat{Y}) = \sum_{k \in S} \frac{1 - \pi_k}{\pi_k^2}(y_k - \widehat{\overline{Y}}_{\text{HAJ}})^2.$$

Exercise 15.3 For the variance,

$$v_k = [(y_k - \overline{Y})^2 - S_y^2]/(N - 1).$$

For the covariance,

$$u_k = [(y_k - \overline{Y})(x_k - \overline{X}) - S_{xy}]/(N - 1).$$

Since $r^2 = S_{xy}^2/(S_y^2 S_x^2)$, the linearized variable of r^2 is

$$w_k = [(x_k - \overline{X})e_k + (y_k - \overline{Y})f_k]r^2/[(N - 1)S_{xy}],$$

where e_k and f_k are, respectively, the residuals of the regression of y in x and of x in y. The determination coefficient can be written as a function of the regression coefficients. In fact, $r^2 = b_1 b_2$, where b_1 and b_2 are, respectively, regression coefficients of y in x and of x in y. The linearized variable of the determination coefficient is

$$a_k = u_k(b_1)b_2 + u_k(b_2)b_1 = [(x_k - \overline{X})e_k + (y_k - \overline{Y})f_k]r^2/[(N - 1)S_{xy}].$$

Both methods give the same result.

Bibliography

Abramowitz, M. & Stegun, I. A. (1964). *Handbook of Mathematical Functions*. New York: Dover.

Aires, N. (1999). Algorithms to find exact inclusion probabilities for conditional Poisson sampling and Pareto πps sampling designs. *Methodology and Computing in Applied Probability* **4**, 457–469.

Aires, N. (2000). Comparisons between conditional Poisson sampling and Pareto πps sampling designs. *Journal of Statistical Planning and Inference* **82**, 1–15.

Aires, N. & Rosén, B. (2005). On inclusion probabilities and relative estimator bias for Pareto pps sampling. *Journal of Statistical Planning and Inference* **128**, 543–566.

Alexander, C. H. (1987). A class of methods for using person controls in household weighting. *Survey Methodology* **13**, 183–198.

Amstrup, S. C., McDonald, T. L. & Manly, B. F. J., eds. (2005). *Handbook of Capture-Recapture Analysis*. Princeton, NJ: Princeton University Press.

Antal, E. & Tillé, Y. (2011). A direct bootstrap method for complex sampling designs from a finite population. *Journal of the American Statistical Association* **106**, 534–543.

Antal, E. & Tillé, Y. (2014). A new resampling method for sampling designs without replacement: the doubled half bootstrap. *Computational Statistics* **29**, 1345–1363.

Ardilly, P. (1991). Échantillonnage représentatif optimum à probabilités inégales. *Annales d'Économie et de Statistique* **23**, 91–113.

Ardilly, P. (1994). *Les techniques de sondage*. Paris: Technip.

Ardilly, P. (2006). *Les techniques de sondage*. Paris: Technip.

Ardilly, P. & Lavallée, P. (2017). *Les sondages pas à pas*. Paris: Technip.

Ardilly, P. & Tillé, Y. (2003). *Exercices corrigés de méthodes de sondage*. Paris: Ellipses.

Ardilly, P. & Tillé, Y. (2006). *Sampling Methods: Exercises and Solutions*. New York: Springer.

Arora, H. R. & Brackstone, G. J. (1977). An investigation of the properties of raking ratio estimator: I with simple random sampling. *Survey Methodology* **3**, 62–83.

Asok, C. & Sukhatme, B. V. (1976). On Sampford's procedure of unequal probability sampling without replacement. *Journal of the American Statistical Association* **71**, 912–918.

Bankier, M. D. (1986). Estimators based on several stratified samples with applications to multiple frame surveys. *Journal of the American Statistical Association* **81**, 1074–1079.

Bankier, M. D. (1988). Power allocations: Determining sample sizes for subnational areas. *The American Statistician* **42**, 174–177.

Barbe, P. & Bertail, P. (1995). *The Weighted Bootstrap*. New York: Springer.

Sampling and Estimation from Finite Populations, First Edition. Yves Tillé.
© 2020 John Wiley & Sons Ltd. Published 2020 by John Wiley & Sons Ltd.

Bartolucci, F. & Montanari, G. E. (2006). A new class of unbiased estimators of the variance of the systematic sample mean. *Journal of Statistical Planning and Inference* **136**, 1512–1525.

Basu, D. (1958). On sampling with and without replacement. *Sankhyā* **20**, 287–294.

Basu, D. (1969). Role of the sufficiency and likelihood principles in sample survey theory. *Sankhyā* **A31**, 441–454.

Basu, D. & Ghosh, J. K. (1967). Sufficient statistics in sampling from a finite universe. In *Proceedings of the 36th Session of the International Statistical Institute.* pp. 850–859.

Bayless, D. L. & Rao, J. N. K. (1970). An empirical study of stabilities of estimators and variance estimators in unequal probability sampling ($n = 3$ or 4). *Journal of the American Statistical Association* **65**, 1645–1667.

Beaumont, J.-F. (2005). Calibrated imputation in surveys under a quasi-model-assisted approach. *Journal of the Royal Statistical Society* **B67**, 445–458.

Beaumont, J.-F. & Haziza, D. (2016). A note on the concept of invariance in two-phase sampling designs. *Survey Methodology* **42**, 319–323.

Beaumont, J.-F., Haziza, D. & Bocci, C. (2007). On variance estimation under auxiliary value imputation in sample surveys. Technical report. Statistics Canada, Ottawa.

Beaumont, J.-F. & Patak, Z. (2012). On the generalized bootstrap for sample surveys with special attention to Poisson sampling. *International Statistical Review* **80**, 127–148.

Bebbington, A. C. (1975). A simple method of drawing a sample without replacement. *Applied Statistics* **24**, 136.

Bellhouse, D. R. (1988a). A brief history of random sampling methods. In *Handbook of Statistics Volume 6: Sampling*, P. R. Krishnaiah & C. R. Rao, eds. New York, Amsterdam: Elsevier/North-Holland, pp. 1–14.

Bellhouse, D. R. (1988b). Systematic sampling. In *Handbook of Statistics Volume 6: Sampling*, P. R. Krishnaiah & C. R. Rao, eds. New York, Amsterdam: Elsevier/North-Holland, pp. 125–145.

Bellhouse, D. R. & Rao, J. N. K. (1975). Systematic sampling in the presence of a trend. *Biometrika* **62**, 694–697.

Benedetti, R., Espa, G. & Lafratta, G. (2008). A tree-based approach to forming strata in multipurpose business surveys. *Survey Methodology* **34**, 195–203.

Berger, Y. G. (1996). Asymptotic variance for sequential sampling without replacement with unequal probabilities. *Survey Methodology* **22**, 167–173.

Berger, Y. G. (1998a). Rate of convergence for asymptotic variance for the Horvitz–Thompson estimator. *Journal of Statistical Planning and Inference* **74**, 149–168.

Berger, Y. G. (1998b). Rate of convergence to normal distribution for the Horvitz-Thompson estimator. *Journal of Statistical Planning and Inference* **67**, 209–226.

Berger, Y. G. (1998c). Variance estimation using list sequential scheme for unequal probability sampling. *Journal of Official Statistics* **14**, 315–323.

Berger, Y. G. (2003). A modified Hájek variance estimator for systematic sampling. *Statistics in Transition* **6**, 5–21.

Berger, Y. G. (2005). Variance estimation with Chao's sampling scheme. *Journal of Statistical Planning and Inference* **127**, 253–277.

Berger, Y. G. (2007). A jackknife variance estimator for unistage stratified samples with unequal probabilities. *Biometrika* **94**, 953–964.

Berger, Y. G., El Haj Tirari, M. & Tillé, Y. (2001). Estimateur optimal par la régression dans les plans équilibrés. In *Enquêtes, modèles et applications*, J.-J. Droesbeke & L. Lebart, eds. Paris: Dunod, pp. 371–375.

Berger, Y. G. & Rao, J. N. K. (2006). Adjusted jackknife for imputation under unequal probability sampling without replacement. *Journal of the Royal Statistical Society* **B68**, 531–547.

Berger, Y. G. & Skinner, C. J. (2005). A jackknife variance estimator for unequal probability sampling. *Journal of the Royal Statistical Society* **B67**, 79–89.

Berger, Y. G. & Tillé, Y. (2009). Sampling with unequal probabilities. In *Handbook of Statistics. Vol 29A, Sampling*, D. Pfeffermann & C. R. Rao, eds. New York: Elsevier, pp. 37–52.

Bertail, P., Chautru, E. & Clémençon, S. (2017). Empirical processes in survey sampling with (conditional) Poisson designs. *Scandinavian Journal of Statistics* **44**, 97–111.

Bertail, P. & Combris, P. (1997). Bootstrap généralisé d'un sondage. *Annals of Economics and Statistics/Annales d'Économie et de Statistique* **46**, 49–83.

Bethlehem, J. G., Cobben, F. & Schouten, B. (2011). *Handbook of Nonresponse in Household Surveys*. New York: Wiley.

Bethlehem, J. G. & Keller, W. J. (1987). Linear weighting of sample survey data. *Journal of Official Statistics* **3**, 141–153.

Bickel, P. J. & Freedman, D. A. (1984). Asymptotic normality and the bootstrap in stratified sampling. *Annals of Statistics* **12**, 470–482.

Binder, D. A. (1983). On the variances of asymptotically normal estimators from complex survey. *International Statistical Review* **51**, 279–292.

Binder, D. A. (1996). Linearization methods for single phase and two-phase samples: a cookbook approach. *Survey Methodology* **22**, 17–22.

Binder, D. A. & Patak, Z. (1994). Use of estimating functions for estimation from complex surveys. *Journal of the American Statistical Association* **89**, 1035–1043.

Binder, D. A. & Théberge, A. (1988). Estimating the variance of raking-ratio estimators. *Canadian Journal of Statistics* **16**, 47–55.

Boistard, H., Lopuhaä, H. P. & Ruiz-Gazen, A. (2017). Functional central limit theorems for single-stage sampling designs. *The Annals of Statistics* **45**, 1728–1758.

Boistard, H., Lopuhaä, H. P., Ruiz-Gazen, A. et al. (2012). Approximation of rejective sampling inclusion probabilities and application to high order correlations. *Electronic Journal of Statistics* **6**, 1967–1983.

Bol'shev, L. N. (1965). On a characterization of the Poisson distribution. *Teoriya Veroyatnostei i ee Primeneniya* **10**, 64–71.

Bondesson, L., Traat, I. & Lundqvist, A. (2006). Pareto sampling versus Sampford and conditional Poisson sampling. *Scandinavian Journal of Statistics* **33**, 699–720.

Bonnéry, D., Breidt, F. J. & Coquet, F. (2012). Uniform convergence of the empirical cumulative distribution function under informative selection from a finite population. *Bernoulli* **18**, 1361–1385.

Booth, J. G., Butler, R. W. & Hall, P. (1994). Bootstrap methods for finite populations. *Journal of the American Statistical Association* **89**, 1282–1289.

Booth, J. G. & Hall, P. (1994). Monte Carlo approximation and the iterated bootstrap. *Biometrika* **81**, 331–340.

Bowley, A. L. (1906). Address to the economic and statistics section of the British Association for the Advancement of Sciences. *Journal of the Royal Statistical Society* **69**, 540–558.

Brackstone, G. J. & Rao, J. N. K. (1979). An investigation of raking ratio estimators. *Sankhyā* **C41**, 97–114.

Brändén, P. & Jonasson, J. (2012). Negative dependence in sampling. *Scandinavian Journal of Statistics* **39**, 830–838.

Breidt, F. J., Claeskens, G. & Opsomer, J. D. (2005). Model-assisted estimation for complex surveys using penalized splines. *Biometrika* **92**, 831–846.

Breidt, F. J. & Opsomer, J. D. (2000). Local polynomial regression estimators in survey sampling. *Annals of Statistics* **28**, 1026–1053.

Brewer, K. R. W. (1963a). A model of systematic sampling with unequal probabilities. *Australian Journal of Statistics* **5**, 5–13.

Brewer, K. R. W. (1963b). Ratio estimation in finite populations: Some results deductible from the assumption of an underlying stochastic process. *Australian Journal of Statistics* **5**, 93–105.

Brewer, K. R. W. (1975). A simple procedure for πpswor. *Australian Journal of Statistics* **17**, 166–172.

Brewer, K. R. W. (1999). Cosmetic calibration for unequal probability sample. *Survey Methodology* **25**, 205–212.

Brewer, K. R. W. (2000). Deriving and estimating an approximate variance for the Horvitz–Thompson estimator using only first order inclusion probabilities. In *Proceedings of the Second International Conference on Establishment Surveys*. Buffalo, New York: American Statistical Association, pp. 1417–1422.

Brewer, K. R. W. (2002). *Combined Survey Sampling Inference, Weighing Basu's Elephants*. London: Arnold.

Brewer, K. R. W. & Donadio, M. E. (2003). The high entropy variance of the Horvitz–Thompson estimator. *Survey Methodology* **29**, 189–196.

Brewer, K. R. W., Early, L. J. & Joyce, S. F. (1972). Selecting several samples from a single population. *Australian Journal of Statistics* **3**, 231–239.

Brewer, K. R. W. & Hanif, M. (1983). *Sampling with Unequal Probabilities*. New York: Springer.

Brown, M. B. & Bromberg, J. (1984). An efficient two-stage procedure for generating random variates from the multinomial distribution. *The American Statistician* **38**, 216–219.

Bülher,W. & Deutler, T. (1975). Optimal stratification and grouping by dynamic programming. *Metrika* **22**, 161–175.

Cao, G. & West, M. (1997). Computing distributions of order statistics. *Communications in Statistics – Theory and Methods* **26**, 755–764.

Caron, N. (1996). Les principales techniques de correction de la non-réponse et les modèles associés. Technical Report 9604. Insee, Paris.

Caron, N. (1999). Le logiciel POULPE, aspects méthodologiques. In *Actes des Journées de Méthodologie Statistique*, vol. 84. pp. 173–200.

Carroll, J. L. & Hartley, H. O. (1964). The symmetric method of unequal probability sampling without replacement. *Biometrics* **20**, 908–909.

Cassel, C. M., Särndal, C.-E. & Wretman, J. H. (1977). *Foundations of Inference in Survey Sampling*. New York: Wiley.

Cassel, C. M., Särndal, C.-E. & Wretman, J. H. (1993). *Foundations of Inference in Survey Sampling*. New York: Wiley.

Causey, B. D. (1972). Sensitivity of raked contingency table totals to change in problem conditions. *Annals of Mathematical Statistics* **43**, 656–658.

Causey, B. D., Cox, L. H. & Ernst, L. R. (1985). Application of transportation theory to statistical problems. *Journal of the American Statistical Association* **80**, 903–909.

Chambers, R. L. (1996). Robust case-weighting for multipurpose establishment surveys. *Journal of Official Statistics* **12**, 3–32.

Chambers, R. L. & Clark, R. G. (2012). *An Introduction to Model-Based Survey Sampling with Applications*. Oxford University Press.

Chang, T. & Kott, P. S. (2008). Using calibration weighting to adjust for nonresponse under a plausible model. *Biometrika* **95**, 555–571.

Chao, M.-T. (1982). A general purpose unequal probability sampling plan. *Biometrika* **69**, 653–656.

Chao, M.-T. & Lo, S.-H. (1985). A bootstrap method for finite population. *Sankhyā* **A47**, 399–405.

Chapman, D. G. (1951). Some properties of the hypergeometric distribution with applications to zoological censuses. *University of California Publications in Statistics* **1**, 131–160.

Chauvet, G. (2009). Stratified balanced sampling. *Survey Methodology* **35**, 115–119.

Chauvet, G. (2012). On a characterization of ordered pivotal sampling. *Bernoulli* **18**, 1320–1340.

Chauvet, G., Deville, J.-C. & Haziza, D. (2011). On balanced random imputation in surveys. *Biometrika* **98**, 459–471.

Chauvet, G. & Tillé, Y. (2005a). De nouvelles macros SAS d'échantillonnage équilibré. In *Actes des Journées de Méthodologie Statistique, Insee*. Paris, pp. 1–20.

Chauvet, G. & Tillé, Y. (2005b). *Fast SAS macros for balancing samples: user's guide*. Software Manual, University of Neuchâtel.

Chauvet, G. & Tillé, Y. (2006). A fast algorithm of balanced sampling. *Journal of Computational Statistics* **21**, 9–31.

Chauvet, G. & Tillé, Y. (2007). Application of the fast SAS macros for balancing samples to the selection of addresses. *Case Studies in Business, Industry and Government Statistics* **1**, 173–182.

Chauvet, G. & Vallée, A.-A. (2018). Consistency of estimators and variance estimators for two-stage sampling. *arXiv preprint arXiv:1808.09758*.

Chen, J. & Qin, J. (1993). Empirical likelihood estimation for finite populations and the effective usage of auxiliary information. *Biometrika* **80**, 107–116.

Chen, J. & Sitter, R. R. (1993). Edgeworth expansion and the bootstrap for stratified sampling without replacement from a finite population. *Canadian Journal of Statistics* **21**, 347–357.

Chen, J. & Sitter, R. R. (1999). A pseudo empirical likelihood approach to the effective use of auxiliary information in complex surveys. *Statistica Sinica* **9**, 385–406.

Chen, J. & Wu, C. (1999). Estimation of distribution function and quantiles using the model-calibrated pseudo empirical likelihood method. *Statistica Sinica* **12**, 1223–1239.

Chen, X.-H., Dempster, A. P. & Liu, J. S. (1994). Weighted finite population sampling to maximize entropy. *Biometrika* **81**, 457–469.

Chikkagoudar, M. S. (1966). A note on inverse sampling with equal probabilities. *Sankhyā* **A28**, 93–96.

Cicchitelli, G., Herzel, A. & Montanari, G. E. (1992). *Il campionamento statistico*. Bologna: Il Mulino.

Cicchitelli, G., Herzel, A. & Montanari, G. E. (1997). *Il campionamento statistico*. Strumenti. Economia. Bologna: Il Mulino.

Clark, J., Winglee, M. & Liu, B. (2007). Handling imperfect overlap determination in a dual-frame survey. In *ASA Proceedings of the Social Statistics Section*. American Statistical Association, pp. 3233–3238.

Cochran, W. G. (1939). The use of the analysis of variance in enumeration by sampling. *Journal of the American Statistical Association* **24**, 492–510.

Cochran, W. G. (1942). Sampling theory when the sampling units are of unequal sizes. *Journal of the American Statistical Association* **37**, 199–212.

Cochran, W. G. (1946). Relative accuracy of systematic and stratified random samples for a certain class of population. *Annals of Mathematical Statistics* **17**, 164–177.

Cochran, W. G. (1953). *Sampling Techniques*. New York: Wiley.

Cochran, W. G. (1961). Comparison of methods for determining stratum boundaries. In *Proceedings of the International Statistical Institute*, vol. **38**. pp. 345–358.

Cochran, W. G. (1963). *Sampling Techniques*. New York: Wiley.

Cochran, W. G. (1977). *Sampling Techniques*. New York: Wiley.

Connor, W. S. (1966). An exact formula for the probability that specified sampling units will occur in a sample drawn with unequal probabilities and without replacement. *Journal of the American Statistical Association* **61**, 384–490.

Conti, P. L. & Marella, D. (2012). *Campionamento da popolazioni finite: Il disegno campionario*. UNITEXT. Milano: Springer.

Cordy, C. B. (1993). An extension of the Horvitz–Thompson theorem to point sampling from a continuous universe. *Statistics and Probability Letters* **18**, 353–362.

Cornfield, J. (1944). On samples from finite populations. *Journal of the American Statistical Association* **39**, 236–239.

Cotton, F. & Hesse, C. (1992). Tirages coordonnés d'échantillons. Document de travail de la Direction des Statistiques Économiques E9206. Technical report. Insee, Paris.

Cumberland, W. G. & Royall, R. M. (1981). Prediction models in unequal probability sampling. *Journal of the Royal Statistical Society* **B43**, 353–367.

Cumberland, W. G. & Royall, R. M. (1988). Does simple random sampling provide adequate balance? *Journal of the Royal Statistical Society* **B50**, 118–124.

da Silva, D. N. & Opsomer, J. D. (2006). A kernel smoothing method of adjusting for unit non-response in sample surveys. *Canadian Journal of Statistics* **34**, 563–579.

da Silva, D. N. & Opsomer, J. D. (2009). Nonparametric propensity weighting for survey nonresponse through local polynomial regression. *Survey Methodology* **35**, 165–176.

Dagpunar, J. (1988). *Principles of Random Numbers Generation*. Oxford: Clarendon.

Dahmstroem, P. & Hagnell, M. (1978). Multivariate stratification of primary sampling units in a general purpose survey. *Scandinavian Journal of Statistics* **5**, 53–56.

Dalenius, T. D. (1950). The problem of optimum stratification. *Skandinavisk Aktuarietidskrift* **33**, 203–213.

Dalenius, T. D. (1957). *Sampling in Sweden*. Stockholm: Almqvist and Wicksell.

Dalenius, T. D. & Hodges Jr., J. L. (1957). The choice of stratification points. *Skandinavisk Aktuarietidskrift* **40**, 198–203.

Dalenius, T. D. & Hodges Jr., J. L. (1959). Minimum variance stratification. *Journal of the American Statistical Association* **54**, 88–101.

Davis, C. S. (1993). The computer generation of multinomial random variables. *Computational Statistics and Data Analysis* **16**, 205–217.

De Ree, S. J. M. (1983). A system of co-ordinated sampling to spread response burden of enterprises. In *Contributed paper, 44th Session of the ISI Madrid.* pp. 673–676.

De Ree, S. J. M. (1999). Co-ordination of business samples using measured response burden. In *Contributed paper, 52th Session of the ISI Helsinki.* pp. 1–4.

De Waal, T., Pannekoek, J. & Scholtus, S. (2011). *Handbook of Statistical Data Editing and Imputation.* New York: Wiley.

Dell, F. & d'Haultfoeuille, X. (2006). Measuring the evolution of complex indicators theory and application to the poverty rate in France. Working Papers 2006-11. Centre de Recherche en Économie et Statistique.

Dell, F., d'Haultfoeuille, X., Février, P. & Massé, E. (2002). Mise en œuvre du calcul de variance par linéarisation. In *Actes des Journées de Méthodologie Statistique.* Paris: Insee-Méthodes, pp. 73–104.

Deming, W. E. (1948). *Statistical Adjustment of Data.* New York: Wiley.

Deming, W. E. (1950). *Some Theory of Sampling.* New York: Dover Publications.

Deming, W. E. (1960). *Sample Design in Business Research.* New York: Wiley.

Deming, W. E. & Stephan, F. F. (1940). On a least square adjustment of sampled frequency table when the expected marginal totals are known. *Annals of Mathematical Statistics* **11**, 427–444.

Demnati, A. & Rao, J. N. K. (2004). Linearization variance estimators for survey data (with discussion). *Survey Methodology* **30**, 17–34.

Deroo, M. & Dussaix, A.-M. (1980). *Pratique et analyse des enquêtes par sondage.* Paris: Presses Universitaires de France.

Desabie, J. (1966). *Théorie et pratique des sondages.* Paris: Dunod.

Devaud, D. & Tillé, Y. (2019). Deville and Särndal's calibration: revisiting a 25 years old successful optimization problem. Institut de Statistique, Université de Neuchâtel.

Deville, J.-C. (1988). Estimation linéaire et redressement sur informations auxiliaires d'enquêtes par sondage. In *Mélanges économiques: Essais en l'honneur de Edmond Malinvaud*, A. Monfort & J.-J. Laffond, eds. Paris: Economica, pp. 915–927.

Deville, J.-C. (1989). Une théorie simplifiée des sondages. In *Les ménages : mélanges en l'honneur de Jacques Desabie*, P. L'Hardy & C. Thélot, eds. Paris: Insee, pp. 191–214.

Deville, J.-C. (1996). Estimation de la variance du coefficient de Gini estimé par sondage. In *Actes des Journées de Méthodologie Statistique*, vol. 69-70-71. Paris: Insee-Méthodes, pp. 269–288.

Deville, J.-C. (1998a). La correction de la non-réponse par calage ou par échantillonnage équilibré. Technical report. Recueil de la Section des méthodes d'enquête, Insee, Paris.

Deville, J.-C. (1998b). La correction de la non-réponse par calage ou par échantillonnage équilibré. In *Recueil de la Section des méthodes d'enquêtes des communications présentées au 26ème congrès de la Société Statistique du Canada.* Sherbrooke, pp. 103–110.

Deville, J.-C. (1998c). Une nouvelle (encore une!) méthode de tirage à probabilités inégales. Technical Report 9804. Méthodologie Statistique, Insee, Paris.

Deville, J.-C. (1999). Variance estimation for complex statistics and estimators: Linearization and residual techniques. *Survey Methodology* **25**, 219–230.

Deville, J.-C. (2000a). Generalized calibration and application to weighting for non-response. In *Compstat – Proceedings in Computational Statistics: 14th Symposium Held in Utrecht, The Netherlands.* New York: Springer, pp. 65–76.

Deville, J.-C. (2000b). Note sur l'algorithme de Chen, Dempster et Liu. Technical report. CREST-ENSAI, Rennes.

Deville, J.-C. (2002). La correction de la nonréponse par calage généralisé. In *Actes des Journées de Méthodologie Statistique.* Paris: Insee-Méthodes, pp. 3–20.

Deville, J.-C. (2004). Calage, calage généralisé et hypercalage. Technical report. Insee, Paris.

Deville, J.-C. (Undated). Cours de sondage, chapitre 3: Les outils de base. Course note. ENSAE, Paris.

Deville, J.-C. & Dupont, F. (1993). Non-réponse: principes et méthodes. In *Actes des Journées de Méthodologie Statistique.* Paris: Insee, pp. 53–70.

Deville, J.-C., Grosbras, J.-M. & Roth, N. (1988). Efficient sampling algorithms and balanced sample. In *COMPSTAT, Proceedings in Computational Statistics.* Heidelberg: Physica Verlag, pp. 255–266.

Deville, J.-C. & Lavallée, P. (2006). Indirect sampling: The foundations of the generalized weight share method. *Survey Methodology* **32**, 165.

Deville, J.-C. & Särndal, C.-E. (1992). Calibration estimators in survey sampling. *Journal of the American Statistical Association* **87**, 376–382.

Deville, J.-C. & Särndal, C.-E. (1994). Variance estimation for the regression imputed Horvitz–Thompson estimator. *Journal of Official Statistics* **10**, 381–394.

Deville, J.-C., Särndal, C.-E. & Sautory, O. (1993). Generalized raking procedure in survey sampling. *Journal of the American Statistical Association* **88**, 1013–1020.

Deville, J.-C. & Tillé, Y. (1998). Unequal probability sampling without replacement through a splitting method. *Biometrika* **85**, 89–101.

Deville, J.-C. & Tillé, Y. (2000). Selection of several unequal probability samples from the same population. *Journal of Statistical Planning and Inference* **86**, 215–227.

Deville, J.-C. & Tillé, Y. (2004). Efficient balanced sampling: The cube method. *Biometrika* **91**, 893–912.

Deville, J.-C. & Tillé, Y. (2005). Variance approximation under balanced sampling. *Journal of Statistical Planning and Inference* **128**, 569–591.

Devroye, L. (1986). *Non-uniform Random Variate Generation.* New York: Springer.

Díaz-García, J. A. & Cortez, L. U. (2008). Multi-objective optimisation for optimum allocation in multivariate stratified sampling. *Survey Methodology* **34**, 215–222.

Dickson, M. M. & Tillé, Y. (2016). Ordered spatial sampling by means of the traveling salesman problem. *Computational Statistics* **31**, 1359–1372.

Doss, D. C., Hartley, H. O. & Somayajulu, G. R. (1979). An exact small sample theory for post-stratification. *Journal of Statistical Planning and Inference* **3**, 235–248.

Droesbeke, J.-J., Fichet, B. & Tassi, P., eds. (1987). *Les Sondages.* Paris: Economica.

Dupont, F. (1996). Calage et redressement de la non-réponse totale : Validité de la pratique courante de redressement et comparaison des méthodes alternatives pour l'enquête sur la consommation alimentaire de 1989. In *Actes des journées de méthodologie statistique, 15 et 16 décembre 1993, Insee-Méthodes No 56-57-58, Complément.* pp. 9–42.

Durieux, B. & Payen, J.-F. (1976). Ajustement d'une matrice sur ses marges : la méthode ASAM. *Annales de l'Insee* 22–23, 311–337.

Dussaix, A.-M. & Grosbras, J.-M. (1992). *Exercices de sondages.* Paris: Economica.

Dussaix, A.-M. & Grosbras, J.-M. (1996). *Les sondages : principes et méthodes*. Paris: Presses Universitaires de France (Que sais-je ?).

Edwards, C. H. (1994). *Advanced calculus of several variables*. New York: Courier Corporation.

Efron, B. (1979). Bootstrap methods: Another look at the jackknife. *Annals of Statistics* **7**, 1–26.

Efron, B. & Tibshirani, R. J. (1993). *An Introduction to the Bootstrap*. Boca Raton, FL.: Chapman and Hall/CRC.

Ernst, L. R. (1986). Maximizing the overlap between surveys when information is incomplete. *European Journal of Operational Research* **27**, 192–200.

Ernst, L. R. & Ikeda, M. M. (1995). A reduced-size transportation algorithm for maximizing the overlap between surveys. *Survey Methodology* **21**, 147–157.

Ernst, L. R. & Paben, S. P. (2002). Maximizing and minimizing overlap when selecting any number of units per stratum simultaneously for two designs with different stratifications. *Journal of Official Statistics* **18**, 185–202.

Estevao, V. M. & Särndal, C.-E. (2000). A functional form approach to calibration. *Journal of Official Statistics* **16**, 379–399.

Estevao, V. M. & Särndal, C.-E. (2006). Survey estimates by calibration on complex auxiliary information. *International Statistical Review* **74**, 127–147.

Fan, C. T., Muller, M. E. & Rezucha, I. (1962). Development of sampling plans by using sequential (item by item) selection techniques and digital computer. *Journal of the American Statistical Association* **57**, 387–402.

Fay, R. E. (1985). A jackknifed chi-squared test for complex samples. *Journal of the American Statistical Association* **80**, 148–157.

Fay, R. E. (1991). A design-based perspective on missing data variance. In *Proceedings of the 1991 Annual Research Conference*. U.S. Census Bureau, pp. 429–440.

Félix-Medina, M. H. (2000). Analytical expressions for Rao–Blackwell estimators in adaptive cluster sampling. *Journal of Statistical Planning and Inference* **84** (1-2), 221–236.

Fellegi, I. P. (1981). Currents topics in survey sampling. In *Should the census counts be adjusted for allocation purposes? Equity considerations*, D. Krewski, R. Platek & J. N. K. Rao, eds. New York: Academic Press, pp. 47–76.

Fienberg, S. E. (1970). An iterative procedure for estimation in contingency tables. *Annals of Mathematical Statistics* **41**, 907–917.

Fitch, D. J. (2002). Studies in multivariate stratification: Similarity analysis vs Friedman-Rubin. In *ASA Proceedings of the Joint Statistical Meetings*. American Statistical Association, pp. 996–999.

Folsom, R. E. & Singh, A. C. (2000). The generalized exponential model for design weight calibration for extreme values, nonresponse and poststratification. In *Section on Survey Research Methods, American Statistical Association*. pp. 598–603.

Friedlander, D. (1961). A technique for estimating a contingency table, given the marginal totals and some supplementary data. *Journal of the Royal Statistical Society* **A124**, 412–420.

Froment, R. & Lenclud, B. (1976). Ajustement de tableaux statistiques. *Annales de l'Insee* **22-23**, 29–53.

Frosini, B. V., Montinaro, M. & Nicolini, G. (2011). *Campionamenti da popolazioni finite*. Torino: Giappichelli.

Fuller, W. A. (1970). Sampling with random stratum boundaries. *Journal of the Royal Statistical Society* **B32**, 209–226.

Fuller, W. A. (1976). *Introduction to Statistical Time Series*. New York: Wiley.

Fuller, W. A. (1987). *Measurement Error Models*. New York: Wiley.

Fuller, W. A. (2002). Regression estimation for survey samples. *Survey Methodology* **28**, 5–23.

Fuller, W. A. (2011). *Sampling Statistics*. John Wiley & Sons.

Fuller, W. A. & Burmeister, L. F. (1972). Estimators for samples selected from two overlapping frames. In *ASA Proceedings of the Social Statistics Section*. American Statistical Association, pp. 245–249.

Funaoka, F., Saigo, H., Sitter, R. R. & Toida, T. (2006). Bernoulli bootstrap for stratified multistage sampling. *Survey Methodology* **32**, 151–156.

Gabler, S. (1981). A comparison of Sampford's sampling procedure versus unequal probability sampling with replacement. *Biometrika* **68**, 725–727.

Gabler, S. (1990). *Minimax Solutions in Sampling from Finite Populations*. New York: Springer.

Gastwirth, J. L. (1972). The estimation of the Lorenz curve and Gini index. *Review of Economics and Statistics* **54**, 306–316.

Gini, C. (1912). *Variabilità e Mutabilità*. Bologna: Tipografia di Paolo Cuppin.

Gini, C. (1914). Sulla misura della concentrazione e della variabilità dei caratteri. *Atti del Reale Istituto Veneto di Scienze, Lettere e Arti, LXXIII* **73**, 1203–1248.

Gini, C. (1921). Measurement of inequality and incomes. *The Economic Journal* **31**, 124–126.

Gini, C. (1928). Une application de la methode représentative aux matériaux du dernier recensement de la population italienne (1er décembre 1921). *Bulletin of the International Statistical Institute* **23**, Liv. 2, 198–215.

Gini, C. & Galvani, L. (1929). Di una applicazione del metodo rappresentativo al censimento italiano della popolazione (1. dicembre 1921). *Annali di Statistica* **Series 6, 4**, 1–107.

Giorgi, G. M., Palmitesta, P. & Provasi, C. (2006). Asymptotic and bootstrap inference for the generalized Gini indices. *Metron* **64**, 107–124.

Godambe, V. P. (1955). A unified theory of sampling from finite population. *Journal of the Royal Statistical Society* **B17**, 269–278.

Godambe, V. P. & Joshi, V. M. (1965). Admissibility and Bayes estimation in sampling finite populations I. *Annals of Mathematical Statistics* **36**, 1707–1722.

Goga, C. (2005). Réduction de la variance dans les sondages en présence d'information auxiliaire : une approche nonparamétrique par splines de régression. *Canadian Journal of Statistics/Revue Canadienne de Statistique* **2**, 1–18.

Goga, C., Deville, J.-C. & Ruiz-Gazen, A. (2009). Use of functionals in linearization and composite estimation with application to two-sample survey data. *Biometrika* **96**, 691–709.

Gouriéroux, C. (1981). *Théorie des sondages*. Paris: Economica.

Graf, É. & Qualité, L. (2014). Sondage dans des registres de population et de ménages en Suisse : coordination d'échantillons, pondération et imputation. *Journal de la Société Française de Statistique* **155**, 95–133.

Graf, M. (2011). Use of survey weights for the analysis of compositional data. In *Compositional Data Analysis: Theory and Applications*, V. Pawlowsky-Glahn & A. Buccianti, eds. Chichester: Wiley, pp. 114–127.

Graf, M. (2014). Tight calibration of survey weights. Technical report. Université de Neuchâtel, Neuchâtel.

Graf, M. (2015). A simplified approach to linearization variance for surveys. Technical report. Institut de statistique, Université de Neuchâtel.

Grafström, A. & Lisic, J. (2016). *BalancedSampling: Balanced and spatially balanced sampling*. R package version 1.5.2.

Grafström, A. & Lundström, N. L. P. (2013). Why well spread probability samples are balanced? *Open Journal of Statistics* **3**, 36–41.

Grafström, A., Lundström, N. L. P. & Schelin, L. (2012). Spatially balanced sampling through the pivotal method. *Biometrics* **68**, 514–520.

Grafström, A. & Tillé, Y. (2013). Doubly balanced spatial sampling with spreading and restitution of auxiliary totals. *Environmetrics* **14**, 120–131.

Gregoire, T. G. & Valentine, H. T. (2007). *Sampling strategies for natural resources and the environment*. Boca Raton, FL: CRC Press.

Grosbras, J.-M. (1987). *Méthodes statistiques des sondages*. Paris: Economica.

Groves, R. M. (2004a). *Survey Errors and Survey Costs*. New York: Wiley.

Groves, R. M. (2004b). *Survey Methodology*. New York: Wiley.

Groves, R. M., Fowler, F. J., Couper, M. P., Lepkowski, J. M. & Singer, E. (2009). *Survey Methodology*. New York: Wiley.

Guandalini, A. & Tillé, Y. (2017). Design-based estimators calibrated on estimated totals from multiple surveys. *International Statistical Review* **85**, 250–269.

Gunning, P. & Horgan, J. M. (2004). A new algorithm for the construction of stratum boundaries in skewed populations. *Survey Methodology* **30**, 159–166.

Gunning, P., Horgan, J. M. & Keogh, G. (2006). Efficient Pareto stratification. *Mathematical Proceedings of the Royal Irish Academy* **106A**, 131–136.

Gunning, P., Horgan, J. M. & Yancey, W. (2004). Geometric stratification of accounting data. *Revista Contaduría y Administración* **214**, 11–21.

Gutiérrez, H. A. (2009). *Estrategias de muestreo diseño de encuestas y estimacion de parametros*. Bogotá: Universidad Santo Tomas.

Haines, D. E. & Pollock, K. H. (1998). Combining multiple frames to estimate population size and totals. *Survey Methodology* **24**, 79–88.

Hájek, J. (1964). Asymptotic theory of rejective sampling with varying probabilities from a finite population. *Annals of Mathematical Statistics* **35**, 1491–1523.

Hájek, J. (1971). Discussion of an essay on the logical foundations of survey sampling, part on by D. Basu. In *Foundations of Statistical Inference*, V. P. Godambe & D. A. Sprott, eds. Toronto: Holt, Rinehart, Winston, p. 326.

Hájek, J. (1981). *Sampling from a Finite Population*. New York: Marcel Dekker.

Hampel, F. R. (1974). The influence curve and its role in robust estimation. *Journal of the American Statistical Association* **69**, 383–393.

Hampel, F. R., Ronchetti, E., Rousseeuw, P. J. & Stahel, W. A. (1985). *Robust Statistics: The Approach Based on the Influence Function*. New York: Wiley.

Hanif, M. & Brewer, K. R. W. (1980). Sampling with unequal probabilities without replacement: A review. *International Statistical Review* **48**, 317–335.

Hansen, M. H. & Hurwitz, W. N. (1943). On the theory of sampling from finite populations. *Annals of Mathematical Statistics* **14**, 333–362.

Hansen, M. H. & Hurwitz, W. N. (1949). On the determination of the optimum probabilities in sampling. *Annals of Mathematical Statistics* **20**, 426–432.

Hansen, M. H., Hurwitz, W. N. & Madow, W. G. (1953a). *Sample Survey Methods and Theory, I*. Reprint in 1993, New York: Wiley.

Hansen, M. H., Hurwitz, W. N. & Madow, W. G. (1953b). *Sample Survey Methods and Theory, II*. Reprint in 1993, New York: Wiley.

Hansen, M. H. & Madow, W. G. (1974). Some important events in the historical development of sample survey. In *On the History of Statistics and Probability*, D. B. Owen, ed. New York: Marcel Dekker, pp. 75–102.

Hansen, M. H., Madow, W. G. & Tepping, B. J. (1983). An evaluation of model dependent and probability-sampling inferences in sample surveys (with discussion and rejoinder). *Journal of the American Statistical Association* **78**, 776–807.

Hanurav, T. V. (1968). Hyper-admissibility and optimum estimators for sampling finite population. *Annals of Mathematical Statistics* **39**, 621–642.

Harford, T. (2014). Big data: A big mistake? *Significance* **11**, 14–19.

Hartley, H. O. (1962). Multiple frame survey. In *ASA Proceedings of the Social Statistics Section*. American Statistical Association, pp. 203–206.

Hartley, H. O. (1974). Multiple frame methodology and selected applications. *Sankhyā* **C36**, 99–118.

Hartley, H. O. & Rao, J. N. K. (1962). Sampling with unequal probabilities and without replacement. *Annals of Mathematical Statistics* **33**, 350–374.

Hasler, C. & Tillé, Y. (2014). Fast balanced sampling for highly stratified population. *Computational Statistics and Data Analysis* **74**, 81–94.

Hasler, C. & Tillé, Y. (2016). Balanced k-nearest neighbor imputation. *Statistics* **105**, 11–23.

Haziza, D. (2005). Inférence en présence d'imputation simple dans les enquêtes : un survol. *Journal de la Société Française de Statistique* **146**, 69–118.

Haziza, D. (2009). Imputation and inference in the presence of missing data. In *Sample surveys: Design, methods and applications*, D. Pfeffermann & C. R. Rao, eds. New York, Amsterdam: Elsevier/North-Holland, pp. 215–246.

Haziza, D. & Beaumont, J.-F. (2017). Construction of weights in surveys: A review. *Statistical Science* **32**, 206–226.

Haziza, D. & Lesage, É. (2016). A discussion of weighting procedures for unit nonresponse. *Journal of Official Statistics* **32**, 129–145.

Haziza, D. & Rao, J. N. K. (2003). Inference for population means under unweighted imputation for missing survey data. *Survey Methodology* **29**, 81–90.

Henderson, T. (2006). *Estimating the variance of the Horvitz-Thompson estimator*. Master Thesis, School of Finance and Applied Statistics, The Australian National University, Canberra.

Hidiroglou, M. A. & Gray, G. B. (1980). Construction of joint probability of selection for systematic PPS sampling. *Applied Statistics* **29**, 107–112.

Hodges Jr., J. L. & Le Cam, L. (1960). The Poisson approximation to the Poisson binomial distribution. *Annals of Mathematical Statistics* **31**, 737–740.

Holmberg, A. (1998). A bootstrap approach to probability proportional-to-size sampling. In *ASA Proceedings of the Section on Survey Research Methods*. American Statistical Association, pp. 378–383.

Holmberg, A. & Swensson, B. (2001). On Pareto πps sampling: reflections on unequal probability sampling strategies. *Theory of Stochastic Processes* **7**, 142–155.

Holt, D. & Smith, T. M. F. (1979). Post-stratification. *Journal of the Royal Statistical Society* **A142**, Part 1, 33–46.

Horgan, J. M. (2006). Stratification of skewed populations: A review. *International Statistical Review* **74**, 67–76.

Horvàth, R. A. (1974). Les idées de Quetelet sur la formation d'une discipline moderne et sur le rôle de la théorie des probabilités. In *Mémorial Adolphe Quetelet*, N° 3. Académie royale des Sciences de Belgique, pp. 57–67.

Horvitz, D. G. & Thompson, D. J. (1952). A generalization of sampling without replacement from a finite universe. *Journal of the American Statistical Association* **47**, 663–685.

Huang, E. T. & Fuller, W. A. (1978). Non-negative regression estimation for sample survey data. In *Proceedings of the Social Statistics Section of the American Statistical Association*. pp. 300–305.

Hyndman, R. J. & Fan, Y. (1996). Sample quantiles in statistical packages. *American Statistician* **50**, 361–365.

Iachan, R. (1982). Systematic sampling: A critical review. *International Statistical Review* **50**, 293–303.

Iachan, R. (1983). Asymptotic theory of systematic sampling. *Annals of Statistics* **11**, 959–969.

Iachan, R. & Dennis, M. L. (1993). A multiple frame approach to sampling the homeless and transient population. *Journal of Official Statistics* **9**, 747–764.

Iannacchione, V. G., Milne, J. G. & Folsom, R. E. (1991). Response probability weight adjustments using logistic regression. In *Proceedings of the Survey Research Methods Section, American Statistical Association*. pp. 637–42.

Ireland, C. T. & Kullback, S. (1968). Contingency tables with given marginals. *Biometrika* **55**, 179–188.

Isaki, C. T. & Fuller, W. A. (1982). Survey design under a regression population model. *Journal of the American Statistical Association* **77**, 89–96.

Isaki, C. T., Tsay, J. H. & Fuller, W. A. (2004). Weighting sample data subject to independent controls. *Survey Methodology* **30**, 35–44.

Jagers, P. (1986). Post-stratification against bias in sampling. *International Statistical Review* **54**, 159–167.

Jagers, P., Odén, A. & Trulsson, L. (1985). Post-stratification and ratio estimation: usages of auxiliary information in survey sampling and opinion polls. *International Statistical Review* **53**, 221–238.

Jessen, R. J. (1978). *Statistical Survey Techniques*. New York: Wiley.

Jewett, R. S. & Judkins, D. R. (1988). Multivariate stratification with size constraints. *SIAM Journal on Scientific and Statistical Computing* **9**, 1091–1097.

Johnson, N. L., Kotz, S. & Balakrishnan, N. (1997). *Discrete Multivariate Distributions*. New York: Wiley.

Johnson, N. L. & Smith, H. (1969). *New Developments in Survey Sampling*. New York: Wiley.

Jonasson, J. et al. (2013). The BK inequality for pivotal sampling a.k.a. the Srinivasan sampling process. *Electronic Communications in Probability* **18**.

Jones, H. L. (1974). Jackknife estimation of functions of stratum means. *Biometrika* **61**, 343–348.

Kabe, D. G. (1976). Inverse moments of discrete distributions. *Canadian Journal Statistics* **4**, 133–141.

Kalton, G. (1983). *Introduction to Survey Sampling*. No. 35 in Quantitative Applications in the Social Sciences. Newbury Park, London, New Delhi: SAGE Publications.

Kalton, G. & Anderson, D. W. (1986). Sampling rare populations. *Journal of the Royal Statistical Society* **A149**, 65–82.

Kauermann, G. & Küchenhoff, H. (2010). *Stichproben: Methoden und praktische Umsetzung mit R*. Berlin Heidelberg: Springer.

Kaufman, S. (1998). A bootstrap variance estimator for systematic PPS sampling. In *Proceedings of the Section on Survey Research Methods*. American Statistical Association, pp. 769–774.

Kaufman, S. (1999). Using the bootstrap to estimate the variance from a single systematic PPS sample. In *Proceedings of the Section on Survey Research Methods*. American Statistical Association, pp. 683–688.

Kemp, C. D. & Kemp, A. W. (1987). Rapid generation of frequency tables. *Applied Statistics* **36**, 277–282.

Keverberg, B. d. (1827). Notes. *Nouveaux mémoires de l'Académie royale des sciences et belles lettres de Bruxelles* **4**, 175–192.

Keyfitz, N. (1951). Sampling with probabilities proportional to size: adjustmentfor changes in the probabilities. *Journal of the American Statistical Association* **46**, 105–109.

Kiær, A. N. (1896). Observations et expériences concernant des dénombrements représentatifs. *Bulletin de l'Institut International de Statistique* **9**, 176–183.

Kiær, A. N. (1899). Sur les méthodes représentatives ou typologiques appliquées à la statistique. *Bulletin de l'Institut International de Statistique* **11**, 180–185.

Kiær, A. N. (1903). Sur les méthodes représentatives ou typologiques. *Bulletin de l'Institut International de Statistique* **13**, 66–78.

Kiær, A. N. (1905). Discours sans intitulé sur la méthode représentative. *Bulletin de l'Institut International de Statistique* **14**, 119–134.

Kim, J. K. (2017). *Survey Sampling (Pyobon josaron, in Korean)*. Paju-si, Gyeonggi-do, Republic of Korea: 2nd edition, Freedom Academy Inc. (Jayu Akademisa).

Kim, J. K., Breidt, F. J. & Opsomer, J. D. (2005). Nonparametric regression estimation of finite population totals under two-stage sampling. Technical report. Department of Statistics, Iowa State University.

Kim, J. K. & Kim, J. J. (2007). Nonresponse weighting adjustment using estimated response probability. *Canadian Journal of Statistics* **35**, 501–514.

Kim, J. K. & Rao, J. N. K. (2009). A unified approach to linearization variance estimation from survey data after imputation for item nonresponse. *Biometrika* **96**, 917–932.

Kim, J. K. & Shao, J. (2013). *Statistical Methods for Handling Incomplete Data*. Boca Raton, FL: Chapman & Hall/CRC.

Kish, L. (1965). *Survey Sampling*. New York: Wiley.

Kish, L. (1976). Optima and proxima in linear sample designs. *Journal of Royal Statistical Society* **A139**, 80–95.

Kish, L. (1989). *Sampling Method for Agricultural Surveys, Statistical Development Series*. Rome: Food and Agriculture Organization of the United Nations.

Kish, L. (1995). *Survey Sampling*. New York: Wiley.

Kish, L. & Anderson, D. W. (1978). Multivariate and multipurpose stratification (Corr: V74 p516). *Journal of the American Statistical Association* **73**, 24–34.

Kish, L. & Scott, A. J. (1971). Retaining units after changing strata and probabilities. *Journal of the American Statistical Association* **66**, 461–470.

Knuth, D. E. (1981). *The Art of Computer Programming (Volume II): Seminumerical Algorithms*. Reading, MA: Addison-Wesley.

Koeijers, C. A. J. & Willeboordse, A. J. (1995). Reference manual on design and implementation of business surveys. Technical report. Statistics Netherlands.

Konijn, H. S. (1973). *Statistical Theory of Sample Survey Design and Analysis*. Amsterdam: North-Holland.

Konijn, H. S. (1981). Biases, variances and covariances of raking ratio estimators for marginal and cell totals and averages of observed characteristics. *Metrika* **28**, 109–121.

Kott, P. S. (2004). Comment on Demnati and Rao: Linearization variance estimators for survey data. *Survey Methodology* **30**, 27–28.

Kott, P. S. (2006). Using calibration weighting to adjust for nonresponse and coverage errors. *Survey Methodology* **32**, 133–142.

Kott, P. S. (2009). Calibration weighting: Combining probability samples and linear prediction models. In *Handbook of Statistics: Volume 29, Part B, Sampling*, D. Pfeffermann & C. R. Rao, eds. New York, Amsterdam: Elsevier/North-Holland, pp. 55–82.

Kott, P. S., Amrhein, J. F. & Hicks, S. D. (1998). Sampling and estimation from multiple list frames. *Survey Methodology* **24**, 3–9.

Kott, P. S. & Chang, T. (2010). Using calibration weighting to adjust for nonignorable unit nonresponse. *Journal of the American Statistical Association* **105**, 1265–1275.

Kovacevic, M. S. (1997). Calibration estimation of cumulative distribution and quantile functions from survey data. In *Proceedings of the Survey Methods Section, Statistical Society of Canada*. pp. 139–144.

Kovacevic, M. S. & Binder, D. A. (1997). Variance estimation for measures of income inequality and polarization – the estimating equations approach. *Journal of Official Statistics* **13**, 41–58.

Kovar, J. G., Rao, J. N. K. & Wu, C. (1988). Bootstrap and other methods to measure errors in survey estimates. *Canadian Journal of Statistics* **16**, 25–45.

Krewski, D. & Rao, J. N. K. (1981a). Inference from stratified samples: properties of linearization, jackknife and balanced repeated replication methods. *Annals of Statistics* **9**, 1010–1019.

Krewski, D. & Rao, J. N. K. (1981b). Inference from stratified samples: properties of the linearization, jackknife and balanced repeated replication methods. *Annals of Statistics* **9**, 1010–1019.

Krishnaiah, P. R. & Rao, C. R. (1994). *Handbook of Statistics: Volume 6, Sampling*. New York: Elsevier Science Publishers.

Kuhn, T. S. (1970). *The Structure of Scientific Revolutions*. Paris: International Encyclopedia of Unified Science.

Kuk, A. Y. C. (1989). Double bootstrap estimation of variance under systematic sampling with probability proportional to size. *Journal of Statistical Computation and Simulation* **31**, 73–82.

Langel, M. & Tillé, Y. (2011a). Corrado Gini, a pioneer in balanced sampling and inequality theory. *Metron* **69**, 45–65.

Langel, M. & Tillé, Y. (2011b). Statistical inference for the quintile share ratio. *Journal of Statistical Planning and Inference* **141**, 2976–2985.

Langel, M. & Tillé, Y. (2013). Variance estimation of the Gini index: Revisiting a result several times published. *Journal of the Royal Statistical Society* **A176**, 521–540.

Lanke, J. (1975). *Some Contributions to the Theory of Survey Sampling*. PhD Thesis, Department of Mathematical Statistics, University of Lund, Sweden.

Laplace, P.-S. (1847). *Théorie analytique des probabilités*. Paris: Imprimerie royale.

Lavallée, P. (2002). *Le sondage indirect ou la méthode généralisée du partage des poids*. Paris: Ellipses.

Lavallée, P. (2007). *Indirect Sampling*. New York: Springer.

Lavallée, P. & Hidiroglou, M. A. (1988). On the stratification of skewed populations. *Survey Methodology* **14**, 35–45.

Le Guennec, J. & Sautory, O. (2002). CALMAR2: une nouvelle version de la macro CALMAR de redressement d'échantillon par calage. In *Actes des Journées de Méthodologie, Insee, Paris*. Paris, pp. 33–38.

Lednicki, B. & Wieczorkowski, R. (2003). Optimal stratification and sample allocation between subpopulations and strata. *Statistics in Transition* **6**, 287–305.

Lee, H., Rancourt, E. & Särndal, C.-E. (2002). Variance estimation from survey data under single imputation. In *In Survey Nonresponse*, R. M. Groves, D. A. Dillman, J. L. Eltinge & R. J. A. Little, eds. New York: Wiley, pp. 315–328.

Leiten, E. (2005). On variance estimation of the Gini coefficient. In *Workshop on Survey Sampling Theory and Methodology*. Vilnius: Statistics Lithuania, pp. 115–118.

Lemaître, G. & Dufour, J. (1987). An integrated method for weighting persons and families. *Survey Methodology* **13**, 199–207.

Lemel, Y. (1976). Une généralisation de la méthode du quotient pour le redressement des enquêtes par sondages. *Annales de l'Insee* **22-23**, 273–281.

Lepkowski, J. M. & Groves, R. M. (1986). A mean squared error model for multiple frame, mixed mode survey design. *Journal of the American Statistical Association* **81**, 930–937.

Lesage, É. (2008). Contraintes d'équilibrage non linéraires. In *Méthodes de sondage : applications aux enquêtes longitudinales, à la santé et aux enquêtes électorales*, P. Guilbert, D. Haziza, A. Ruiz-Gazen & Y. Tillé, eds. Paris: Dunod, pp. 285–289.

Lesage, É., Haziza, D. & D'Haultfoeuille, X. (2018). A cautionary tale on instrumental calibration for the treatment of nonignorable unit nonresponse in surveys. *Journal of the American Statistical Association* **114**, 906–915.

Little, R. J. A. & Rubin, D. B. (1987). *Statistical Analysis with Missing Data*. New York: Wiley.

Little, R. J. A. & Rubin, D. B. (2014). *Statistical Analysis with Missing Data*. New York: Wiley.

Lohr, S. L. (1999). *Sampling: Design and Analysis*. Belmont, CA: Duxbury Press.

Lohr, S. L. (2000). *Muestreo: diseño y análisis*. Matemáticas Thomson. México: S.A. Ediciones Paraninfo.

Lohr, S. L. (2007). Recent developments in multiple frame surveys. In *ASA Proceedings of the Section on Survey Research Methods*. American Statistical Association, pp. 3257–3264.

Lohr, S. L. (2009a). Multiple frame surveys. In *Handbook of Statistics: Volume 29, Sample Surveys: Theory, Methods and Inference*. Amsterdam: North Holland, pp. 71–88.

Lohr, S. L. (2009b). *Sampling: Design and Analysis*. Boston: Brooks/Cole.

Lohr, S. L. & Rao, J. N. K. (2000). Inference in dual frame surveys. *Journal of the American Statistical Association* **95**, 271–280.

Lohr, S. L. & Rao, J. N. K. (2006). Estimation in multiple-frame surveys. *Journal of the American Statistical Association* **101**, 1019–1030.

Lomnicki, Z. A. (1952). The standard error of Gini's mean difference. *Annals of Mathematical Statistics* **23**, 635–637.

Lorenz, M. O. (1905). Methods of measuring the concentration of wealth. *Publications of the American Statistical Association* **9**, 209–219.

Loukas, S. & Kemp, C. D. (1983). On computer sampling from trivariate and multivariate discrete distribution. *Journal of Statistical Computation and Simulation* **17**, 113–123.

Lundström, S. & Särndal, C.-E. (1999). Calibration as a standard method for treatment of nonresponse. *Journal of Official Statistics* **15**, 305–327.

Mac Carthy, P. J. & Snowden, C. B. (1985). The bootstrap and finite population sampling. Technical report. Data Evaluation and Methods Research, Series 2, No. 95. U.S. Department of Health and Human Services, Washington.

Madow, W. G. (1948). On the limiting distribution based on samples from finite universes. *Annals of Mathematical Statistics* **19**, 535–545.

Madow, W. G. (1949). On the theory of systematic sampling, II. *Annals of Mathematical Statistics* **20**, 333–354.

Mandallaz, D. (2008). *Sampling Techniques for Forest Inventories*. Boca Raton, FL: Chapman & Hall/CRC.

Marazzi, A. & Tillé, Y. (2017). Using past experience to optimize audit sampling design. *Review of Quantitative Finance and Accounting* **49**, 435–462.

Matei, A. & Skinner, C. J. (2009). Optimal sample coordination using controlled selection. *Journal of Statistical Planning and Inference* **139**, 3112–3121.

Matei, A. & Tillé, Y. (2005a). Evaluation of variance approximations and estimators in maximum entropy sampling with unequal probability and fixed sample size. *Journal of Official Statistics* **21**, 543–570.

Matei, A. & Tillé, Y. (2005b). Maximal and minimal sample co-ordination. *Sankhyā* **67**, 590–612.

Matei, A. & Tillé, Y. (2007). Computational aspects of order πps sampling schemes. *Computational Statistics and Data Analysis* **51**, 3703–3717.

McDonald, T. (2016). *SDraw: Spatially Balanced Sample Draws for Spatial Objects*. R package version 2.1.3.

McLeod, A. I. & Bellhouse, D. R. (1983). A convenient algorithm for drawing a simple random sampling. *Applied Statistics* **32**, 182–184.

Mecatti, F. (2007). A single frame multiplicity estimator for multiple frame surveys. *Survey Methodology* **33**, 151–157.

Merkouris, T. (2004). Combining independent regression estimators from multiple surveys. *Journal of the American Statistical Association* **99**, 1131–1139.

Merkouris, T. (2010). Combining information from multiple surveys by using regression for efficient small domain estimation. *Journal of the Royal Statistical Society* **B72**, 27–48.

Mészáros, P. (1999). A program for sample co-ordination: Salomon. In *Proceedings of the International Seminar on Exchange of Technology and Knowhow*. Prague, pp. 125–130.

Montanari, G. E. (1987). Post sampling efficient QR-prediction in large sample survey. *International Statistical Review* **55**, 191–202.

Moran, P. A. P. (1950). Notes on continuous stochastic phenomena. *Biometrika* **37**, 17–23.

Murthy, M. N. (1967). *Sampling Theory and Methods*. Calcutta: Statistical Publishing Society.

Murthy, M. N. & Rao, T. J. (1988). Systematic sampling with illustrative examples. In *Handbook of Statistics: Volume 6, Sampling*, P. R. Krishnaiah & C. R. Rao, eds. New York, Amsterdam: Elsevier/North-Holland, pp. 147–185.

Narain, R. D. (1951). On sampling without replacement with varying probabilities. *Journal of the Indian Society of Agricultural Statistics* **3**, 169–174.

Nedyalkova, D., Pea, J. & Tillé, Y. (2006). A review of some current methods of coordination of stratified samples. Introduction and comparison of new methods based on microstrata. Technical report. Université de Neuchâtel.

Nedyalkova, D., Pea, J. & Tillé, Y. (2008). Sampling procedures for coordinating stratified samples: Methods based on microstrata. *International Statistical Review* **76**, 368–386.

Nedyalkova, D., Qualité, L. & Tillé, Y. (2009). General framework for the rotation of units in repeated survey sampling. *Statistica Neerlandica* **63**, 269–293.

Nedyalkova, D. & Tillé, Y. (2008). Optimal sampling and estimation strategies under linear model. *Biometrika* **95**, 521–537.

Neyman, J. (1934). On the two different aspects of the representative method: The method of stratified sampling and the method of purposive selection. *Journal of the Royal Statistical Society* **97**, 558–606.

Ng, M. P. & Donadio, M. E. (2006). Computing inclusion probabilities for order sampling. *Journal of Statistical Planning and Inference* **136**, 4026–4042.

Nieuwenbroek, N. J. (1993). An integrated method for weighting characteristics of persons and households using the linear regression estimator. Technical report. Central Bureau of Statistics, The Netherlands.

Nieuwenbroek, N. J. & Boonstra, H. J. (2002). Bascula 4. 0 for weighting sample survey data with estimation of variances. *The Survey Statistician, Software Reviews* **46**, 6–11.

Nigam, A. K. & Rao, J. N. K. (1996). On balanced bootstrap for stratified multistage samples. *Statistica Sinica* **6**, 199–214.

Ohlsson, E. (1990). Sequential Poisson sampling from a business register and its application to the Swedish Consumer Price Index. R&d report. Statistics Sweden, Stockholm.

Ohlsson, E. (1995). Coordination of samples using permanent random numbers. In *Business Survey Methods*, B. G. Cox, D. A. Binder, B. N. Chinnappa, A. Christianson, M. J. Colledge & P. S. Kott, eds. New York: Wiley, pp. 153–169.

Ohlsson, E. (1998). Sequential Poisson sampling. *Journal of Official Statistics* **14**, 149–162.

Osier, G. (2009). Variance estimation for complex indicators of poverty and inequality using linearization techniques. *Survey Research Methods* **3**, 167–195.

Owen, D. B. & Cochran, W. G., eds. (1976). *On the History of Statistics and Probability, Proceedings of a Symposium on the American Mathematical Heritage, to celebrate the bicentennial of the United States of America, held at Southern Methodist University*, New York: Marcel Dekker.

Pathak, P. K. (1961). On simple random sampling with replacement. *Sankhyā* **A24**, 415–420.

Pathak, P. K. (1962). On the evaluation of moments of distinct units in a sample. *Sankhyā* **A23**, 287–302.

Pathak, P. K. (1988). Simple random sampling. In *Handbook of Statistics: Volume 6, Sampling*, P. R. Krishnaiah & C. R. Rao, eds. New York, Amsterdam: Elsevier/North-Holland, pp. 97–109.

Patterson, H. D. (1950). Sampling on successive occasions with partial replacement of units. *Journal of the Royal Statistical Society* **B12**, 241–255.

Pea, J., Qualité, L. & Tillé, Y. (2007). Systematic sampling is a minimal support design. *Computational Statistics & Data Analysis* **51**, 5591–5602.

Pebesma, E. J. & Bivand, R. S. (2005). Classes and methods for spatial data in R. *R News* **5**, 9–13.

Pérez López, C. (2000). *Técnicas de muestreo estadístico: teoría, práctica y aplicaciones informáticas*. México: Alfaomega Grupo Editor.

Pfeffermann, D. & Rao, C. R. eds. (2009a). Sample surveys: inference and analysis. In *Handbook of Statistics: Volume 29A*. Elsevier.

Pfeffermann, D. & Rao, C. R. eds. (2009b). *Sample surveys: inference and analysis*. In *Handbook of Statistics: Volume 29A*. Elsevier.

Pollock, K. H. (1981). Capture-recapture models: A review of current methods, assumptions and experimental design. In *Estimating Numbers of Terrestrial Birds, Studies in Avian Biology*, J. M. Scott & C. J. Ralph, eds., vol. 6. Oxford: Pergamon, pp. 426–435.

Pollock, K. H. (2000). Capture-recapture models. *Journal of the American Statistical Association* **95**, 293–296.

Pollock, K. H., Nichols, J. D., Brownie, C. & Hines, J. E. (1990). *Statistical Inference for Capture-recapture Experiments*. Bethesda, MD: Wildlife Society.

Qualité, L. (2009). *Unequal probability sampling and repeated surveys*. PhD Thesis, Université de Neuchâtel, Neuchâtel.

Quenouille, M. H. (1949). Problems in plane sampling. *The Annals of Mathematical Statistics* **20**, 355–375.

Quételet, A. (1846). *Lettres à S. A. R. le Duc régnant de Saxe-Cobourg et Gotha sur la théorie des probabilités appliquées aux sciences morales et politiques*. Bruxelles: M. Hayez.

Raghunathan, T. E., Lepkowski, J. M., Van Hoewyk, J. & Solenberger, P. W. (2001). A multivariate technique for multiply imputing missing values using a sequence of regression models. *Survey Methodology* **27**, 85–95.

Raghunathan, T. E., Solenberger, P. W. & Van Hoewyk, J. (2002). IVEware: Imputation and variance estimation software. Technical report. Survey Methodology Program, Survey Research Center, Institute for Social Research, University of Michigan, Ann Arbor.

Raghunathan, T. E., Solenberger, P. W. & Van Hoewyk, J. (2016). IVEware: Imputation and variance estimation software. 2007. Technical report. Survey Methodology Program, Institute for Social Research, University of Michigan, Ann Arbor.

Raj, D. (1968). *Sampling Theory*. New York:McGraw-Hill.

Raj, D. & Khamis, S. D. (1958). Some remarks on sampling with replacement. *Annals of Mathematical Statistics* **29**, 550–557.

Rao, J. N. K. (1963). On three procedures of unequal probability sampling without replacement. *Journal of the American Statistical Association* **58**, 202–215.

Rao, J. N. K. (1965). On two simple schemas of unequal probability sampling without replacement. *Journal of the Indian Statistical Association* **3**, 173–180.

Rao, J. N. K. (1990). Variance estimation under imputation for missing data. Technical report. Statistics Canada, Ottawa.

Rao, J. N. K. (1994). Estimating totals and distribution functions using auxiliary information at the estimation stage. *Journal of Official Statistics* **10**, 153–165.

Rao, J. N. K. (2003). *Small Area Estimation*. New York: Wiley.

Rao, J. N. K. & Bayless, D. L. (1969). An empirical study of the stabilities of estimators and variance estimators in unequal probability sampling of two units per stratum. *Journal of the American Statistical Association* **64**, 540–549.

Rao, J. N. K. & Molina, I. (2015). *Small Area Estimation*. New York: Wiley.

Rao, J. N. K. & Shao, J. (1992). Jackknife variance estimation with survey data under hot-deck imputation. *Biometrika* **79**, 811–822.

Rao, J. N. K. & Sitter, R. R. (1995). Variance estimation under two-phase sampling with application to imputation for missing data. *Biometrika* **82**, 453–460.

Rao, J. N. K. & Skinner, C. J. (1996). Estimation in dual frame surveys with complex designs. In *Proceedings of the Survey Methods Section of the Statistical Society of Canada*. Statistical Society of Canada, pp. 63–68.

Rao, J. N. K. & Tausi, M. (2004). Estimating function jackknife variance estimators under stratified multistage sampling. *Communications in Statistics: Theory and Methods* **33**, 2087–2095.

Rao, C. R. (1971). Some aspects of statistical inference in problems of sampling from finite population. In *Foundations of Statistical Inference*, V. P. Godambe & D. A. Sprott, (dir.). Toronto, Montréal : Holt, Rinehart and Winston, pp. 177–202.

Rebecq, A. (2017). *icarus: Calibrates and Reweights Units in Samples, likelihood, estimation*. R package version 0.3.0.

Rebecq, A. (2019). *Méthodes de sondage pour les données massives*. PhD Thesis, Université Paris Nanterre.

Ren, R. (2000). *Utilisation d'information auxiliaire par calage sur fonction de répartition*. PhD Thesis, Université Paris Dauphine.

Ren, R. & Ma, X. (1996). *Sampling Survey Theory and Its Applications*. Kaifeng, China: Henan University Press. In Chinese.

Rivest, L.-P. (2002). A generalization of the Lavellée-Hidiroglou algorithm for stratification in business surveys. *Survey Methodology* **28**, 191–198.

Rivière, P. (1998). Description of the chosen method: Deliverable 2 of supcom 1996 project (part "co-ordination of samples"). Report. Eurostat.

Rivière, P. (1999). Coordination of samples: the microstrata methodology. In *13th International Roundtable on Business Survey Frames*. Paris: Insee.

Rivière, P. (2001a). Coordinating samples using the microstrata methodology. In *Proceedings of Statistics Canada Symposium 2001, Achieving Data Quality in a Statistical Agency: A Methodological Perspective*. Ottawa, pp. 1–10.

Rivière, P. (2001b). Random permutations of random vectors as a way of co-ordinating samples. Report. University of Southampton.

Rosén, B. (1972a). Asymptotic theory for successive sampling I. *Annals of Mathematical Statistics* **43**, 373–397.

Rosén, B. (1972b). Asymptotic theory for successive sampling II. *Annals of Mathematical Statistics* **43**, 748–776.

Rosén, B. (1995). Asymptotic theory for order sampling. R&D report. Statistics Sweden, Stockholm.

Rosén, B. (1997a). Asymptotic theory for order sampling. *Journal of Statistical Planning and Inference* **62**, 135–158.

Rosén, B. (1997b). On sampling with probability proportional to size. *Journal of Statistical Planning and Inference* **62**, 159–191.

Rosén, B. (1998). On inclusion probabilities of order sampling. R&D Report, Research – Methods – Development.

Rosén, B. (2000). On inclusion probabilties for order πps sampling. *Journal of Statistical Planning and Inference* **90**, 117–143.

Roy, G. & Vanheuverzwyn, A. (2001). Redressement par la macro CALMAR : applications et pistes d'amélioration. *Traitements des fichiers d'enquêtes, éditions PUG*, 31–46.

Royall, R. M. (1970a). Finite population sampling – On labels in estimation. *Annals of Mathematical Statistics* **41**, 1774–1779.

Royall, R. M. (1970b). On finite population sampling theory under certain linear regression models. *Biometrika* **57**, 377–387.

Royall, R. M. (1971). Linear regression models in finite population sampling theory. In *Foundations of Statistical Inference*, V. P. Godambe & D. A. Sprott, eds. Toronto, Montréal: Holt, Rinehart et Winston, pp. 259–279.

Royall, R. M. (1976a). Likelihood functions in finite population sampling theory. *Biometrika* **63**, 605–614.

Royall, R. M. (1976b). The linear least squares prediction approach to two-stage sampling. *Journal of the American Statistical Association* **71**, 657–664.

Royall, R. M. (1988). The prediction approach to sampling theory. In *Sampling*, P. R. Krishnaiah & C. R. Rao, eds., vol. 6 of *Handbook of Statistics*. Amsterdam: Elsevier, pp. 399–413.

Royall, R. M. (1992a). The model based (prediction) approach to finite population sampling theory. In *Current issues in statistical inference: Essays in honor of D. Basu*, M. Ghosh & P. K. Pathak, eds., vol. 17 of *Lecture Notes, Monograph Series*. Institute of Mathematical Statistics, pp. 225–240.

Royall, R. M. (1992b). Robustness and optimal design under prediction models for finite populations. *Survey Methodology* **18**, 179–185.

Royall, R. M. & Cumberland, W. G. (1978). Variance estimation in finite population sampling. *Journal of the American Statistical Association* **73**, 351–358.

Royall, R. M. & Cumberland, W. G. (1981a). An empirical study of the ratio estimator and its variance. *Journal of the American Statistical Association* **76**, 66–77.

Royall, R. M. & Cumberland, W. G. (1981b). The finite population linear regression estimator and estimators of its variance – an empirical study. *Journal of the American Statistical Association* **76**, 924–930.

Royall, R. M. & Herson, J. (1973a). Robust estimation in finite populations I. *Journal of the American Statistical Association* **68**, 880–889.

Royall, R. M. & Herson, J. (1973b). Robust estimation in finite populations II: Stratification on a size variable. *Journal of the American Statistical Association* **68**, 890–893.

Rubin, D. B. (1976). Inference and missing data. *Biometrika* **63**, 581–592.

Rubin-Bleuer, S. (2002). Report on Rivière's random permutations method of sampling co-ordination. Report 2002-07-26. Statistics Canada.

Rubin-Bleuer, S. (2011). The proportional hazards model for survey data from independent and clustered super-populations. *Journal of Multivariate Analysis* **102**, 884–895.

Saavedra, P. J. (1995). Fixed sample size PPS approximations with a permanent random number. In *ASA Proceedings of the Section on Survey Research Methods*. American Statistical Association, pp. 697–700.

Saegusa, T. (2015). Variance estimation under two-phase sampling. *Scandinavian Journal of Statistics* **42**, 1078–1091.

Saigo, H., Shao, J. & Sitter, R. R. (2001). A repeated half-sample bootstrap and balanced repeated replications for randomly imputed data. *Survey Methodology* **27**, 189–196.

Sampford, M. R. (1967). On sampling without replacement with unequal probabilities of selection. *Biometrika* **54**, 499–513.

Sandström, A., Wretman, J. H. & Waldén, B. (1985). Variance estimators of the Gini coefficient: Simple random sampling. *Metron* **43**, 41–70.

Särndal, C.-E. (1980). On π-inverse weighting versus best linear unbiased weighting in probability sampling. *Biometrika* **67**, 639–650.

Särndal, C.-E. (1982). Implication of survey design for generalized regression estimation of linear functions. *Journal of Statistical Planning and Inference* **7**, 155–170.

Särndal, C.-E. (1984). Design-consistent versus model dependent estimation for small domains. *Journal of the American Statistical Association* **68**, 880–889.

Särndal, C.-E. (1992). Methods for estimating the precision of survey estimates when imputation has been used. *Survey Methodology* **18**, 241–252.

Särndal, C.-E. (2007). The calibration approach in survey theory and practice. *Survey Methodology* **33**, 99–119.

Särndal, C.-E. & Lundström, S. (2005). *Estimation in Surveys with Nonresponse*. New York: Wiley.

Särndal, C.-E. & Swensson, B. (1987). A general view of estimation for two phases of selection with applications to two-phase sampling and non-response. *International Statistical Review* **55**, 279–294.

Särndal, C.-E., Swensson, B. & Wretman, J. H. (1989). The weighted residual technique for estimating the variance of the general regression estimator of the finite population total. *Biometrika* **76**, 527–537.

Särndal, C.-E., Swensson, B. & Wretman, J. H. (1992). *Model Assisted Survey Sampling*. New York: Springer.

Särndal, C.-E. & Wright, R. L. (1984). Cosmetic form of estimators in survey sampling. *Scandinavian Journal of Statistics* **11**, 146–156.

Sautory, O. & Le Guennec, J. (2003). La macro CALMAR2 : redressement d'un échantillon par calage sur marges – documentation de l'utilisateur. Technical report. Insee, Paris.

Schwarz, C. J. & Seber, G. A. F. (1999). Estimating animal abundance: Review III. *Statistical Science* **14**, 427–456.

Seber, G. A. F. (1970). The effects of trap response on tag recapture estimates. *Biometrics* **26**, 13–22.

Seber, G. A. F. (1986). A review of estimating animal abundance. *Biometrics* **42**, 267–292.

Seber, G. A. F. (1992). A review of estimating animal abundance II. *International Statistical Review* **60**, 129–166.

Seber, G. A. F. (2002). *The Estimation of Animal Abundance and Related Parameters*. London: Blackburn Press.

Sen, A. R. (1953). On the estimate of the variance in sampling with varying probabilities. *Journal of the Indian Society of Agricultural Statistics* **5**, 119–127.

Serfling, R. J. (1980). *Approximation Theorems of Mathematical Statistics*. New York: Wiley.

Sethi, V. K. (1963). A note on the optimum stratification of populations for estimating the population means. *Australian Journal of Statistics* **5**, 20–33.

Shao, J. (2003). Impact of the bootstrap on sample surveys. *Statistical Science* **18**, 191–198.

Shao, J. & Steel, P. (1999). Variance estimation for survey data with composite imputation and nonneglible sampling fractions. *Journal of the American Statistical Association* **94**, 254–265.

Shao, J. & Tu, D. (1995). *The Jacknife and Bootstrap*. New York: Springer-Verlag.

Shaw, R., Sigman, R. S. & Zayatz, L. (1995). Multivariate stratification and sample allocation satisfying multiple constraints for surveys of post-office mail volumes. In *ASA Proceedings of the Section on Survey Research Methods*. American Statistical Association, pp. 763–768.

Sheynin, O. B. (1986). Adolphe Quételet as a statistician. *Archive for History of Exact Science* **36**, 282–325.

Sitter, R. R. (1992a). Comparing three bootstrap methods for survey data. *Canadian Journal of Statistics* **20**, 135–154.

Sitter, R. R. (1992b). A resampling procedure for complex survey data. *Journal of the American Statistical Association* **87**, 755–765.

Skinner, C. J. (1991). On the efficiency of raking estimation for multiple frame surveys. *Journal of the American Statistical Association* **86**, 779–784.

Skinner, C. J. (2011). Inverse probability weighting for clustered nonresponse. *Biometrika* **98**, 953–966.

Skinner, C. J., Holmes, D. J. & Holt, D. (1994). Multiple frame sampling for multivariate stratification. *International Statistical Review* **62**, 333–347.

Skinner, C. J., Holt, D. & Smith, T. M. F. (1989). *Analysis of Complex Surveys*. New York: Wiley.

Skinner, C. J. & Rao, J. N. K. (2001). Jackknife variance estimation for multivariate statistics under hot deck imputation from common donors. *Journal of Statistical Planning and Inference* **79**, 149–167.

Srinivasan, A. (2001). Distributions on level-sets with applications to approximation algorithms. In *Proceedings of the 42nd IEEE Symposium on Foundations of Computer Science, 2001*. IEEE, pp. 588–597.

Steel, D. G. & Clark, R. G. (2007). Person-level and household-level regression estimation in household surveys. *Survey Methodology* **33**, 51–60.

Stenger, H. (1985). *Stichproben*. Würzburg: Physica-Verlag.

Stephan, F. F. (1942). An iterative method of adjusting sample frequency data tables when expected marginal totals are known. *Annals of Mathematical Statistics* **13**, 166–178.

Stephan, F. F. (1945). The expected value and variance of the reciprocal and other negative powers of a positive Bernoullian variate. *Annals of Mathematical Statistics* **16**, 50–61.

Stephan, F. F. (1948). History of the uses of modern sampling procedures. *Journal of the American Statistical Association* **43**, 12–49.

Stevens Jr., D. L. & Olsen, A. R. (1999). Spatially restricted surveys over time for aquatic resources. *Journal of Agricultural, Biological, and Environmental Statistics* **4**, 415–428.

Stevens Jr., D. L. & Olsen, A. R. (2003). Variance estimation for spatially balanced samples of environmental resources. *Environmetrics* **14**, 593–610.

Stevens Jr., D. L. & Olsen, A. R. (2004). Spatially balanced sampling of natural resources. *Journal of the American Statistical Association* **99**, 262–278.

Stigler, S. M. (1986). *The History of Statistics*. Cambridge-London: Harvard University Press.

Stukel, D. M., Hidiroglou, M. A. & Särndal, C.-E. (1996). Variance estimation for calibration estimators: A comparison of jackknifing versus Taylor linearization. *Survey Methodology* **22**, 117–125.

Sukhatme, P. V. (1954). *Sampling Theory of Surveys with Applications*. New Delhi: Indian Society of Agricultural Statistics.

Sukhatme, P. V. & Sukhatme, B. V. (1970). *Sampling Theory of Surveys with Applications*. Calcutta, India: Asian Publishing House.

Sunter, A. B. (1977). List sequential sampling with equal or unequal probabilities without replacement. *Applied Statistics* **26**, 261–268.

Theobald, D. M., Stevens Jr., D. L., White, D. E., Urquhart, N. S., Olsen, A. R. & Norman, J. B. (2007). Using GIS to generate spatially balanced random survey designs for natural resource applications. *Environmental Management* **40**, 134–146.

Thionet, P. (1953). *La théorie des sondages.* Paris: Institut National de la Statistique et des Études Économiques, Études théoriques vol. 5, Imprimerie nationale.

Thionet, P. (1959). L'ajustement des résultats des sondages sur ceux des dénombrements. *Revue de l'Institut International de Statistique* **27**, 8–25.

Thionet, P. (1976). Construction et reconstruction de tableaux statistiques. *Annales de l'Insee* **22-23**, 5–27.

Thompson, M. E. (1997). *Theory of Sample Surveys.* London: Chapman and Hall.

Thompson, S. K. (1988). Adaptive sampling. In *Proceedings of the Section on Survey Research Methods of the American Statistical Association.* pp. 784–786.

Thompson, S. K. (1990). Adaptive cluster sampling. *Journal of the American Statistical Association* **85**, 1050–1059.

Thompson, S. K. (1991a). Adaptive cluster sampling: Designs with primary and secondary units. *Biometrics* **47**, 1103–1115.

Thompson, S. K. (1991b). Stratified adaptive cluster sampling. *Biometrika* **78**, 389–397.

Thompson, S. K. (1992). *Sampling.* New York: Wiley.

Thompson, S. K. (2012). *Sampling.* New York: Wiley.

Thompson, S. K. & Ramsey, F. L. (1983). Adaptive sampling of animal populations. Technical report. Oregon State University, Corvallis, Oregon, USA.

Thompson, S. K. & Seber, G. A. F. (1996). *Adaptive Sampling.* New York: Wiley.

Tillé, Y. (1996a). An elimination procedure of unequal probability sampling without replacement. *Biometrika* **83**, 238–241.

Tillé, Y. (1996b). Some remarks on unequal probability sampling designs without replacement. *Annales d'Économie et de Statistique* **44**, 177–189.

Tillé, Y. (1998). Estimation in surveys using conditional inclusion probabilities: simple random sampling. *International Statistical Review* **66**, 303–322.

Tillé, Y. (1999). Estimation in surveys using conditional inclusion probabilities: complex design. *Survey Methodology* **25**, 57–66.

Tillé, Y. (2002). Unbiased calibrated estimation on a distribution function in simple random sampling. *Survey Methodology* **28**, 77–85.

Tillé, Y. (2006). *Sampling Algorithms.* New York: Springer.

Tillé, Y. (2010). *Muestreo equilibrado eficiente: el método del cubo.* Vitoria-Gasteiz: Euskal Estatistika Erakundea/Instituto Vasco de Estadística.

Tillé, Y. (2011). Ten years of balanced sampling with the cube method: an appraisal. *Survey Methodology* **37**, 215–226.

Tillé, Y. (2018). Fast implementation of Fuller's unequal probability sampling method. Technical report. University of Neuchâtel.

Tillé, Y., Dickson, M. M., Espa, G. & Giuliani, D. (2018). Measuring the spatial balance of a sample: A new measure based on the Moran's *I* index. *Spatial Statistics* **23**, 182–192.

Tillé, Y. & Favre, A.-C. (2004). Co-ordination, combination and extension of optimal balanced samples. *Biometrika* **91**, 913–927.

Tillé, Y. & Matei, A. (2016). *Sampling: Survey Sampling.* R package version 2.8.

Tillé, Y., Newman, J. A. & Healy, S. D. (1996). New tests for departures from random behavior in spatial memory experiments. *Animal Learning and Behavior* **24**, 327–340.

Tillé, Y. & Vallée, A.-A. (2017a). Balanced imputation for Swiss cheese nonresponse. Invited Conference at the 2017 Annual Meeting of the Canadian Statistical Society.

Tillé, Y. & Vallée, A.-A. (2017b). Revisiting variance decomposition when independent samples intersect. *Statistics and Probability Letters* **130**, 71–75.

Tillé, Y. & Wilhelm, M. (2017). Probability sampling designs: Balancing and principles for choice of design. *Statistical Science* **32**, 176–189.

Traat, I., Bondesson, L. & Meister, K. (2004). Sampling design and sample selection through distribution theory. *Journal of Statistical Planning and Inference* **123**, 395–413.

Tschuprow, A. (1923). On the mathematical expectation of the moments of frequency distributions in the case of correlated observation. *Metron* **2**, 461–493, 646–680.

Vallée, A.-A. & Tillé, Y. (2019). Linearization for variance estimation by means of sampling indicators: application to nonresponse. *International Statistical Review*, 1–25, to appear.

Valliant, R. (1993). Poststratification and conditional variance estimation. *Journal of the American Statistical Association* **88**, 89–96.

Valliant, R., Dever, J. A. & Kreuter, F. (2013). *Practical Tools for Designing and Weighting Survey Samples*. New York: Springer.

Valliant, R., Dorfman, A. H. & Royall, R. M. (2000). *Finite Population Sampling and Inference: A Prediction Approach*. New York: Wiley.

Van Huis, L. T., Koeijers, C. A. J. & de Ree, S. J. M. (1994a). EDS, sampling system for the central business register at statistics netherlands. Technical report. Statistics Netherlands.

Van Huis, L. T., Koeijers, C. A. J. & de Ree S. J. M. (1994b). Response burden and co-ordinated sampling for economic surveys. Technical report. Statistics Netherlands, Volume 9.

Vanderhoeft, C. (2001). Generalised calibration at Statistics Belgium SPSS module gCALIBS and current practices. Technical report. Statistics Belgium Working Paper no. 3.

Vanderhoeft, C., Waeytens, E. & Museux, J.-M. (2001). Generalised calibration with SPSS 9. 0 for Windows baser. In *Enquêtes, Modèles et Applications*, J.-J. Droesbeke & L. Lebart, eds. Paris: Dunod, pp. 404–415.

Vitter, J. S. (1985). Random sampling with a reservoir. *ACM Transactions on Mathematical Software* **11**, 37–57.

Wolf, C., Joye, D., Smith, T. W. & Fu, Y.-c., eds. (2016). *The SAGE Handbook of Survey Methodology*. SAGE Publications.

Wolter, K. M. (1984). An investigation of some estimators of variance for systematic sampling. *Journal of the American Statistical Association* **79**, 781–790.

Wolter, K. M. (1985). *Introduction to Variance Estimation*. New York: Springer.

Woodruff, R. S. (1971). A simple method for approximating the variance of a complicated estimate. *Journal of the American Statistical Association* **66**, 411–414.

Wright, R. L. (1983). Finite population sampling with multivariate auxiliary information. *Journal of the American Statistical Association* **78**, 879–884.

Wu, C. & Rao, J. N. K. (2006). Pseudo-empirical likelihood ratio confidence intervals for complex surveys. *Canadian Journal of Statistics/La revue canadienne de statistique* **34**, 359–376.

Wynn, H. P. (1977). Convex sets of finite population plans. *Annals of Statistics* **5**, 414–418.

Yates, F. (1946). A review of recent statistical developments in sampling and sampling surveys. *Journal of the Royal Statistical Society* **A109**, 12–43.

Yates, F. (1949). *Sampling Methods for Censuses and Surveys*. London: Charles Griffin.

Yates, F. (1960). *Sampling Methods for Censuses and Surveys*. London: Charles Griffin, 3rd edition.

Yates, F. (1979). *Sampling Methods for Censuses and Surveys*. London: Charles Griffin, 4th edition.

Yates, F. & Grundy, P. M. (1953). Selection without replacement from within strata with probability proportional to size. *Journal of the Royal Statistical Society* **B15**, 235–261.

Yung, W. & Rao, J. N. K. (1996). Jackknife linearization variance estimators under stratified multi-stage sampling. *Survey Methodology* **22**, 23–31.

Yung, W. & Rao, J. N. K. (2000). Jackknife variance estimation under imputation for estimators using poststratification information. *Journal of the American Statistical Association* **95**, 903–915.

Zieschang, K. D. (1986). A generalized least square weighting system for the consumer expenditure survey. *Section of Survey Research Methods, Journal of the American Statistical Association*, 64–71.

Zieschang, K. D. (1990). Sample weighting methods and estimation of totals in the consumer expenditure survey. *Journal of the American Statistical Association* **85**, 986–1001.

Author Index

Sampling and Estimation from Finite Populations, First Edition. Yves Tillé.
© 2020 John Wiley & Sons Ltd. Published 2020 by John Wiley & Sons Ltd.

Subject Index

a

admissible   24, 88, 184

algorithm
   for Bernoulli sampling   51
   of the cube method   130, 132
   for Poisson sampling   93
   for simple random sampling with
      replacement   56
   for systematic sampling
      with equal probabilities   57
      with unequal probabilities   90
   for unequal probability sampling designs
      86

alternative estimator   35

approach
   design-based   1, 13, 272, 273, 275, 277
   model-assisted   1
   model-based   1, 8, 11, 231, 263–280
   reverse   152, 347

approximately unbiased   203, 237

approximation
   bias   49
   of the variance   111

asymptotic statistic   295

auxiliary
   information   2, 9, 10, 32, 65, 72, 77, 83,
      111, 120, 143, 195, 196, 200, 201,
      209–211, 217, 225, 226, 237,
      255–257
   variable   83, 84, 100, 111, 123, 128, 131,
      135, 141, 195, 199, 200, 202, 331

b

balanced
   sample   119–142
   sampling   119–142
balancing   119–142, 172
   constraint   131
   equation   121, 122, 126–130, 132, 138,
      169
   variable   127, 130, 133, 134, 136, 137,
      141, 142

Bernoulli   32
   design   32
   sampling   32, 51

between-strata variance   66

bias   17, 19
   conditional   235
   of the expansion estimator   19
   of the Hájek estimator   359

biased   19, 20, 81, 197
   Hájek estimator   359
   ratio estimator   197
   sample   19

Bienaymé–Chebyshev inequality   298

bootstrap   331

bounded sequence   296
   in probability   296

bound management   253

Brewer method   109

burden   174

c

calibration   10, 141, 190, 218–220, 222,
      223, 226, 236–261, 277, 313, 315,
      318, 321, 376
   bound management   253
   equation   238
   estimator   190, 328
   existence of a solution   254

Sampling and Estimation from Finite Populations, First Edition. Yves Tillé.
© 2020 John Wiley & Sons Ltd. Published 2020 by John Wiley & Sons Ltd.